Texts and Monographs in Symbolic Computation

A Series of the
Research Institute for Symbolic Computation,
Johannes-Kepler-University, Linz, Austria

Edited by
B. Buchberger and G. E. Collins

**B. F. Caviness and
J. R. Johnson (eds.)**

**Quantifier Elimination and
Cylindrical Algebraic
Decomposition**

SpringerWienNewYork

Dr. Bob F. Caviness
Department of Computer and Information Sciences
University of Delaware, Newark, Delaware, U.S.A.

Dr. Jeremy R. Johnson
Department of Mathematics and Computer Science
Drexel University, Philadelphia, Pennsylvania, U.S.A.

Graphic design: Ecke Bonk
Printed on acid-free and chlorine-free bleached paper

With 20 Figures

Library of Congress Cataloging-in-Publication Data

Quantifier elimination and cylindrical algebraic decomposition / B. F.
    Caviness, J. R. Johnson (eds.).
        p.   cm. — (Texts and monographs in symbolic computation, ISSN
    0943-853X)
        Includes bibliographical references (p.   –   ) and index.
        ISBN-13: 978-3-211-82794-9   e-ISBN-13: 978-3-7091-9459-1
        DOI: 10.1007/978-3-7091-9459-1
        1. Algebra—Data processing. 2. Decomposition method—Data
    processing. 3. Algorithms. I. Caviness, Bob F. II. Johnson, J.
    R. (Jeremy R.) III. Series.
    QA155.7.E4Q36 1998
    512—dc21
                                                                96-43590
                                                                   CIP

ISSN 0943-853X
ISBN-13: 978-3-211-82794-9

Preface

A symposium on Quantifier Elimination and Cylindrical Algebraic Decomposition was held October 6–8, 1993, at the Research Institute for Symbolic Computation (RISC) in Linz, Austria. The symposium celebrated the 20th anniversary of George Collins' discovery of Cylindrical Algebraic Decomposition (CAD) as a method for Quantifier Elimination (QE) for the elementary theory of real closed fields (Collins 1973b)[1], and was devoted to the many advances in this subject since Collins' discovery.

The symposium was preceded by a workshop presented by George Collins and his former students Hoon Hong, Jeremy Johnson, and Scott McCallum. The workshop surveyed the CAD algorithm and its application to QE. In addition the participants were introduced to and experimented with **qepcad**, an interactive program, written by Hong with improvements by Johnson, M. Encarnación, and Collins for solving quantifier elimination problems using CAD (see the survey by Collins in this volume).

The symposium itself centered on four invited talks by George Collins, Hoon Hong, Daniel Lazard, and James Renegar. Collins described 20 years of progress of CAD based quantifier elimination, Hong showed how CAD can be used in the investigation of Hilbert's 16th problem, Lazard spoke on arithmetic support needed for real geometry, and Renegar surveyed recent theoretical improvements in algorithmic quantifier elimination.

Several additional invited speakers spoke on some of Collins' many contributions to computer algebra. Erich Kaltofen spoke on polynomial factorization, Werner Krandick spoke on algorithms for polynomial complex zero calculation, Jeremy Johnson spoke on polynomial real root isolation, David Musser spoke on Collins' contributions to the design of computer algebra software and library design, and David Saunders spoke on symbolic linear algebra. Finally, Rüdiger Loos gave an appropriate historical presentation of Collins' discovery of cylindrical algebraic decomposition and his work on the many computer algebra subalgorithms required by CAD.

In addition to the invited presentations surveying CAD/QE and Collins' contributions to computer algebra, there were contributed talks on current research in CAD, QE, and related subalgorithms. These talks were selected by

---

[1]For references, see the bibliography at the end of this book.

a refereeing process carried out by the program committee listed at the end of the preface.

The contributed papers included new QE algorithms, CAD in domains other than the real numbers, contributions to important subalgorithms needed by CAD and various QE algorithms, and applications of QE. L. González-Vega presented a combinatorial algorithm for solving a subclass of quantifier elimination problems. V. Weispfenning presented a new approach to real quantifier elimination based on methods for counting real zeros of multivariate polynomial systems and the computation of comprehensive Gröbner bases. D. Richardson spoke on local theories and cylindrical decomposition. D. Dubhashi spoke on quantifier elimination in p-adic fields. R. Liska presented joint work with S. Steinberg on the application of quantifier elimination to stability analysis of difference schemes. M.-F. Roy discussed several theorems for counting the number of real roots of a polynomial. Her presentation was based on joint work with L. González-Vega, H. Lombardi, T. Recio. R. Pollack presented joint work with M.-F. Roy on the number of cells defined by a family of polynomials. H. Hong presented joint work with J. R. Sendra on the computation of two variants of the resultant. G. Hagel described various characterizations of the Macaulay matrix and their algorithmic impact. Yu. A. Blinkov presented joint work with A. Yu. Zharkov on involutive bases of zero-dimensional ideals.

Dedication

The symposium and these proceedings are dedicated to George Edwin Collins as a belated gift on the occasion of his 65th birthday. It is through these means that those of us in his research community express our respect and admiration for his continuing work in computer algebra, mathematics, and computer science.

George Collins was born on January 10, 1928, on a farm near Stuart, Iowa, the seventh child of Martin Wentworth Collins and Linnie Fry Collins. It was not long before his interest in things mathematical became apparent. When he was seven or eight years old, he found a mathematics problem book that his father had used when he attended a business college. Soon George was able to do the problems much faster than his father even though his father had above average mathematical ability. At about the same time, George began to entertain himself by adding or multiplying large numbers in his head. Thus his father became the first of many who have marveled at his remarkable mental agilities in doing calculations and in remembering details that most of us never noticed in the first place!

George went to several different schools, including four high schools. The last one, from which he graduated in May, 1945, was Washington Township Consolidated High School located in the country more than five miles from the nearest town, Minburn, Iowa. There were six students in his graduating class. At this high school was a teacher who had the good sense to encourage

George to find a way to attend college. She complimented him on his "stick-to-it-iveness." He found a way by first serving in the U.S. Navy, which he did from the day after his eighteenth birthday until November 12, 1947.

In February, 1948, George entered the University of Iowa from which he graduated in February, 1951, with a B.A. in Mathematics with Highest Distinction. After finishing his bachelor's degree, he stayed at the University of Iowa to study for a master's degree during which time he was supported by a Sanxay Fellowship. Prof. E. W. Chittenden encouraged his interest in mathematical logic and suggested that he go to Cornell to study with J. Barkley Rosser. He matriculated at Cornell in the Fall of 1953 supported by a National Science Foundation Graduate Fellowship. He received his Ph.D. degree under Prof. Rosser in June, 1955. In the latter part of his studies at Cornell, Dr. Lisl Gaal brought George's attention to Tarski's method on the theory of elementary algebra and geometry with the suggestion that he might like to generalize it.

After receiving his Ph.D., George went to work at the IBM Scientific Computer Center in New York City. Soon thereafter, Dr. F. S. Beckman again suggested Tarski's method to George with the additional thought that he might like to program it. This sparked George's interest in potential applications and in the problem of obtaining a more efficient method, interests that have remained with him throughout his career. It seems that his high school teacher's observation about his "stick-to-it-iveness" was very insightful!

In tackling a difficult problem Collins, like others, found that his research spun-off a variety of problems (and solutions) of fundamental importance. One of the first was the need for efficient symbolic processing. In 1960 this led him to begin work on the development of the PM system (Collins 1966a), an early computer algebra system with built-in list processing procedures that made use of reference counts for storage management, an idea that he introduced in (Collins 1960a). PM was developed for efficient calculation with polynomials and Collins quickly realized that the calculation of polynomial greatest common divisors (GCDs) was a significant bottleneck. This led to some of his most important contributions to algebraic computation – namely efficient GCD algorithms. His work on the relationships between polynomial remainder sequences and subresultants (Collins 1966b, 1967) has been fundamental. His early work (along with that of others such as Stanley Brown) on methods for GCD computation pioneered work on a whole range of algorithms that now are used routinely in computer algebra systems for a variety of tasks.

In 1973, George made a breakthrough when he generalized his cylindrical algebraic decomposition method from two variables to n variables, $n \geq 0$. His new method was a dramatic improvement over the original method of Tarski (1951) and subsequent methods suggested by Seidenberg (1954) and Cohen (1969). The CAD method itself has been improved subsequently by Collins and several of his doctoral students, namely Arnon (1981), McCallum (1984), and Hong (1990a).

When asked how he came to discover his cylindrical algebraic decomposition method, Collins responded,

> I first thought about the best way to do quantifier elimination when
> there is only one variable. Then I generalized this to two variables.
> But when I considered three variables, I couldn't prove that my
> projection method for two variables would still work (nor could I
> prove that it wouldn't). This finally led me to the subresultant
> projection method, which would work for any number of variables.
> Scott McCallum later proved (modulo a few details) that my two-
> variable projection method would work in general. I want to thank
> Rüdiger Loos, who shared my interest and encouraged me to persist
> in 1973 at Stanford University, where I finally succeeded.

Perhaps the most striking feature of Collins' CAD research is that he has
developed the scientific tools that are basic in computer algebra: concrete computing time analysis, descriptions and implementations of algorithms, research
on better bounds for both the analysis and implementation of algorithms, better
mathematics for better algorithms, and he used important research problems
as yardsticks of progress. Without his work, contemporary computer algebra
would not be as far advanced and would, undoubtedly, be very different.

George is a master of clarifying often overlooked, but crucially important
details, is not afraid to pursue research paths independent of prevailing opin-
ions, and is an enthusiastic programmer. All his talents result in fast and
masterful actual programs. Collins is known to have admired Rosser's fresh
way of looking at old problems from an algorithmic point of view and his way
of working, often with nothing more than a note pad in his lap. Collins has kept
these attributes alive. His entire scientific career reflects a fresh approach to old
problems via an algorithmic approach, and his ability to work without many
aids is well-known. He seldom visits a library since many things others have to
look up, he solves faster from scratch, often in his head. He has a strong power
of concentration. During long car trips he has developed algorithms completely
in his head (as Mozart did with his music).

His writing, as Bruno Buchberger has noted, is a fine balance between formal precision and a more descriptive style. He is never imprecise but also never
needlessly formal. Don Knuth, in his wonderful little paperback on mathematical writing (Knuth et al. 1989), paid George a supreme compliment. We
quote from p. 32 where the refereeing of scientific papers is being discussed.

> Authors, thought Don, should always be encouraged to do better;
> he could recall only a single occasion when, as a referee or editor, he
> could recommend no improvements at all. (The author in this case
> was George Collins writing for the ACM journal.) Let us publish
> journals to be proud of, he said.

Collins' writing, indeed all his work, has the delightful attribute that you
find yourself drawn into studying it in detail and, as you do, you discover that

in all aspects it is "polished, significant, and worth reprinting in textbooks," to quote (Knuth et al. 1989) once more.

Collins also has high praise for Knuth. When asked about important influences in his career, he first mentioned his high school teacher and Professors Chittenden and Rosser. Then he said, "Since then probably no one has influenced me as much as Donald Knuth, through his monumental books on *The Art of Computer Programming*." Collins also acknowledges the importance of a series of grants from the National Science Foundation that supported his research at the University of Wisconsin and at the Ohio State University as well as an IBM Fellowship at California Institute of Technology for the 1963–64 year.

George has supervised thirteen Ph.D. students to date. They are, in order of completion: Ellis Horowitz, Lee E. Heindel , Michael T. McClellan, David R. Musser, James R. Pinkert, Cyrenus M. Rubald, Stuart C. Schaller, Tsu-Wu J. Chou, Dennis S. Arnon, Scott McCallum, Hoon Hong, Jeremy R. Johnson, and Werner Krandick. Mark J. Encarnación, a current student, is the likely fourteenth[2]. As a Ph.D. advisor, Collins notes "I am indebted to all of my former and current doctoral students, from each of which I have learned."

As of this symposium (October, 1993), a non-exhaustive search identified 63 additional Ph.D. academic descendants in the second through fifth generations from Collins, a total of 76 descendants indicating that the Collins branch of the academic genealogy tree is healthy and flourishing! Collins' academic ancestors are a distinguished group: Rosser, Ph.D. Princeton, 1934; Alonzo Church, Ph.D. Princeton, 1927; Oswald Veblen, Ph.D. Chicago, 1903; E. H. Moore, Ph.D. Yale, 1885, and H. A. Newton, a Yale faculty member who did not have a Ph.D. Prof. Newton, the root of this tree of Ph.D.'s, earned his BA degree from Yale in 1850 and was elected a professor at the commencement of 1855 whereby he took a one-year leave of absence for advanced study in Paris.

In December, 1953, while in graduate school at Cornell, George and his fellow graduate student in mathematical logic, Elliot Mendelson, went to a reception hosted by the University's female secretaries. Dorothy Day Guise always claimed that she found George standing in a fireplace talking to Mendelson. George and Dorothy were married on September 4, 1954. They are the parents of three daughters, Cynthia Day, Nancy Helen and Rebecca Lynn. "Three beautiful, talented and loving daughters" as their proud father aptly describes them. Nancy is the mother of George's two grandchildren.

Many of us have had the privilege of being the recipients of the wonderful hospitality of Dorothy and George in their home, where Dorothy's dining room table often reflected her love and talent for cooking. George and Dorothy complemented one another in a satisfying way. Dorothy was gregarious, animated, and energetic – actively involved in the life of her family and community. George is quiet and reflective. There was a subtle, gentle humor that flowed

---

[2]M. J. Encarnación finished his Ph.D. degree in 1995 under the supervision of Prof. Collins.

between them. Dorothy died much too soon in August, 1986.

George has not only dedicated himself to his family and his research. From 1970–72 he served as Chair of the Department of Computer Sciences at the University of Wisconsin where he was a faculty member for twenty years. From 1975–77, he served as the sixth Chair of the ACM Special Interest Group on Symbolic and Algebraic Manipulation (SIGSAM), a leading professional organization in algebraic computation. At the age of 50 he became a dedicated runner and completed many marathons. In the decade of his 60s he has started running in ultra marathons, 50-mile trail runs. In 1994 he made his third attempt to complete the Ice Age 50 Mile Trail Run in Wisconsin. Stick-to-it-iveness? George continues to run on a regular basis just as he continues his research at an age that many others are sitting in a rocking chair. His best results in running and in research may still be ahead of him, an awesome thought, but completely in character for him! With the warmth and respect of his many friends and admirers, we honor George Edwin Collins with this symposium.

Proceedings Overview

The pre-symposium plans were for this book to be a collection of the contributed and invited papers. Afterwards, Bruno Buchberger suggested a more ambitious goal for the conference proceedings. His vision was that this book together with the collection of papers in the book *Algorithms in Real Algebraic Geometry* edited by Arnon and Buchberger (1988) would contain most of the important papers on Cylindrical Algebraic Decomposition and Quantifier Elimination. These two books would serve as an introduction and survey of the field of algorithmic quantifier elimination. This volume would emphasize Cylindrical Algebraic Decomposition and Collins' contributions to the field. Therefore, in addition to material from the symposium, it was decided to include important, previously published papers from the field with the proviso that they had not already appeared in (Arnon and Buchberger 1988). It was further decided that we would not reprint papers that appeared in the special issue of *The Computer Journal*, volume 36, number 5, 1993, edited by Hoon Hong that was devoted to work on quantifier elimination. Finally, a bibliography of work on QE, CAD, and important subalgorithms, including all of Prof. Collins' papers, was prepared. All of the references from the individual papers in this monograph are contained in this bibliography.

The papers selected for reprinting in this book were chosen in consultation with George Collins. They are Tarski's original monograph (Tarski 1951), Collins' original paper on CAD (Collins 1975), a paper by Fischer and Rabin (1974a) that presented the first result on the theoretical complexity of QE, two survey papers on CAD by Arnon, Collins, and McCallum (1984a, 1984b), two papers by Hong (1990a, 1992e) containing important practical improvements to CAD based QE, a joint paper by Collins and Hong (1991) introducing the concept of partial CAD, and a survey paper by Renegar (1991) on recent progress in the complexity of algorithms for QE. All of these papers have been reformatted to conform to the style of this monograph. In addition, several typographical errors have been removed and an error in (Collins 1975) has been corrected by Professor Collins.

The monograph begins with an introduction to the elementary theory of real closed fields (also known as elementary algebra and geometry), QE, and CAD. The introduction is followed by the survey paper by Collins outlining the developments in CAD based QE that have taken place in the last twenty years. The following nine papers are those that were selected for reprinting. The remaining papers were contributed to the symposium and appear for the first time.

The first five contributed papers are devoted to important subalgorithms of CAD and other QE algorithms. The paper by McCallum discusses improvements in the projection phase of CAD. The papers by Johnson and González-Vega et al. discuss algorithms for finding and/or counting the real zeros of polynomials, a special case of QE for real closed fields, and an important subalgorithm of CAD. The paper by Hagel discusses the Macaulay matrix and resultant that plays an important role in recent QE algorithms (see Renegar 1991). The paper by Hong and Sendra discusses two variants of the resultant that play an important role in the specialized QE algorithms developed by (1993d, 1993e) for formulas involving quadratic polynomials.

The remaining contributed papers present new QE algorithms. The paper by Basu et al. presents an algorithm, with improved theoretical computing time, for finding a point in each connected component of each non-empty sign condition of a set of polynomials. The paper by Richardson presents an algorithm for finding a cylindrical decomposition of a small neighborhood of the origin that is sign invariant for a set of multivariate polynomials. The paper by González-Vega presents a special purpose QE algorithm for formulas with one quantifier and one polynomial sign condition. Weispfenning presents a new QE algorithm based on the method for counting real zeros of multivariate polynomial systems in (Becker and Wörmann 1991; Pedersen, Roy, and Szpirglas 1993) and the computation of comprehensive Gröbner bases.

Three papers presented at the symposium were published elsewhere and were not selected for republication. The papers by Dubhashi (1993) and Liska and Steinberg (1993) were published in the aforementioned special issue of *The Computer Journal* (Hong 1993a). The invited presentation by Hong (1996) was published in the *Journal of IMACS*. A fourth paper presented by Zharkov and Blinkov (1993) was sub itted for publication elsewhere. The survey talks given by Kaltofen, Krandick, Lazard, Loos, and Saunders were not prepared in written form and hence do not appear in this volume.

Acknowledgments

We gratefully acknowledge financial support from the Austrian Fonds zur Förderung der Wissenschaftlichen Forschung (FWF), the American National Science Foundation (NSF Grant INT–9224304), and RISC-Linz without which support this symposium would have been impossible.

To all the following we owe our gratitude for various jobs well done:

Bruno Buchberger was an enthusiastic originator, the General Chair of the Symposium, the person responsible for obtaining the financial support of the Austrian FWF, and was responsible for committing RISC-Linz to host the conference. Prof. Collins has expressed his personal appreciation for Buchberger's efforts.

Rüdiger Loos, a long time collaborator of Collins, provided valuable comments and suggestions for the preface for which we are especially appreciative.

Betina Curtis, secretary at RISC-Linz, handled all the local arrangements expertly and expeditiously. Without her efforts, it is difficult to imagine having had everything go so smoothly.

The Program Committee consisted of:

Dennis Arnon, Xerox PARC, USA
B. Caviness, University of Delaware, USA (Co-Chair)
H. Hong, Johannes Kepler University, Austria
J. Johnson, Drexel University, USA (Co-Chair)
M. Kalkbrener, Eidgenössische Technische Hochschule Zürich, Switzerland
E. Kaltofen, Rensselaer Polytechnic Institute, USA
D. Kozen, Cornell University, USA
W. Krandick, Johannes Kepler University, Austria
D. Lazard, University of Paris VI, France
S. McCallum, Macquarie University, Australia
J. Renegar, Cornell University, USA
V. Weispfenning, University of Passau, Germany

To all these individuals we are appreciative for their role in shaping the program and in helping to evaluate the contributed papers.

Kudos to D. Arnon, G. Collins, B. Mishra, M.-F. Roy, and V. Weispfenning for assistance with the references.

B. F. Caviness
Newark
Jeremy R. Johnson
Philadelphia
1995

Contents

List of Contributors

Dennis Arnon
Frame Technology Corporation
333 W. San Carlos Street
San Jose, CA 95110, USA.

Saugata Basu
Courant Institute of Mathematical
Sciences
New York University
New York, NY 10012 USA.

B. F. Caviness
Department of Computer and
Information Sciences
University of Delaware
Newark, DE 19716 USA.

George E. Collins
Research Institute for Symbolic
Computation
Johannes Kepler University
A-4040 Linz, Austria.

Michael J. Fischer
Department of Computer Science
Yale University, P. O. Box 208285
New Haven, CT 06520-8285 USA.

L. González-Vega
Departamento de Matemáticas,
Estadística y Computación
Universidad de Cantabria
E-39005 Santander, Spain.

Georg Hagel
Wilhelm-Schickard-Institut
Eberhard-Karls-Universität
D-72076 Tübingen, Germany.

Hoon Hong
Research Institute for Symbolic
Computation
Johannes Kepler University
A-4040 Linz, Austria.

J. R. Johnson
Department of Mathematics and
Computer Science
Drexel University
Philadelphia, PA 19104 USA.

H. Lombardi
Laboratoire de Mathematiques
Université de Franche-Comté
F-25030 Besancon, France.

Scott McCallum
Department of Computing
School of MPCE
Macquarie University
Macquarie NSW 2109, Australia.

Richard Pollack
Courant Institute of Mathematical
Sciences
New York University
New York, NY 10012 USA.

Michael O. Rabin
Computer Science
Harvard University
Cambridge, MA 02138 USA.

T. Recio
Departamento de Matemáticas
Universidad de Cantabria
E-39005 Santander, Spain.

James Renegar
School of Operations Research and
Industrial Engineering
Frank Rhodes Hall
Cornell University
Ithaca, NY 14853 USA.

Daniel Richardson
Department of Mathematics
University of Bath
Bath BA2 7AY, UK.

Marie-Françoise Roy
IRMAR (URA CNRS 305)
Université de Rennes
Campus de Beaulieu
F-35042 Rennes cedex, France.

J. Rafael Sendra
Departamento de Matemáticas
Universidad de Alcalá
E-28871 Madrid, Spain.

Alfred Tarski

V. Weispfenning
Lehrstuhl für Mathematik
Universität Passau
D-94030 Passau, Germany.

Introduction

This chapter provides a brief introduction to the primary topic of these proceedings and some indication of the subject's importance.

1   Introduction to the Method

The elementary theory of real-closed fields (RCF) is expressed in a formal language in which one can create atomic formulas of the forms $A = B$ and $A > B$ where A and B are arbitrary polynomials in any number of variables with integer coefficients. These atomic formulas may be combined with the Boolean connectives, conjunction, $\wedge$, disjunction, $\vee$, or negation, $\neg$, and some or all of the variables in these resulting formulas may be quantified over the RCF by universal and existential quantifiers, $(\forall x)$, $(\exists x)$. If all of the variables are quantified, the resulting formula is called a sentence, and it has a definite truth value. In our applications we are interested in the ordered field of the real numbers, a particular RCF.

In 1930, Alfred Tarski proved the existence of a decision procedure for sentences in RCF. More generally, Tarski presented a quantifier elimination method for RCF. A QE method for RCF is an algorithmic method that accepts as input any formula of RCF and outputs an equivalent formula containing the same free (unquantified) variables and no quantifiers. The equivalent formula is true for the same values of its free variables as the input formula. In particular, if the input formula is a sentence then the output formula is trivially reducible to the true formula $0 = 0$ or the false formula $1 = 0$. If the input formula contains free variables then the output formula can be thought of as expressing a necessary and sufficient algebraic condition for the input formula to hold.

The formula $a \neq 0 \wedge (\exists x)(ax^2 + bx + c = 0)$ of RCF states that the quadratic polynomial has a real root. QE applied to this formula easily produces the well-known necessary and sufficient condition $b^2 - 4ac \geq 0$. Here the parameters a, b and c are free variables.

A less trivial example is provided by a problem posed by Kahan (1975). Find necessary and sufficient conditions on an ellipse, with axes parallel to the coordinate axes, that it lies wholly within a circle of radius one centered at the origin. Expressed in the language of RCF this becomes

$$a > 0 \land b > 0 \land (\forall x)(\forall y)((x - c)^2/a^2 + (y - d)^2/b^2 = 1 \Rightarrow x^2 + y^2 \leq 1).$$

The divisions in this formula are not permitted in our language but the formula can obviously be replaced with the equivalent formula

$$a > 0 \land b > 0 \land (\forall x)(\forall y)(b^2(x - c)^2 + a^2(y - d)^2 = a^2 b^2 \Rightarrow x^2 + y^2 \leq 1).$$

This is a problem with four free variables and two quantified variables that is sufficiently difficult that it has thus far defied unaided solution by any quantifier elimination method. In fact, Lazard (1988) has shown that any solution must contain a certain polynomial $T(a, b, c, d)$ of total degree 12 and containing 104 terms; in this paper Lazard gives a solution containing this polynomial.

The CAD method for QE can be briefly described as a method that extracts the polynomials occurring in the input formula (having first trivially reduced all atomic formulas to the forms $A = 0$ and $A > 0$) and then constructs a decomposition of real r-dimensional space, where r is the number of variables in the formula, into a finite number of connected regions, called cells, in which each polynomial is invariant in sign. Moreover these cells are arranged in a certain cylindrical manner. From this cylindrical decomposition it is then quite straightforward to apply the quantifiers by using a sample point from each cell to determine the (invariant) truth value of the input formula in that cell. This application of quantifiers reveals which cells in the subspace of the free variables are true. It then remains to construct an equivalent quantifier-free formula from this knowledge. In Collins' original QE method this problem was solved by a method called augmented projection that provided a quantifier-free formula for each of the true cells. Hong (1990a) has recently devised a much superior method that appears to work in most cases.

2   Importance of QE and CAD Algorithms

The language of RCF is rich enough to permit the expression of many difficult and important mathematical problems. Therefore any QE algorithm that does not require too much computation is extremely valuable. The 1973 CAD method of Collins was an enormous stride forward and much additional progress has been made since then. In the following we describe some of the potential applications.

Solutions of systems of polynomial equalities and inequalities In these problems there are no quantifiers. The true cells of the CAD contain all solutions of the system. The number of solutions is finite just in case each true cell is zero-dimensional. Otherwise an algebraic-geometric description is obtained for each component of the solution set.

Polynomial optimization This is the problem of finding the maximum value of a polynomial subject to polynomial inequality constraints. This problem is easily expressed as a QE problem. For example, to find the maximum value of $P(x,y)$ subject to $Q(x,y) > 0$, we would input the formula $Q(x,y) > 0 \land (\forall u)(\forall v)(Q(u,v) > 0 \Rightarrow P(u,v) \leq P(x,y))$.

Approximation An example is to find the best approximation, in the uniform norm, on a finite interval of a polynomial of degree n by a polynomial of degree $n - 2$ or less. This is known as Solotareff's second problem (Achieser 1956). Schematically this problem can be expressed as $(\forall \bar{Q})(\forall x)(a \leq x \leq b \Rightarrow |P(x) - Q(x)| \leq |P(x) - \bar{Q}(x)|)$. Here P is the polynomial of degree n, $[a,b]$ is the interval of approximation, Q, a polynomial of degree at most $n - 2$, is the best approximation and $\bar{Q}$ is any polynomial of degree $\leq n - 2$. It remains only to introduce indeterminates for the coefficients of Q and $\bar{Q}$ and to replace the $\bar{Q}$ quantifier by quantifiers for the corresponding indeterminates. Technically absolute value is not a part of the elementary theory of RCF, however, it is possible to systematically replace formulas using absolute values with equivalent formulas from RCF. For example, $|x| > |y|$ is equivalent to $(x \geq 0 \land y \geq 0 \land x > y) \lor (x < 0 \land y \geq 0 \land -x > y) \lor (x \geq 0 \land y < 0 \land x > -y) \lor (x < 0 \land y < 0 \land -x > -y)$.

Topology of semi-algebraic sets The subsets of n-dimensional real space that are definable by a quantifier-free formula in n variables are called semi-algebraic sets. Because of the existence of QE methods these are the same as the sets of n-tuples of real numbers that satisfy some arbitrary formula of elementary algebra. By construction of a CAD, each true cell is a connected subset of the semi-algebraic set. The true cells can be grouped into components (maximal connected subsets) if a method is available to decide whether any two cells are adjacent (meaning that the union of the two cells is connected). Adjacency of cells is itself describable as a QE problem, however application of QE does not provide a sufficiently efficient adjacency algorithm. Efficient alternative methods for $n \leq 3$ have been developed by Arnon, Collins, and McCallum (1984b, 1988). Other alternative methods have been proposed by Prill (1986) and by Schwartz and Sharir (1983b). Recent progress on this problem has been reported in (Canny, Grigor'ev, and Vorobjov 1992). Efficient adjacency algorithms have been used by Arnon (1988b) as a means of improving the CAD algorithm through the use of clustering.

Algebraic curve display Since a curve is a special case of a semi-algebraic set, we can completely determine its topology by means of a CAD and an adjacency determination of its cells. This also yields some non-topological geometric information about the curve since it tells us whether the curve is bounded and also reveals the number of branches of the curve in each cylinder. Such topological and geometric information is an important first step in *reliably* displaying the curve or any portion of it. Further geometric information is obtainable. Suppose, for example, that the curve is bivariate. If the curve is bounded then we immediately have from the CAD the minimum and maximum values of x for all points on the curve and QE applied to the formula $(\exists x)(A(x,y) = 0)$ will provide minimum and maximum values for y. If we are interested in dividing the curve into monotonic pieces for display purposes we can compute a CAD for the formula $A(x,y) = 0 \land A_x(x,y) \neq 0 \land A_y(x,y) \neq 0$, where A_x and

A_y are the first partial derivatives, obtaining a CAD in which each branch is monotonic in each cylinder.

Surface intersection By QE we can decide whether two algebraic surfaces intersect. If so, we can analyze and display the space curve of the intersection. Besides the methods described above that easily generalize to space curves, we can project the space curve onto any of the coordinate planes, or indeed onto any plane, providing arbitrary perspectives of the curve.

Algebraic curve implicitization If an algebraic curve is given in parametric form we can use QE to obtain an implicit representation, that is a bivariate polynomial whose real variety defines the curve. We need merely to existentially quantify the argument of the parametric functions over their common argument domain.

Termination proof of term-rewriting systems Lankford (1979) proposed the use of quantifier elimination for proving termination of term rewriting systems. Under Lankford's scheme as further developed by others (see Bündgen 1991; Küchlin 1982; Dershowitz 1987), the constants of the rewrite system are interpreted as integers and the k-ary operators as k-variate integral polynomials. This induces an interpretation of all terms of the system as n-variate integral polynomials where n is the maximum arity of any operator. Then to prove the termination property for the rewrite system it suffices to prove for each rule of the system that the interpretation of the left-hand side, say $L(x_1, \ldots, x_n)$, is greater than the interpretation of the right-hand side, say $R(x_1, \ldots, x_n)$, meaning that

$$(\forall x_1) \ldots (\forall x_n)(x_1 > h \wedge \cdots \wedge x_n > h \Rightarrow L(x_1, ..., x_n) > R(x_1, ..., x_n)),$$

where h is the largest interpretation of any of the constants. QE can be used to decide these statements. Furthermore QE can also be used to decide whether there is any interpretation of this type in which the polynomials that interpret the operators satisfy some degree bound by quantifying over the finitely many coefficients of these polynomials.

Geometry theorem proving and discovery By the introduction of coordinates for points and of equations for lines and circles, Kutzler (Kutzler 1988; Kutzler and Stifter 1986a) using the method of Gröbner bases and Chou (1984, 1988) using the method of Wu have proved algorithmically many classical theorems of geometry as universally quantified statements of elementary algebra. For these theorems no polynomial inequalities are required and all quantifiers are universal. In such cases the special methods of Wu and Groebner bases are usually applicable and are generally much faster than the CAD QE method. But there are many other geometry problems where polynomial inequalities are required and also both universal and existential quantifiers are required, in which case the Wu and Groebner basis methods are not applicable. Also, if one wishes to decide the truth of a geometric conjecture rather than merely confirm a geometric theorem, then these methods are not adequate. Thus these methods, while of much interest as the best known examples of mechanical geometric theorem proving, are of limited value in advancing our geometric knowledge whereas the CAD QE method does have this potential. Chou and McPhee are currently investigating the possibilities for combining in some way the Wu and CAD QE methods in order to extend the scope of Chou's geometry prover (Chou, Gao, and McPhee 1989; McPhee 1990). For a survey of the use of algebraic methods in geometry theorem proving see the article by Buchberger, Collins, and Kutzler (1988).

Geometric modeling As an example of geometric modeling, suppose that we are given a finite set of solid objects, each defined by a semi-algebraic description and a container, also semi-algebraically defined. We wish to decide whether the objects can be packed into the container and, if so, to find a configuration of the objects for such a packing. The configuration can include a suitable orientation, as well as a position, for each object. As a special case this includes the sphere packing problem, namely to determine the maximum number of spheres of radius r that can be packed into a unit cube for any specified value of r.

Robot motion planning In motion planning we are given several algebraic objects in some initial configuration, some of the objects being movable and others not. We wish to determine whether the movable objects can be continuously moved, without collisions, in order to arrive at a specified final configuration and, if so, to find motions that will achieve the goal. The motions of the movable objects can be subjected to various limitations reflecting properties of the objects such as hinges or objects too heavy to lift, for example. Every configuration of the entire set of objects is represented as a point in a higher dimension space containing one dimension for each parameter describing the position and orientation of one of the objects. The solution to the problem, if it exists, is then a path in this higher dimension space. A path exists just in case the two points representing the initial and final configurations are in the same connected component of the set of configuration points that are collision-free (no two objects are occupying the same point of physical space). The set of collision-free points is a semi-algebraic set that can be determined by QE. It is then necessary to decide all cell adjacencies in the CAD in order to determine whether the two points are in the same component. In case they are in the same component then a path of adjacent cells is produced and a path of configurations can then be found by choosing a suitable path in each of those cells.

Stability analysis QE can be used in the stability analysis of difference schemes used for numerically solving time dependent partial differential equations. The von Neuman condition for the stability of a difference scheme can be converted into a real QE problem. Liska and Steinberg (1993) have successfully used the **qepcad** program to analyze several sample difference schemes.

Miscellaneous problems QE has been successfully applied to many other problems. For example, Collins and Johnson (1989b) used QE to study properties of Descartes' rule of signs as it relates to the coefficient sign variation method for real root isolation. QE has also been used to investigate ill-conditioned polynomials (polynomials where a small perturbation in the coefficients leads to a large change in the roots) (Johnson 1991). Renegar has used general results from semialgebraic geometry to theoretically study this problem. There are many problems dealing with polynomials for which QE and CAD can construct interesting examples, find counterexamples, and suggest general theorems by proving specific cases.

3   Alternative Approaches

Collins' work on quantifier elimination in real closed field stimulated various research efforts (see the references in this volume). Some of these efforts have dealt with the theoretical complexity of QE and algorithms for performing QE.

Collins (1975) gave an upper bound for the worst case time complexity of his QE algorithm, that is doubly-exponential in the number of variables. Davenport and Heintz (1988) showed that this doubly-exponential behavior is intrinsic to the problem: they found a lower bound for a worst case time complexity that is doubly-exponential in the number of quantifiers. Independently Weispfenning (1988) showed that the double exponential lower bound holds for the simpler problem of QE for formulas that are linear in the bound variables. Weispfenning's result relies on an earlier result of Fischer and Rabin (1974a), that proved an exponential lower bound for the decision problem of the reals.

Ben-Or, Kozen, and Reif (1986) presented a new decision algorithm that is well suited to parallel computation. Recently Grigor'ev (1988), Fitchas, Galligo, and Morgenstern (1990a),Heintz, Roy, and Solernó (1989a), and Renegar (1992a, 1992b, 1992c) devised still newer QE or decision algorithms that behave better than Collins' for problems with few quantifier alternations. Renegar (1991) surveys these recent developments. Another method, which may benefit from a few quantifier alternations, is presented in this volume by Weispfenning.

Arnborg and Feng (1988) are investigating an idea that might lead to a CAD construction method that is fast when applied to a set of "regular" polynomials and whose cost increases with the number and the complexity of the singularities in the polynomials.

Recently several researchers have concentrated on QE algorithms specialized to particular types of input formulas. Weispfenning (1988) and Loos and Weispfenning (1993) have developed and implemented QE algorithms for input formulas that are linear in their bound variables. Hong in a series of papers has designed an algorithm for quadratic QE problems. Continuing in this direction, Weispfenning has designed an algorithm for cubic problems. In these proceedings González-Vega has developed an algorithm for inputs with a single quantifier.

4 Practical Issues

Many important practical improvements to CAD based QE have been made since Arnon's original implementation. These improvements have enabled the current implementation (Hong 1990a) to solve many non-trivial problems, some of which were previously unsolvable by computer methods. For example, consider the problem of finding conditions on the coefficients of a quartic polynomial so that the polynomial is positive semi-definite. This can be stated as the following QE problem.

$$(\forall x)(x^4 + px^2 + qx + r \geq 0) \tag{1}$$

This problem was originally solved in several days of computation by Arnon's program. On today's computers Arnon's program can solve this problem in several hours. The current implementation completes in several seconds.

Hong's program relies on the important concept of a partial CAD (Collins and Hong 1991; Hong 1990a) that in many cases allows QE to be performed without computing the entire CAD. In many important practical problems this can lead to improved computing times of several orders of magnitude. Another significant improvement has been achieved with an alternative approach to solution formula construction (Hong 1992e). This improvement allows the construction of an equivalent quantifier free formula, in many cases, without using the time consuming augmented projection. Furthermore, the new approach typically produces much shorter formulas that are accessible to human interpretation. As an example of the success of this approach, the quantifier free formula,

$$[256r^3 - 128p^2r^2 + 144pq^2r + 16p^4r - 27q^4 - 4p^3q^2 \geq 0 \wedge$$
$$8pr - 9q^2 - 2p^3 \leq 0] \vee [27q^2 + 8p^3 \geq 0 \wedge 8pr - 9q^2 - 2p^3 \geq 0] \wedge$$
$$r \geq 0$$

for the quartic problem stated in equation (1) contains five atomic formulas whereas the solution formula, produced by the original approach has 401 occurrences of atomic formulas and requires several pages to write down.

The current implementation also includes improvements in the projection phase of CAD. These improvments are due to Hong (1990b) and McCallum (1984). Further practical improvements have been achieved by improved subalgorithms for polynomial real root isolation and real algebraic number computation (Johnson 1991; Langemyr 1990). Recently, several researchers have investigated the use of parallelism to further reduce computing times (Hong 1993c; Saunders, Lee, and Abdali 1989).

Acknowledgments

This chapter is based on material provided by G. E. Collins, H. Hong, and J. R. Johnson.

Quantifier Elimination by Cylindrical Algebraic Decomposition – Twenty Years of Progress

George E. Collins

1 Introduction

The CAD (cylindrical algebraic decomposition) method and its application to QE (quantifier elimination) for ERA (elementary real algebra) was announced by the author in 1973 at Carnegie Mellon University (Collins 1973b). In the twenty years since then several very important improvements have been made to the method which, together with a very large increase in available computational power, have made it possible to solve in seconds or minutes some interesting problems. In the following we survey these improvements and present some of these problems with their solutions.

As a framework for the survey we provide a brief outline of the original method in the next section. Following sections are about the improvements resulting from adjacency and clustering, an improved projection method, partial CADs, an interactive implementation, solution formula construction, and equational constraints. The final two sections describe improvements which have been made to subalgorithms, and improvements conceived but not yet implemented.

2 Original Method

In the original method, as described in the first full paper on the method (Collins 1975), the CAD and the QE were distinct algorithms. The input to the CAD algorithm is a set of integral polynomials in, say r, variables and the output is a CAD of $\mathbb{R}^r$, r-dimensional real space, in each cell of which each input polynomial is sign-invariant. Input to the QE algorithm is a prenex formula. The QE algorithm applies the CAD algorithm to the set of polynomials occurring in this formula, then uses the output of the CAD algorithm to determine which cells of the CAD satisfy the matrix (unquantified part) of the input formula. The CAD of r-dimensional space induces (or implicitly determines) a CAD of k-dimensional space, where k is the number of free variables. Because of the cylindrical arrangement of the cells, applying the quantifiers then determines which cells of $\mathbb{R}^k$ satisfy the quantified formula. In order to produce a quantifier-free formula satisfied by just the true cells the CAD algorithm must be used in a mode called augmented projection.

The CAD algorithm has two phases. In the first phase a *projection* operation is repeatedly applied. Applied to the set of input polynomials the operation eliminates one variable, producing a set of integral polynomials in the remaining $r - 1$ variables. It is applied to this set to eliminate another variable. Continuing, the projection phase ends with a set of polynomials in only one variable.

The second phase is the *lifting*, or *stack construction*, phase. The univariate polynomials produced by the last projection determine a minimal CAD of one-dimensional space in the cells of which they are all sign-invariant. Its cells are the real zeros of these polynomials, which are called *sections*, and the open intervals between consecutive zeros (also the semi-infinite open intervals preceding and following all zeros), called *sectors*. These sections and sectors in their natural order comprise a *stack*, and also a CAD of one-dimensional space. The CAD algorithm also computes a *sample point* for each cell, which may be any point belonging to the cell. It is clear that for $\mathbb{R}^1$ some sample points are necessarily irrational algebraic, and this is true more generally. The CAD of $\mathbb{R}^1$ is then extended to a CAD of $\mathbb{R}^2$ by constructing a stack over each cell in the CAD of $\mathbb{R}^1$, using the set of bivariate polynomials which were produced by projection. By the properties of the projection operation, and by the underlying theory, the real zeros of these bivariate polynomials intersected with the cylinder whose base is any cell of 1-space are linearly ordered continuous functions defined on the base, called sections. The connected sets between successive sections (also below all sections and above all sections) are called sectors. These sections and sectors together comprise a stack and all such stacks comprise a sign-invariant CAD of $\mathbb{R}^2$. Sample points for the cells of this CAD are obtained by substituting the sample points of the base cells into the bivariate polynomials and computing the real zeros of the resulting univariate algebraic polynomials. The terminology of sections, sectors and stacks is due to Arnon, see (Arnon et al. 1984a), which also provides an alternative exposition of the method.

Every cell in a CAD has an *index*. The index of a cell of $\mathbb{R}^k$ is a k-tuple of positive integers. The cells in a stack are given consecutive positive integers in accordance with their position in the stack. For example, the cell with index (3,6) is the sixth cell (from the bottom) in the stack constructed in the cylinder over the third cell (from the left) in a CAD of $\mathbb{R}^1$ (as pictured in $\mathbb{R}^2$).

3   Adjacency and Clustering

The first complete implementation of the CAD method was carried out by Arnon in 1979 and 1980. Examples of its use were included in his doctoral thesis (Arnon 1981). Arnon's task was quite formidable, involving also the programming of numerous subalgorithms needed for the projection phase and also many needed for the algebraic number computations of the lifting phase.

Concurrently Arnon, Collins and McCallum were devising algorithms for computing cell adjacencies. Each cell is a connected set and two cells are said to be *adjacent* in case their union is also connected. The resulting algorithms, for cells in two- and three-dimensional spaces, were ultimately published in (Arnon et al. 1984b and 1988 respectively).

Cell adjacency has several important applications. One application, related to Hilbert's 16th problem, is to determine the topological type of the plane curve given by any bivariate integral polynomial. Arnon and McCallum (1988) developed a method and a program, using CAD and adjacency, for this purpose. More recently Hong (1996, 1994) has developed improved algorithms for this problem. Schwartz and Sharir (1983b) showed the potential applicability of QE and CAD along with adjacency calculation to robot motion planning.

Arnon recognized the applicability of adjacency calculation to CAD construction itself. A stack can be constructed over sets which are larger than single cells; it is only necessary that the set used as a base of the stack be connected and that the projection polynomials are sign-invariant in the set. Therefore it is possible to combine cells belonging to the same topological component and in which each projection polynomial is sign-invariant. Maximal sets of cells with these properties were given the name *clusters* by Arnon. The practice of clustering reduces the number of stack constructions which must be performed. Stack constructions generally require many computations with algebraic numbers and are therefore very time-consuming. Also, the sample point of a cluster can be chosen to be the sample point of any cell in the cluster. So if any cell of a cluster has a sample point whose coordinates are all rational, as will be the case whenever the cell is full-dimensional (that is, has the same dimension as the space in which it resides) then all algebraic number computations for all cells in the cluster are avoided. Thus potentially great time savings are possible, but these savings must be weighed against the time cost of the adjacency calculations. Arnon (1988b) reported comparisons for several examples. For the example consisting of a sphere $z^2 + y^2 + x^2 - 1$ and a catastrophe surface polynomial $z^3 + xz + y$, he reported that his program, when run without adjacency and clustering, failed to complete the CAD in 827 minutes, whereas with clustering the CAD was completed in 282 minutes. The latter time includes the time, not separately given, to cluster the cells in $\mathbb{R}^3$, which is not needed to obtain the CAD of $\mathbb{R}^3$. Using a new program called **qepcad** (quantifier elimination by partial cylindrical algebraic decomposition), described in following sections of this paper, a CAD for this example was computed without adjacency and clustering in 11.0 seconds[1], of which 9.2 seconds were required for the 157 stack constructions in $\mathbb{R}^3$. **qepcad** does not yet have a facility for adjacency computation and clustering. However an auxiliary program was used to do the required adjacency computations for this problem in $\mathbb{R}^2$ which completed in 0.47 seconds. The 157 cells in $\mathbb{R}^2$ produced 36 clusters and **qepcad** then did the stack constructions for these clusters in 2.5 seconds. Thus the investment of 0.47 seconds produced a gross saving of 6.7 seconds and a net saving of about 6.2 seconds out of the previous 9.2 seconds.

It is not easy to see how the adjacency algorithm of (Arnon et al. 1988) for $\mathbb{R}^3$ could be extended to higher dimension spaces. However Collins and McCallum are currently developing a new algorithm for $\mathbb{R}^3$ which is probably more efficient and which probably can be generalized to higher dimensions. Prill (1986) presented a method for arbitrary dimension. However the method requires a linear change of variables, which makes it useless for quantifier elimination. Also the method employs a priori bounds for algebraic numbers so there is reason for doubt about the practical efficiency of his method.

4 Improved Projection

For the discussion of this section we need to elaborate the outline of projection given in Section 2. In the original method the projection of a set $A_1, \ldots, A_n$ consisted, in general, of:

- each coefficient of each A_i;
- each principal subresultant coefficient $\mathrm{psc}_k(B_i, B_i')$ where B_i is a reductum of A_i, B_i' is the derivative of B_i and $k \geq 0$;
- each principal subresultant coefficient $\mathrm{psc}_k(B_i, B_j)$ where B_i is a reductum of A_i, B_j is a reductum of A_j, $i < j$ and $k \geq 0$.

The phrase 'in general' above is intended to imply that in practice many of these polynomials may be omitted, namely, whenever certain subsets of these polynomials can be shown to have only a finite number of common real zeros. The variable k above ranges up to, but not including, the minimum of the degrees of the two polynomial arguments. Thus the projection set defined is very large. This has the consequence that the resulting decomposition is very fine, consisting of very many stacks and cells, and therefore requires much time to compute.

In the case $k = 0$ above, $\mathrm{psc}_0(B_i, B_i')$ is just the leading coefficient of B_i times the discriminant of B_i and $\mathrm{psc}_0(B_i, B_j)$ is just the resultant of B_i and B_j. Prior to the discovery of the projection scheme above the author attempted without success to prove that these discriminants and resultants alone suffice for a projection. Later, in 1984, McCallum (1984 and 1988) proved that, apart

---

[1]All new times reported in this paper were obtained using a DECstation 5000/240 with a 40MHz R3400 risc-processor.

from a slight technicality, this is indeed the case. In fact, he even proved the much stronger result that only the discriminants of the A_i and the resultants of the pairs (A_i, A_j) are needed, along with the coefficients of the A_i.

As one example of the efficacy of McCallum's improved projection we consider again an example used previously by Arnon and Mignotte (1988). The problem is to determine necessary and sufficient conditions on the coefficients for a general quartic polynomial to be positive semi-definite. Thus the input formula is

$$(\forall x)[x^4 + px^2 + qx + r \geq 0].$$

Using the original projection method **qepcad** computed 17 projection polynomials and required 6.2 seconds to produce a solution. Using McCallum's projection only 4 projection polynomials were produced and the total computing time was only 1.1 seconds. Unfortunately, however, **qepcad** was unable to produce an equivalent quantifier-free formula when using the McCallum projection. This will be discussed further in Section 7.

Hong proved in 1989 that the original projection could be restricted by including $\mathrm{psc}_k(B_i, B_j)$ only in the case that B_j is A_j (see Hong 1990a). Although the resulting projection is still a superset of McCallum's projection, there may be examples where Hong's projection could be useful for two reasons. First, if clustering is used with McCallum's projection, the clusters must be order-invariant, not merely sign-invariant, and the union of the cells in a cluster must constitute a smooth manifold. These requirements entail extra computation and possibly smaller clusters. Second, McCallum's projection method may fail to be applicable in rare cases in which a projected polynomial is zero everywhere in a cylinder.

In 1990 Lazard (see Lazard 1994) announced a projection which would further reduce McCallum's projection to the extent that not all coefficients of each polynomial need to be included, but only the leading and trailing coefficients. Unfortunately a gap was subsequently discovered in Lazard's proof which has not been filled. Lazard's projection has nevertheless been used in a number of examples without finding a case where it fails.

5   Partial CADs

Hong (1990b) (see also Collins and Hong 1991) made the important observation that quantifier elimination can proceed concurrently with the stack construction phase of the CAD algorithm. If there are at least two quantifiers this can make many stack constructions unnecessary. A very simple example will illustrate. Suppose the quantifier elimination problem is

$$(\exists y)(\exists z)[x^2 + y^2 + z^2 < 1].$$

The first projection will consist of only $x^2 + y^2 - 1$ and the second projection will consist of only $x + 1$ and $x - 1$. Suppose we first construct a stack over the cell which is the open interval $(-1, +1)$. This stack contains 5 cells, one of which is the interior of the unit circle centered at the origin. Suppose we next construct a stack over this cell. Again the resulting stack consists of 5 cells and one of these is the interior of the sphere, in which the formula $x^2 + y^2 + z^2 < 1$ is true. Applying the quantifier $(\exists z)$ to this stack therefore shows that the formula $(\exists z)[x^2 + y^2 + z^2 < 1]$ is true in the base of the stack, namely the interior of the unit circle. Now applying the quantifier $(\exists y)$ shows that the input formula is true in the cell $(-1, +1)$. We have thereby avoided constructing stacks over the other 4 cells in the first-constructed stack. This illustrates what is called *truth value propagation*.

In partial CAD construction, stack construction is also avoided in another way. Some of the polynomials occurring in the input formula may not contain all of the variables of the problem. Suppose a cell c has already been constructed in $\mathbb{R}^k$ and that one or more of the input polynomials contain no variables other than the first k. The signs of the values of these polynomials in c are computed from the sample point of c and thereby the truth values of the atomic formulas in which they occur are determined. These truth values may suffice to determine the truth value of the quantifier-free part of the input formula. If they do then a stack need not be constructed over the cell c. This device is called *trial evaluation*. Let us consider a modification of the example above:

$$(\exists y)(\exists z)[x > 0 \wedge x^2 + y^2 + z^2 < 1].$$

In this example stacks would not be constructed over any of the four cells in $\mathbb{R}^1$ containing points less than or equal to 0.

In order to incorporate these ideas and others, Hong completely reprogrammed CAD and QE. The resulting program, **qepcad**, with various subsequent improvements, is the one used for all examples reported in this paper. These subsequent improvements are variously due to Hong, J. R. Johnson, M. J. Encarnación and the author. The program uses SACLIB (Buchberger et al. 1993) and will itself be made available to SACLIB users after further development.

As an example of the tremendous improvement resulting from partial CAD construction, we consider one arising in solving a special case of the Solotareff approximation problem (Achieser 1956). The problem is to obtain the best approximation, in the sense of the uniform norm on the interval $[-1, +1]$, to a polynomial of degree n by one of degree $n - 2$ or less. Here we consider only a portion of the case $n = 3$, namely, where the best approximation to $x^3 + rx^2$ is $ax + b$, to obtain a as a function of r for $0 \leq r < 1$. The input formula is

$$(\exists b)(\exists u)(\exists v)[r \geq 0 \wedge r < 1$$
$$\wedge -1 < u \wedge u < v \wedge v < 1 \wedge 3u^2 + 2ru - a = 0$$
$$\wedge 3v^2 + 2rv - a = 0 \wedge u^3 + ru^2 - au + a - r - 1 = 0$$
$$\wedge v^3 + rv^2 - av - 2b - a + r + 1 = 0].$$

Using the partial CAD methodology the solution to this problem requires 1.7 seconds, of which 1.3 seconds are devoted to stack construction. But when

qepcad was forced to construct a complete CAD, the program spent 577 seconds on stack construction prior to aborting due to insufficient memory. (It needed more than 4,000,000 cells of list memory.)

6   Interactive Implementation

Hong's new implementation is interactive; the user has a repertoire of commands and he can step the program through from beginning to end, using his expertise to assist the program in obtaining a solution. Interaction with the program has also suggested several ways to further improve the program.

For purposes of interaction the program is divided into sequential phases: the normalization phase, the projection phase, the stack construction phase, and the solution formula phase. The normalization phase extracts the polynomials from the input formula, factors them, and rewrites the input formula in terms of these irreducible factors. The projection phase and the stack construction phase are further divided into a variable number of steps. In the projection phase each projection is a step and each subsequent factorization of the resulting projection polynomials is a step. In the stack construction phase, each stack construction is a step. The user may instruct the program to do all steps in a phase, or to do a specified number of steps of the phase.

The user also has at his disposal a number of commands which modify program behavior. As one example the user can choose whether to use a modular or non-modular algorithm for computing resultants and discriminants. There are cases where either method may be much faster than the other. Especially important is the user's ability to control the order in which stack constructions are performed, since this will affect the number of stack constructions which must be performed. During the stack construction phase a number of stacks have already been constructed and now any of the cells in any of these stacks can be chosen as the base of a new stack provided that (a) such stack has not already been constructed, (b) the cell belongs to some $\mathbb{R}^k$ with $k < r$ and (c) the cell has not been assigned a truth value by trial evaluation. For purposes of choosing the next base for stack construction such eligible cells have a linear order imposed on them and this order is under user control by means of various cell attributes such as cell *level* (the dimension of the space in which it resides), the cell dimension, the degree of the cell's sample coordinates, and its index. If all irreducible factors of the input polynomials contain all variables then there is no chance that trial evaluation will reduce the required number of stack constructions so it will generally be best to build stacks over highest level cells first in order to benefit maximally from truth propagation. But if some of these irreducible factors have missing variables then trial evaluation may be more effective, so that it would be better to first choose lower level cells. Different stack constructions will require different amounts of computation and the amount depends quite strongly on the degree of the sample point. Choosing an optimal ordering for a given problem is difficult but by experimentation one can sometimes obtain a solution to a problem which would not have been obtainable otherwise.

Perhaps most important of all is a command which imposes further necessary conditions on any cell over which a stack construction is to be performed. For example, the condition can be imposed that the cell be full-dimensional, that is, that its dimension is the same as that of the space in which it resides. This is useful for solving a system of strict inequalities, as was first noticed by McCallum (1987). McCallum (1993) includes an example for which the solution time was reduced from more than 12644 seconds to only 6 seconds by this device. Perhaps equally important, the command can be used to solve a system of polynomial equations if it is known, as is often the case, that the system has only finitely many solutions. In this case the command is used to restrict stack constructions to zero-dimensional cells. As an example we used the following system of equations:

$$-7xyz + 6yz - 14xz + 9z - 3xy - 12y - x + 1 = 0$$

$$\wedge\ 2xyz - yz + 14z + 15xy + 14y - 15x = 0$$

$$\wedge\ -8xyz + 11yz - 12xz - 5z + 15xy + 2y + 10x - 14 = 0.$$

Without imposing the zero-dimensional condition **qepcad** required 24.8 seconds to find the four solution points and to approximate their coordinates to 10 decimal places. This included 17.8 seconds to do 119 stack constructions. With the condition imposed the total time was reduced to 18.5 seconds, including 12.9 seconds for only 25 stack constructions. Thus the avoided stack constructions, though numerous, were relatively fast.

7 Solution Formula Construction

In Section 2 it was explained that in the original algorithm the CAD computation had to be performed with an "augmented" projection operator in order to obtain a solution formula for each individual true cell and thereby a solution formula for the collection of all true cells. This provided a method which always succeeds but at the cost of a much finer decomposition, which takes much longer to produce. Also, the resulting formula becomes so long as to often be unintelligible and useless. Hong (1990b and 1992e) developed a new method for attempting to construct a solution formula without using augmented projection. In fact, in numerous examples which have been tried his method has usually been successful, although it is easy to construct simple examples where the method will fail. One especially simple example is

$$(\exists y)[x^2 + y^2 < 1 \wedge x + y > 0].$$

In this example the projection polynomials are $x^2 - 1$, the discriminant of the circle, and $2x^2 - 1$, the resultant of the circle and the line. Hence the projection factors are $x - 1$, $x + 1$ and $2x^2 - 1$, from which the program attempts to construct a solution formula. But this is impossible. The true cells in $\mathbb{R}^1$ are the open interval $(-\sqrt{1/2}, +\sqrt{1/2})$, the one-point set $\sqrt{1/2}$ and the open interval $(\sqrt{1/2}, 1)$. There is no formula containing only the three projection factors which is true for $\sqrt{1/2}$ but false for $-\sqrt{1/2}$.

The augmented projection of the original method includes resultants and discriminants of all derivatives of the set of polynomials being projected. By Rolle's theorem this guarantees that different roots of the same polynomial can be distinguished as needed in constructing a solution formula. In the example above the derivative of $x^2 + y^2 - 1$ with respect to y, namely $2y$, would have been computed and then the resultant of $2y$ and $x + y$, namely $2x$. The resulting projection factor x suffices to distinguish the true cells from the false cells. It is planned to improve **qepcad** by introducing derivatives only as needed during the solution formula construction phase rather than during projection. In this example the polynomial x would be obtained directly from the derivative of $2x^2 - 1$.

A more natural and less simple example is provided by the so-called quartic problem described in Section 4. Arnon and Mignotte reported that for this problem the original program, using augmented projection, required 19 hours to compute a CAD and a solution formula, the resulting formula being 80 lines long. They then showed how it was possible to interact with the program to reduce the computation time to an hour or less and by analysis of displayed information to deduce three different solution formulas, each no more than two or three lines long. Hong (1992e) reported that the original method required almost 3 hours and produced a solution formula containing 401 occurrences of atomic formulas whereas **qepcad** required 18.5 seconds to compute the required partial CAD and then an additional 2.3 seconds to construct the simple solution formula

$$[[D \geq 0 \wedge S \leq 0] \vee [E \geq 0 \wedge S \geq 0]] \wedge r \geq 0,$$

where D is the discriminant of the quartic polynomial, S is the first subdiscriminant of the quartic polynomial and E is a factor (of multiplicity three) of the discriminant of D.

$$D = 256r^3 - 128p^2r^2 + 144pq^2r + 16p^4r - 27q^4 - 4p^3q^2,$$

$$S = 8pr - 9q^2 - 2p^3,$$

$$E = 27q^2 + 8p^3.$$

As is implied by the presence in the solution formula of the first subdiscriminant, S, Hong used the original method for the first projection. Later the author did the problem again using the McCallum projection instead and found that the program was unable to construct a solution formula. **qepcad** used 1.0 seconds to compute the partial CAD and then only 17 milliseconds in attempting to construct a solution formula. The only projection factors were D, E, p and q. It is necessary to introduce both the first derivative D' and the second derivative D'' in order to obtain a solution formula. By careful analysis Scott McCallum and the author found the simple solution formula

$$[D \geq 0 \wedge D' \geq 0 \wedge D'' \geq 0] \vee [p \geq 0 \wedge D \geq 0].$$

This author can foresee that in the near future **qepcad** will be improved so that it will automatically introduce these two derivatives and obtain a solution formula similar to this one. Actually McCallum and the author also found the even simpler solution formula

$$D \geq 0 \wedge [4r - p^2 \geq 0 \vee p \geq 0].$$

8   Equational Constraints

This section gives a brief account of a recent improvement by the author which has still been only partially implemented in **qepcad**. A *constraint* of a quantifier elimination problem is an atomic formula which is logically implied by the quantifier-free matrix of the prenex input formula. If the atomic formula is an equation it is called an *equational constraint*, and the polynomial in the equation is an *equational constraint polynomial*.

Quantifier elimination problems quite often have equational constraints and they can be used to reduce the projection sets. Since the matrix can only be true in the sections of an equational constraint polynomial, other polynomials occurring in the formula need only to be sign-invariant in those sections and do not need to be delineable. Hence, by restriction of McCallum's projection, if A is an equational constraint polynomial and $B_1, \ldots, B_n$ are the remaining polynomials then the projection set may consist of only

- the discriminant and (enough of) the coefficients of A, and
- for each i, the resultant of A and B_i.

There may be several equational constraint polynomials. Then any one of them may be chosen as A in this description of the projection set. We call this one the *pivot*. It is usually a good strategy to choose for the pivot a polynomial of least degree in the projection variable, but currently the program chooses a pivot arbitrarily and this choice can be overridden by the user.

Since polynomials which are to be projected are always made irreducible, it is important to realize that the equational constraint polynomial A may be a product of several irreducible polynomials, $A_1, \ldots, A_m$. Then the discriminant of A has as factors

- the discriminant of each A_i, and
- the square of each resultant $\mathrm{res}(A_i, A_j), i < j$,

and in the projection set the coefficients of A may be replaced by appropriate coefficients of the A_i.

If A_1 and A_2 are both equational constraint polynomials then so is their resultant, $res(A_1, A_2)$, since $A_1 = 0 \wedge A_2 = 0 \Rightarrow res(A_1, A_2) = 0$. Currently the user must declare the equational constraint polynomials which occur in the input formula, but the program will then propagate these equational resultants by application of this 'resultant rule'. This is a non-trivial task since in general A is a product of several irreducible factors and each resultant arising from a factor of A may itself factor into several irreducible polynomials.

As an example to illustrate the use of equational constraints we again use the problem of Solotareff (see Section 5). It is known that the best approximation is unique. Clearly there is no loss of generality in assuming that $A(x) = x^n + rx^{n-1}$ with $r \geq 0$. Solotareff showed that for $0 \leq r \leq S_n = n \tan^2(\pi/2n)$ the coefficients of $B(x)$ are rational functions of r which are easy to compute in terms of Chebyshev polynomials. For $r > S_n$ he obtained a characterization of $B(x)$ which is helpful in solving the problem, for any particular n, by quantifier elimination. The characterization is that there exist points u_i, $-1 = u_0 < u_1 < \cdots < u_{n-2} < +1$, such that, if $E(x) = A(x) - B(x)$ is the error polynomial then $\| E(x) \| = E(1) = (-1)^{i-1} E(u_{n-i})$, $i = 2, \ldots, n$. It is important to observe that we must have $E'(u_i) = 0$ for $1 \leq i \leq n-2$.

We now treat the case $n = 4$. Then $S_4 = 12 - 8\sqrt{2}$, the smallest root of $x^2 - 24x + 16$. Let $B(x) = ax^2 + bx + c$. From $E(1) = -E(-1)$ we obtain $c = 1 - a$ and $E(1) = r - b$. Substituting u for u_1, v for u_2 and $1 - a$ for c, we obtain the following quantified formula for a as a function of r.

$$(\exists b)(\exists u)(\exists v)[[r^2 - 24r + 16 < 0 \vee r >]$$

$$\wedge\, r - b > 0 \wedge -1 < u \wedge u < v \wedge v < 1$$

$$\wedge\, u^4 + ru^3 - au^2 - bu - (1 - a) = r - b$$

$$\wedge\, v^4 + rv^3 - av^2 - bv - (1 - a) = -r + b$$

$$\wedge\, 4u^3 + 3u^2 - 2au - b = 0 \wedge 4v^3 + 3rv^2 - 2av - b = 0].$$

Notice that there are four equational constraint polynomials, two of them with v as last variable, two with u as last variable. Propagation through resultants produces two equational constraint polynomials with b as main variable and one with a as main variable. (In the variable ordering r precedes a.) When the projection phase is run for this problem without the use of equational constraints, 442 projection polynomials having a total of 288 distinct irreducible factors are produced, and the time required is 23.8 seconds. When it is run with the use of equational constraints this is reduced to 80 projection polynomials having 76 irreducible factors, and the time required is only 3.0 seconds.

Equational constraints can also be used very effectively during stack construction. Whenever a stack has already been constructed in which there are sections of an equational constraint polynomial all other cells in the stack can be marked false and it is unnecessary to construct stacks over them. This may reduce enormously the number of stacks which must be constructed, as it indeed does in our Solotareff problem. Although this aspect of equational constraints has not yet been implemented, we have simulated its effect in this problem by means of user control commands. First we need to explain that in solving this example problem we have proved that a must be a continuous function of r and we used this also to reduce the computation. Because of the continuity it is only necessary to construct stacks over those cells in $\mathbb{R}^1$ which are sectors. This eliminates about half of the needed stack constructions and, moreover, the ones which are eliminated are much more costly because they have algebraic sample points of higher degrees. There are 21 sections and 22 sectors over which stacks would have to be constructed otherwise. We simulated the application of equational constraints for just one of the sectors; the results would be quite similar for the others. Without the use of equational constraints there were 286 stack constructions needed over the sample sector and they required 24.5 seconds exclusive of garbage collection time. With the use of equational constraints only 37 stack constructions were needed and they required only 2.7 seconds exclusive of garbage collection time. We conclude that in this problem the use of equational constraints during stack construction would likely reduce computing time from about 539 seconds to about 59 seconds. Actually we solved this problem with the aid of user commands based on mathematical conjecture using only 6.9 seconds of stack construction time. We did not use the solution formula construction part of the program but merely displayed the resulting partial CAD to infer the result. We found that the polynomial

$$324a^4 + 324r^2a^3 - 2016a^3 + 108r^4a^2 - 1128r^2a^2 +$$
$$4576a^2 + 12r^6a - 224r^4a + 1392r^2a - 4480a -$$
$$15r^6 + 112r^4 - 608r^2 + 1600,$$

which is one of seven factors of the propagated bivariate equational constraint, has two real roots and that a is the greater of these for every $r \geq S_4$. As a function of r, a asymptotically approaches $5/4$ from below as r approaches ∞.

9 Subalgorithms

Besides improvements in the method itself and increases in available computational power the potential of the CAD method has been enhanced by improved algorithms for the arithmetic and algebraic operations that are required. This twenty-year progress report would not be complete without some mention of these.

For cylindrical algebraic decomposition the most crucial subalgorithms are for operations with algebraic numbers and algebraic polynomials (polynomials with coefficients from an algebraic number field). The required operations include computing the greatest common divisors of univariate algebraic polynomials, isolating the real roots of univariate algebraic polynomials, and converting a double extension of the rational numbers to a simple extension.

For the problem of computing the gcd of two univariate algebraic polynomials, a significant advance was made in 1987 by Langemyr and McCallum (Langemyr and McCallum 1989) with a new modular algorithm. Their paper showed impressive speedups relative to the classical non-modular algorithm for some randomly generated examples. However the method proved to be disappointing when applied in CAD construction; it was slower than the classical method in so many cases that its use was abandoned. The method of Langemyr and McCallum requires that the field generator be an algebraic integer. As a result the input algebraic polynomials must be converted to satisfy this requirement, which greatly increases the sizes of integer coefficients occurring in them and therewith the computation time. Recently Encarnación (1994) has discovered and proved that these conversions are unnecessary, enabling a large speedup. In addition he has devised a version which achieves a further speedup by using modular residue to rational conversion in lieu of a certain precomputed bound for the denominators in the gcd. This version achieves speedups relative to the Langemyr–McCallum method in the range of 10 to 20 for randomly generated examples of modest size. It has replaced the classical method in the CAD-QE program, resulting in significant overall performance improvement, since many algebraic polynomial gcds are needed to compute squarefree bases for polynomials arising from substitution of sample points during stack constructions.

Johnson (1991) studied in detail various algorithms for isolating the real roots of both univariate integral polynomials and univariate algebraic polynomials. His comparisons showed that for both cases the method based on Descartes' method, interval bisections and linear fractional transformations (the coefficient sign variation method) was, with only a few exceptions, faster than other methods, usually much faster. In particular this method was found to be appreciably faster for polynomials arising in CAD computations. **qepcad** would be much slower if it used either the method of Sturm sequences or the derivative sequence method.

An especially big gain was made by a change in the method for converting double extensions to simple extensions. A frequently occurring problem in CAD construction is that we have a real algebraic number α given by its minimal integral polynomial $A(x)$ and an isolating interval I, and another real algebraic number β given by a squarefree algebraic polynomial $B(y)$ over $\mathbb{Q}(\alpha)$ and an isolating interval J. We wish to compute a minimal polynomial $C(x)$ and an isolating interval K for γ such that $\mathbb{Q}(\alpha, \beta) = \mathbb{Q}(\gamma)$. Originally we first computed the norm, B^*, of B and obtained the minimal polynomial, $\bar{B}$, of β as the appropriate irreducible factor of B^*. Then we computed γ by the algorithm SIMPLE of (Loos 1982a), computing as a resultant a polynomial $C^*(x)$ having $\gamma = \alpha + t\beta$, t a small positive integer such that C^* is squarefree, as a root, and factoring C^* to obtain the minimal polynomial C of γ. If the degrees of A and B are m and n respectively, it may be that $\bar{B}$ has degree mn, and C then has degree m^2n although the degree of C is at most mn. The high degree of C^* makes this an expensive method. A much more efficient method was already described by Trager (1976) and was discovered independently by J. R. Johnson and incorporated into **qepcad**. This method instead computes as a resultant the norm, $C^*(x)$, of $B(x - s\alpha)$, s a small non-negative integer chosen so that C^* is squarefree. Then the minimal polynomial C of $\gamma = s\alpha + \beta$ is obtained as the appropriate irreducible factor of C^*, the degree of C^* being only mn. Trager required that $B(x)$ be irreducible over $\mathbb{Q}(\alpha)$ but this requirement is not essential. This method has the further advantage that if $\mathbb{Q}(\alpha, \beta) = \mathbb{Q}(\beta)$ the value obtained for γ will be β rather than $\alpha + t\beta$ for some $t > 0$. This results in much simpler representations for α and β as elements of $\mathbb{Q}(\gamma)$. In addition Johnson adds a test of whether $\deg(C) = \deg(A)$. If so, his method computes β as an element of $\mathbb{Q}(\alpha)$ as the root of the linear polynomial $\gcd(B, C)$. Again this gives much simpler representations than using $\gamma = s\alpha + \beta$. As an indication of the enormous importance of this change, a **qepcad** application which had previously taken 589 seconds, of which 456 seconds were used for converting double to simple extensions, required, after the new method was installed, only 67 seconds, of which only 5.3 seconds were used for double to simple extensions.

Ultimately all of the algebraic subalgorithms depend on algorithms for integer arithmetic. Recently the SACLIB programs for integer arithmetic have been given new implementations which are faster. The improvements are based in part on increased use of arrays. Also, Jebelean's (1993) method for exact integer division, applicable when the remainder is known in advance to be zero, has been implemented and substituted for division with remainder wherever possible. It was found that this change produced a typical reduction of about 10% in the total computing time for several quantifier elimination problems.

10   Future Improvements

Improvement of the CAD method is a continuing process. This final section of our progress report will describe further improvements which are already planned but not yet carried out.

We have already discussed one such improvement in Section 8, namely the use of equational constraints during lifting. The necessary changes to **qepcad** have already been analyzed and programming is about to begin.

Another improvement which will be made involves the use of multiple pivots for propagating equational constraints. Suppose that we have three equational constraints of level $k \geq 3$, say $A_1 = 0$, $A_2 = 0$ and $A_3 = 0$. If we select $A_1 = 0$ as pivot and perform two projections we will obtain $A_{12} = \mathrm{res}(A_1, A_2)$, $A_{13} = \mathrm{res}(A_1, A_3)$ and $A_{123} = \mathrm{res}(A_{12}, A_{13})$ as equational constraint polynomials. As already observed in (Collins 1975), A_{123} will, with few exceptions, be reducible. In fact, A_{123} will have a "genuine" factor, say G, and an "extraneous" factor, say E. (We do not rule out the possibility that either G or E is 1, nor do we exclude the possibility that either G or E is reducible.) G has the property that every solution of $G = 0$ is extendable to a solution of $A_1 = 0 \wedge A_2 = 0 \wedge A_3 = 0$ and E has the property that every solution of $E = 0$ which is so extendable is a solution of $G = 0$. It follows from this definition that $G = 0$ is an equational constraint, in fact a stronger equational constraint than $A_{123} = 0$, which should therefore replace the latter. The problem is to identify the irreducible factors of G or of E. Let us rename A_{123} as B_1 in recognition of the pivot A_1. Using A_2 and A_3 as pivots in place of A_1 we will obtain B_2 and B_3. Let B be the product of those irreducible factors common to B_1, B_2 and B_3. Let $C_i = B_i/B, i = 1, 2, 3$. If it can be shown that $C_1 = 0 \land C_2 = 0 \land C_3 = 0$ has no solution, or even that every solution is already a solution of $B = 0$ then we can conclude that $B = 0$ is an equational constraint. In particular it will follow that every irreducible factor of C_1 is a factor of E.

In the case $k = 3$, the C_i will be univariate polynomials, so it is likely, we believe, that only two pivots will be needed, that is, the system $C_1 \land C_2$ will already be inconsistent. Notice that this is equivalent to $\deg(\gcd(C_1, C_2)) = 0$. This is indeed the case for the system of three equations in three variables considered in Section 6. In this example we obtain $B = 5801340x^6 - 31075609x^5 + 107500x^4 + 49420395x^3 - 23556366x^2 - 18379208x + 16877004$, $C_1 = 35x^2 + 154x - 114$ and $C_2 = 30x^2 + 195x + 196$. Since C_1 and C_2 are both irreducible they have no common solution. In the case $k = 4$, the C_i will be bivariate polynomials so one will usually need three of them to obtain an inconsistent system. But the benefit of removing extraneous factors will be greater since the extraneous factors would otherwise generate additional extraneous univariate factors. More generally, for $k \geq 4$ one will probably need to use $k - 1$ pivots in order to remove the extraneous factor. Lacking that many equational constraints one could nevertheless when $k = 4$, for example, replace the constraint $B = 0 \lor C_1 = 0$ by the more effective constraint $B = 0 \lor D = 0$, where $D = \mathrm{res}(C_1, C_2) = 0$. Also, one could generalize the notion of an equational constraint to allow the use of $B = 0 \lor C_1 = 0 \land C_2 = 0$ as an equational constraint during lifting.

Another method which we plan to introduce uses the "genealogy" of projection factors during lifting in order to reduce the cost of squarefree basis computations for sets of univariate algebraic polynomials. The genealogy of a projection factor specifies the projection polynomials of which it is a factor and specifies for each such projection factor its two "parent" projection factors if it is a resultant, its single "parent" projection factor if it is a discriminant. If a projection polynomial is a factor of an input polynomial or a factor of a coefficient of a projection factor then that will be indicated. Suppose that a projection factor B is a factor of only some discriminants, say $\mathrm{discr}(A_i)$, $i = 1, \ldots, m$, where $A_1, \ldots, A_n$ are all of the level k projection factors. If we construct a stack over a cell whose sample point is a zero of B we will substitute the sample point into each A_i, obtaining univariate algebraic polynomials A_i^*, $i = 1, \ldots, n$. Then we know that the polynomials A_i^* for $i > m$ are squarefree and we avoid the cost of computing $\gcd(A_i^*, A_i^{*\prime})$ for $i > m$. Typically n will be much larger than m. Similarly, if the projection factor is a factor of only a few resultants $\mathrm{res}(A_i, A_j)$ then we know that most of the pairs (A_i^*, A_j^*) are relatively prime. Although gcd's which are 1 are relatively fast to compute, the abundance of such gcd computations that will be avoided may amount to a significant portion of squarefree basis computation time.

We can reduce the time that is spent on isolating the real roots of a square-free basis of algebraic polynomials by using interval arithmetic. Let $\mathbb{Q}[\alpha]$ be the algebraic number field in which the coefficients of the polynomials lie. We first compute a small isolating interval for α. Then, using interval arithmetic, we evaluate the univariate polynomials representing the coefficients, obtaining a set of univariate polynomials with interval coefficients. Next we apply the coefficient sign variation method to these polynomials to try to obtain disjoint isolating intervals for their zeros. This will be possible provided that our isolating interval for α is sufficiently small and that we use interval arithmetic of sufficiently high precision. Otherwise the attempt may fail because we obtain a polynomial having a coefficient that is an interval containing zero, positive and negative numbers, so that its number of coefficient sign variations is undetermined. Then the precision must be increased. This method, and other related methods, were described by Johnson (1992). However at that time the method had only been implemented using interval arithmetic programs which performed exact arithmetic on the rational endpoints of the intervals, rendering the method relatively ineffective. Subsequently Krandick and Johnson (1994) have developed highly efficient SACLIB programs for floating-point interval arithmetic of arbitrary specified precision.

Although Hong's method of solution formula construction succeeds in many cases, it needs to be modified to produce a solution formula in all cases. It is easy to detect, for any given partial CAD, whether a formula can be constructed from the projection factors. We know that when this isn't possible one can obtain a solution formula by also using some derivatives of projection factors; the derivatives need not be introduced during projection and only a few derivatives are likely to suffice. If there are two cells, one true and one false, for which all projection factors have the same sign, we can compare their indices. There must be a least level k such that the k-th elements of their indices differ but all level k projection factors have the same signs in the two cells. There must then be some level k projection factor which has at least two real roots, multiplicities counted, in the level k stack to which both cells belong. The two cells can be distinguished with the use of one or more derivatives of this projection factor. The choices which are made, namely, which two cells are considered first and which projection factors are chosen for differentiation, will lead to different solution formulas, some undoubtedly much better than others. So an algorithm based on this approach which is both reasonably fast and produces simple solution formulas may be a challenge, but one which will be pursued.

Acknowledgments

The author's work on quantifier elimination is currently supported in part by FWF Grant P8572-PHY.

A Decision Method for Elementary Algebra and Geometry[1]

Alfred Tarski
Prepared for publication with the assistance of J. C. C. McKinsey

This work has a long history. Its main results were found in 1930 and first mentioned in print a year later. It took nine years, however, before the material in its full development was prepared for publication. In fact, a monograph embodying those results was scheduled to appear in 1939 under the title *The completeness of elementary algebra and geometry* in the collection *Actualités scientifiques et industrielles*, Hermann & Cie, Paris. As a result of the war, the publication did not materialize; two existing sets of page proofs are probably the only trace of this venture.

Naturally, I did not abandon the hope of seeing the results in print. However, as often happens with authors, the original version ceased to satisfy me. The specific character of the results seemed to call for a more formal presentation, and the addition of new observations and conclusions seemed to be desirable. The perplexity of the war and postwar periods resulted in postponing the revision from year to year. Hence I was happy when, in the beginning of 1948, the RAND Corporation, Santa Monica, California, became interested in my results and offered to publish them. It was especially fortunate that Professor J. C. C. McKinsey (Stanford University), who was at the time working with the RAND Corporation, was entrusted with the task of preparing the work for publication. With my collaboration, he actually prepared a new draft of the monograph and, aside from his competent editorial work, he contributed to a simplification of the development. Within a few months the monograph was published. As was to be expected, it reflected the specific interests which

---

the RAND Corporation found in the results. The decision method for elementary algebra and geometry – which is one of the main results of the work – was presented in a systematic and detailed way, thus bringing to the fore the possibility of constructing an actual decision machine. Other, more theoretical aspects of the problems discussed were treated less thoroughly, and only in notes.

The present edition is a photographic reprint of that published by the RAND Corporation, and hence no extensive changes could be introduced in the text. However, all known errors have been corrected, the introduction has been partly changed, and some supplementary notes have been added. In addition to new bibliographical references (mostly to recent literature), these supplementary notes contain elaborations of the original text which, in at least one case, result in enriching the content of the monograph by new theoretical material.

I should like to thank the readers and reviewers who drew my attention to misprints and minor flaws in the original edition, in particular, Professor L. A. Henkin (University of Southern California), Dr. G. F. Rose (University of Wisconsin), and Professor B. L. van der Waerden (University of Zurich, Switzerland). I also want to express my warm gratitude to Mr. Solomon Feferman and Mr. Frederick B. Thompson (both of the University of California) for their assistance in preparing the present edition. Mr. Feferman's work in this connection was done within the framework of a project sponsored by the Office of Naval Research.

Alfred Tarski
Berkeley, California
May 1951

1 Introduction

By a *decision method* for a class K of sentences (or other expressions) is meant a method by means of which, given any sentence θ, one can always decide in a finite number of steps whether θ is in K; by a *decision problem* for a class K we mean the problem of finding a decision method for K. A decision method must be like a recipe, which tells one what to do at each step so that no intelligence is required to follow it; and the method can be applied by anyone so long as he is able to read and follow directions.

The importance of the decision problem for the whole of mathematics (and for various special mathematical theories) was stressed by Hilbert, who considered this as the main task of a new field of mathematical research for which he suggested the term "metamathematics". The most important kind of decision problems is that in which K is defined to be the class of true sentences of a certain theory. When we say that there is a decision method for a certain theory, we mean that there is a decision method for the class of true sentences of the theory (see Note 1 in Section 5).

Some decision methods have been known for a very long time. For example, Euclid's algorithm provides (among other things) a decision method for the class of all true sentences of the form "p and q are relatively prime," where p and q are integers (or polynomials with constant coefficients). And Sturm's theorem enables one to decide how many roots a given polynomial has and thus to decide on the truth of sentences of the form, "the polynomial p has exactly k roots."

Other decision methods are of more recent date. Löwenheim (1915) gave a decision method for the class of correct formulas of the lower predicate calculus involving only one variable. Post (1921) gave an exact proof of the validity of the familiar decision method (the so-called "truth-table method") for ordinary sentential calculus. Langford (1927a, 1927b) gave a decision method for an elementary theory of linear order. Presburger (1930) gave a decision method for the part of the arithmetic of integers which involves only the operation of addition. Tarski (1940) found a decision method for the elementary theory of Boolean algebra. McKinsey (1943) gave a decision method for the class of true universal sentences of elementary lattice theory. Mrs. Szmielew has recently found a decision method for the elementary theory of Abelian groups (see Note 2 in Section 5).

There are also some important negative results in this connection. From the fundamental results of Gödel (1931) and subsequent improvements of them obtained by Church (1936) and Rosser (1936), it follows that there does not exist a decision method for any theory to which belong all the sentences of elementary number theory (i.e., the arithmetic of integers with addition and multiplication) – and hence no decision method for the whole of mathematics is possible. A similar result has been obtained recently by Mrs. Robinson for theories to which belong all the sentences of the arithmetic of rationals. It is also known that there do not exist decision methods for various parts of modern algebra – in fact, for the elementary theory of rings (Mostowski and Tarski), the elementary theories of groups and lattices (Tarski), and the elementary theory of fields (Mrs. Robinson) (see Note 3 in Section 5).

In this monograph we present a method (found in 1930 but previously unpublished, see Note 4 in Section 5) for deciding on the truth of sentences of the elementary algebra of real numbers – and hence also of elementary geometry.

By elementary algebra we understand that part of the general theory of real numbers in which one uses exclusively variables representing real numbers, constants denoting individual numbers, like "0" and "1", symbols denoting elementary operations on and elementary relations between real numbers, like "$+$", "$\cdot$", "$-$", "$<$", "$>$", and "$=$", and expressions of elementary logic such as "and", "or", "not", "for some x", and "for all x". Among formulas of elementary algebra we find algebraic equations and inequalities; and by combining equations and inequalities by means of the logical expressions listed above, we obtain arbitrary sentences of elementary algebra. Thus, for example, the following are sentences of elementary algebra:

$0 > (1 + 1) + (1 + 1)$;

For every a, b, c, and d, where $a \neq 0$, there exists an x such that

$$ax^3 + bx^2 + cx + d = 0.$$

The first sentence is false, and the second sentence is true.

On the other hand, in elementary algebra we do not use variables standing for arbitrary sets or sequences of real numbers, for arbitrary functions of real numbers, and the like. (When in this monograph we attach the qualifier "elementary" to the name of a theory, we refer to this abstention from the use of set-theoretical notions.) Hence those algebraic concepts whose definitions in terms of the fundamental notions listed above would require some set-theoretical devices cannot be represented in our system of elementary algebra. This applies, for instance, to the general notion of a polynomial, to the notion of solvability of an equation by means of radicals, and the like. For this reason it is not possible, for example, to consider as a sentence of elementary algebra the sentence:

Every polynomial has at least one root.

On the other hand, one can formulate in elementary algebra the sentences:

Every polynomial of degree 1 has a root;
Every polynomial of degree 2 has a root;
Every polynomial of degree 3 has a root;

and so on. Since we are dealing with real – not complex – algebra, the above sentences are true for odd degree but false for even degree.

It should be emphasized that the general notion of an integer (as well as that of a rational, or of an algebraic number) also belongs to those notions which cannot be represented in our system of elementary algebra – and this in spite of the fact that each individual integer can easily be represented; e.g., 2 as $1 + 1$, 3 as $1 + 1 + 1$, etc. (see Note 5 in Section 5). The variables in elementary algebra always stand for arbitrary real numbers and cannot be supposed to assume only integers as values. For such a supposition would imply that the class of all sentences of elementary algebra contains all sentences of elementary number theory; and, by results mentioned above, there could be no universal method for deciding on the truth of sentences of such a class. Thus, the following is not a sentence of elementary algebra:

The equation

$$x^3 + y^3 = z^3$$

has no solution in positive integral x, y, z.

This gives, we hope, an adequate idea of what is understood here by ·a sentence of elementary algebra. Turning now to geometry, we can say roughly that by a sentence of elementary geometry we understand one which can be translated into a sentence of elementary algebra by fixing a coordinate system. It is well know that most sentences of elementary geometry in the traditional meaning are of this kind. There are, however, exceptions. There are, for instance, statements which involve explicitly or implicitly the general notion of a natural number: for instance, statements regarding polygons with an arbitrary number of sides – such as, that in every polygon each side is shorter than the sum of the remaining sides. It goes without saying that statements which involve the general notion of a point set – of an arbitrary geometrical figure – are also not elementary in our sense, but they would hardly be regarded as elementary in the everyday understanding of the term.

On the other hand, there are sentences which are elementary according to our definition but which are not ordinarily so considered. Most sentences of analytic geometry concerning algebraic curves of any definite degree belong here: for example, the theorem that any two ellipses intersect in at most four points.

It is important to realize that only the nature of the concepts involved, and not the character of the means of proof, determines whether a geometrical theorem is a sentence of elementary geometry. For instance, the statement that every angle can be divided into three congruent angles is an elementary sentence in our sense, and of course a true elementary sentence – despite the fact that the usual proofs of this statement make essential use of the axiom of continuity. On the other hand, the general notion of constructibility by rule and compass cannot be defined in elementary geometry, and therefore the statement that an angle in general cannot be trisected by rule and compass is not an elementary sentence – although we can express in elementary geometry the facts that, say, an angle of 30° cannot be trisected by $1, 2, \ldots$, or in general any fixed number n of applications of rule and compass.

If we now compare the theories treated in this monograph (i.e., elementary algebra and geometry) with the other theories mentioned above for which decision methods have been found, we see at once that although the logical structure in both cases is equally elementary, the theories investigated here have a considerably richer mathematical content. It would be possible to mention numerous problems which can be formulated in these theories, and which played in the past an important role in the development of mathematics. In the solution of these problems, and in general in the development of the theories considered, a great variety of modes of inference have been applied – some of them of a rather intricate nature (to mention only one example: the proof of the theorem that a triangle is isosceles if the bisectors of two of its angles are congruent). Thus the fact that there exists a universal decision method for elementary algebra and geometry could hardly have been regarded as a foregone conclusion.

In light of these remarks one should not expect that the mathematical basis for the decision method to be discussed will be of a quite obvious and trivial nature. In fact by analyzing this decision method the reader will easily see that in its mathematical content it is closely related to a classical algebraic result –

namely, the theorem of Sturm previously mentioned – and it even provides an extension of this theorem to arbitrary systems of equations and inequalities in many unknowns.

Since a decision method, by its very nature, requires no intelligence for its application, it is clear that, whenever one can give a decision method for a class K of sentences, one can also devise a machine to decide whether an arbitrary sentence belongs to K. It often happens in mathematical research, both pure and applied, that problems arise as to the truth of complicated sentences of elementary algebra or geometry. The decision method presented in this work gives the mathematician the assurance that he will be able to solve every such problem by working at it long enough. Once the machine is devised, his task will reduce to explaining the problem to the machine – or to its operator. It may be instructive to illustrate, by means of an example, the more specific ways in which a decision machine could prove helpful in the study of unsolved problems.

As is well known, any two polygons of equal area, P and Q, can be decomposed into the same finite number n of non-overlapping triangles in such a way that each triangle in P is congruent to the corresponding triangle in Q. We are interested in determining the smallest number for which such a decomposition is possible. We assume in the following that P is the unit square and Q is a rectangle of unit area whose base has x units. Now the smallest number n depends exclusively on x and is denoted by $d(x)$; our problem reduces to describing the behavior of the function d for all positive values of x.

In particular, given any x_0, we can ask what the value of $d(x_0)$ is. In most cases, even the answer to this simple question presents difficulty; e.g., it is not easily seen whether or not $d(7/2) = 8$. However, we can easily establish, by means of a direct geometrical argument, an upper bound for $d(x_0)$; in fact, if $1 \leq x_0 \leq n$, where n is an integer, we have $d(x_0) \leq 2n$. Consequently, just one of the sentences "$d(x_0) = 1$", "$d(x_0) = 2$", ..., "$d(x_0) = 2n$" is true. If, moreover, x_0 is an algebraic number, all of these sentences prove to be expressible in elementary geometry. Hence, by setting the machine in motion at most $2n$ times, we could check which of the sentences is true and thus find the value of $d(x_0)$.

In turn we may consider hypotheses regarding the behavior of the function d in some intervals. For instance, offhand, it seems plausible that $5 \leq d(x) \leq 6$ whenever $2 < x \leq 3$. This hypothesis is still expressible in elementary geometry, and hence could be confirmed or rejected by means of a machine. The situation changes when we consider hypotheses of a more general character concerning the behavior of the function in its whole domain, e.g., the following one: for any real x and integral n, if $x > n$, then $d(x) > 2n$. This hypothesis has not yet been confirmed even for small values of n. In its general form, the hypothesis cannot be formulated in elementary geometry, and hence cannot be tested by means of the machine suggested here. However, the machine would permit us to test the hypothesis for any special value of n. We could carry out such tests for a sequence of consecutive values, $n = 2, 3, \ldots$, up to, say, $n = 100$.

If the result of at least one test were negative, the hypothesis would prove to be false; otherwise, our confidence in the hypothesis would increase, and we should feel encouraged to attempt establishing the hypothesis (by means of a normal mathematical proof), instead of trying to construct a counterexample.

As is seen from the last remarks, the machine envisaged may prove useful in connection with certain problems which cannot be formulated in elementary algebra (or geometry). The most typical in this class of problems are those of the form "Is it the case that, for every integer n, the condition C_n holds?" where C_n is expressible in elementary algebra for each fixed value of n. The machine could be used to solve mechanically this sort of problem for a series of consecutive values of n; in consequence, either we would learn that the solution of the problem in its general form is negative or else the plausibility of a positive solution would increase. Many important and difficult problems belong to this class, and the applicability of the machine to such problems may greatly enhance its value for mathematical research. (The results of this work have further implications, independent of the use of the machine, for the class of problems discussed; see Supplementary Note 7.)

It will be seen later, from the detailed description of the decision method, that the machine could serve some further purposes. We are often concerned, not with a sentence of elementary algebra, but with a condition involving parameters a, b, c, ..., and formulated in terms of elementary algebra; the condition may be very involved, and we are interested in simplifying it – and, in fact, in reducing it to a standard form, in which it appears as a combination of algebraic equations and inequalities in a, b, c, ... To give an example, consider the condition satisfied by the numbers a, b, and c if and only if there are exactly two (real) solutions of the equation:

$$ax^2 + bx + c = 0.$$

In this case, the reduction is very simple and is well known from high-school algebra; the condition can be given the standard form:

$$a \neq 0 \text{ and } b^2 - 4ac > 0.$$

The decision method developed below will give the assurance that such a reduction is always possible; and the decision machine would perform the reduction mechanically.

This monograph is divided into three sections. The next section contains a description of the system of algebra to which the decision method applies. In Section 3, the decision method itself is developed in a detailed way. In Section 4, some extensions of the results obtained as well as some related open problems are discussed. The notes at the end of the monograph contain, in addition to historical and bibliographical references, the discussion of various points of theoretical interest which are not directly related to the question of constructing a decision machine. A short bibliography following the notes[1] lists

---

[1]Editors note: All bibliographic references are given at the end of the book.

the works which are referred to in the monograph (see Note 1 in Section 5).

2   The System of Elementary Algebra

In this section we want to describe a formal system of elementary algebra – and in particular to define in a precise way the class of sentences of this system(see Note 7 in Section 5).

By a variable we shall mean any one of the following symbols:

$$x, x_1, x_2, \ldots ; y, y_1, \ldots ; z, z_1, z_2, \ldots .$$

We suppose that there are infinitely many variables and that they are arranged in a sequence, so that we can speak of the variable occupying the 1st, 2nd,..., n-th place in the sequence. These variables are to be thought of intuitively as ranging over the set of real numbers.

By an *algebraic constant* we shall mean one of the following three symbols:

$$1, 0, -1.$$

By an *algebraic operation-sign* we shall mean one of the following two symbols:

$$+, \cdot .$$

The first is called the *addition sign*, and the second the *multiplication sign*.

By an *algebraic term* we understand any meaningful expression built up from variables and algebraic constants by means of the elementary operation-signs. Thus for example,

$$x, \quad x_1 + y, \quad -1 \cdot x, \quad [(x_1 \cdot -1) \cdot x_1] + x_2$$

are algebraic terms. But

$$x+, \quad \sqrt{2} + x$$

are not algebraic terms: the first, because it is meaningless; the second, because it involves the sign "$\sqrt{2}$", which is neither a variable nor an algebraic constant (in the restricted meaning we have given to the latter term).

If one wants a precise definition of algebraic terms, they can be defined recursively as follows: An algebraic term of first order is simply a variable or one of the three algebraic constants. If α and β are algebraic terms of order at most k, and if the maximum of the orders of α and β is k, then $(\alpha \cdot \beta)$ and $(\alpha + \beta)$ are algebraic terms of order $k + 1$. An expression is called an algebraic term if, for some k, it is an algebraic term of order k.

According to the above definition, one should inclose in parentheses the results of performing operations on terms. Thus one should write, for example, always

$$(x + y) \text{ and } (x \cdot y)$$

instead of simply
$$x + y \text{ and } x \cdot y.$$

We shall often omit these parentheses, however, when no ambiguity will result from doing so; we shall use, in general, the ordinary conventions as to omitting parentheses in writing algebraic terms. Thus we write

$$x + y \cdot z$$

instead of

$$[x + (y \cdot z)].$$

It is convenient to introduce the operation of subtraction as follows (see Note 8 in Section 5): if α and β are any terms, then we set

$$(\alpha - \beta) \equiv [\alpha + (-1 \cdot \beta)].$$

We have used here the symbol "$\equiv$" to indicate that two formulas are identically the same – in the present case by definition. We shall use this symbol throughout the rest of this report. When we write

$$\alpha \equiv \beta$$

we mean that α and β are composed of exactly the same symbols, written in exactly the same order. Thus, for example, it is true that

$$0 \equiv 0,$$

and that

$$(0 = 1) \equiv (0 = 1),$$

but not that

$$(0 + 0) \equiv 0,$$

nor that

$$(0 = 1) \equiv (1 = 0).$$

It is also convenient to introduce notation for sums and products of arbitrary finite length. Let $\alpha_1, \alpha_2, \ldots$ be a sequence of terms. Then we set

$$\sum_{i=1}^{1} \alpha_i \equiv \alpha_1$$

$$\sum_{i=1}^{k+1} \alpha_i \equiv \left(\sum_{i=1}^{k} \alpha_i + \alpha_{k+1} \right),$$

and similarly,

$$\prod_{i=1}^{1} \alpha_i \equiv \alpha_1$$

$$\prod_{i=1}^{k+1} \alpha_i \equiv \left(\prod_{i=1}^{k} \alpha_i \cdot \alpha_{k+1} \right).$$

Instead of

$$\sum_{i=1}^{n} \alpha_i$$

we shall also sometimes use the notation

$$\alpha_1 + \alpha_2 + \cdots + \alpha_n,$$

or simply,

$$\alpha_1 + \cdots + \alpha_n;$$

and instead of

$$\prod_{i=1}^{n} \alpha_i$$

we shall sometimes write

$$\alpha_1 \cdot \alpha_2 \cdot \cdots \cdot \alpha_n,$$

or

$$\alpha_1 \cdot \cdots \cdot \alpha_n.$$

If $\alpha_1, \ldots, \alpha_n$ are all the same, and equal, say to α, then instead of

$$\prod_{i=1}^{n} \alpha_i$$

we sometimes write simply

$$\alpha^n.$$

Thus, for example,

$$\xi^3$$

has the same meaning as

$$[(\xi \cdot \xi) \cdot \xi].$$

Moreover, we shall sometimes write α^0 for 1.

By an *algebraic relation-symbol* we shall mean one of the two symbols:

$$=, >,$$

called, respectively, the *equality sign* and the *greater-than sign* (see Note 8 in Section 5).

By an *atomic formula* we shall mean an expression of one of the forms

$$(\alpha = \beta), \ (\alpha > \beta)$$

where α and β stand for arbitrary algebraic terms; according to our previous remarks, parentheses will sometimes be omitted. The first kind of expression is called an *equality*, and the second an *inequality*. Thus, for example, the following are atomic formulas:

$$1 = 1 + 1$$
$$0 + x = x$$
$$x \cdot (y + x) = 0$$
$$[x \cdot (1 + 1)] + (y \cdot y) > 0$$
$$x > (y \cdot y) + x.$$

By a *sentential connective* we shall mean one of the following three symbols:

$$\sim, \wedge, \vee.$$

The first is called the *negation sign* (and is to be read "not"), the second is called the *conjunction sign* (and is to be read "and"), and the third is called the *disjunction sign* (and is to be read "or" – in the nonexclusive sense).

By the *(existential) quantifier* we understand the symbol "E". If ξ is any variable, then $(E\xi)$ is called a *quantifier expression*. The expression $(E\xi)$ is to be read "there exists a ξ such that."

By a *formula* we shall mean an expression built up from atomic formulas by use of sentential connectives and quantifiers. Thus, for example, the following are formulas:

$$0 = 0,$$
$$(Ex)\,(x = 0),$$
$$(x = 0) \vee (Ey)\,(x > y),$$
$$(Ex) \sim (Ey) \sim [(x = y) \vee (x > 1 + y)],$$
$$\sim (x > 1) \wedge (Ey)\,(x = y \cdot y).$$

If one wants a precise definition of formulas, they can be defined recursively as follows: A formula of first order is simply an atomic formula. If Θ is a formula of order k, then $\sim \Theta$ is a formula of order $k + 1$. If Θ is a formula of order k and ξ is any variable, then $(E\xi)\Theta$ is a formula of order $k + 1$. If Θ and Φ are formulas of order at most k, and one of them is of order k, then $(\Theta \wedge \Phi)$ and $(\Theta \vee \Phi)$ are formulas of order $k + 1$. An expression is a formula, if, for some n, it is a formula of order n.

Among the variables occurring in a formula, it is for some purposes convenient to distinguish the so-called "free" variables. We define this notion recursively in the following way: If Φ is an atomic formula, then ξ is *free* in Φ if and only if ξ occurs in Φ; ξ is free in $(E\eta)\Theta$ if and only if η is not the same variable as ξ, and ξ is free in Θ; ξ is free in $\sim \Theta$ if and only if ξ is free in Θ; ξ is free in $(\Theta \wedge \Phi)$, and in $(\Theta \vee \Phi)$, if and only if ξ is free in at least one of the two formulas Θ and Φ. Thus, for example, x is free in the formulas

$$x = 1$$

$$x = x$$
$$(Ey)\,(y = x)$$
$$(x = 1) \vee (Ex)\,(x = 2)$$

but not in the formulas

$$y = 1$$
$$(Ex)\,(x = x)$$
$$(Ex)(Ey)\,(y = x)\,.$$

(For certain purposes a more subtle notion is needed: that of a variable's being free, or not free, at a certain occurrence in a formula. Thus, for instance, in the formula

$$(Ey)\,(y = x) \wedge (Ex)\,[x + y = x \cdot w]\,,$$

the variable x is free at its first occurrence – reading from left to right – but not in its other occurrences. This notion is not necessary for our discussion, however, so we shall not give a more exact explanation of it.)

It is convenient to introduce some abbreviated notation (see Note 8 in Section 5). If Θ and Φ are any formulas, then we regard

$$(\Theta \rightarrow \Phi)$$

as an abbreviation for

$$(\sim \Theta \vee \Phi)\,,$$

and

$$(\Theta \leftrightarrow \Phi)$$

as an abbreviation for

$$(\Theta \rightarrow \Phi) \wedge (\Phi \rightarrow \Theta)\,.$$

The sign $\rightarrow$ is called the *implication sign*, and the sign $\leftrightarrow$ is called the *equivalence sign*. If Θ and Φ are any formulas, then the formula $\Theta \rightarrow \Phi$ is called an *implication*; we call Θ the *antecedent* or *hypothesis* and Φ the *consequent* or *conclusion* of this implication.

If Θ is any formula, and ξ is any variable, then

$$(A\xi)\Theta$$

is an abbreviation for

$$\sim (E\xi) \sim \Theta.$$

We also introduce notation to represent disjunctions and conjunctions of arbitrarily many formulas. In the simplest case the formulas in question are arranged in a finite sequence $\Theta_1, \Theta_2, \ldots, \Theta_n$; we then denote the disjunction of the formulas by

$$\bigvee_{1 \leq i \leq n} \Theta_i$$

or

$$\Theta_1 \vee \Theta_2 \vee \cdots \vee \Theta_n$$

and their conjunctions by

$$\bigwedge_{1 \leq i \leq n} \Theta_i$$

or

$$\Theta_1 \wedge \Theta_2 \wedge \cdots \wedge \Theta_n.$$

A recursive definition of these symbolic expressions hardly needs to be formulated explicitly here. Sometimes we are confronted with more involved cases: for instance, we may have a finite set S of ordered couples (n, p), a formula $\Theta_{n,p}$ being correlated with each member (n, p) of S. To denote the disjunction and conjunction of all such formulas $\Theta_{n,p}$ we use the symbolic expressions

$$\bigvee_{(n,p) \in S} \Theta_{n,p}$$

and

$$\bigwedge_{(n,p) \in S} \Theta_{n,p}$$

(where "$(n, p) \in S$" may be replaced by formulas defining the set S). To ascribe to these symbolic expressions an unambiguous meaning we have of course to specify the order in which the formulas $\Theta_{n,p}$ are taken in forming the disjunction or conjunction. The way in which this order is specified is immaterial for our purposes; we can, if we wish, specify once and for all that the formulas are taken in lexicographical order of their indices; thus, for instance, the symbolic expression

$$\bigvee_{\substack{n+p \leq 4 \\ 1 \leq n,p}} \Theta_{n,p}$$

will denote the disjunction

$$\left(\left[\left\{ \left[(\Theta_{1,1} \vee \Theta_{1,2}) \vee \Theta_{1,3} \right] \vee \Theta_{2,1} \right\} \vee \Theta_{2,2} \right] \vee \Theta_{3,1} \right).$$

Analogous notations are used for disjunctions and conjunctions of finite systems of formulas which are correlated, not with couples, but with triples, quadruples,..., or even arbitrary finite sequences, of integers.

We also need a symbolism to denote arbitrarily long sequences of quantifier expressions. For this purposes we shall use exclusively the "three-dot" notation:

$$(E\xi_1) \cdots (E\xi_n)\Theta$$

and

$$(A\xi_1) \cdots (A\xi_n)\Theta,$$

where $\xi_1, \ldots, \xi_n$ are arbitrary variables, and Θ is an arbitrary formula.

A formula is called a *sentence*, if it contains no free variables. Thus, for example, the following are sentences:

$$0 = 0$$
$$0 = 1$$
$$(0 = 0) \wedge (1 = 1)$$
$$(Ex) (0 = 0)$$
$$(Ex) (x = 0) \wedge (Ey) \sim (y = 0)$$
$$(Ex)(Ey) (y > x).$$

On the other hand, the following are not sentences since they contain free variables:

$$x > 0$$
$$(Ey) (x > 0)$$
$$(Ex) (x > [(x + 1) + (y \cdot y)]).$$

It should be noticed that while a sentence is either true or false, this is not the case for a formula with free variables, which in general will be satisfied by some values of the free variables and not satisfied by others.

The notion of the truth of a sentence will play a fundamental role in our further discussion. It will occur either explicitly or, more often, implicitly; in fact, in terms of the notion of truth we shall define that of the equivalence of two formulas, and the latter notion will be involved in numerous theorems of Section 3, which are essential for establishing the decision method. We shall use the notion of truth intuitively, without defining it in a formal way. We hope, however, that the correctness of the theorems involving the notion of truth will be apparent to anyone who grasps the mathematical content of our arguments. No one will doubt, for instance, that a sentence of elementary algebra like

$$(Ax)(Ay) [(x + y) = (y + x)]$$

is true, and that the sentence

$$(Ax)(Ay) [(x - y) = (y - x)]$$

is false (see Note 9 in Section 5).

As examples of general laws involving the notion of truth, we give the following:

If Θ is a sentence, then $\sim \Theta$ is true if and only if Θ is not true. If Θ and Φ are sentences, then $(\Theta \wedge \Phi)$ is true if and only if Θ and Φ are both true. If Θ and Φ are sentences, then $(\Theta \vee \Phi)$ is true if and only if at least one of the sentences Θ and Φ is true; $\Theta \to \Phi$ is true if and only if either Θ is not true, or Φ is true; and $\Theta \leftrightarrow \Phi$ is true if and only if Θ and Φ are both true or both false.

Let Θ and Φ be any formulas, and let $\xi_1, \xi_2, \ldots, \xi_n$ be the totality of free variables that occur in Θ and Φ or both.

Then if the sentence

$$(A\xi_1) \cdots (A\xi_n)\, (\Theta \leftrightarrow \Phi)$$

is true, we say that Θ and Φ are *equivalent*.

Thus, for example, the following two formulas are equivalent:

$$(x > 0) \vee (x = 0), \ \ (Ey)\,(x = y \cdot y).$$

(Notice that neither of these formulas is equivalent to

$$(z > 0) \vee (z = 0)$$

since the latter contains "z" instead of "x".)

We now have some simple but very useful theorems regarding this notion of equivalence; they will be used in the subsequent discussion without explicit references.

A. *The relation of equivalence is symmetric, reflexive, and transitive.*
B. *Let Θ_1 and Θ_2 be equivalent formulas, and suppose that the formula Ψ_2 arises from the formula Ψ_1 by replacing Θ_1 by Θ_2 at one or more places. Then Ψ_1 is equivalent to Ψ_2.*

The proof of A and B presents no difficulty; in establishing B we apply induction with respect to the order of Ψ_1.

It is also convenient to define an equivalence relation for terms. Let α and β be any terms, and let $\xi_1, \xi_2, \ldots, \xi_n$ be the variables which occur in α or β or both. Then if the sentence

$$(A\xi_1) \cdots (A\xi_n)\, (\alpha = \beta)$$

is true, we say that α and β are *equivalent*.

The fundamental theorems regarding the equivalence of terms are analogous to those concerning equivalence of formulas. In fact we have:

C. *The relation of equivalence of terms is symmetric, reflexive, and transitive.*
D. *If α_1 and α_2 are equivalent terms, and if the term β_2 arises from the term β_1 by replacing α_1 by α_2 at one or more places, then β_1 is equivalent to β_2.*
E. *If α_1 and α_2 are equivalent terms, and if the formula Ψ_2 arises from the formula Ψ_1 by replacing α_1 by α_2 at one or more places, then Ψ_1 is equivalent to Ψ_2.*

3   Decision Method for Elementary Algebra

The decision method for elementary algebra which will be explained in this section can be properly characterized as the "method of eliminating quantifiers" (see Notes 10 and 11 in Section 5). It falls naturally into two parts. The first, essential, part consists in a procedure by means of which, given any formula Θ, one can always find in a mechanical way an equivalent formula which involves no quantifiers, and no free variables besides those already occurring in Θ; in particular, this procedure enables us, given any sentence, to find an equivalent sentence without quantifiers. Mathematically, this part of the decision method coincides with the extension of Sturm's theorem mentioned in the Introduction. The second part consists in a procedure by means of which, given any sentence Θ without quantifiers, one can always decide in a mechanical way whether Θ is true. It is obvious that these two procedures together provide the desired decision method.

In order to establish the first half of the decision method, we proceed by induction on the order of a formula. As is easily seen (using the elementary properties of equivalence of formulas mentioned in Section 2) it suffices to describe a procedure by means of which, given a formula $(E\xi)\Theta$, where Θ contains no quantifiers, one can always find an equivalent formula Φ, without quantifiers, and such that every variable in Φ is free in $(E\xi)\Theta$; i.e., to give a method of eliminating the quantifier from $(E\xi)\Theta$. Actually, it turns out to be convenient to do slightly more: i.e., to give a method of eliminating the quantifier from $(E_k\xi)\Theta$, where the prefix "$(E_k x)$" is to be read "there exist exactly k values of x such that."

Definition 1 *Let* $\alpha_0, \alpha_1, \ldots, \alpha_n$ *be terms which do not involve* ξ. *Then the term*

$$\alpha_0 + \alpha_1 \cdot \xi + \cdots + \alpha_n \cdot \xi^n$$

is called a polynomial in ξ. *We say that the degree of this polynomial is* n, *and that* $\alpha_0, \ldots, \alpha_n$ *are its coefficients:* α_n *is called the leading coefficient.*

Remark 1 *Our definition of the degree of polynomials differs slightly from the one usually given in algebra, in that we do not require that the leading coefficient be different from zero. Thus we call*

$$1 + (1+1)\,x + (1-1)\,x^2$$

a polynomial of the second degree, not of the first degree.

Definition 2 *Let* α *and* β *be polynomials in* ξ *of degrees* m *and* n *respectively: i.e., let*

$$\begin{aligned}
\alpha &\equiv \alpha_0 + \alpha_1 \cdot \xi + \cdots + \alpha_m \cdot \xi^m, \\
\beta &\equiv \beta_0 + \beta_1 \cdot \xi + \cdots + \beta_n \cdot \xi^n,
\end{aligned}$$

where $\alpha_0, \ldots, \alpha_m$ *and* $\beta_0, \ldots, \beta_n$ *are terms which do not involve* ξ. *Let* r *be the minimum of the integers* m *and* n, *and let* s *be their maximum. Let*

$$\gamma_i \equiv \alpha_i + \beta_i \text{ for } i \leq r.$$

If $m < n$, let

$$\gamma_i \equiv \beta_i \ for \ r < i \le s.$$

If $m > n$, let

$$\gamma_i \equiv \alpha_i \ for \ r < i \le s.$$

Then we set

$$\alpha +_\xi \beta \equiv \gamma_0 + \gamma_1 \cdot \xi + \cdots + \gamma_s \cdot \xi^s.$$

Definition 3 *Let*

$$\alpha \equiv \alpha_0 + \alpha_1 \cdot \xi + \cdots + \alpha_m \cdot \xi^m$$

be a polynomial in ξ. Then by the first reductum (or, simply, the reductum) of α we mean the polynomial obtained by leaving off the term $\alpha_m \cdot \xi^m$: i.e., we set

$$Rd_\xi(\alpha) \equiv \alpha_0 + \alpha_1 \cdot \xi + \cdots + \alpha_{m-1} \cdot \xi^{m-1};$$

if $m = 0$ (so that α does not involve ξ at all) we set

$$Rd_\xi(\alpha) \equiv 0.$$

We define reducta of all orders recursively, by setting

$$
\begin{aligned}
Rd_\xi^0(\alpha) &\equiv \alpha \\
Rd_\xi^{k+1}(\alpha) &\equiv Rd_\xi \left[Rd_\xi^k(\alpha) \right].
\end{aligned}
$$

The following theorem is easily established by an induction on the degree of α.

Theorem 1 *If α is a polynomial in ξ, then $Rd_\xi(\alpha)$ is also a polynomial in ξ (whose coefficients, of course, are the same as certain of the coefficients of α – and hence contain no variables except those occurring in the coefficients of α). If α is of degree $m > 0$, then $Rd_\xi(\alpha)$ is of degree $m - 1$.*

We make use of Theorem 1 in defining recursively the product of two polynomials:

Definition 4 *Let*

$$
\begin{aligned}
\alpha &\equiv \alpha_0 + \alpha_1 \cdot \xi + \cdots + \alpha_m \cdot \xi^m \\
\beta &\equiv \beta_0 + \beta_1 \cdot \xi + \cdots + \beta_n \cdot \xi^n
\end{aligned}
$$

be polynomials in ξ of degrees m and n respectively. If $m = 0$ then we set

$$\alpha \cdot_\xi \beta \equiv (\alpha \cdot \beta_0) + (\alpha \cdot \beta_1) \cdot \xi + \cdots + (\alpha \cdot \beta_n) \cdot \xi^n.$$

If $m > 0$, let

$$\begin{array}{rcl} \gamma_i & \equiv & 0 \ for \ i < m \\ \gamma_m & \equiv & \alpha_m \cdot \beta_0 \\ \gamma_{m+1} & \equiv & \alpha_m \cdot \beta_1 \\ & \vdots & \\ \gamma_{m+n} & \equiv & \alpha_m \cdot \beta_n, \end{array}$$

and we set

$$\alpha \cdot_\xi \beta \equiv [Rd_\xi(\alpha) \cdot_\xi \beta] +_\xi \left(\gamma_0 + \gamma_1 \cdot \xi + \cdots + \gamma_{m+n} \cdot \xi^{m+n}\right).$$

Definition 5 *If α and β are polynomials in ξ, then we set*

$$\alpha -_\xi \beta \equiv \alpha +_\xi [(-1) \cdot_\xi \beta].$$

Theorem 2 *If α and β are polynomials in ξ, then $\alpha +_\xi \beta$, $\alpha \cdot_\xi \beta$, and $\alpha -_\xi \beta$ are polynomials in ξ.*

Proof. Obvious from the definitions. ∎

Definition 6 *If $\alpha \equiv \xi$, then we set*

$$P_\xi(\alpha) \equiv 0 + 1 \cdot \xi.$$

If α is a constant $(0,1,$ or $-1)$, or a variable different from ξ, then we set

$$P_\xi(\alpha) \equiv \alpha.$$

If α and β are arbitrary terms, then we set

$$\begin{array}{rcl} P_\xi(\alpha + \beta) & \equiv & P_\xi(\alpha) +_\xi P_\xi(\beta) \\ P_\xi(\alpha \cdot \beta) & \equiv & P_\xi(\alpha) \cdot_\xi P_\xi(\beta). \end{array}$$

Theorem 3 *If α is any term, and ξ is any variable, then $P_\xi(\alpha)$ is a polynomial in ξ, and is equivalent to α.*

Proof. By an induction on the order of α, making use of Theorem 2. ∎

Remark 2 *It will be seen that if α is any term, then $P_\xi(\alpha)$ is the polynomial which results from "multiplying out" and "arranging in increasing powers of ξ." It is convenient to extend the definition of P_ξ so that it will be defined not only for all terms but for all formulas without quantifiers. This is done in our next definition. The intuitive significance of $P_\xi(\Theta)$, when Θ is a formula, will become clear in Theorem 4.*

Definition 7 *For all terms α and β, and for all formulas Θ and Φ, we set*

(i) $P_\xi(\alpha = \beta) \equiv P_\xi(\alpha - \beta) = 0$

(ii) $P_\xi(\alpha > \beta) \equiv P_\xi(\alpha - \beta) > 0$

(iii) $P_\xi[\sim (\alpha = \beta)] \equiv [P_\xi(\alpha > \beta) \vee P_\xi(\beta > \alpha)]$

(iv) $P_\xi[\sim (\alpha > \beta)] \equiv [P_\xi(\alpha = \beta) \vee P_\xi(\beta > \alpha)]$

(v) $P_\xi(\Theta \vee \Phi) \equiv [P_\xi(\Theta) \vee P_\xi(\Phi)]$

(vi) $P_\xi(\Theta \wedge \Phi) \equiv [P_\xi(\Theta) \wedge P_\xi(\Phi)]$

(vii) $P_\xi[\sim (\Theta \vee \Phi)] \equiv P_\xi(\sim \Theta) \wedge P_\xi(\sim \Phi)]$

(viii) $P_\xi[\sim (\Theta \wedge \Phi)] \equiv P_\xi(\sim \Theta) \vee P_\xi(\sim \Phi)]$

(ix) $P_\xi(\sim\sim \Theta) \equiv P_\xi(\Theta).$

Theorem 4 *Let Θ be any formula without quantifiers, and let ξ be any variable. Then $P_\xi(\Theta)$ is equivalent to Θ. Moreover, $P_\xi(\Theta)$ is a formula built up by means of conjunction and disjunction signs (but without using negation signs) from atomic formulas of the form*

$$\alpha = 0$$

and

$$\alpha > 0,$$

where α is a polynomial in ξ.

Proof. We prove that $P_\xi(\Theta)$ is equivalent to Θ by an induction on the order of Θ. If Θ is of first order, the theorem is obvious by 7 (i), 7 (ii), 3, and the following facts: $\alpha = \beta$ is equivalent to $\alpha - \beta = 0$; and $\alpha > \beta$ is equivalent to $\alpha - \beta > 0$. In order to carry out the recursive step, we make use of the facts: that $\sim (\alpha = \beta)$ is equivalent to $(\alpha > \beta) \vee (\beta > \alpha)$; that $\sim (\alpha > \beta)$ is equivalent to $(\alpha = \beta) \vee (\beta > \alpha)$; that $\sim (\Theta \vee \Phi)$ is equivalent to $\sim \Theta \wedge \sim \Phi$; that $\sim (\Theta \wedge \Phi)$ is equivalent to $\sim \Theta \vee \sim \Phi$; and that $\sim\sim \Theta$ is equivalent to Θ.

The second part of the theorem can also be proved by an induction on the order of Θ, making use of Theorem 3. ∎

Given any formula Θ without quantifiers, we have thus obtained an equivalent formula $\Psi \equiv P_\xi(\Theta)$ which contains no quantifiers and no negation signs. We are now going to define an operator Q which subjects any formula Ψ of this kind to further transformations by applying mainly the distributive law of sentential calculus, so as to bring Ψ to the so-called "disjunctive normal form."

Definition 8 *If Φ is an atomic formula, then we set*

$$Q(\Phi) \equiv \Phi.$$

If

$$Q(\Phi_1) \equiv \bigvee_{i \leq m} \bigwedge_{j \leq m_i} \Psi_{i,j},$$

and

$$Q(\Phi_2) \equiv \bigvee_{m < i \leq m+n} \bigwedge_{j \leq m_i} \Psi_{i,j}$$

where $\Psi_{i,j}$ (for $i \leq m + n$ and $j \leq m_i$) is an atomic formula, then we set

$$Q(\Phi_1 \vee \Phi_2) \equiv \bigvee_{i \leq m+n} \bigwedge_{j \leq m_i} \Psi_{i,j}$$

and

$$Q(\Phi_1 \wedge \Phi_2) \equiv \bigvee_{\substack{i \leq m \\ m < j \leq m+n}} \left(\Psi_{i,1} \wedge \cdots \wedge \Psi_{i,m_i} \wedge \Psi_{j,1} \wedge \cdots \wedge \Psi_{j,m_j} \right).$$

Theorem 5 *If Φ is any formula which involves no negation signs or quantifiers, then $Q(\Phi)$ is a disjunction of conjunctions of atomic formulas. Moreover, $Q(\Phi)$ is equivalent to Φ.*

Proof. By induction on the order of Φ, making use of the following fact: for any formulas Φ, Θ, and Ψ, the formulas $\Phi \wedge (\Theta \vee \Psi)$ and $(\Phi \wedge \Theta) \vee (\Phi \wedge \Psi)$ are equivalent, as are also the formulas $\Phi \wedge (\Theta \wedge \Psi)$ and $(\Phi \wedge \Theta) \wedge \Psi$, as well as the formulas $\Phi \vee (\Theta \vee \Psi)$ and $(\Phi \vee \Theta) \vee \Psi$. ∎

Theorem 6 *Let Φ be any formula without quantifiers, and let ξ be any variable. Then $QP_\xi(\Phi)$ is a disjunction of conjunctions of atomic formulas – each of the atomic formulas in question having a polynomial in ξ for its left member and 0 for its right member. Moreover, $Q\Phi_\xi(\Phi)$ is equivalent to Φ.*

Proof. By Theorems 4 and 5; $QP_\xi(\Phi)$ is of course used here to mean $Q[P_\xi(\Phi)]$. ∎

We now introduce the notion of a derivative (with respect to a given variable).

Definition 9 *If*

$$\alpha \equiv \alpha_0 + \alpha_1 \cdot \xi + \cdots + \alpha_n \cdot \xi^n$$

is a polynomial in ξ, of degree $n > 0$, then we put (writing "2" for "1 + 1", etc.)

$$D_\xi(\alpha) \equiv \alpha_1 + (2 \cdot \alpha_2) \cdot \xi + \cdots + (n \cdot \alpha_n) \cdot \xi^{n-1}.$$

If α is of degree zero in ξ, we set

$$D_\xi(\alpha) \equiv 0.$$

Remark 3 *The notion of a derivative can of course be extended to arbitrary terms which are not formally polynomials in ξ according to Definition 1 by putting*

$$D_\xi(\alpha) \equiv D_\xi P_\xi(\alpha).$$

Theorem 7 *If α is a polynomial in ξ, so is $D_\xi(\alpha)$.*

Proof. By Definitions 1 and 9. ∎

We also define derivatives of arbitrary order as follows:

Definition 10 *If α is any term, and ξ is any variable, we set*

$$
\begin{aligned}
D_\xi^0(\alpha) &\equiv \alpha \\
D_\xi^{k+1}(\alpha) &\equiv D_\xi[D_\xi^k(\alpha)].
\end{aligned}
$$

Theorem 8 *If α is a polynomial in ξ, and k is a non-negative integer, then $D_\xi^k(\alpha)$ is a polynomial in ξ.*

Proof. By Theorem 7 and Definition 10. ∎

The operator M which will be introduced next correlates, with every polynomial α, every variable ξ, and every non-negative integer n, a formula $M_\xi^n(\alpha)$, which in intuitive interpretation means that ξ is a root of α of order n. In case $n = 0$, this formula means simply that ξ is not a root of α. In this connection the formula $M_\xi^n(\alpha)$ will be read "the number ξ is of order n in the polynomial α," independent of whether n is positive or equal to zero.

Definition 11 *Let α be any polynomial in ξ. If n is any positive integer, we set*

$$M_\xi^n(\alpha) \equiv \left\{ \left(\bigwedge_{1 \le i \le n} \left[D_\xi^{i-1}(\alpha) = 0 \right] \right) \wedge \sim \left[D_\xi^n(\alpha) = 0 \right] \right\}.$$

We set, in addition,

$$M_\xi^0(\alpha) \equiv \sim (\alpha = 0).$$

We now introduce by definition a new kind of existential quantifier, which may be called the *numerical* existential quantifier. If n is any non-negative integer, ξ any variable, and Φ any formula, then $(E_n\xi)\Phi$ is to be interpreted intuitively as meaning that there exist exactly n values of ξ which make Φ true.

Definition 12 *Let ξ be any variable, and let Φ be any formula. We set*

$$(E_0\xi)\Phi \equiv (A\xi) \sim \Phi.$$

Let n be any positive integer, and let $\eta_1, \ldots, \eta_n$ be the first n variables (in the sequence of all variables) which do not occur in Φ and are different from ξ. Then we set

$$(E_n\xi)\Phi \;\equiv\; \Big\{ (E\eta_1)\cdots(E\eta_n)\Big(\bigwedge_{1\leq i<j\leq n} \sim (\eta_i = \eta_j) \wedge$$

$$(A\xi)\big[\Phi \leftrightarrow \bigvee_{1\leq i\leq n} (\eta_i = \xi)\big]\Big)\Big\}.$$

We next introduce an operator F with a more complicated and technical interpretation. If n is an integer, ξ a variable, and α and β any polynomials, then $F_\xi^n(\alpha, \beta)$ is a formula to be intuitively interpreted as meaning that there are exactly n numbers ξ which satisfy the following conditions: (1) ξ is a root of higher order of α than of β, and the difference between these two orders is an odd integer; (2) there exists an open interval, whose right end-point is ξ, within which α and β have the same sign. The exact form of the symbolic expression used to define $F_\xi^n(\alpha, \beta)$ will probably seem strange at first glance even to those acquainted with logical symbolism; we have chosen this form so as to avoid the necessity of introducing a notation to indicate the result of replacing one variable by another in a given term. (An analogous remark applies to some other symbolic formulations given elsewhere in this work -- in particular, in Note 9.) It will be noticed that the variable ξ is not free in $F_\xi^n(\alpha, \beta)$.

Definition 13 *Let α be a polynomial of degree p in ξ, and let β be a polynomial of degree q in ξ. Let η_1 and η_2 be the first two variables which are different from ξ and which do not occur in α or β. Then we set*

$$F_\xi^n(\alpha, \beta) \;\equiv\; (E_n\xi)\Bigg\{ \bigvee_{\substack{0\leq k\leq q \\ 0\leq 2m\leq p-k-1}} \Big[M_\xi^{k+2m+1}(\alpha) \wedge M_\xi^k(\beta)\Big] \wedge$$

$$(E\eta_1)(E\eta_2)\Big[(\eta_1 = \xi) \wedge (\xi > \eta_2) \wedge$$

$$(A\xi)\big\{ [(\xi > \eta_2) \wedge (\eta_1 > \xi)] \rightarrow (\alpha \cdot \beta > 0)\big\}\Big]\Bigg\}.$$

The operator G which will now be defined is closely related to the operator F. In fact, α and β being polynomials in ξ, $G_\xi^n(\alpha, \beta)$ has the following meaning: if n_1 is the integer for which $F_\xi^{n_1}(\alpha, \beta)$ holds and n_2 is the integer for which $F_\xi^{n_2}(\alpha, \beta')$ holds (where β' is the negative of β -- i.e., the polynomial obtained by multiplying β by -1), then $n = n_1 - n_2$; the integer n may be positive, zero, or negative. Remembering the intuitive meaning of $F_\xi^n(\alpha, \beta)$, the intuitive meaning of $G_\xi^n(\alpha, \beta)$ now becomes clear.

Definition 14 *Let n be any integer (positive, negative, or zero), and let α and β be any polynomials in ξ. Let k be the maximum of the degrees of α and β. Then we set*

$$G_\xi^n(\alpha,\beta) \equiv \bigvee_{\substack{0 \le m \le k \\ 0 \le m+n \le k}} \left[F_\xi^{n+m}(\alpha,\beta) \wedge F_\xi^m(\alpha,(-1)\cdot_\xi \beta) \right].$$

We need also the notion of the *remainder* obtained by dividing one polynomial by another. For our purposes, however, it turns out to be slightly more convenient to introduce a notation for the *negative of the remainder*.

Definition 15 *Let ξ be a variable. Let α be a polynomial of degree m in ξ, whose leading coefficient is α_m. Let β be a polynomial of degree n in ξ, whose leading coefficients is β_n. If $m < n$, we set*

$$R_\xi(\alpha,\beta) \equiv (-1)\cdot_\xi \alpha.$$

If $m = n$, we set

$$R_\xi(\alpha,\beta) \equiv Rd_\xi P_\xi \left(\alpha_m \cdot \beta_n \cdot \beta - \beta_n^2 \cdot \alpha \right).$$

If $m > n$, we set

$$R_\xi(\alpha,\beta) \equiv R_\xi \left\{ Rd_\xi P\xi \left(\beta_n^2 \cdot \alpha - \alpha_m \cdot \beta_n \cdot \xi^{m-n} \cdot \beta \right), \beta \right\}.$$

Theorem 9 *If α and β are two polynomials in a variable ξ, then $R_\xi(\alpha,\beta)$ is again a polynomial in ξ whose coefficients contain no variables except those occurring in the coefficients of α and β. If β is a polynomial of degree $n > 0$, then $R_\xi(\alpha,\beta)$ is of degree less than n. If β is of degree zero, then $R_\xi(\alpha,\beta) \equiv 0$.*

It should be noticed that our definition of the negative of the remainder diverges somewhat from that which would normally be given in a textbook of algebra. According to the usual definition of the remainder, the negative of the remainder obtained by dividing a polynomial α of degree m by a polynomial β of degree n – both in the variable ξ – is a polynomial δ of a degree lower than n, such that, for some polynomial γ, the equation

$$\alpha = \beta \cdot \gamma - \delta$$

is satisfied identically. The coefficients of γ and δ can be obtained from those of α and β be means of the four rational operations, division included. We have modified this definition so as to eliminate division, which as we know, is not available in our system. In consequence, we cannot construct for the negative remainder in our sense a polynomial γ which satisfies the above equation. We have instead:

Theorem 10 *Let α and β be any polynomials in a variable ξ, of degrees m and n, respectively, and let β_n be the leading coefficient of β. We set $q = 0$ in case $m < n$, and $q = m - n + 1$ in case $m \geq n$. Then there is a polynomial γ in ξ, whose coefficients contain no variables except those occurring in the coefficients of α and β, and for which $\alpha \cdot \beta_n^{2q}$ and $\beta \cdot \gamma - R_\xi(\alpha, \beta)$ are equivalent.*

Proof. By induction on the difference of the degrees of α and β. ∎

One rather undesirable consequence of our modification of the notion of a negative remainder is that, in case the leading coefficient β_n of β is 0, all of the coefficients of $R_\xi(\alpha, \beta)$ prove to be terms equivalent to 0. No difficulty will arise from this fact, however, since we shall never use $R_\xi(\alpha, \beta)$ except when conjoined with the hypothesis $\sim (\beta_n = 0)$.

It should be pointed out that our negative remainder $R_\xi(\alpha, \beta)$ is still a polynomial of lower degree than β – except when β is of degree zero, in which case $R_\xi(\alpha, \beta) \equiv 0$. This circumstance, together with the analogous property of the reductum of a polynomial (see Theorem 1) will be the basis for some recursive definitions given in our later discussion.

The following three definitions (16, 17, and 18) and the theorems which follow them (11, 12, and 13) are of crucial importance for the decision method under discussion. In these definitions we introduce three operators S, T, and U which correlate, with certain formulas Φ, new formulas $S(\Phi)$, $T(\Phi)$, and $U(\Phi)$ containing no quantifiers; and in the subsequent theorems we show that the correlated formulas are always equivalent to the original ones. The operator S is defined only for rather special formulas – in fact, for those of the form $G_\xi^k(\alpha, \beta)$. The operator T, which is constructed with the help of S, is defined for a rather extensive class of formulas, which contains formulas like

$$(E_k \xi)(\alpha = 0) \ , \ (E_k \xi)\left[(\alpha = 0) \wedge (\beta > 0)\right]$$

(where α and β are any polynomials in ξ), and some related but more complicated types of formulas. The operator U, finally, constructed in terms of T, is defined for all possible formulas; hence Theorem 13, which establishes the equivalence of Φ and $U(\Phi)$, provides us with a universal method of eliminating quantifiers. It may be pointed out that the operator U, though constructed with the help of T, and thus indirectly of S, is not an extension of either of these operators; thus, if Φ is a formula for which S is defined, then $S(\Phi)$ and $U(\Phi)$ are in general formally different, though equivalent, formulas.

Definition 16 *Let k be an integer, and let α and β be polynomials in a variable ξ of degrees m and n respectively and having leading coefficients α_m and β_n; and let*

$$\Phi \equiv G_\xi^k(\alpha, \beta).$$

(i) *If α or β is the polynomial 0, we set*

$$S(\Phi) \equiv (0 = 0), \; \text{for } k = 0$$

and

$$S(\Phi) \equiv (0 = 1), \; \text{for } k \neq 0.$$

(ii) *If neither α nor β is the polynomial 0, and if $m + n$ is even, we set*

$$
\begin{aligned}
S(\Phi) \; \equiv \; & \Big\{ \big[(\alpha_m = 0) \wedge SG_\xi^k \left(Rd_\xi(\alpha), \beta \right) \big] \vee \\
& \big[(\beta_n = 0) \wedge SG_\xi^k \left(\alpha, Rd_\xi(\beta) \right) \big] \vee \\
& \big[\sim (\alpha_m \cdot \beta_n \doteq 0) \wedge SG_\xi^k \left(\beta, R_\xi(\alpha, \beta) \right) \big] \Big\}.
\end{aligned}
$$

(iii) *If neither α nor β is the polynomial 0, and if $m + n$ is odd, we set*

$$
\begin{aligned}
S(\Phi) \; \equiv \; & \Big\{ \big[(\alpha_m = 0) \wedge SG_\xi^k \left(Rd_\xi(\alpha), \beta \right) \big] \vee \\
& \big[(\beta_n = 0) \wedge SG_\xi^{\prime k} \left(\alpha, Rd_\xi(\beta) \right) \big] \vee \\
& \big[(\alpha_m \cdot \beta_n > 0) \wedge SG_\xi^{k+1} \left(\beta, R_\xi(\alpha, \beta) \right) \big] \vee \\
& \big[(0 > \alpha_m \cdot \beta_n) \wedge SG_\xi^{k-1} \left(\beta, R_\xi(\alpha, \beta) \right) \big] \Big\}.
\end{aligned}
$$

Theorem 11 *Let Φ be one of the formulas for which the operator S is defined (by 16). Then $S(\Phi)$ is a formula which contains no quantifiers, and no variables except those that occur free in Φ. Moreover, Φ is equivalent to $S(\Phi)$.*

Proof. The first part follows immediately from Definition 16.

To show the second part we consider the recursive definition for S given in 16. In view of this definition, we easily see that it suffices to establish what follows (for arbitrary polynomials α and β of degrees m and n respectively in a variable ξ, with leading coefficients α_m and β_n, and for an arbitrary integer k):

(1) if α or β is the polynomial 0, then $G_\xi^k(\alpha, \beta)$ is equivalent to $(0 = 0)$ for $k = 0$, and to $(0 = 1)$ for $k \neq 0$;

(2) if $m + n$ is even, then $G_\xi^k(\alpha, \beta)$ is equivalent to

$$
\begin{aligned}
& \Big\{ \big[(\alpha_m = 0) \wedge G_\xi^k \left(Rd_\xi(\alpha), \beta \right) \big] \vee \big[(\beta_n = 0) \wedge G_\xi^k \left(\alpha, Rd_\xi(\beta) \right) \big] \vee \\
& \big[\sim (\alpha_m \cdot \beta_n = 0) \wedge G_\xi^k \left(\beta, R_\xi(\alpha, \beta) \right) \big] \Big\};
\end{aligned}
$$

(3) if $m + n$ is odd, then $G_\xi^k(\alpha, \beta)$ is equivalent to

$$
\begin{aligned}
& \Big\{ \big[(\alpha_m = 0) \wedge G_\xi^k \left(Rd_\xi(\alpha), \beta \right) \big] \vee \big[(\beta_n = 0) \wedge G_\xi^k \left(\alpha, Rd_\xi(\beta) \right) \big] \vee \\
& \big[(\alpha_m \cdot \beta_n > 0) \wedge G_\xi^{k+1} \left(\beta, R_\xi(\alpha, \beta) \right) \big] \vee \\
& \big[(0 > \alpha_m \cdot \beta_n) \wedge G_\xi^{k-1} \left(\beta, R_\xi(\alpha, \beta) \right) \big] \Big\}.
\end{aligned}
$$

Let $\xi_1, \ldots, \xi_s$ be all the variables which occur in the coefficients of α or β or both.

It is easily seen that the proof of (1) reduces to showing that, in case $\alpha \equiv 0$ or $\beta \equiv 0$, the following sentences are true:

$$(A\xi_1) \cdots (A\xi_s) G_\xi^k(\alpha, \beta), \text{ for } k = 0;$$
$$(A\xi_1) \cdots (A\xi_s) \sim G_\xi^k(\alpha, \beta), \text{ for } k \neq 0.$$

Now, we notice by Definition 9 that, for every non-negative integer p, $D_\xi^p(0) \equiv 0$. Hence we conclude by Definition 11 that, in case $\alpha \equiv 0$ or $\beta \equiv 0$, the formula $F_\xi^k(\alpha, \beta)$ is satisfied by all values of $\xi_1, \ldots, \xi_s$ if $k = 0$, and by no such values if $k \neq 0$. From Definition 14 we then easily see that the same applies to the formula $G_\xi^k(\alpha, \beta)$; and this is just what we wanted to show.

Analogously, by means of easy transformations, we see that the proof of (2) and (3) reduces to showing that the following sentences are true:

(4) $(A\xi_1) \cdots (A\xi_s) \left\{ \sim (\alpha_m \cdot \beta_n = 0) \rightarrow [G_\xi^k(\alpha, \beta) \leftrightarrow G_\xi^k(\beta, R_\xi(\alpha, \beta))] \right\}$, for $m + n$ even;

(5) $(A\xi_1) \cdots (A\xi_s) \left\{ (\alpha_m \cdot \beta_n > 0) \rightarrow [G_\xi^k(\alpha, \beta) \leftrightarrow G_\xi^{k+1}(\beta, R_\xi(\alpha, \beta))] \right\}$, for $m + n$ odd;

(6) $(A\xi_1) \cdots (A\xi_s) \left\{ (0 > \alpha_m \cdot \beta_n) \rightarrow [G_\xi^k(\alpha, \beta) \leftrightarrow G_\xi^{k-1}(\beta, R_\xi(\alpha, \beta))] \right\}$, also for $m + n$ odd.

Actually it turns out to be more convenient to establish, instead of (4), (5) and (6), certain stronger statements. For this purpose we introduce the formula $H_\xi^p(\alpha, \beta)$ expressing the fact that there are just p numbers ξ such that the difference between the order of ξ in α and the order of ξ in β is an odd integer, not necessarily positive. A precise formal definition of $H_\xi^p(\alpha, \beta)$ hardly needs to be given here. The sentences whose truth we want to establish can now be formulated as follows (letting γ and δ be arbitrary polynomials in ξ, whose coefficients involve no variables except $\xi_1, \ldots, \xi_s$, and letting p and q be arbitrary non-negative integers):

(7) $(A\xi_1) \cdots (A\xi_s)$
$$\left\{ [H_\xi^p(\alpha, \beta) \wedge (A\xi)(\alpha \cdot \beta_n^{2q} = \beta \cdot \gamma - \delta) \wedge \sim (\alpha_m \cdot \beta_n = 0)] \rightarrow \right.$$
$$\left. [G_\xi^k(\alpha, \beta) \leftrightarrow G_\xi^k(\beta, \delta)] \right\}, \text{ for } p \text{ even};$$

(8) $(A\xi_1) \cdots (A\xi_s)$
$$\left\{ [H_\xi^p(\alpha, \beta) \wedge (A\xi)(\alpha \cdot \beta_n^{2q} = \beta \cdot \gamma - \delta) \wedge (\alpha_m \cdot \beta_n > 0)] \rightarrow \right.$$
$$\left. [G_\xi^k(\alpha, \beta) \leftrightarrow G_\xi^{k+1}(\beta, \delta)] \right\}, \text{ for } p \text{ odd};$$

(9) $(A\xi_1)\cdots(A\xi_s)$

$$\left\{\left[H_\xi^p(\alpha,\beta)\wedge(A\xi)\left(\alpha\cdot\beta_n^{2q}=\beta\cdot\gamma-\delta\right)\wedge(0>\alpha_m\cdot\beta_n)\right]\right.\to$$

$$\left.\left[G_\xi^k(\alpha,\beta)\leftrightarrow G_\xi^{k-1}(\beta,\delta)\right]\right\},\ \text{also for } p \text{ odd.}$$

It is easily seen that the truth of (7), (8), and (9) implies that of (4), (5), and (6) respectively. We shall sketch the proof of this for the cases (7) and (4). Thus assume (7) to be true and $m+n$ to be even. Consider any fixed but arbitrary set of values of $\xi_1,\ldots,\xi_s$ and suppose the hypothesis of (4) to be satisfied. Let p be the (uniquely determined) integer for which $H_\xi^p(\alpha,\beta)$ is satisfied (by the given values of $\xi_1,\ldots,\xi_s$). An elementary algebraic argument shows that p is congruent to $m+n$ modulo two; therefore p is even, and (7) may be applied. We now set $q=0$ if $m<n$, and $q=m-n+1$ otherwise; and we construct a polynomial γ in ξ, with coefficients involving no variables except those occurring in the coefficients of α or β, and such that

$$\alpha\cdot\beta_n^{2q}=\beta\cdot\gamma-R_\xi(\alpha,\beta)$$

holds for every value of ξ. (Regarding the possibility of constructing such a γ, see Theorem 10.) We then see that the hypothesis of (7) is satisfied with δ replaced by $R_\xi(\alpha,\beta)$. Hence the conclusion of (7) is also satisfied. This conclusion, however, with the indicated replacement, coincides with the conclusion of (4). The proof now reduces to establishing the truth of (7), (8), and (9). It is convenient in this part of the proof to avail ourselves of customary mathematical language and symbolism. Also, we shall not be too meticulous in trying avoid possible confusions between mathematical and metamathematical formulations.

Given a polynomial α and a number λ, we shall denote by $f(\lambda,\alpha)$ the order of λ in α: i.e., the uniquely determined non-negative integer r such that $M_\lambda^r(\alpha)$ holds. The function f is thus defined for every number λ, and for every polynomial α that does not vanish identically.

Similarly, for any given polynomials α and β in ξ we denote by $g(\alpha,\beta)$ the integer k for which $G_\xi^k(\alpha,\beta)$ holds. From the definition of $G_\xi^k(\alpha,\beta)$ (see Definition 14) it follows that such an integer always exists and is uniquely determined. It can be computed in the following way. We consider all these numbers λ for which $f(\lambda,\alpha)-f(\lambda,\beta)$ is positive and odd, and we divide them into two sets P and N; λ belongs to P (or N) if there is an open interval whose right-hand end-point is λ, within which the values of α and β have always the same sign (or always different signs). Both sets P and N are clearly finite, and the difference between the number of elements in P and the number of elements in N is just $g(\alpha,\beta)$. Thus $g(\alpha,\beta)$ can be positive, negative, or zero; in case α or β vanishes identically, $g(\alpha,\beta)=0$.

Finally we introduce the symbol $h(\alpha,\beta)$ to denote the integer p for which $H_\xi^p(\alpha,\beta)$ holds; in other words, $h(\alpha,\beta)$ is the number of all those numbers λ for which $f(\lambda,\alpha)-f(\lambda,\beta)$ is odd – though not necessarily positive.

For later use we state here without proof (which would be quite elementary) the following property of the function f defined above:

(10) Let α, β, γ, and δ be polynomials in ξ, such that

$$\alpha \cdot \beta_n^{2q} = \gamma \cdot \beta - \delta$$

holds for every value of ξ, β_n being the (nonvanishing) leading coefficient of β and q some integer. If, for any given number λ, $f(\lambda, \beta) > f(\lambda, \alpha)$ – so that α, as well as β, does not vanish identically – then $f(\lambda, \alpha) = f(\lambda, \delta)$ (so that δ does not vanish identically either). Similarly, if $f(\lambda, \beta) > f(\lambda, \delta)$, then $f(\lambda, \alpha) = f(\lambda, \delta)$.

We now take up the proof of (7), (8), and (9), which will be done by a simultaneous induction on $h(\alpha, \beta) = p$. The reader can easily verify that (7), (8), and (9) hold in case the polynomial δ vanishes identically; therefore we shall assume henceforth that δ does not vanish identically.

Assume first that $h(\alpha, \beta) = 0$. Thus there are no numbers λ such that $f(\lambda, \alpha) - f(\lambda, \beta)$ is odd. A *fortiori* there are no numbers λ such that $f(\lambda, \alpha) - f(\lambda, \beta)$ is positive and odd; and hence $g(\alpha, \beta) = 0$. Furthermore, there are no numbers λ such that $f(\lambda, \beta) - f(\lambda, \delta)$ is positive and odd; for if such a number λ existed, we should have $f(\lambda, \alpha) = f(\lambda, \delta)$ by (10), and hence $f(\lambda, \alpha) - f(\lambda, \beta)$ would be odd. Consequently, $g(\beta, \delta) = 0$ and therefore $g(\alpha, \beta) = g(\beta, \delta)$. Thus in this case (7) proves to hold, while (8) and (9) are of course vacuously satisfied.

Assume that (7), (8), and (9) have been established for arbitrary polynomials α and β with $h(\alpha, \beta) = p$ (p any given integer). Consider any polynomials α and β in ξ with nonvanishing coefficients α_m and β_n, and with

(11) $h(\alpha, \beta) = p + 1$,

as well as two further polynomials γ and δ in ξ such that

(12) $\alpha \cdot \beta_n^{2q} = \gamma \cdot \beta - \delta$ holds identically for some non-negative integer q.

Two cases can be distinguished here, according as $\alpha_m \cdot \beta_n > 0$ or $0 > \alpha_m \cdot \beta_n$; since the arguments are entirely analogous in both cases, however, we restrict ourselves to the case

(13) $\alpha_m \cdot \beta_n > 0$.

Our assumption (11) implies that there are exactly $p + 1$ numbers λ for which $f(\lambda, \alpha) - f(\lambda, \beta)$ is odd. Let

(14) $\lambda_0 =$ the largest λ such that $f(\lambda, \alpha) - f(\lambda, \beta)$ is odd.

Condition (13) implies that for sufficiently large numbers $\xi > \lambda_0$ the values of α and β are of the same sign. This can be extended to every number $\xi > \lambda_0$ (not a root of α or β), since, by (14), there is no number $\xi > \lambda_0$ for which $f(\xi, \alpha) - f(\xi, \beta)$ is odd (and therefore at which one of the polynomials α and β changes sign while the other does not). Hence, and from the fact that $f(\lambda_0, \alpha) - f(\lambda_0, \beta)$ is odd, we conclude:

(15) There is an open interval whose right-hand end-point is λ_0, within which the values of α and β are everywhere of different signs.

We now introduce three new polynomials α', γ', and δ' by stipulating that the equations

(16) $\alpha' = \alpha \cdot (\lambda_0 - \xi)$, $\quad \gamma' = \gamma \cdot (\lambda_0 - \xi)$, $\quad \delta' = \delta \cdot (\lambda_0 - \xi)$

hold identically. By (12), (13), and (16) we obviously have:

(17) $\alpha' \cdot \beta_n^{2q} = \gamma' \cdot \beta - \delta'$ holds identically for some non-negative integer q;

(18) $\alpha'_{m+1} \cdot \beta_n < 0$, where α'_{m+1} is the leading coefficient of α';

(19) $f(\lambda_0, \alpha') = f(\lambda_0, \alpha) + 1$, and also $f(\lambda_0, \delta') = f(\lambda_0, \delta) + 1$ (since δ and δ' do not vanish identically);

(20) $f(\xi, \alpha') = f(\xi, \alpha)$ for every $\xi \neq \lambda_0$, and similarly $f(\xi, \delta') = f(\xi, \delta)$ for every $\xi \neq \lambda_0$ (since δ and δ' do not vanish identically).

From (19) and (20) we conclude that the set of numbers λ such that $f(\lambda, \alpha') - f(\lambda, \beta)$ is odd differs from the analogous set for α and β only by the absence of λ_0; thus, using also (11),

$$h(\alpha', \beta) = h(\alpha, \beta) - 1 = p.$$

Consequently, our inductive premise applies to the polynomials α' and β; i.e., sentences (7), (8), and (9) are true if α is replaced by α'. Remembering the meaning of $g(\alpha, \beta)$, and taking account of (17) and (18), we conclude:

(21) $g(\alpha', \beta) = g(\beta, \delta')$ in case p is even;
(22) $g(\alpha', \beta) = g(\beta, \delta') + 1$ in case p is odd.

We now want to show that

(23) $g(\alpha, \beta) - g(\beta, \delta) = g(\alpha', \beta) - g(\beta, \delta') - 1$.

To do this, we first notice that by (16):

(24) The values of α and α' are of the same sign for every $\xi < \lambda_0$ (not a root of α); similarly for the values of δ and δ'.

We also observe that:

(25) There is no $\xi > \lambda_0$ such that $f(\xi, \beta) - f(\xi, \delta)$ is positive and odd; similarly for β and δ'.

For, if $f(\xi, \beta) - f(\xi, \delta)$ were positive and odd for some $\xi > \lambda_0$, then, by (10) and (12), $f(\xi, \alpha) - f(\xi, \beta)$ would be odd for the same $\xi > \lambda_0$, and this would contradict (14). The argument for β and δ' is analogous; instead of (12) we use (17), and when stating the final contradiction we refer to (14) combined with first part of (20), instead of merely to (14).

We now distinguish two cases, dependent on the sign of $f(\lambda_0, \alpha) - f(\lambda_0, \beta)$. In view of (20), (24), and (25), the only number which can cause a difference in the values of $g(\alpha, \beta)$ and $g(\alpha', \beta)$, or in the values of $g(\beta, \delta)$ and $g(\beta, \delta')$, is the number λ_0. If now

(A) $f(\lambda_0, \alpha) - f(\lambda_0, \beta) > 0$,

then by (14) and (15) the number λ_0 effects a decrease of $g(\alpha, \beta)$ by 1; while, as a result of (19), it has no effect on the value of $g(\alpha', \beta)$. Hence

(26) $g(\alpha, \beta) = g(\alpha', \beta) - 1$.

Furthermore, in the case (A) considered, $f(\lambda_0, \beta) - f(\lambda_0, \delta)$ cannot be positive by (10); and therefore $f(\lambda_0, \beta) - f(\lambda_0, \delta')$ cannot a fortiori be positive by (19). Thus in this case the number λ_0 proves to have no effect on the values of $g(\beta, \delta)$ and $g(\beta, \delta')$; and consequently

(27) $g(\beta, \delta) = g(\beta, \delta')$.

Equations (26) and (27) at once imply (23).

Turning to the case

(B) $f(\lambda_0, \alpha) - f(\lambda_0, \beta) < 0$,

we first notice that under this assumption λ_0 does not affect the value of $g(\alpha, \beta)$. Nor does it affect the value of $g(\alpha', \beta)$, since, by (14) and (19), $f(\lambda_0, \alpha') - f(\lambda_0, \beta)$ is even. Therefore,

(28) $g(\alpha, \beta) = g(\alpha', \beta)$.

Moreover, in the case (B) under consideration, we see from (10) that $f(\lambda_0, \alpha) = f(\lambda_0, \delta)$, and hence, using (14), that

(29) $f(\lambda_0, \beta) - f(\lambda_0, \delta)$ is positive and odd.

Let

$$f(\lambda_0, \alpha) = f(\lambda_0, \delta) = r.$$

Thus λ_0 is of order r in α and δ, and of a higher order in β. Consequently there are three polynomials α'', β'', and δ'' such that the equations

(30) $\alpha = \alpha'' \cdot (\lambda_0 - \xi)^r, \quad \beta = \beta'' \cdot (\lambda_0 - \xi)^r, \quad \delta = \delta'' \cdot (\lambda_0 - \xi)^r,$

hold identically; λ_0 is a root of β'', but not of α'' or δ''. We obtain from (12) and (30):
$$\alpha'' \cdot \beta_n^{2q} = \gamma \cdot \beta'' - \delta''.$$

Consequently, the values of α'' and δ'', for $\xi = \lambda_0$, have different signs. Therefore there is an open interval, whose right-hand end-point is λ_0, within which the values of α'' and δ'' have different signs; and, by (30), this applies also to α and δ. By comparing this result with (15), we conclude that there is an open interval whose right-hand end-point is λ_0, within which the values of β and δ have the same sign. Hence, and by (29), λ_0 contributes to the increase of $g(\beta, \delta)$ by 1. On the other hand, by (19) and (29), $f(\lambda_0, \beta) - f(\lambda_0, \delta')$ is even, so that λ_0 has no effect on the value of $g(\beta, \delta')$. Thus, finally,

(31) $g(\beta, \delta) = g(\beta, \delta') + 1.$

Equations (28) and (31) again imply (23). Hence (23) holds in both the cases (A) and (B).

From (21), (22), and (23) we obtain at once:

$$
\begin{aligned}
g(\alpha, \beta) &= g(\beta, \delta) \text{ in case } p + 1 \text{ is even};\\
g(\alpha, \beta) &= g(\beta, \delta) - 1 \text{ in case } p + 1 \text{ is odd}.
\end{aligned}
$$

Thus we have shown that (7), (8), and (9) hold for polynomials α and β with $h(\alpha, \beta) = p + 1$; and hence by induction they hold for arbitrary polynomials α and β. This completes the proof. ∎

Definition 17 *Let*

$$
\begin{aligned}
\alpha &\equiv \alpha_0 + \alpha_1 \xi + \cdots + \alpha_m \xi^m\\
\beta &\equiv \beta_0 + \beta_1 \xi + \cdots + \beta_n \xi^n\\
\gamma_1 &\equiv \gamma_{1,0} + \gamma_{1,1} \xi + \cdots + \gamma_{1,n_1} \xi^{n_1}\\
&\vdots\\
\gamma_r &\equiv \gamma_{r,0} + \gamma_{r,1} \xi + \cdots + \gamma_{r,n_r} \xi^{n_r}
\end{aligned}
$$

be arbitrary polynomials in ξ. We define the function T as follows:

(i) If Φ is a formula of the form

$$(E_k \xi) \, [\alpha = 0] \, ,$$

   then we set

$$T(\Phi) \equiv \left[\sim (\alpha_0 = 0) \vee \cdots \vee \sim (\alpha_m = 0) \right] \wedge SG_\xi^{-k} \, (\alpha, D_\xi(\alpha))$$

(ii) *If Φ is a formula of the form*

$$\left[\sim (\alpha_0 = 0) \vee \cdots \vee \sim (\alpha_m = 0)\right] \wedge (E_k \xi)\left[(\alpha = 0) \wedge (\beta > 0)\right],$$

then we set

$$T(\Phi) \equiv \left\{\left[\sim (\alpha_0 = 0) \vee \cdots \vee \sim (\alpha_m = 0)\right] \wedge \right.$$

$$\left. \bigvee_{\substack{2k=r_1-r_2+r_3 \\ 0 \leq r_1, r_2 \leq m \\ -m \leq r_3 \leq m}} \left(\begin{array}{l} SG_\xi^{-r_1}[\alpha, D_\xi(\alpha)] \wedge \\ SG_\xi^{-r_2}[P_\xi(\alpha^2 + \beta^2), D_\xi P_\xi(\alpha^2 + \beta^2)] \wedge \\ SG_\xi^{-r_3}[\alpha, D_\xi(\alpha) \cdot_\xi \beta] \end{array} \right) \right\}.$$

(iii) *If Φ is a formula of the form*

$$\left[\sim (\alpha_0 = 0) \vee \cdots \vee \sim (\alpha_m = 0)\right] \wedge$$
$$(E_k \xi)\left[(\alpha = 0) \wedge (\gamma_1 > 0) \wedge \cdots \wedge (\gamma_r > 0)\right],$$

where $r \geq 2$, then we set

$$T(\Phi) \equiv \bigvee_{\substack{2k=r_1+r_2-r_3 \\ 0 \leq r_1, r_2, r_3 \leq m}} \left\{ T(\Phi_1) \wedge T(\Phi_2) \wedge T(\Phi_3) \right\},$$

where

$$\Phi_1 \equiv \left\{ \left[\sim (\alpha_0 = 0) \vee \cdots \vee \sim (\alpha_m = 0)\right] \wedge \right.$$
$$(E_{r_1} \xi)\left[(\alpha = 0) \wedge (\gamma_1 > 0) \wedge \cdots \wedge (\gamma_{r-2} > 0) \wedge \right.$$
$$\left. \left. P_\xi(\gamma_{r-1} \cdot \gamma_r^2) > 0\right] \right\},$$

$$\Phi_2 \equiv \left\{ \left[\sim (\alpha_0 = 0) \vee \cdots \vee \sim (\alpha_m = 0)\right] \wedge \right.$$
$$(E_{r_2} \xi)\left[(\alpha = 0) \wedge (\gamma_1 > 0) \wedge \cdots \wedge (\gamma_{r-2} > 0) \wedge \right.$$
$$\left. \left. P_\xi(\gamma_{r-1}^2 \cdot \gamma_r) > 0\right] \right\},$$

$$\Phi_3 \equiv \left\{ \left[\sim (\alpha_0 = 0) \vee \cdots \vee \sim (\alpha_m = 0)\right] \wedge \right.$$
$$(E_{r_3} \xi)\left[(\alpha = 0) \wedge (\gamma_1 > 0) \wedge \cdots \wedge (\gamma_{r-2} > 0) \wedge \right.$$
$$\left. \left. P_\xi \left((-1) \cdot \gamma_{r-1} \cdot \gamma_r\right) > 0\right] \right\};$$

in the case $r = 2$ we omit the expression

$$(\gamma_1 > 0) \wedge \cdots \wedge (\gamma_{r-2} > 0) \wedge$$

from the formulas defining Φ_1, Φ_2, and Φ_3.

(iv) *If Φ is a formula of the form*

$$\sim (\alpha_m = 0) \wedge (E_k\xi)\,[(\alpha = 0) \wedge (\gamma_1 > 0) \wedge \cdots \wedge (\gamma_r > 0)],$$

then we set

$$
\begin{aligned}
T(\Phi) \;\equiv\; \Big(&\sim (\alpha_m = 0) \wedge \\
&T\Big\{ \big[\sim (\alpha_0 = 0) \vee \cdots \vee \sim (\alpha_m = 0)\big] \wedge \\
&\quad (E_k\xi)\,[(\alpha = 0) \wedge (\gamma_1 > 0) \wedge \cdots \wedge (\gamma_r > 0)]\Big\}\Big).
\end{aligned}
$$

(v) *If Φ is a formula of the form*

$$
\begin{aligned}
&\big[\sim (\gamma_{1,n_1} = 0) \wedge \sim (\gamma_{2,n_2} = 0) \wedge \cdots \wedge \sim (\gamma_{r,n_r} = 0)\big] \wedge \\
&(E_k\xi)\,[(\gamma_1 > 0) \wedge \cdots \wedge (\gamma_r > 0)],
\end{aligned}
$$

then, if $k > 0$, we set
$$T(\Phi) \equiv (0 = 1);$$
while if $k = 0$ and $n_1 + \cdots + n_r = 0$, we set

$$
\begin{aligned}
T(\Phi) \;\equiv\; \Big\{ &\big[\sim (\gamma_{1,0} = 0) \wedge \cdots \wedge \sim (\gamma_{r,0} = 0)\big] \wedge \\
&\big[(0 > \gamma_{1,0}) \vee \cdots \vee (0 > \gamma_{r,0})\big]\Big\};
\end{aligned}
$$

and if $k = 0$ and $n_1 + \cdots + n_r > 0$, we set

$$
\begin{aligned}
T(\Phi) \;\equiv\; \Big\{ &\sim (\gamma_{1,n_1} = 0) \wedge \cdots \wedge \sim (\gamma_{r,n_r} = 0) \wedge \\
&\big[(0 > \gamma_{1,n_1}) \vee \cdots \vee (0 > \gamma_{r,n_r})\big]\Big\} \wedge \\
&\Big\{ \big[0 > (-1)^{n_1} \cdot \gamma_{1,n_1}\big] \vee \cdots \vee \big[0 > (-1)^{n_r} \cdot \gamma_{r,n_r}\big]\Big\} \wedge \\
&T\Big\{ \sim (\delta = 0) \wedge (E_0\xi)\Big(\big[D_\xi P_\xi(\gamma_1 \cdots \gamma_r) = 0\big] \wedge \\
&\qquad\qquad\qquad\qquad\quad (\gamma_1 > 0) \wedge \cdots \wedge (\gamma_r > 0)\Big)\Big\}
\end{aligned}
$$

where δ is the leading coefficient of $D_\xi P_\xi(\gamma_1 \cdots \gamma_r)$.

(vi) *If Φ is a formula of the form*

$$(E_k\xi)\,[(\gamma_1 > 0) \wedge \cdots \wedge (\gamma_r > 0)],$$

and if $k \neq 0$, we set
$$T(\Phi) \equiv (0 = 1);$$

while if $k = 0$, we set

$$T(\Phi) \equiv \Big\{ [(\gamma_{1,0} = 0) \wedge \cdots \wedge (\gamma_{1,n_1} = 0)] \vee \cdots \vee$$

$$[(\gamma_{r,0} = 0) \wedge \cdots \wedge (\gamma_{r,n_r} = 0)] \Big\} \vee$$

$$\bigvee_{(s_1,\ldots,s_r) \in S} \Big\{ \Psi_{1,s_1} \wedge \cdots \wedge \Psi_{r,s_r} \wedge$$

$$T\Big([\sim (\gamma_{1,s_1} = 0) \wedge \cdots \wedge \sim (\gamma_{r,s_r} = 0)] \wedge$$

$$(E_0\xi)[(Rd_\xi^{n_1 - s_1}(\gamma_1) > 0) \wedge \cdots \wedge$$

$$(Rd_\xi^{n_r - s_r}(\gamma_r) > 0)] \Big) \Big\}.$$

where S is the set of all ordered r-tuples $(s_1, \ldots, s_r)$ with $0 \le s_1 \le n_1, \ldots, 0 \le s_r \le n_r$,

$$\Psi_{k,l} \equiv [(\gamma_{k,l+1} = 0) \wedge \cdots \wedge (\gamma_{k,n_k} = 0)]$$

for $0 \le l < n_k$, and $\Psi_{k,l} \equiv (0 = 0)$ for $l = n_k$.

(vii) *If Φ is a formula of the form*

$$(E_k\xi)[(\alpha = 0) \wedge (\gamma_1 > 0) \wedge \cdots \wedge (\gamma_r > 0)]$$

then we set

$$T(\Phi) \equiv T\Big\{ [\sim (\alpha_0 = 0) \vee \cdots \vee \sim (\alpha_m = 0)] \wedge \Phi \Big\} \vee$$

$$\Big([(\alpha_0 = 0) \wedge \cdots \wedge (\alpha_m = 0)] \wedge$$

$$T\Big\{ (E_k\xi)[(\gamma_1 > 0) \wedge \cdots \wedge (\gamma_r > 0)] \Big\} \Big).$$

(viii) *If Φ is a formula of the form*

$$(E_k\xi)[(\gamma_1 = 0) \wedge \cdots \wedge (\gamma_r = 0)],$$

then we set

$$T(\Phi) \equiv T\Big\{ (E_k\xi)[P_\xi(\gamma_1^2 + \cdots + \gamma_r^2) = 0] \Big\}.$$

(ix) *If Φ is a formula of the form*

$$(E_k\xi)[(\gamma_1 = 0) \wedge \cdots \wedge (\gamma_s = 0) \wedge (\gamma_{s+1} > 0) \wedge \cdots \wedge (\gamma_r > 0)]$$

where $1 < s < r$, then we set

$$T(\Phi) \equiv T\Big\{ (E_k\xi)[P_\xi(\gamma_1^2 + \cdots + \gamma_s^2) = 0 \wedge$$

$$(\gamma_{s+1} > 0) \wedge \cdots \wedge (\gamma_r > 0)] \Big\}.$$

(x) *If Φ is a formula, not of any of the previous forms, but such that*

$$\Phi \equiv (E_k \xi)(\Phi_1 \wedge \Phi_2 \wedge \cdots \wedge \Phi_r),$$

where each Φ_i is of one of the forms $\gamma_i = 0$ or $\gamma_i > 0$, and if $j_1, \ldots, j_u$ are the values of i (in increasing order) for which $\Phi_i \equiv (\gamma_i = 0)$; and if $j_{u+1}, \ldots, j_r$ are the values of i (in increasing order) for which $\Phi_i \equiv (\gamma_i > 0)$, then we set

$$T(\Phi) \equiv T(E_k \xi)(\Phi_{j_1} \wedge \cdots \wedge \Phi_{j_r}).$$

Theorem 12 *Let ξ be any variable, and let Φ be any formula such that $T(\Phi)$ is defined (by Definition 17). Then $T(\Phi)$ is a formula which contains no quantifiers, and no variables except those that occur free in Φ. Moreover, Φ is equivalent to $T(\Phi)$.*

Proof. The first part follows immediately from Theorem 11 and Definition 17.

We shall prove the second part by considering separately the ten possible forms Φ can have according to Definition 17. As in the proof of Theorem 11, we shall use here partially ordinary mathematical modes of expression, without taking any great pains to distinguish sharply mathematical from metamathematical notions.

Suppose first, then, that Φ is of the form 17 (i): i.e., that $\Phi \equiv (E_k \xi)(\alpha = 0)$, where α is a polynomial in ξ. We are to show that Φ is equivalent to the formula

$$\left[\sim (\alpha_0 = 0) \vee \cdots \vee \sim (\alpha_m = 0) \right] \wedge SG_\xi^{-k}(\alpha, D_\xi(\alpha)),$$

where $\alpha_0, \ldots, \alpha_m$ are the coefficients of α. Since by Theorem 11 the latter formula is equivalent to

$$\left[\sim (\alpha_0 = 0) \vee \cdots \vee \sim (\alpha_m = 0) \right] \wedge G_\xi^{-k}(\alpha, D_\xi(\alpha)),$$

we see that our task reduces to establishing the following: if α is any polynomial in ξ which is not identically zero, then α has k distinct roots if and only if $G_\xi^{-k}(\alpha, D_\xi(\alpha))$ holds. Let s_1 be the number of numbers ξ such that: (1) the order of ξ in α is by a positive odd integer higher than its order in $D_\xi(\alpha)$; (2) there exists an open interval whose right-hand end-point is ξ, within which the values of α and $D_\xi(\alpha)$ have the same sign. Let s_2 be the number of numbers ξ which satisfy condition (1) above and moreover the condition: (3) there exists an open interval whose right-hand end-point is ξ, within which the values of α and $D_\xi(\alpha)$ have different signs. By the remark preceding Definition 14, we see that $G_\xi^{-k}(\alpha, D_\xi(\alpha))$ is true if and only if $-k = s_1 - s_2$. Moreover, it is readily seen that $s_1 = 0$, and that s_2 is simply the number of distinct roots of α. Thus $G_\xi^{-k}(\alpha, D_\xi(\alpha))$ holds if and only if k is the number of distinct roots of α, as was to be shown.

In order to treat the case where Φ is of the form 17 (ii), it is convenient first to establish the following: Let α and β be polynomials in ξ, let t_1 be the number of roots of α at which β is positive, and let t_2 be the number of roots of α at which β is negative; then $SG_\xi^{-c}(\alpha, D_\xi(\alpha) \cdot_\xi \beta)$ is true if and only if $c = t_1 - t_2$. This can easily be done by the sort of argument applied in the preceding paragraph – making use of Theorem 11 and the remark preceding Definition 14. We notice also that, since we are dealing with the algebra of real numbers, the common roots of two polynomials α and β coincide with the roots of $\alpha^2 + \beta^2$. Now let Φ be a formula of the form given in 17 (ii). To show that Φ is equivalent to $T(\Phi)$, it suffices to show that, if k is a non-negative integer, and α and β are polynomials, where α is of degree m and not identically zero, then the following conditions are equivalent: (1) there are exactly k roots of α at which β is positive; (2) there are integers r_1, r_2, and r_3 satisfying $2k = r_1 - r_2 + r_3$, $0 \le r_1 \le m$, $0 \le r_2 \le m$, $-m \le r_3 \le m$, such that r_1 is the number of roots of α, r_2 is the number of roots common to α and β, and r_3 is the difference between the number of roots of α at which β is positive and the number of roots of α at which β is negative. Now α has at most m roots; hence, if r_1, r_2, and r_3 have the meanings indicated in (2) – i.e., if r_1 is the number of roots of α, etc. – we obviously have

$$0 \le r_1 \le m, \ 0 \le r_2 \le m, \ \text{and} \ -m \le r_3 \le m.$$

Let k be the number of roots of α at which β is positive, and let r_4 be the number of roots of α at which β is negative. We see immediately from the definitions of r_1, r_2, r_3, k, and r_4 that

$$r_1 - r_2 = k + r_4$$
$$r_3 = k - r_4.$$

Eliminating r_4 between these two equations, we obtain

$$2k = r_1 - r_2 + r_3.$$

Thus (1) implies (2); the proof in the opposite direction is almost obvious.

To prove our theorem for formulas of the form 17 (iii), it suffices to show that if α is a polynomial of degree m, and not identically zero, and if $\gamma_1, \ldots, \gamma_r$ are any polynomials, then the following conditions are equivalent: (1) there are exactly k roots of α at which $\gamma_1, \ldots, \gamma_r$ are all positive; and (2) there are three integers r_1, r_2, r_3 satisfying $2k = r_1 + r_2 - r_3$, $0 \le r_1, r_2, r_3 \le m$, such that r_1 is the number of roots of α at which $\gamma_1, \ldots, \gamma_{r-2}$, and $\gamma_{r-1} \cdot \gamma_r^2$ are all positive, r_2 is the number of roots of α at which $\gamma_1, \ldots, \gamma_{r-2}$ and $\gamma_{r-1}^2 \cdot \gamma_r$ are all positive, and r_3 is the number of roots of α at which $\gamma_1, \ldots, \gamma_{r-2}$, and $-1 \cdot \gamma_{r-1} \cdot \gamma_r$ are all positive. In fact, if r_1, r_2, and r_3 have the meanings just indicated, we obviously have

$$0 \le r_1, r_2, r_3 \le m.$$

Let r_4 be the number of roots of α at which $\gamma_1, \ldots, \gamma_{r-2}, \gamma_{r-1}$ are all positive, and γ_r is negative. Let r_5 be the number of roots of α at which $\gamma_1, \ldots, \gamma_{r-2}$

and γ_r are all positive, and γ_{r-1} is negative. Let k be the number of roots of α at which $\gamma_1, \ldots, \gamma_r$ are all positive. From the definitions of r_1, r_2, r_3, r_4, r_5, and k we see that

$$
\begin{aligned}
k + r_4 &= r_1 \\
k + r_5 &= r_2 \\
r_4 + r_5 &= r_3 .
\end{aligned}
$$

Eliminating r_4 and r_5 from these equations, we obtain

$$2k = r_1 + r_2 - r_3,$$

which completes this part of the proof.

To prove our theorem for formulas of the form 17 (iv), we need only notice that the formula

$$\sim (\alpha_m = 0) \wedge \left[\sim (\alpha_0 = 0) \vee \cdots \vee \sim (\alpha_m = 0) \right]$$

is equivalent to

$$\sim (\alpha_m = 0) .$$

Now suppose that Φ is of the form 17 (v): i.e., that Φ is

$$\left[\sim (\gamma_{1,n_1} = 0) \wedge \cdots \wedge \sim (\gamma_{r,n_r} = 0) \right] \wedge (E_k \xi) \left[(\gamma_1 > 0) \wedge \cdots \wedge (\gamma_r > 0) \right].$$

We notice first that if $k > 0$, then the formula

$$(E_k \xi) \left[(\gamma_1 > 0) \wedge \cdots \wedge (\gamma_r > 0) \right]$$

is never satisfied (i.e., is satisfied by no values of the free variables occurring in it), so that Φ is never satisfied either – and hence is equivalent to $(0 = 1)$. If $k = 0$, and $n_1 + \cdots + n_r = 0$, then $n_1 = n_2 = \cdots = n_r = 0$, and hence Φ reduces to

$$\left[\sim (\gamma_{1,0} = 0) \wedge \cdots \wedge \sim (\gamma_{r,0} = 0) \right] \wedge (E_0 \xi) \left[(\gamma_{1,0} > 0) \wedge \cdots \wedge (\gamma_{r,0} > 0) \right],$$

where $\gamma_{1,0}, \ldots, \gamma_{r,0}$ are terms which do not involve ξ; since

$$(E_0 \xi) \left[(\gamma_{1,0} > 0) \wedge \cdots \wedge (\gamma_{r,0} > 0) \right]$$

is equivalent to

$$\sim \left[(\gamma_{1,0} > 0) \wedge \cdots \wedge (\gamma_{r,0} > 0) \right],$$

we see that Φ is equivalent to $T(\Phi)$, as was to be shown.

Thus we are left with the case that Φ is of the form 17 (v) where $k = 0$ and $n_1 + \cdots + n_r > 0$. To establish in this case that Φ is equivalent to $T(\Phi)$, it suffices to prove: If $\gamma_1, \ldots, \gamma_r$ are polynomials in ξ not all of which are of degree zero, and whose leading coefficients are all different from zero, then a necessary and sufficient condition that there exists no value of ξ which makes all these polynomials positive is that the following three conditions hold: (1) at least one of the polynomials has a negative leading coefficient; (2) at least one of the polynomials satisfies $(-1)^{n_i}\gamma_{i,n_i} < 0$ (where n_i is the degree of the polynomial, and γ_{i,n_i} its leading coefficient); (3) there exists no value of ξ which is a root of the derivative of the product of the polynomials, and which makes them all positive. To see that the condition is necessary, suppose that $\gamma_1, \ldots, \gamma_r$ are polynomials which are never all positive for the same value of ξ; then it is immediately apparent that (3) is satisfied; to see that (1) is satisfied, we remember that, if the leading coefficient of a polynomial is positive, then we can find a number μ such that the polynomial is positive for all values of the variable ξ greater then μ; the proof of (2) is similar, by considering large negative values of the variable. Now suppose, if possible, that the condition is not sufficient: i.e., suppose that (1), (2), and (3) are satisfied, and that there exists some ξ which makes all the polynomials positive. Let λ be a value of ξ at which $\gamma_1 > 0, \ldots, \gamma_r > 0$. Then we see that, for $\xi = \lambda$,

$$\gamma_1 \cdot \gamma_2 \cdots \gamma_r > 0.$$

On the other hand, since (1) is true, there exists an i such that γ_i has a negative leading coefficient. Hence we can find a λ' which is larger then λ and sufficiently large that γ_i is negative at λ'. Then γ_i is positive at λ and negative at λ' and hence has a root between these numbers. Since every root of γ_i is also a root of $\gamma_1 \cdot \gamma_2 \cdots \gamma_r$, we conclude that $\gamma_1 \cdot \gamma_2 \cdots \gamma_r$ has a root to the left of λ. Now let μ_1 be the largest root of $\gamma_1 \cdot \gamma_2 \cdots \gamma_r$ to the left of λ and let μ_2 be the smallest root of $\gamma_1 \cdot \gamma_2 \cdots \gamma_r$ to the right of λ. Then $\gamma_1 \cdot \gamma_2 \cdots \gamma_r$ is positive in the open interval (μ_1, μ_2) and zero at its end-points. We see that no γ_i can have a root within the open interval (μ_1, μ_2); since each γ_i is positive at λ, which lies within this interval, we conclude that each γ_i is positive throughout the whole open interval. On the other hand, since $\gamma_1 \cdot \gamma_2 \cdots \gamma_r$ is zero at μ_1 and μ_2, we see by Rolle's theorem that there is a point ν within (μ_1, μ_2) at which the derivative of $\gamma_1 \cdot \gamma_2 \cdots \gamma_r$ vanishes. Since this contradicts (3), we conclude that the condition is also sufficient.

Now suppose that Φ is of the form 17 (vi). If $k \neq 0$, it is obvious that Φ is equivalent to $T(\Phi)$. Hence suppose that $k = 0$. Two cases are logically possible: either one of the polynomials $\gamma_1, \ldots, \gamma_r$ vanishes identically, or none of them does. In the first case, Φ obviously holds. In the second case, let s_i, for $i = 1, \ldots, r$ be the largest index j such that $\gamma_{i,j}$ (i.e. the j-th coefficient of γ_i) does not vanish. Then Φ is obviously equivalent to the formula

$$\begin{aligned}
\Psi \;\equiv\; & \left[\sim (\gamma_{1,s_1} = 0) \wedge \cdots \wedge \sim (\gamma_{r,s_r} = 0) \right] \wedge \\
& (E_0\xi)\Big(\left[Rd_\xi^{n_1-s_1}(\gamma_1) > 0 \right] \wedge \cdots \wedge \left[Rd_\xi^{n_r-s_r}(\gamma_r) > 0 \right] \Big).
\end{aligned}$$

However, Ψ is of the form 17 (v), and hence, as we have shown above, is equivalent to $T(\Psi)$. In view of these remarks, by looking at the formula defining $T(\Phi)$ in 17 (vi), we see at once that Φ and $T(\Phi)$ are equivalent.

If Φ is of the form 17 (vii), our theorem follows from the obvious fact that Φ is equivalent to

$$\left\{\left[\sim (\alpha_0 = 0) \vee \cdots \vee \sim (\alpha_m = 0)\right] \wedge \Phi\right\} \vee \left\{\left[(\alpha_0 = 0) \wedge \cdots \wedge (\alpha_m = 0)\right] \wedge \Phi\right\}.$$

If Φ is of the form 17 (viii), or of the form 17 (ix), our theorem follows from the fact – which was already used in discussing 17 (ii) – that the common roots of r polynomials $\gamma_1, \ldots, \gamma_r$ coincide with the roots of $\gamma_1^2 + \cdots + \gamma_r^2$.

If Φ is of the form 17 (x), our theorem follows from the associative and commutative laws for conjunction, familiar from elementary logic. ∎

Definition 18 *Let Φ, Ψ, and Θ be any formulas, and ξ any variable.*

(i) *If Φ is an atomic formula, we set*

$$U(\Phi) \equiv \Phi.$$

(ii) *If $\Phi \equiv (\Psi \vee \Theta)$, then we set*

$$U(\Phi) \equiv [U(\Psi) \vee U(\Theta)].$$

(iii) *If $\Phi \equiv (\Psi \wedge \Theta)$, then we set*

$$U(\Phi) \equiv [U(\Psi) \wedge U(\Theta)].$$

(iv) *If $\Phi \equiv\, \sim \Psi$, then we set*
$$U(\Phi) \equiv\, \sim U(\Psi).$$

(v) *If $\Phi \equiv (E\xi)\Psi$, and*

$$QP_\xi U(\Psi) \equiv \Psi_1 \vee \Psi_2 \vee \cdots \vee \Psi_n,$$

where Ψ_i, for $i = 1, \ldots, n$ is a conjunction of atomic formulas, then we set

$$U(\Phi) \equiv\, \sim T\left[(E_0\xi)\Psi_1\right] \vee\, \sim T\left[(E_0\xi)\Psi_2\right] \vee \cdots \vee\, \sim T\left[(E_0\xi)\Psi_n\right].$$

Theorem 13 *If Φ is any formula, then $U(\Phi)$ is a formula which contains no quantifiers, and no free variables except variables which occur free in Φ. Moreover, Φ is equivalent to $U(\Phi)$.*

Proof. By induction on the order of Φ, making use of Theorems 6 and 12. ∎

Corollary 1 *If Φ is any sentence, then $U(\Phi)$ is an equivalent sentence without any variables or quantifiers.*

The first part of our task as outlined at the beginning of this section has thus been completed. We have established a general procedure which permits us to transform every formula (and in particular every sentence) into an equivalent formula (or sentence) without quantifiers (see Notes 12 and 13 in Section 5). Before continuing the discussion, we should like to give a few relatively simple examples in which such a transformation has actually been carried out. The equivalent transformations $U'(\Phi)$ which are given below for some formulas Φ do not coincide with $U(\Phi)$ but can be obtained from the latter by means of elementary simplifications.

Let ξ be any variable, and α_0, α_1, α_2, α_3, β_0, β_1, and β_2 any terms which do not involve ξ. If

$$\Phi \equiv (E\xi) \left[\alpha_0 + \alpha_1 \cdot \xi + \alpha_2 \xi^2 + \alpha_3 \cdot \xi^3 = 0 \right]$$

we obtain an equivalent formula by setting

$$U'(\Phi) \equiv \left\{ (\alpha_0 = 0) \vee [\sim (\alpha_1 = 0) \wedge (\alpha_2 = 0)] \vee \right.$$
$$\left. [\sim (\alpha_2 = 0) \wedge \sim (4 \cdot \alpha_0 \cdot \alpha_2 > \alpha_1^2)] \vee \sim (\alpha_3 = 0) \right\}$$

(where, as can easily be guessed, 4 stands for $1+1+1+1$). If

$$\Phi \equiv (E\xi) \left[\alpha_0 + \alpha_1 \cdot \xi + \alpha_2 \cdot \xi^2 + \alpha_3 \cdot \xi^3 > 0 \right],$$

we can put

$$U'(\Phi) \equiv \left\{ (\alpha_0 > 0) \vee (\alpha_1^2 > 4 \cdot \alpha_0 \cdot \alpha_2) \vee (\alpha_2 > 0) \vee \sim (\alpha_3 = 0) \right\}.$$

If, finally,

$$\Phi \equiv (E\xi) \left[(\alpha_0 + \alpha_1 \cdot \xi + \alpha_2 \cdot \xi^2 = 0) \wedge (\beta_0 + \beta_1 \cdot \xi + \beta_2 \cdot \xi^2 > 0) \right],$$

we can put

$$U'(\Phi) \equiv \left\{ \left[(\alpha_0 = 0) \wedge (\alpha_1 = 0) \wedge (\alpha_2 = 0) \wedge \right. \right.$$
$$\left. ((\beta_0 > 0) \vee (\beta_1^2 > 4 \cdot \beta_0 \cdot \beta_2) \vee (\beta_2 > 0)) \right] \vee$$
$$\left[\sim (\alpha_1 = 0) \wedge (\alpha_2 = 0) \wedge (\alpha_0^2 \cdot \beta_2 + \alpha_1^2 \cdot \beta_0 > \alpha_0 \cdot \alpha_1 \cdot \beta_1) \right] \vee$$
$$\left[\sim (\alpha_2 = 0) \wedge \sim (4 \cdot \alpha_0 \cdot \alpha_2 > \alpha_1^2) \wedge \right.$$
$$\left. (\alpha_1^2 \cdot \beta_2 + 2 \cdot \alpha_2^2 \cdot \beta_0 > 2 \cdot \alpha_0 \cdot \alpha_2 \cdot \beta_2 + \alpha_1 \cdot \alpha_2 \cdot \beta_1) \right] \vee$$
$$\left[\sim (\alpha_2 = 0) \wedge (\alpha_1^2 > 4 \cdot \alpha_0 \cdot \alpha_2) \wedge \right.$$
$$(\alpha_0^2 \cdot \beta_2^2 + \alpha_0 \cdot \alpha_2 \cdot \beta_1^2 + \alpha_1^2 \cdot \beta_0 \cdot \beta_2 + \alpha_2^2 \cdot \beta_0^2 >$$
$$\left. \left. \alpha_0 \cdot \alpha_1 \cdot \beta_1 \cdot \beta_2 + 2 \cdot \alpha_0 \cdot \alpha_2 \cdot \beta_0 \cdot \beta_2 + \alpha_1 \cdot \alpha_2 \cdot \beta_0 \cdot \beta_1) \right] \right\}.$$

We now turn to the second part of our task. We want to correlate, with every sentence Φ which contains no variables or quantifiers, an equivalent sentence of a very special form: in fact, one of the two sentences

$$0 = 0$$

and

$$0 = 1.$$

We first consider terms which occur in such sentences. As is easily seen, every such term is obtained from the algebraic constants 0, 1, and -1 by combining them by means of addition and multiplication. Hence we can correlate with every such term α an integer $n(\alpha)$ in the following way.

Definition 19 *We set*

$$
\begin{aligned}
n(1) &= 1, \\
n(-1) &= -1, \\
n(0) &= 0.
\end{aligned}
$$

If $\alpha \equiv \beta + \gamma$, then we set

$$n(\alpha) = n(\beta) + n(\gamma).$$

If $\alpha \equiv (\beta \cdot \gamma)$, then we set

$$n(\alpha) = n(\beta) \cdot n(\gamma).$$

Remark 4 *It should be emphasized that the above definition correlates integers, not expressions, with terms. It is for this reason that we have written, for example,*

$$n(1) = 1,$$

instead of

$$n(1) \equiv 1;$$

$n(1)$ is the integer 1, not a name of that integer. In the equation

$$n(\alpha) = n(\beta) + n(\gamma)$$

the addition sign indicates the sum of the two integers $n(\beta)$ and $n(\gamma)$. $n(\alpha)$ is what would ordinarily be called the "value" of the expression α; thus, if

$$\alpha \equiv 1 + (1 + 1) \cdot (1 + 1),$$

then $n(\alpha) = 5$.

On the other hand, we could use for our purposes, instead of integers, certain expressions of our formal system of algebra – in fact one of the terms of the following sequence

$$\ldots, (-1) + (-1), -1, 0, 1, 1 + 1, \ldots.$$

We can use these special terms since they can obviously be put in one-to-one correspondence with arbitrary integers. As a result of this modification, however, Definition 19 and the subsequent Definition 20 would assume a more complicated form.

Definition 20 *Let α and β be terms, and Φ, Ψ, and Θ be formulas, none of which contain any variables.*

(i) *If $\Phi \equiv (\alpha = \beta)$, we set*

$$W(\Phi) \equiv (0 = 0)$$

in case $n(\alpha) = n(\beta)$, and otherwise

$$W(\Phi) \equiv (0 = 1).$$

(ii) *If $\Phi \equiv (\alpha > \beta)$, we set*

$$W(\Phi) \equiv (0 = 0)$$

in case $n(\alpha) > n(\beta)$, and otherwise

$$W(\Phi) \equiv (0 = 1).$$

(iii) *If $\Phi \equiv (\Psi \vee \Theta)$, we set*

$$W(\Phi) \equiv (0 = 0)$$

in case either $W(\Psi) \equiv (0 = 0)$ or $W(\Theta) \equiv (0 = 0)$, and otherwise

$$W(\Phi) \equiv (0 = 1).$$

(iv) *If $\Phi \equiv (\Psi \wedge \Theta)$, we set*

$$W(\Phi) \equiv (0 = 0)$$

in case both $W(\Psi) \equiv (0 = 0)$ and $W(\Theta) \equiv (0 = 0)$, and otherwise

$$W(\Phi) \equiv (0 = 1).$$

(v) *If $\Phi \equiv \sim \Psi$, we set*

$$W(\Phi) \equiv (0 = 0)$$

in case $W(\Psi) \equiv (0 = 1)$, and otherwise

$$W(\Phi) \equiv (0 = 1).$$

Theorem 14 *If Φ is any sentence which involves no quantifiers or variables, then $W(\Phi)$ is one or the other of the two sentences $0 = 0$ and $0 = 1$. Moreover, Φ is equivalent to $W(\Phi)$.*

Proof. By induction on the order of Φ. ∎

Theorem 15 *If Φ is any sentence, then $WU(\Phi)$ is one or the other of the two sentences $0 = 0$ or $0 = 1$. Moreover, Φ is equivalent to $WU(\Phi)$.*

Proof. By 1 and 14. ∎

Now by analyzing the definitions of W, U, and the preceding functions, we notice that for any given sentence Φ we can actually find the value of $WU(\Phi)$ in a finite number of steps (see Note 14 in Section 5). By combining this with the result stated in Theorem 15, we obtain

Theorem 16 *There is a decision method for the class of all true sentences of elementary algebra (see Note 15 in Section 5).*

In concluding this section we should like to remark that the minimum number of steps which are necessary for the evaluation of $WU(\Phi)$ is of course a function of the form of Φ – in particular, this number depends on the length of Φ, on the number of quantifiers occurring in it, and so on. The problem of estimating the order of increase of this function is of primary importance in connection with the question of the feasibility of constructing a decision machine for elementary algebra.

4 Extensions to Related Systems

In this section we shall discuss some applications to other systems of the results obtained in Section 3, as well as some problems that are still open.

The decision method found for the algebra of real numbers can be extended to various algebraic systems built upon real numbers – thus to the elementary algebra of complex numbers, that of quaternions, and that of n-dimensional vectors. We can think of the elementary algebra of complex numbers, for example, as a formal system very closely related to that described in Section 2: variables are now thought of as representing arbitrary complex numbers; the logical and mathematical constants remain unchanged; but now the greater-than relation is thought of as holding exclusively between real numbers – thus we can define real numbers within the system by saying that x is real if, for some y, $x > y$. If one wishes, one can enrich the system by a new predicate $\mathrm{Rl}(x)$, agreeing that $\mathrm{Rl}(x)$ will mean that x is real (see Note 16 in Section 5).

The results obtained can furthermore be extended to the elementary systems of n-dimensional Euclidean geometry. Since the methods of extending the results to the algebraic systems and the geometric systems are essentially the same, we shall consider a little more closely the case of 2-dimensional Euclidean geometry.

We first give a sketchy description of the formal system of 2-dimensional Euclidean geometry. We use infinitely many *variables*, which are to be thought of as representing arbitrary points of the Euclidean plane. We use three constants denoting relations between points: the binary relation of *identity*, symbolized by "$=$"; the ternary relation of *betweenness*, symbolized by "B", so that "$B(x, y, z)$" is to be read "y is between x and z" (i.e., y lies between x and z on the straight line connecting them; it is not necessary that the three points all be distinct; $B(x, y, z)$ is always true if $x = y$ or if $y = z$; but we cannot have $x = z$ unless $x = y = z$); and the quaternary *equidistance* relation, symbolized by "D", so that "$D(x, y; x', y')$" is to be read "x is just as far from y as x' is from y'" (or, "the distance from x to y equals the distance from x' to y'") (see Note 17 in Section 5). The only *terms* of this system are variables. An *atomic formula* is an expression of one of the forms

$$\xi = \eta, \ B(\xi_1, \xi_2, \xi_3), \ D(\xi, \eta; \xi', \eta'),$$

where

$$\xi, \eta, \xi_1, \xi_2, \xi_3, \xi', \text{ and } \eta'$$

are arbitrary variables. As in the formal system of elementary algebra we build up *formulas*, from atomic formulas by means of negation, conjunction, disjunction, and the application of quantifiers; we also introduce here as abbreviations the symbols $\rightarrow$ and $\leftrightarrow$.

Sentences of elementary geometry, in our formulation, express certain facts about points and relations between them. On the other hand, most theorems which one finds in high-school textbooks on this subject involve also such notions as triangle, plane, circle, line, and the like. It is easy, however, to convince oneself that a considerable part of these notions can be translated into the language of our system. Thus, for example, the theorem that the medians of a triangle are concurrent can be expressed as in Figure 1.

On the other hand, it would not be difficult to enrich our system of geometry so as to enable us to refer to these elementary figures directly. Regarding more essential limitations of our system, see the remarks in the Introduction.

In order to obtain a decision procedure for elementary geometry, we correlate with every sentence Φ of elementary geometry a sentence Φ^* of elementary algebra in the sense of Section 2. The construction of Φ^* can be roughly described in the following way. With every (geometric) variable ξ in Φ we correlate two different (algebraic) variables $\overline{\xi}$ and $\overline{\overline{\xi}}$, in such a way that if ξ and η are two different variables in Φ, then $\overline{\xi}, \overline{\overline{\xi}}, \overline{\eta},$ and $\overline{\overline{\eta}}$ are all distinct. Next we replace in Φ every quantifier expression $(E\xi)$ by $(E\overline{\xi})(E\overline{\overline{\xi}})$; every partial formula $\xi = \eta$ by $(\overline{\xi} = \overline{\eta}) \wedge (\overline{\overline{\xi}} = \overline{\overline{\eta}})$; every formula $B(\xi, \eta, \mu)$ by

$$\left[(\overline{\overline{\eta}} - \overline{\overline{\xi}}) \cdot (\overline{\mu} - \overline{\eta}) = (\overline{\overline{\mu}} - \overline{\overline{\eta}}) \cdot (\overline{\eta} - \overline{\xi}) \right] \wedge$$

$$\left[((\overline{\xi} - \overline{\eta}) \cdot (\overline{\eta} - \overline{\mu}) > 0) \vee ((\overline{\xi} - \overline{\eta}) \cdot (\overline{\eta} - \overline{\mu}) = 0) \right] \wedge$$

$$\left[((\overline{\overline{\xi}} - \overline{\overline{\eta}}) \cdot (\overline{\overline{\eta}} - \overline{\overline{\mu}}) > 0) \vee ((\overline{\overline{\xi}} - \overline{\overline{\eta}}) \cdot (\overline{\overline{\eta}} - \overline{\overline{\mu}}) = 0) \right];$$

and every partial formula $D(\xi, \eta; \mu, \nu)$ by

$$(\overline{\xi} - \overline{\eta})^2 + (\overline{\overline{\xi}} - \overline{\overline{\eta}})^2 = (\overline{\mu} - \overline{\nu})^2 + (\overline{\overline{\mu}} - \overline{\overline{\nu}})^2.$$

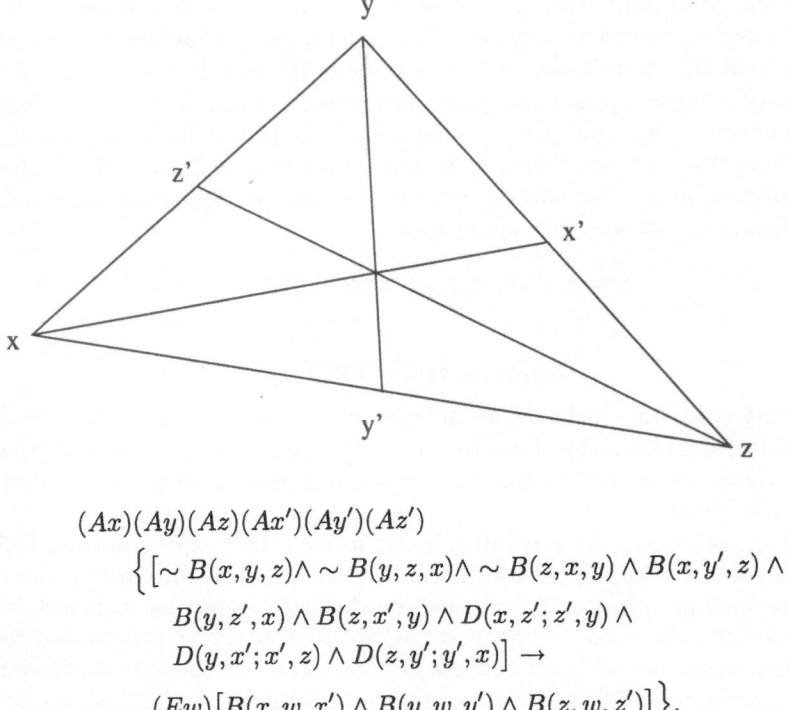

$$(Ax)(Ay)(Az)(Ax')(Ay')(Az')$$
$$\big\{[\sim B(x,y,z) \wedge \sim B(y,z,x) \wedge \sim B(z,x,y) \wedge B(x,y',z) \wedge$$
$$B(y,z',x) \wedge B(z,x',y) \wedge D(x,z';z',y) \wedge$$
$$D(y,x';x',z) \wedge D(z,y';y',x)] \rightarrow$$
$$(Ew)[B(x,w,x') \wedge B(y,w,y') \wedge B(z,w,z')]\big\}.$$

Figure 1: Theorem on the medians of a triangle

It is now obvious to anyone familiar with the elements of analytic geometry that whenever Φ is true then Φ^* is true, and conversely. And since we can always decide in a mechanical way about the truth of Φ^*, we can also do this for Φ.

The decision method just outlined applies with obvious changes to Euclidean geometry of any number of dimensions (see Note 18 in Section 5). And, since it depends exclusively on the possibility of introducing into geometry a system of real coordinates, it will apply as well to various systems of non-Euclidean and projective geometry (see Note 19 in Section 5).

We can attempt to extend the results concerning elementary algebra in still another way: in fact, by introducing into the system of algebra new mathematical terms which cannot be defined by means of those occurring in the original system. The new terms may denote certain properties of numbers, certain relations between numbers, or certain operations on numbers (in particular, unary operations – i.e., functions of one real variable). In consequence of any such extension of the original system we are presented with a new decision problem. In some cases, from the results known in the literature it easily follows that the solution of the problem is negative – i.e., that no decision method for the enlarged system can ever be found, and that no decision machine can be constructed. In view of the Gödel-Church-Rosser result mentioned in the Introduction, this applies, for instance, if we introduce into the system of real algebra the predicate In, to denote the property of being an integer (so that In(x) is read: "x is an integer"); and, by the result of Mrs. Robinson, the same applies to the predicate Rt, denoting the property of being rational. The situation is still the same if we introduce a symbol for some periodic function, for instance, sine; this is seen if only from the fact that the notion of an integer and of a rational number can easily be defined in terms of sine and the notions of our original system; thus we can say that x is a rational if and only if it satisfies the formula

$$(Ey)(Ez)\big[(x \cdot y = z) \wedge \sim (y = 0) \wedge (\sin y = 0) \wedge (\sin z = 0)\big].$$

In other cases, by introducing a new symbol we arrive at a system for which the decision problem is open. This applies, for instance, to the system obtained by introducing the operation of exponentiation (of course restricted to the cases where it yields a definite real result), or – what amounts essentially to the same thing – the symbol Exp to denote an exponential with a fixed base, for example, 2 (see Note 20 in Section 5). The decision problem for the system just mentioned is of great theoretical and practical interest. But its solution seems to present considerable difficulties. These difficulties appear, however, to be of a purely mathematical (not logical) nature: they arise from the fact that our knowledge of conditions for the solvability of equations and inequalities in the enlarged system is far from adequate (see Note 21 in Section 5).

In this connection it may be worthwhile to mention that by introducing the operation of exponentiation into the system of elementary complex algebra, we arrive at a system for which the solution of the decision problem is negative. In fact it is well known that the exponential function in the complex domain is periodic, and hence, like the function sine in the real domain, it allows us to define the notion of an integer.

5 Notes

1. When dealing with theories presented as formal axiomatized systems, one often uses the term "decision method" for a theory in a different sense, by referring it to the class, not of all true sentences, but of all theorems of the theory; i.e., of all sentences of the theory which can be derived from the axioms by means of certain prescribed rules of inference.
2. See (Löwenheim 1915), (Post 1921), (Langford 1927a, 1927b), (Presburger 1930), and (McKinsey 1943). The results of Tarski and Mrs. Szmielew are unpublished.
3. See (Gödel 1931), (Church 1936), and (Rosser 1936). The results of Mostowski, Tarski, and Mrs. Robinson are unpublished.

4. This result was mentioned, though in an implicit form and without proof, in (Tarski 1931, pp. 233 and 234); see also (Tarski 1933). Some partial results tending in the same direction – e.g., decision methods for elementary algebra with addition as the only operation, and for the geometry of the straight line – are still older, and were obtained by Tarski and presented in his university lectures in the years 1926–1928; cf. (Presburger 1930, p. 95, footnote 4) and (Tarski 1931, p. 324, footnote 53).

5. In this connection A. Mostowski has pointed out the following. Although the general concept of an integer is lacking in our system of elementary algebra, it can easily be shown that a "general arithmetic" in the sense of (Carnap 1937, p. 206), is "contained" in this system. Since the language in question is consistent and decidable (again in the sense of (Carnap 1937, pp. 207 and 209)), it provides an example against Carnap's assertion that "every consistent language which contains a general arithmetic is irresoluble" (Carnap 1937, p. 210). Carnap's definition of the phrase "contains a general arithmetic" is therefore certainly too wide.

6. Among the works listed in the Bibliography, Bernays and Hilbert (1939) may be consulted for various logical and metamathematical notions and results involved in our discussion, and van der Waerden (1937) will provide necessary information in the domain of algebra.

7. In this monograph we establish certain results concerning various mathematical theories, such as elementary algebra and elementary geometry. Hence our discussion belongs to the general theory of mathematical theories: i.e., to what is called "metamathematics". To give our discussion a precise form we have to use various metamathematical symbols and notions. Since, however, we do not want to create any special difficulties for the reader, we apply the following method: when referring to individual symbols of the mathematical theory being discussed, or to expressions involving these symbols, we use the symbols and expressions themselves. We could thus say that the symbols and expressions play in our discussion the role of metamathematical constants. On the other hand, when referring to arbitrary symbols and expressions, or to arbitrary expressions of a certain form, we use special metamathematical variables. In fact, small Greek letters, as for instance "α", "β", "γ", are used to represent arbitrary terms, and in particular the letters "ξ", "η", "λ", "μ", "ν", will be used to represent arbitrary variables; on the other hand, Greek capitals "Φ", "Θ", "Ψ" will be used to represent arbitrary formulas and sentences. With these exceptions we do not introduce any special metamathematical symbolism. Various metamathematical notions whose intuitive meaning is clear will be used without any explanation; this applies, for instance, to such a phrase as "the variable ξ occurs in the formula Φ." Also, we do not consider it necessary to set up an axiomatic foundation for our metamathematical discussion, and we avoid a strictly formal exposition of metamathematical arguments. We assume that we can avail ourselves in metamathematics of elementary number theory; we use variables "m", "n", "p", and so on

to represent arbitrary integers; and we employ the ordinary notation for individual integers, arithmetical relations between integers, and operations on them.

The reader who is interested in the deductive foundations, and a precise development, of metamathematical discussion, may be referred to (Carnap 1937, part II, pp. 55 ff.), (Gödel 1931), (Tarski 1936, Section 2, pp. 279 ff., in particular p. 289), and (Tarski 1933, especially p. 100).

8. In choosing symbols for the formalized system of algebra, we have been interested in presenting the metamathematical results in the simplest possible form. For this reason we have not introduced into the system various mathematical and logical symbols which are ordinarily used in expressing mathematical theorems: such as the subtraction symbol "$-$", the symbol "$<$", the implication sign "$\rightarrow$", the equivalence sign "$\leftrightarrow$", and the universal quantifier "A". Nevertheless, some of these symbols are made available for our use, since they are introduced as metamathematical abbreviations. If we wished, we could reduce the number of symbols still further; we could, for instance, dispense with the "$>$" sign, by treating

$$x > y$$

merely as an abbreviation for

$$(Ez) \left[\sim (z = 0) \wedge (x = y + z^2) \right].$$

In an analogous way we could dispense with the symbols 0, 1, and -1, and with one of the two logical connectives $\vee$ and $\wedge$. But such a reduction in the number of symbols would hardly be advantageous from our point of view.

It should be pointed out that, in order to increase the efficiency of the decision machine which may be constructed on the basis of this monograph, it might very well turn out to be useful to enrich the symbolism of our system, even if this carried with it certain complications in the description of the decision method.

9. A formal definition of truth can be found in (Tarski 1936). It should be pointed out that we can eliminate the notion of truth from our whole discussion by subjecting the system of elementary algebra to the process of axiomatization. For this purpose, we single out certain sentences of our system which we call "axioms". They are divided into logical and algebraic axioms. The logical axioms (or rather, axiom schemata) are those of the sentential calculus and the lower predicated calculus with identity; they can be found, for instance in (Bernays and Hilbert 1939, see sections 3, 4 and 5 in vol.1, and supplement 1 in vol.2). Among algebraic axioms we find, in the first place, those which characterize the set of real numbers as a commutative ordered field with the operations $+$ and $\cdot$ and the relation $>$, and which single out in a familiar way the three special elements 0, 1, and -1. These axioms are supplemented by one additional axiom schema comprehending all sentences of the form

(i) $(A\xi_1)\ldots(A\xi_n)(A\eta)(A\zeta)$

$$\left\{\left[(\eta > \zeta) \wedge (E\xi)((\xi = \eta) \wedge (\alpha > 0)) \wedge (E\xi)((\xi = \zeta) \wedge (0 > \alpha))\right]\right.$$

$$\left.\to (E\xi)((\eta > \xi) \wedge (\xi > \zeta) \wedge (\alpha = 0))\right\},$$

where $\xi_1,\ldots,\xi_n$, η, ζ are arbitrary variables, ξ is any variable different from η and ζ, and α is any term – which, in the non-trivial cases, of course involves the variable ξ. Intuitively speaking, this axiom schema expresses the fact that every function which is represented by a term of our symbolism (i.e., every rational integral function) and which is positive at one point and negative at another, vanishes at some point in between.

From what can be found in the literature (see van der Waerden (1937, in particular pp. 235 f.)), it is seen that this axiom schema can be equivalently replaced by the combination of an axiom expressing the fact that every positive number has a square root, with an axiom schema comprehending all sentences to the effect that every polynomial of odd degree has a zero: i.e., all sentences of the form

(ii) $(A\eta_0)(A\eta_1)\ldots(A\eta_{2n+1})$
$$[\sim (\eta_{2n+1} = 0) \to (E\xi)(\eta_0 + \eta_1\xi + \cdots + \eta_{2n+1}\xi^{2n+1} = 0)],$$

where η_0, η_1, ..., η_{2n+1} are arbitrary variables, and ξ is any variable different from all of them. It is also possible to use, instead of (ii), a schema comprehending all sentences to the effect that every polynomial of degree at least three has a quadratic factor. Finally, it turns out to be possible to replace equivalently schema (i) by the seemingly much stronger axiom schema comprehending all those particular cases of the continuity axiom which can be expressed in our symbolism. (By the continuity axiom we may understand the statement that every set of real numbers which is bounded above has a least upper bound; when expressing particular cases of this axiom in our symbolism, we speak, not of elements of a set, but of numbers satisfying a given formula.) The possibility of this last replacement, however, is a rather deep result, which is the by-product of other results presented in this work; in fact, of those discussed below in Note 15.

After having selected the axioms, we describe the operations by means of which new sentences can be derived from given ones. These operations are expressed in the so-called "rules of inference" familiar from mathematical logic. A sentence which can be derived from axioms by repeated applications of the rules of inference is called a provable sentence. In our further discussion – in particular, in defining the notions of equivalence of terms and equivalence of formulas – we replace everywhere the notion of a true sentence by that of a provable one. Hence, when establishing certain of the results given later – in particular, the theorems about equivalent

formulas – we have to show that the sentences involved are formally deriv-
able from the selected axioms (and not that they are true in any intuitive
sense); otherwise the discussion does not differ from that in the text.

10. We use the term "decision method" here in an intuitive sense, without giving a formal definition. Such a procedure is possible because our result is
of a positive character; we are going to establish a decision method, and
no one who understands our discussion will be likely to have any doubt
that this method enables us to decide in a finite number of steps whether
any given sentence of elementary algebra is true. The situation changes
radically, however, if one intends to obtain a result of a negative character
– i.e., to show for a given theory that no decision method can be found; a
precise definition of a decision method then becomes indispensable. The
way in which such a definition is to be given is of course known from the
contemporary literature. Using one of the familiar methods – for instance
the method due to Gödel – one establishes a one-to-one correspondence
between expressions of the system and positive integers, and one agrees to
treat the phrase "there exists a decision method for the class A of expressions" as equivalent with the phrase "the set of numbers correlated with
the expressions of A is general recursive." (When the set of numbers correlated with a class A is general recursive, we sometimes say simply that A is
general recursive.) For a discussion of the notion of general recursiveness,
see (Bernays and Hilbert 1939) and (Kleene 1936).

11. The method of eliminating quantifiers occurs in a more or less explicit
form in the papers (Löwenheim 1915, section 3), (Skolem 1919, section 4),
(Langford 1927a, 1927b), and (Presburger 1930). In Tarski's university
lectures for the years 1926–1928 this method was developed in a general
and systematic way; cf. (Presburger 1930, p. 95, footnote 4, and p. 97,
footnote 1).

12. The results obtained in Theorems 11 and 12, and culminating in Theorem 13, seem to deserve interest even from the purely mathematical point
of view. They are closely related to the well-known theorem of Sturm, and
in proving them we have partly used Sturm's methods.

The theorem most closely related to Sturm's ideas is Theorem 11. In
fact, by analyzing, and slightly generalizing, the proof of this theorem we
arrive at the following formulation. Let α and β be any two polynomials in
a variable ξ, and κ and μ any two real numbers with $\kappa < \mu$. We construct
a sequence of polynomials $\gamma_1, \gamma_2, \ldots, \gamma_n$ – which may be called the Sturm
chain for α and β – by taking α for γ_1, β for γ_2, and assuming that γ_i,
with $i > 2$, is the negative remainder of γ_{i-2} and γ_{i-1}; we discontinue the
construction when we reach a polynomial γ_n which is a divisor of γ_{n-1}. Let
$\kappa_1, \ldots, \kappa_n$ and $\mu_1, \ldots, \mu_n$ the sequence of values of $\gamma_1, \ldots, \gamma_n$ at $\xi = \kappa$ and
$\xi = \mu$ respectively; let k be the number of changes of sign in the sequence
$\kappa_1, \ldots, \kappa_n$, and let m be the number of changes in sign of the sequence
$\mu_1, \ldots, \mu_n$. Then it turns out that $k - m$ is just the number $g(\alpha, \beta)$ defined
as in the proof of Theorem 11, but with the roots assumed to lie between

κ and μ. (In Theorem 11 we were dealing, not with the arbitrary interval (κ, μ), but with the interval $(-\infty, +\infty)$.) Sturm himself considered two particular cases of this general theorem: the case where β is the derivative of α – when the number $k - m$ proves to be simply the number of distinct roots of α in the interval (κ, μ); and the case where β is arbitrary but α is a polynomial without multiple roots – when $k - m$ proves to be the difference between the number of roots of α at which β agrees in sign with the derivative of α, and the number of roots of α at which β disagrees in sign with the derivative of α – the roots being taken from the interval (κ, μ). These two special cases easily follow from the theorem, and we have made an essential use of this fact in the proof of Theorem 12. The general formulation was found recently by J. C. C. McKinsey; it contributed to a simplification, not of the original decision method itself, but of its mathematical description.

Apart, however, from technicalities connected with the notion and construction of Sturm chains, the mathematical content of Sturm's theorem essentially consists in the following: given any algebraic equation in one variable x, and with coefficients $a_0, a_1, \ldots, a_n$, there is an elementary criterion for this equation to have exactly k real solutions (which may be in addition subjected to the condition that they lie in a given interval): such a criterion is obtained by constructing a certain finite sequence of systems, each consisting of finitely many equations and inequalities which involve the coefficients $a_0, a_1, \ldots, a_n$ of the given equation (and possibly the endpoints b and c of the interval); it is shown that the equation has exactly k roots if and only if its coefficients satisfy all the equations and inequalities of at least one of these systems. (When applied to an equation with constant coefficients, the criterion enables us actually to determine the number of roots of the equation, but this is only a by-product of Sturm's theorem.) By applying Sturm's theorem we obtain in particular an elementary condition for an algebraic equation in one unknown to have at least one real solution. Theorem 13 gives directly an extension of this special result to an arbitrary system of algebraic equations and inequalities with arbitrarily many unknowns. It is easily seen, however, that from our theorem one can obtain stronger consequences: in fact, criteria for such systems to have exactly k real solutions. To clear up this point, let us consider the simple case of a system consisting of one equation in two unknowns

(i) $F(x, y) = 0$.

We form the following system of equations and inequalities
(ii)

$$\begin{cases} F(x, y) = 0 \\ F(x', y') = 0 \\ (x - x')^2 + (y - y')^2 > 0. \end{cases}$$

By Theorem 13 we have an elementary criterion for the system (ii) to have at least one solution. But it is obvious that this criterion is at the same time a criterion for (i) to have at least two solutions. In the same way, we can obtain criteria for (i) to have at least $3, 4, \ldots, k$ real solutions. Hence we can also obtain a criterion for (i) to have exactly k solutions (since an equation has exactly k solutions if it has at least k, but not at least $k + 1$, solutions).

The situation does not change if the solutions are required to satisfy additional conditions – namely, to lie within given bounds. We can thus say that *Theorem 13 constitutes an extension of Sturm's theorem* (or, at least, of the essential part of this theorem) *to arbitrary systems of equations and inequalities with arbitrarily many unknowns.*

It may be noticed that by Sturm's theorem a criterion for solvability (in the real domain) involves systems which contain inequalities as well as equations. Hence, to obtain an extension of this theorem to systems of equations in many unknowns, it seemed advisable to consider inequalities from the beginning, and in the first step to extend the theorem to arbitrary systems of equations and inequalities in one unknown. As a result of this preliminary extension, the subsequent induction with respect to the number of unknowns becomes almost trivial.

In its most general form the mathematical result obtained above seems to be new, although, in view of the extent of the literature involved, we have not been able to establish this fact with absolute certainty. At any rate some precedents are known in the literature. From what can be found in Sturm's original paper, the extension of his result to the case of one equation and one inequality with one unknown can easily be obtained; Kronecker, in his theory of characteristics, concerned himself with the case of n (independent) equations with n unknowns. It seems, on the other hand, that such a simple problem as that of finding an elementary criterion for the solvability in the real domain of one equation in two unknowns has not been previously treated; the same applies to the case of a system of inequalities (without equations) in one unknown – although this case is essential for the subsequent induction. (Cf. in this connection (Weber 1895, pp. 271 ff.) and (Runge 1904, pp. 416 ff.) where further references to the literature are also given.)

13. The result established in Theorem 13 and discussed in the preceding note has various interesting consequences. To formulate them, we can use, for instance, a geometric language and refer the result to n-dimensional analytic space with real coordinates -- or, what is slightly more convenient, to the infinite-dimensional space S_ω, in which, however, every point has only finitely many coordinates different from zero. By an elementary algebraic domain in S_ω we understand the set of all points $\langle x_0, x_1, \ldots, x_n, \ldots \rangle$ in which the coordinates $x_{k_1}, x_{k_2}, \ldots, x_{k_m}$ satisfy a given algebraic equation or inequality

$$F(x_{k_1}, \ldots, x_{k_m}) = 0$$

or

$$F(x_{k_1}, \ldots, x_{k_m}) > 0,$$

and the remaining coordinates are zeros. Let $\mathcal{F}$ be the smallest family of
point sets in S_ω which contains among its elements all elementary algebraic
domains, and is closed under the operations of finite set-addition, finite set-
multiplication, set-complementation, and projection parallel to any axis.
(The projection of a set A parallel to the n-th axis is the set obtained by
replacing by zero the n-th coordinate of every point of A.) Now Theorem 13
in geometric formulation implies that the family $\mathcal{F}$ consists of those and
only those sets in S_ω which are finite sums of finite products of elementary
algebraic domains. The possibility of passing from the original formulation
to the new one is a consequence of the known relations between projection
and existential quantifiers.

Theorem 13 has also some implications concerning the notion of arithmetical (or elementary) definability. A set A of real numbers is called
arithmetically definable if there is a formula Φ in our system containing
one free variable and such that A consists of just those numbers which
satisfy Φ. In a similar way we define an arithmetically definable binary,
ternary, and in general an n-ary, relation between real numbers. Now
Theorem 13 gives us a characterization of those sets of real numbers, and
relations between real numbers, which are arithmetically definable. We see,
for instance, that a set of real numbers is arithmetically definable if and
only if it is a set-theoretical sum of a finite number of intervals (bounded,
or unbounded; closed, open, or half-closed, half-open) with algebraic endpoints; in particular, a real number (i.e., the set consisting of this number
alone) is arithmetically definable if and only if it is algebraic. Hence it follows that an arithmetically definable set of real numbers which is bounded
above has an arithmetically definable least upper bound – a consequence
which is relevant in connection with a result mentioned near the end of
Note 9. As further consequences we conclude that the sets of all integers,
of all rationals, etc., are not arithmetically definable, which justifies some
remarks made in the Introduction.

As a simple corollary of Theorem 13 we obtain: For every formula
Φ there is an equivalent formula Ψ with the same free variables of the
following form:

$$\Psi = (E\xi_1) \ldots (E\xi_n)[\alpha = 0].$$

This corollary can also be interpreted geometrically.

For the notions used in this note, cf. (Tarski 1931) and (Kuratowski
and Tarski 1931).

14. In other words, using terminology introduced in Note 10, we state that
the number-theoretic function correlated with WU is general recursive.
Actually this function is easily seen to be a general recursive function of a
very simple type – what is called a "primitive" recursive function.

15. If we take the axiomatic point of view outlined in Note 9 and replace in our whole discussion the notion of truth by that of provability, then the meaning and extent of the fundamental results obtained in Section 3 undergo some essential changes. In the new interpretation, Theorem 15 implies that every sentence of elementary algebra is provably equivalent to one of the sentences $0 = 0$ or $0 = 1$. In addition, we can easily show that $WU(\Phi) \equiv (0 = 0)$ if and only if $WU(\sim \Phi) \equiv (0 = 1)$, and that for any provable sentence Φ we have $WU(\Phi) \equiv (0 = 0)$. By combining these results, we arrive at the conclusion that the axiomatic system of elementary algebra is consistent and complete, in the sense that one and only one of any pair Φ and $\sim \Phi$ of contradictory sentences is provable. The proof of this fact has what is called a constructive character. The completeness of the system implies by itself the existence of a decision method for the class of all provable sentences (even without the knowledge that the number-theoretical function correlated with WU is general recursive); cf. (Kleene 1936).

We further notice that all the axioms listed in Note 9 are satisfied, not only by real numbers, but by the elements of any real closed field in the sense of Artin and Schreier, cf. van der Waerden (1937, chapter IX). Thus all the results just mentioned can be extended to the elementary theory of real closed fields. From the fact that this theory is complete it follows that there is no sentence expressible in our formal system of elementary algebra which would hold in one real closed field and fail in another. In other words, any arithmetically definable property (in the sense of Note 13) which applies to one real closed field also applies to all other such fields: i.e., any two real closed fields are arithmetically indistinguishable.

In general, when applied to axiomatized theories, the notions of truth and provability do not have the same extension. Usually it can be shown only that every provable sentence is true. Since, however, in the case of elementary algebra the class of provable sentences turns out to be complete, we conclude that in this particular case the converse holds, and hence that the two classes coincide. Thus in the case of elementary algebra three equivalent definitions of a true sentence are available: (i) the definition of a true sentence as a sentence Φ such that $WU(\Phi) \equiv (0 = 0)$; (ii) the definition of a true sentence as a provable sentence; (iii) the definition based on the general method of defining truth developed in (Tarski 1936). Correspondingly, when starting to develop elementary algebra, we have three methods of stipulating which sentences will be accepted in this algebra – i.e., recognized as true. Apart from any educational and psychological considerations, the first method has in principle a great advantage; it implies directly that the class of sentences recognized as true is general recursive. Hence it provides us from the beginning with a mechanical device to decide in each particular case whether a sentence should be accepted, and serves as a basis for the construction of a decision machine. The second method – which is the usual axiomatic method - is less advantageous: it has as a direct consequence only the fact that the class of sentences recognized as true is what is called recursively enumerable (not necessarily general recursive). It leads to the construction of a machine which would be much less useful – to a machine which would construct, so to speak, blindly, the infinite sequence of all sentences accepted as true, without being able to tell in advance whether a given sentence would ever appear in the sequence. The third method, though very important for certain theoretical considerations, is even less advantageous than the second. It does not show that the class of accepted sentences is recursively enumerable; it can hardly be applied to a practical construction of a theory unless it is combined on a metamathematical level with the first or the second method. It goes without saying that in the particular case with which we are concerned – that is, in the case of elementary algebra – by establishing the equivalence of these possible definitions of truth we have *eo ipso* shown that in this case the three methods determine eventually the same class of sentences.

16. We can also consider a more restricted elementary system of complex algebra, from which the symbols $>$ and Rl have been eliminated. The decision method applies to such a system as well, and even becomes much simpler. By taking the axiomatic point of view and basing the discussion on the notion of provability, we can carry over to this restricted system of complex algebra all the results pointed out in Note 15. Since the axioms of this system prove to be satisfied by elements of an arbitrary algebraic closed field with characteristic zero (thus, in particular, by the complex algebraic numbers), the results apply to the general elementary theory of such fields; in particular, any two algebraic close fields with characteristic zero turn out to be arithmetically indistinguishable. A slight change in the argument permits us further to extend the results just mentioned to algebraic closed fields with any given characteristic p. For these notions, cf. van der Waerden (1937, chapter 5).

   On the other hand, as was mentioned in the Introduction, no decision method can be given for the arithmetic of rationals, nor for the elementary theory of arbitrary fields. For most special fields the decision problem still remains open. This applies, for instance, to finite algebraic extensions of the field of rational numbers and to the field of all numbers expressible by means of radicals. It would be interesting to solve the decision problem for some of these special fields, or even to obtain a simple mathematical characterization of all those fields for which the solution of the decision problem is positive.

17. As in the case of elementary algebra (see Note 8), some of the symbols listed could be eliminated from the system of elementary geometry and treated merely as abbreviations. It is known, for example, that in n-dimensional geometry with $n \geq 2$ the symbol "B" of the betweenness relation can be defined in terms of the symbol "D" of the equidistance relation.

18. Exactly as in the case of elementary algebra, we can treat the system of elementary geometry in an axiomatic way, and base our discussion of the

decision problem on the notion of provability. If we restrict ourselves to the case of two dimensions, we can take, for instance (in addition to the general logical axioms mentioned in Note 9), the following geometrical axioms:

(i) $(Ax)(Ay)B(x, y, y)$;

(ii) $(Ax)(Ay)[B(x, y, x) \rightarrow (x = y)]$;

(iii) $(Ax)(Ay)(Az)[B(x, y, z) \rightarrow B(z, y, x)]$;

(iv) $(Ax)(Ay)(Az)(Au)$
$$\left\{[B(x, y, u) \wedge B(y, z, u)] \rightarrow B(x, y, z)\right\};$$

(v) $(Ax)(Ay)(Az)(Au)$
$$\left\{[B(x, y, z) \wedge B(y, z, u) \wedge \sim (y = z)] \rightarrow B(x, y, u)\right\};$$

(vi) $(Ax)(Ay)(Az)(Au)$
$$\left\{[B(x, y, u) \wedge B(x, z, u)] \rightarrow [B(x, y, z) \vee B(x, z, y)]\right\};$$

(vii) $(Ax)(Ay)(Az)(Au)\left\{[B(x, y, z) \wedge B(x, y, u) \wedge \sim (x = y)] \rightarrow \right.$
$$\left. [B(x, z, u) \vee B(x, u, z)]\right\};$$

(viii) $(Ex)(Ey)(Ez)[\sim B(x, y, z) \wedge \sim B(y, z, x) \wedge \sim B(z, x, y)]$;

(ix) $(Ax)(Ay)(Az)(Az')(Au)\left\{[B(x, z', z) \wedge B(y, z, u)] \rightarrow \right.$
$$\left. (Ey')[B(x, y', y) \wedge B(y', z', u)]\right\};$$

(x) $(Ax)(Ay)(Az)(Az')(Au)$
$$\left\{[B(x, z, z') \wedge B(y, z, u) \wedge \sim (x = z)] \rightarrow \right.$$
$$\left. (Ey')(Eu')[B(x, y, y') \wedge B(x, u, u') \wedge B(y', z', u')]\right\};$$

(xi) $(Ax)(Ay)(Az)(Au)(Ev)$
$$\left\{[(B(x, u, v) \vee B(u, v, x) \vee B(v, x, u)) \wedge B(y, v, z)] \vee \right.$$
$$[(B(y, u, v) \vee B(u, v, y) \vee B(v, y, u)) \wedge B(z, v, x)] \vee$$
$$\left. [(B(z, u, v) \vee B(u, v, z) \vee B(v, z, u)) \wedge B(x, v, y)]\right\};$$

(xii) $(Ax)(Ay)D(x, y; y, x)$;

(xiii) $(Ax)(Ay)(Az)[D(x, y; z, z) \rightarrow (x = y)]$;

(xiv) $(Ax)(Ay)(Az)(Au)(Av)(Aw)$
$$\left\{[D(x, y; z, u) \wedge D(x, y; v, w)] \rightarrow D(z, u; v, w)\right\};$$

(xv) $(Ax)(Ay)(Az)(Az')(Au)$

$$\Big\{\big[\sim (x = y) \land D(x,z;x,z') \land D(y,z;y,z') \land$$

$$B(y,u,z') \land (B(x,u,z) \lor B(x,z,u))\big] \to (z = z')\Big\};$$

(xvi) $(Ax)(Ax')(Ay)(Ay')(Az)(Az')(Au)(Au')$

$$\Big\{\big[D(x,y;x',y') \land D(y,z;y',z') \land$$

$$D(x,u;x',u') \land D(y,u;y',u') \land$$

$$B(x,y,z) \land B(x',y',z') \land$$

$$\sim (x = y) \land \sim (y = z)\big] \to D(z,u;z',u')\Big\};$$

(xvii) $(Ax)(Ay)(Ay')(Az')(Ez)\big\{B(x,y,z) \land D(y,z;y',z')\big\};$

(xviii) $(Ax)(Ax')(Ay)(Ay')(Az')(Av)\Big\{D(x,y;x',y') \to$

$$(Ez)(Eu)\big[D(x,z;x',z') \land D(y,z;y',z') \land B(z,u,v) \land$$

$$(B(x,y,u) \lor B(y,u,x) \lor B(u,x,y))\big]\Big\}.$$

To these is added the axiom schema which comprehends all particular cases of the axiom of continuity (e.g., in the Dedekind form) that are expressible in our system: i.e., all sentences of the following form:

(xix) $(A\xi_1)\dots(A\xi_n)$

$$\Big\{(E\mu)(A\eta_1)(A\eta_2)\big[(\Phi \land \Psi) \to B(\mu,\eta_1,\eta_2)\big] \to$$

$$(E\mu)(A\eta_1)(A\eta_2)\big[(\Phi \land \Psi) \to B(\eta_1,\mu,\eta_2)\big]\Big\},$$

where neither μ nor η_2 is free in the formula Φ, and neither μ nor η_1 is free in the formula Ψ.

The reader will notice the formal simplicity of most of the axioms just given – which we have tried to put into evidence by avoiding (contrary to the prevailing custom) the use of any defined terms in formulating the axioms. On the other hand, however, the reader will easily recognize a close similarity between our axiom system and various systems which can be found in the comprehensive literature of the foundations of geometry; see, e.g., (Hilbert 1930).

By means of some obvious changes in (viii) and (xi) one can obtain from this axiom system a system of axioms for elementary geometry of any number of dimensions.

Again as in the case of algebra, one of the achievements attained by the axiomatic treatment of the subject is a constructive consistency proof for the whole of elementary geometry. This improves a result to be found in (Bernays and Hilbert 1939, vol. 2, pp. 38 ff.). It may also be mentioned that in (Hilbert 1930, section 35, pp. 96–98) a result is given which is closely connected with the decision method for elementary geometry, but which has a rather restricted character.

19. As is known, ordinary projective geometry can be treated as a specialized branch of lattice theory – more specifically, of the theory of modular lattices: see (Birkhoff 1948), where references to earlier papers of Menger can also be found. The decision method applies to this branch of the theory of modular lattices as well.

20. In the axiomatic presentation, the introduction of the symbol Exp would require the addition of new axioms. The following three axioms can be used, for instance, for this purpose:

$$(Ax)(Ay)\big[(x > y) \rightarrow (\mathrm{Exp}(x) > \mathrm{Exp}(y))\big]$$
$$(Ax)(Ay)\big[(\mathrm{Exp}(x) \cdot \mathrm{Exp}(y)) = \mathrm{Exp}(x + y)\big]$$
$$\mathrm{Exp}(1) = 1 + 1.$$

21. Similar decision problems arise if we introduce into our system of elementary algebra the symbol Al to denote the property of being an algebraic number, or the symbol Cn to denote the property of being a constructible number (i.e., a number which can be obtained from the number 1 by means of the rational operations, together with the operation of extracting square roots.) If the solution of the decision problem for elementary algebra with the addition of the symbol Cn were positive, this result would have an interesting application for geometry: in fact, we should obtain a decision method which would enable us, not only to decide on the truth of every sentence of elementary geometry, but also – in the case of existential sentences (like the sentence stating the possibility of trisecting an arbitrary angle) – to decide whether the truth of such a sentence can be established using only the so-called elementary constructions: i.e., constructions by means of rule and compass. It seems unlikely, however, that the solution of the problem in question is indeed positive; probably we shall be able to show that such a sharper decision method for elementary geometry cannot be found.

6 Supplementary Notes

1. The references to decision methods previously established given on p. 26 and in Note 2, p. 69, were not intended to be complete. For some further results and additional references compare the series of abstracts by Mostowski, Mrs. Szmielew, and Tarski in the Bulletin of the American Mathematical Society, vol. 55, pp. 63–66 and 1192, 1949, as well as the following papers:
Gentzen, G., "Untersuchungen über das logische Schliessen". Mathematische Zeitschrift, vol. 39, pp. 176–210, 1934.
McKinsey, J. C. C., "A solution of the decision problem for the Lewis systems S2 and S4, with an application to topology". Journal of Symbolic Logic, vol. 6, pp. 117–134, 1941.
McKinsey, J. C. C. and Tarski, A., "The algebra of topology". Annals of

Mathematics, vol. 45, pp. 141–191, 1944.

Skolem, T., Über einige Satzfunktionen in der Arithmetik. [Skrifter utgitt av det Norske Videnskaps-Akademi i Oslo, I. klasse 1930, no. 7] Oslo, 1931.

Szmielew, W., "Decision problem in group theory". Proceedings of the Tenth International Congress of Philosophy, fasc. 2, pp. 763–766, Amsterdam, 1949.

2. The results of Mrs. Robinson, Mostowski, and the author mentioned in the first paragraph of p. 26 appeared in print (some only in outline form) after the first edition of this monograph. See the series of abstracts in the Journal of Symbolic Logic, vol. 14, pp. 75–78, 1949, as well as the article:

Robinson, J., "Definability and decision problems in arithmetic". Journal of Symbolic Logic, vol. 14, pp. 98–114, 1949.

Some related results can be found in the article:

Robinson, R. M., "Undecidable rings". Transactions of the American Mathematical Society, vol. 70, pp. 137–159, 1951.

3. The following remarks refer to the discussion on pp. 29 and 30. Many examples of open problems in elementary algebra and geometry are known; one comes across discussions of such problems by looking through any issue of the American Mathematical Monthly. However, the problem of describing the behavior of the function d does not seem to have been previously treated in the literature. For a discussion of a related problem – involving the decomposition of P and Q, not in triangles, but in arbitrary polygons – see the following article (where references to earlier papers of Moese and the author can also be found):

Tarski, A., "Uwagi o stopniu równoważności wielokatów". (Remarks on the degree of equivalence of polygons, in Polish.) Parametr, vol. 2, 1932.

It may be interesting to mention that some conclusions concerning the function d can be derived from the general results stated in Note 13, p. 75. In fact, it can be shown that every bounded interval (a, b) can be divided into finitely many subintervals such that the function d is constant within each of these subintervals; all the end-points of these subintervals are algebraic, with the possible exceptions of a and b.

4. The statement in Note 12, p. 73, to the effect that the case of a system of inequalities in one unknown was not previously treated in the literature, seems to be correct when applied to the situation which existed at the time when the results of this work were found and first mentioned in print (1931), as well as for many years thereafter. However, the author's attention has been called to the fact that this case has recently been treated in the paper:

Meserve, B. E., "Inequalities of higher degree in one unknown". American Journal of Mathematics, vol. 49, pp. 357–370, 1947.

5. The discussion in Note 13, p. 75, may convey the impression that the notions considered in the first paragraph have but little in common with those considered in the second paragraph. Actually, these notions are very closely related to each other. In fact, if the notion of arithmetical definability is applied to arbitrary sets of sequences of real numbers, i.e., to point sets in

S_ω, then the family of all arithmetically definable point sets simply coincides with the family $\mathcal{F}$.

6. It was stated in Note 13, p. 75, that every real number which is arithmetically definable is algebraic. An interesting application of this result to the theory of games has recently been found by O. Gross and is discussed in his paper "On certain games with transcendental values" (to appear in the American Mathematical Monthly).

7. As was pointed out in Note 15, p. 76, the completeness theorem for elementary algebra leads to the following result: every arithmetically definable property which applies to one real closed field also applies to all other such fields. It is important to realize that the result just mentioned extends to a comprehensive class of properties which are not arithmetically definable (i.e., which are not expressible in our formal system of elementary algebra). This class includes in particular all the properties expressed by sentences of the form $(Am)\Phi_m$, $(Am)(En)\Psi_{m,n}, \ldots$, where $m, n, \ldots$ are variables assumed to range over all positive integers and where $\Phi_m, \Psi_{m,n}, \ldots$ are formulas which involve $m, n, \ldots$ (as free variables) and which, for any particular values of $m, n, \ldots$, are equivalent in any real closed field to sentences of elementary algebra. In fact, consider a sentence of this kind, say $(Am)\Phi_m$. If this sentence holds in a given real closed field, the same obviously applies to all the particular sentences of the form Φ_m, i.e., to $\Phi_1, \Phi_2, \Phi_3, \ldots$. Each of these particular sentences is equivalent to a sentence of elementary algebra and hence, by the result discussed, it holds in every real closed field. Consequently, the universal sentence $(Am)\Phi_m$ also holds in every real closed field. Various theorems of these types are known which were originally established for the field of real numbers using essentially the continuity of this field (sometimes with the help of difficult topological methods) and whose extension to arbitrary real closed fields presented a new and difficult problem; in view of our general result such an extension now becomes automatic. As examples the following three theorems may be mentioned.

    I. Let R be an m-dimensional region defined as the set of all points

$$\langle x_0, x_1, \ldots, x_{m-1} \rangle$$

satisfying a finite system of inequalities

$$P_i(x_0, x_1, \ldots, x_{m-1}) \geq 0$$

where the P_i's for $i = 0, 1, \ldots, n-1$ are polynomials of degree at most p; let F be a rational function whose denominator does not vanish on R. Then there is a positive integer q (dependent exclusively on m, n, and p) such that the set S of all function values of F on R is a sum of at most q closed intervals; if R is bounded, then all these intervals are also bounded, and hence F reaches a maximum and minimum on R.

    II. For every system of m polynomials $P_0, P_1, \ldots, P_{m-1}$ in m variables there are real numbers $c \geq 0, x_0, x_1, \ldots, x_{m-1}$ such that $P_i(x_0, x_1, \ldots, x_{m-1}) = c \cdot x_i$ for all $i = 0, 1, \ldots, m-1$.

III. Every commutative division algebra – whether associative or not – over the field of real numbers is of order 1 or 2; if it has a unit, it coincides either with the field of real numbers or with the algebraic closure of this field (i.e., with the field of complex numbers).

While I is simply a particular case of a familiar theorem concerning continuous functions, and the same applies to the "eigenvalue theorem" II, Theorem III has a specifically algebraic character; it was proved, with the help of topology, in the article:

Hopf, H., "Systeme symmetrischer Bilinearformen und euklidische Modelle der projectiven Räume". Vierteljahrsschrift der Naturforschenden Gesellschaft in Zürich, vol. 85, supplement No. 32, pp. 165–177, 1940.

The research to extend these and similar results, obtained by means of topological methods, to arbitrary real closed fields was initiated by H. Hopf. Compare the following papers where partial results in this direction (in particular, extension of some special cases of Theorem I) have been achieved directly, without the help of our general method:

Behrend, F., "Über Systeme algebraischer Gleichungen". Compositio Mathematica, vol. 7, pp. 1–19, 1939.

Habicht, W., "Ein Existenzsatz über reelle definite Polynome". Commentarii Mathematici Helvetici, vol. 18, pp. 331–348, 1946.

Habicht, W., "Über die Lösbarkeit gewisser algebraischer Gleichungssysteme". Commentarii Mathematici Helvetici, vol. 18, pp. 154–175, 1946.

Kaplansky, I., "Polynomials in topological fields". Bulletin of the American Mathematical Society, vol. 54, pp. 909–916, 1948.

8. In view of the results stated in Note 16, p. 78, the remarks made in the preceding note will still hold if, instead of real closed fields, we consider the class of algebraically closed fields with a given characteristic.

Quantifier Elimination for Real Closed Fields by Cylindrical Algebraic Decomposition[1]

George E. Collins

1 Introduction

Tarski in 1948, (Tarski 1951) published a quantifier elimination method for the elementary theory of real closed fields (which he had discovered in 1930). As noted by Tarski, any quantifier elimination method for this theory also provides a decision method, which enables one to decide whether any sentence of the theory is true or false. Since many important and difficult mathematical problems can be expressed in this theory, any computationally feasible quantifier elimination algorithm would be of utmost significance.

However, it became apparent that Tarski's method required too much computation to be practical except for quite trivial problems. Seidenberg (1954) described another method which he thought would be more efficient. A third method was published by Cohen (1969). Some significant improvements of Tarski's method have been made by W. Boege (private communication in 1973), which are described in a thesis by Holthusen (1974).

This paper describes a completely new method which I discovered in February 1973. This method was presented in a seminar at Stanford University in March 1973 and in abstract form at a symposium at Carnegie/Mellon University in May 1973. In August 1974 a full presentation of the method was delivered at the EUROSAM 74 Conference in Stockholm, and a preliminary version of the present paper was published in the proceedings of that conference (Collins 1974b).

The method described here is much more efficient than the previous methods, and therefore offers renewed hope of practical applicability. In fact, it will

---

[1]Reproduced from *Automata Theory and Formal Languages* (Lecture Notes in Computer Science, vol. 33), edited by H. Brakhage, with permission of Springer-Verlag, with corrections by the author.

be shown that, for a prenex input formula ϕ, the maximum computing time of the method is dominated, in the sense of (Collins 1971a), by $(2n)^{2^{2r+8}} m^{2^{r+6}} d^3 a$, where r is the number of variables in ϕ, m is the number of polynomials occurring in ϕ, n is the maximum degree of any such polynomial in any variable, d is the maximum length of any integer coefficient of any such polynomial, and a is the number of occurrences of atomic formulas in ϕ. Thus, for fixed r, the computing time is dominated by a polynomial function $P_r(m, n, d, a)$. In contrast, it can be shown that the maximum computing times of the methods of Tarski and Seidenberg are exponential in both m and n for every fixed r, including even $r = 1$, and this is likely the case for Cohen's method also. (In fact, Cohen's method is presumably not intended to be efficient.) Boege's improvement of Tarski's method eliminates the exponential dependency on m, but the exponential dependency on n remains.

Fischer and Rabin (1974b) have recently shown that every decision method, deterministic or non-deterministic, for the first order theory of the additive group of the real numbers, a fortiori for the elementary theory of a real closed field, has a maximum computing time which dominates 2^{cN} where N is the length of the input formula and c is some positive constant. Since m, n, d, r and a are all less than or equal to N (assuming that x^n must be written as $x \cdot x \cdot \ldots \cdot x$), the method of this paper has a computing time dominated by $2^{2^{kN}}$ where in fact $k \leq 8$. The result of Fischer and Rabin suggests that a bound of this form is likely the best achievable for any deterministic method.

In a letter received from Leonard Monk in April 1974, I was informed that he and R. Solovay had found a decision method, but not a quantifier elimination method, with a maximum computing bound of the form $2^{2^{2^{2^{kN}}}}$. However, the priority and superiority of the method described below are easily established.

The most essential observation underlying the method to be described is that if $\mathcal{A}$ is any finite set of polynomials in r variables with real coefficients, then there is a decomposition of r-dimensional real space into a finite number of disjoint connected sets called cells, in each of which each polynomial in $\mathcal{A}$ is invariant in sign. Moreover, these cells are cylindrically arranged with respect to each of the r variables, and they are algebraic in the sense that their boundaries are the zeros of certain polynomials which can be derived from the polynomials in $\mathcal{A}$. Such a decomposition is therefore called an $\mathcal{A}$-invariant cylindrical algebraic decomposition. The sign of a polynomial in $\mathcal{A}$ in a cell of the decomposition can be determined by computing its sign at a sample point belonging to the cell. In the application of cylindrical algebraic decomposition to quantifier elimination, we assume that we are given a quantified formula ϕ in prenex form, and we take $\mathcal{A}$ to be the set of all polynomials occurring in ϕ. From a set of sample points for a decomposition, we can decide in which cells the unquantified matrix of the formula ϕ is true. The decomposition of r-dimensional space induces, and is constructed from, a decomposition of each lower-dimension space. Each cylinder is composed of a finite number of cells, so universal and existential quantifiers can be treated like conjunctions and disjunctions, and one can decide in which cells of a lower-dimension space the quantified formula is true. The quantifier elimination can be completed by constructing formulas which define these cells.

The polynomials whose zeros form the boundaries of the cells are the elements of successive "projections" of the set $\mathcal{A}$. The projection of a set of polynomials in r variables is a set of polynomials in $r - 1$ variables. The cylindrical arrangement of cells is ensured by a condition called delineability of roots, which is defined in Section 2. Several theorems giving sufficient conditions for delineability are proved, culminating in the definition of projection and the fundamental theorem that if each element of the projection of a set $\mathcal{A}$ is invariant on a connected set S then the roots of A are delineable on S. This theorem implicitly defines an $\mathcal{A}$-invariant cylindrical algebraic decomposition. Section 2 also defines the "augmented projection", a modification of the projection which is applied in certain contexts in order to facilitate the construction of defining formulas for cells. Section 2 is concluded with the specification of the main algebraic algorithms which are required as subalgorithms of the quantifier elimination algorithm described in Section 3. These algebraic algorithms include algorithms for various operations on real algebraic numbers and on polynomials with rational integer or real algebraic number coefficients.

Section 3 describes the quantifier elimination algorithm, ELIM, and its subalgorithms, which do most of the work. ELIM invokes successively its two subalgorithms, DECOMP (decomposition) and EVAL (evaluation). DECOMP produces sample points and cell definitions, given a set of polynomials. EVAL uses the sample points and cell definitions, together with the prenex formula ϕ, to produce a quantifier-free formula equivalent to ϕ. DECOMP itself uses a subalgorithm, DEFINE, to aid in the construction of defining formulas for cells. These algorithms are described in a precise but informal style with extensive interspersed explanatory remarks and assertions in support of their validity.

Section 4 is devoted to an analysis of the computing time of the quantifier elimination algorithm. Since some of the required algebraic subalgorithms have not yet been fully analyzed, and since in any case improved subalgorithms are likely to be discovered from time to time, the analysis is carried out in terms of a parameter reflecting the computing times of the subalgorithms.

Section 5 is devoted to further discussion of the algorithm, including possible modifications, examples, special cases, and the observed behavior of the method.

It should be noted that the definition of the projection operator has been changed in an important way since the publication of the preliminary version of this paper. This change is justified by a new definition of delineability in Section 3 and a new proof of what is now Theorem 5. This change in the projection operator contributes greatly to the practical feasibility of the algorithm.

2   Algebraic Foundations

In this section we make some needed definitions, prove the basic theorems which provide a foundation for the quantifier elimination algorithm to be presented in Section 3, and define and discuss the main subalgorithms which will be required.

By an *integral polynomial* in r variables we shall mean any element of the ring $\mathbb{Z}[x_1,\ldots,x_r]$, where $\mathbb{Z}$ is the ring of the rational integers. As observed by Tarski, any atomic formula of elementary algebra can be expressed in one of the two forms $A = 0$, $A > 0$, where A is an integral polynomial. Also, any quantifier-free formula can be easily expressed in disjunctive normal form as a disjunction of conjunctions of atomic formulas of these types. However, for the quantifier elimination algorithm to be presented in this paper, there is no reason to be so restrictive, and we define a *standard atomic formula* as a formula of one of the six forms $A = 0$, $A > 0$, $A < 0$, $A \neq 0$, $A \geq 0$, and $A \leq 0$. A *standard formula* is any formula which can be constructed from standard atomic formulas using propositional connectives and quantifiers. A *standard prenex formula* is a standard formula of the form

$$(Q_k x_k)(Q_{k+1} x_{k+1}) \ldots (Q_r x_r) \phi(x_1,\ldots,x_r), \tag{1}$$

where $\phi(x_1,\ldots,x_r)$ is a quantifier-free standard formula, $1 \leq k \leq r$, and each $(Q_i x_i)$ is either an existential quantifier $(\exists x_i)$ or a universal quantifier $(\forall x_i)$.

The variables x_i range over the ordered field $\mathbb{R}$ of all real numbers, or over any other real closed field. For additional background information on elementary algebra, the reader is referred to Tarski's (1951) excellent monograph and van der Waerden (1970) has an excellent chapter on real closed fields.

The quantifier elimination algorithm to be described in the next section accepts as input any standard prenex formula of the form (1), with $1 \leq k \leq r$, and produces as output an equivalent standard quantifier-free formula $\psi(x_1,\ldots,x_{k-1})$.

$\mathcal{R}$ will denote an arbitrary commutative ring with identity. Unless otherwise specified, we will always regard a polynomial $A(x_1,\ldots,x_r) \in \mathcal{R}[x_1,\ldots,x_r]$ as an element of $\mathcal{R}[x_1,\ldots,x_{r-1}][x_r]$; that is, A is regarded as a polynomial in its *main variable*, x_r, with coefficients in the polynomial ring $\mathcal{R}[x_1,\ldots,x_{r-1}]$. Thus, for example, the *leading coefficient* of A, denoted by $\mathrm{ldcf}(A)$, is an element of $\mathcal{R}[x_1,\ldots,x_{r-1}]$. Similarly, $\deg(A)$ denotes the degree of A in x_r. If $A(x_1,\ldots,x_r) = \sum_{i=0}^{n} A_i(x_1,\ldots,x_{r-1}) \cdot x_r^i$ and $\deg(A) = n$ then $\mathrm{ldcf}(A) = A_n$ and $\mathrm{ldt}(A) = A_n(x_1,\ldots,x_{r-1}) \cdot x_r^n$, the *leading term* of A. Following Tarski, $\mathrm{red}(A)$, the *reductum* of A, is the difference $A - \mathrm{ldt}(A)$. By convention, $\deg(0) = \mathrm{ldcf}(0) = 0$, and hence also $\mathrm{ldt}(0) = \mathrm{red}(0) = 0$. A' will denote the derivative of A.

$\mathbb{R}^k$ will denote the k-fold Cartesian product $\mathbb{R} \times \cdots \times \mathbb{R}$, $k \geq 1$. If f and g are real-valued functions defined on a set $S \subseteq \mathbb{R}^k$, we write $f > 0$ *on* S in case $f(x) > 0$ for all $x \in S$; $f < 0$ *on* S, $f \neq 0$ *on* S, $f < g$ *on* S and other such relations are similarly defined. We say that f *is invariant on* S in case $f > 0$

on S, $f = 0$ on S, or $f < 0$ on S. These definitions are also applied to real polynomials, which may be regarded as real-valued functions.

The field of complex numbers will be denoted by $\mathbb{C}$. We will regard $\mathbb{R}$ as a subset, and hence as a subfield, of $\mathbb{C}$. A polynomial $A(x_1, \ldots, x_r)$ belonging to $\mathbb{R}[x_1, \ldots, x_r]$ will be called a *real polynomial*.

Let $A(x_1, \ldots, x_r)$ be a real polynomial, $r \geq 2$, S a subset of $\mathbb{R}^{r-1}$. We will say that the roots of A are delineable on S, and that $f_1, \ldots, f_k$, $k \geq 0$, *delineate the real roots* of A on S in case the following conditions are all satisfied:

(1) There are $m \geq 0$ positive integers e_i such that if $(a_1, \ldots, a_{r-1}) \in S$ then $A(a_1, \ldots, a_{r-1}, x)$ has exactly m distinct roots, with multiplicities $e_1, \ldots, e_m$.

(2) $f_1 < f_2 \ldots < f_k$ are continuous functions from S to $\mathbb{R}$.

(3) If $(a_1, \ldots, a_{r-1}) \in S$ then $f_i(a_1, \ldots, a_{r-1})$ is a root of $A(a_1, \ldots, a_{r-1}, x_r)$ of multiplicity e_i for $1 \leq i \leq k$ and $(a_1, \ldots, a_{r-1}) \in S$,

(4) If $(a_1, \ldots, a_{r-1}) \in S$, $b \in \mathbb{R}$ and $A(a_1, \ldots, a_{r-1}, b) = 0$ then $b = f_i(a_1, \ldots, a_{r-1})$ for some $i, 1 \leq i \leq k$.[1]

e_i will be called the multiplicity of f_i.

Note that if the roots of A are delineable on S then $A(a_1, \ldots, a_{r-1}, x)$ is a non-zero polynomial for $(a_1, \ldots, a_{r-1}) \in S$ and the number of distinct roots of $A(a_1, \ldots, a_{r-1}, x)$ is independent of the choice of $(a_1, \ldots, a_{r-1})$ in S. The number of roots, multiplicities counted, $\sum_{i=1}^{m} e_i = n$, must also be invariant on S. Hence, $\deg(A) = n$ is also invariant on S, so $\mathrm{ldcf}(A) \neq 0$ on S. The following basic theorem shows that if these necessary conditions are satisfied and additionally S is connected, then the roots of A are delineable on S.

Theorem 1 *Let $A(x_1, \ldots, x_r)$ be a real polynomial, $r \geq 2$. Let S be a connected subset of $\mathbb{R}^{r-1}$. If $\mathrm{ldcf}(A) \neq 0$ on S and the number of distinct roots of A is invariant on S, then the roots of A are delineable on S.*

Proof. We may assume S is non-empty and $\deg(A) = n > 0$ since otherwise the theorem is trivial. Let $(a_1, \ldots, a_{r-1}) = a \in S$ and let $\alpha_1, \ldots, \alpha_m$ be the distinct roots of $A(a_1, \ldots, a_{r-1}, x)$. We may assume that $\alpha_1 < \alpha_2 < \cdots < \alpha_k$ are real and $a_{k+1}, \ldots, a_m$ are non-real. If $m = 1$ let $\delta = 1$ and otherwise let $\delta = \frac{1}{2} \min_{i<j} |\alpha_i - \alpha_j|$. Let C_i be the circle with center α_i and radius δ. Let $A(x_1, \ldots, x_r) = \sum_{i=0}^{n} A_1(x_1, \ldots, x_{r-1}) x_r^i$. Since the A_i are continuous functions on S and $A_n \neq 0$ on S, by theorem (1,4) of (Marden 1949) there exists

---

[1] In the original paper there was an additional condition in the definition of delineability that required the non-real roots of A to also be definable by continuous functions on S. However this condition was superfluous for CAD construction and there was a flaw in the attempted proof of the existence of such functions in Theorem 1 below. To remedy this, some changes have been made in the proofs of several theorems of this section. Otherwise only a few corrections of typographical errors have been made to the original paper.

$\epsilon > 0$ such that if $a' = (a'_1, \ldots, a'_{r-1}) \in S$ and $|a - a'| < \epsilon$ then $A(a', x)$ has exactly e_i roots, multiplicities counted, inside C_i, where e_i is the multiplicity of α_i. Since by hypothesis $A(a', x)$ has exactly m distinct roots and the interiors of the m circles C_i are disjoint, each circle must contain a unique root of $A(a', x)$, whose multiplicity is e_i. Since the non-real roots of $A(a, x)$ occur in conjugate pairs, the interiors of the circles $C_{k+1}, \ldots, C_m$ contain no real numbers and hence the roots of $A(a', x)$ in $C_{k+1}, \ldots, C_m$ are non-real. If $i \leq k$ and C_i contained a non-real root of $A(a', x)$ then its conjugate would also be a non-real root of $A(a', x)$ in C_i since the center of C_i is real. So the roots of $A(a', x)$ in $C_1, \ldots, C_k$ are real.

Let $N = \{a' : a' \in S \,\& \, |a - a'| < \epsilon\}$. For $a' \in N$ define $f_1(a')$ to be the unique root of $A(a', x)$ inside C_1. Then $f_1 < f_2 < \cdots < f_k$ are real functions and $f_{k+1}, \ldots, f_m$ are non-real valued. By another application of Theorem $(1,4)$ of (Marden 1949), the f_i are continuous functions on N, which is an open neighborhood of a in S. Hence if $0 \leq k \leq m$ and S_k is the set of all $a \in S$ such that $A(a, x)$ has exactly k distinct real roots, then S_k is open in S. Since a connected set is not a union of two disjoint non-empty subsets, there is a unique k such that $S = S_k$.

We can now define $f_i(a)$ to be the i-th real root of $A(a, x)$ for $a \in S$ and $1 \leq i \leq k$, so that $f_1 < f_2 < \cdots < f_k$ on S. By the preceding paragraph it is immediate that $f_1, \ldots, f_r$ are continuous. By another application of the connectivity, the multiplicity of f_i as a root of A is an invariant e_i throughout S since, as we have already shown, the multiplicity is locally invariant. ∎

We say that the polynomials $A, B \in \mathcal{R}[x]$ are *similar*, and write $A \approx B$, in case there exist non-zero $a, b \in \mathcal{R}$ such that $aA = bB$.

We define $\mathrm{red}^k(A)$, the k-th *reductum* of the polynomials A, for $k \geq 0$, by induction on k as follows: $\mathrm{red}^0(A) = A$ and $\mathrm{red}^{k+1}(A) = \mathrm{red}(\mathrm{red}^k(A))$ for $k \geq 0$. We say that B is a *reductum* of A in case $B = \mathrm{red}^k(A)$ for some $k \geq 0$.

We repeat some definitions from (Collins 1967). Let A and B be polynomials over $\mathcal{R}$ with $\deg(A) = m$ and $\deg(B) = n$. The *Sylvester matrix* of A and B is the $m + n$ by $m + n$ matrix M whose successive rows contain the coefficients of the polynomials $x^{n-1}A(x), \ldots, A(x), x^{m-1}B(x), \ldots, xB(x), B(x)$, with the coefficients of x^i occurring in column $m + n - i$. We allow either $m = 0$ or $n = 0$. As is well known, $\mathrm{res}(A, B)$, the resultant of A and B, is $\det(M)$, the determinant of M. (We adopt the convention $\det(N) = 0$ in case N is a zero by zero determinant.) For $0 \leq i \leq j \leq \min(m, n)$ let $M_{j,i}$ be the matrix obtained from M by deleting the last j rows of A coefficients, the last j rows of B coefficients, and all of the last $2j + 1$ columns except column $m + n - i - j$. The j-th *subresultant* of A and B is the polynomial $S_j(A, B) = \sum_{i=0}^{j} \det(M_{j,i}) x^i$, a polynomial of degree j or less. We define also the j-th *principal subresultant coefficient* of A and B by $\mathrm{psc}_j(A, B) = \det(M_{j,j})$. Thus $\mathrm{psc}_j(A, B)$ is the coefficient of x^j in $S_j(A, B)$. We note, for subsequent application, that if $\deg(A) = m > 0$ then $\mathrm{psc}_{m-1}(A, A') = m \cdot \mathrm{ldcf}(A)$.

Theorem 2 *Let A and B be non-zero polynomials over a unique factorization domain. Then $\deg(\gcd(A, B)) = k$ if and only if k is the least j such that $\mathrm{psc}_j(A, B) \neq 0$.*

Proof. Let $k = \deg(\gcd(A, B))$. By the fundamental theorem of polynomial remainder sequences (Brown and Traub 1971) $S_j(A, B) = 0$ for $0 \leq j < k$, and $\gcd(A, B) \approx S_k(A, B)$. Hence $\mathrm{psc}_j(A, B) = 0$ for $0 \leq j < k$, $\deg(S_k(A, B)) = k$, and $\mathrm{psc}_k(A, B) \neq 0$. ∎

Theorem 3 *Let $A(x)$ be a real univariate polynomial with $\deg(A) = m \geq 1$ and let $k = \deg(\gcd(A, A'))$. Then $m - k$ is the number of distinct roots of A.*

Proof. Let A have the distinct roots $\alpha_1, \ldots, \alpha_n$ with respective multiplicities $e_1, \ldots, e_n$. By a familiar argument, α_i is a root of A' with multiplicity $e_i - 1$ (meaning that α_i is not a root of A' if $e_i = 1$). Hence α_i is a root of $\gcd(A, A')$ with multiplicity $e_i - 1$. Since every root of $\gcd(A, A')$ is some α_i, $k = \deg(\gcd(A, A')) = \sum_{i=0}^{h}(e_i - 1) = \sum_{i=0}^{h} e_i - h = m - h$ and $m - k = h$. ∎

Using reducta and principal subresultant coefficients, we now obtain a more useful sufficient condition for the delineability of the roots of a polynomial.

Theorem 4 *Let $A(x_1, \ldots, x_r)$ be a real polynomial, $r \geq 2$, S a connected subset of $\mathbb{R}^{r-1}$. Let $\mathcal{B} = \{\mathrm{red}_k(A) : k \geq 0 \,\&\, \deg(\mathrm{red}^k(A)) \geq 1\}$, $\mathcal{L} = \{\mathrm{ldcf}(B) : B \in \mathcal{B}\}$. $\mathcal{S} = \{\mathrm{psc}_k(B, B') : B \in \mathcal{B} \,\&\, 0 \leq k < \deg(B')\}$ and $\mathcal{P} = \mathcal{L} \cup \mathcal{S}$. If every element of $\mathcal{P}$ is invariant on S, then the roots of A are delineable on S.*

Proof. If $\deg(A) \leq 1$ then the theorem is obvious, so let $A(x_1, \ldots, x_r) = \sum_{i=0}^{n} A_i(x_1, \ldots, x_{r-1}) x_r^i$ with $\deg(A) = n \geq 1$. If $i \geq 1$ and $A_i \neq 0$ then $A_i \in \mathcal{L}$ so A_i is invariant on S for $1 \leq i \leq n$. If $A_i = 0$ on S for $1 \leq i \leq n$ then the theorem is obvious, so let $m \geq 1$ be maximal such that $A_m \neq 0$ on S and let k be such that $\mathrm{red}_k(A) = \sum_{i=0}^{m} A_i(x_1, \ldots, x_{r-1}) x_r^i = B$. Then $A = B$ on S so it suffices to show that the roots of B are delineable on S. $B \in \mathcal{B}$ so $\mathrm{psc}_j(B, B')$ is invariant on S for $0 \leq j < m - 1$. Also $\mathrm{psc}_{m-1}(B, B') = m A_m \neq 0$ on S. By Theorem 2, $\deg(\gcd(B, B'))$ is invariant on S, that is, for some k, $\deg(\gcd(B(a, x), B'(a, x))) = k$ for all $a \in S$. By Theorem 3, the number of distinct roots of B on S is the invariant $m - k$. By Theorem 1, the roots of B are delineable on S. ∎

Let $\mathcal{A}$ be a set of real polynomials in r variables, $r \geq 2$. Let $\mathcal{B} = \{\mathrm{red}^k(A) : A \in \mathcal{A} \,\&\, k \geq 0 \,\&\, \deg(\mathrm{red}^k(A)) \geq 1\}$, $\mathcal{L} = \{\mathrm{ldcf}(B) : B \in \mathcal{B}\}$, $\mathcal{S}_1 = \{\mathrm{psc}_k(B, B') : B \in \mathcal{B} \,\&\, 0 \leq k < \deg(B')\}$, $\mathcal{S}_2 = \{\mathrm{psc}_k(B_1, B_2) : B_1, B_2 \in \mathcal{B} \,\&\, 0 \leq k < \min(\deg(B_1), \deg(B_2))\}$ and $\mathcal{P} = \mathcal{L} \cup \mathcal{S}_1 \cup \mathcal{S}_2$. Then $\mathcal{P}$ will be called the *projection* of $\mathcal{A}$. If $\mathcal{A} = \{A_1, \ldots, A_n\}$ is a non-empty finite set of non-zero polynomials, we will say that the roots of $\mathcal{A}$ are *delineable* on a set S in case the roots of the product $A = \prod_{i=1}^{n} A_i$ are delineable on S. Note that the roots of each A_i could be delineable on S without the roots of $\mathcal{A}$ being delineable on S. The next theorem shows how the inclusion of the set $\mathcal{S}_2$ in the projection $\mathcal{P} = \mathrm{proj}(\mathcal{A})$ helps to ensure the delineability of the roots of $\mathcal{A}$.

Theorem 5 *Let $\mathcal{A} = \{A_1, \ldots, A_n\}$ be a non-empty set of non-zero real polynomials in r real variables, $r \geq 2$. Let S be a connected subset of $\mathbb{R}^{r-1}$. Let $\mathcal{P}$ be the projection of $\mathcal{A}$. If every element of $\mathcal{P}$ is invariant on S, then the roots of $\mathcal{A}$ are delineable on S.*

Proof. By Theorem 1 and the inclusion in $\mathcal{P}$ of each $\text{proj}(A_1)$, the roots of each A_i are delineable. By Theorem 2 and the inclusion in $\mathcal{P}$ of $\mathcal{S}_2$, the number of common roots of A_i and A_j is invariant in S for all i and j, $i \neq j$. For any fixed $a \in S$ and any two distinct products B and C of some of the A_i let $\text{nd}(B)$ denote the number of distinct roots of B at a and let $\text{nc}(B, C)$ denote the number of common roots of B and C at a. Then we have the recurrence relations $\text{nc}(A_1 \cdots A_{k-1}, A_k) = \text{nc}(A_1 \cdots A_{k-2}, A_k) + \text{nc}(A_{k-1}, A_k) - \text{nc}(A_1 \cdots A_{k-2}, A_{k-1})$ and $\text{nd}(A_1 \cdots A_n) = \text{nd}(A_1 \cdots A_{n-1}) + \text{nd}(A_n) - \text{nc}(A_1 \cdots A_{n-1}, A_n)$. Repeated application of these relations results in an equation for $\text{nd}(A_1 \cdots A_n)$ in terms of the invariants $\text{nd}(A_i)$ and $\text{nc}(A_i, A_j)$, thereby showing that $\text{nd}(A_1 \cdots A_n)$ is also invariant on S. This implies the invariance of the multiplicities of the roots of $A_1 \cdots A_n$, thus establishing condition (1) of the definition of delineability. Let $f_1, \ldots, f_k$ be all of the delineating functions for all of the A_i. From condition (1) it follows that if $i \neq j$ then $f_i(a) \neq f_j(a)$ for all $a \in S$, and hence the f_i can be arranged so that condition (2) is satisfied. Conditions (3) and (4) then follow easily. ∎

Let us write $\text{der}(A)$ for A', the derivative of A. We define $\text{der}^0(A) = A$ and, inductively, $\text{der}^{k+1}(A) = \text{der}(\text{der}^k(A))$ for $k \geq 0$.

Let $\mathcal{A}$ be a set of real polynomials in r variables, $r \geq 2$. Let $\mathcal{B} = \{\text{red}^k(A) : A \in \mathcal{A} \,\&\, k \geq 0 \,\&\, \deg(\text{red}^k(A)) \geq 1\}$, $\mathcal{D} = \{\text{der}^k(B) : B \in \mathcal{B} \,\&\, 0 < k < \deg(B)\}$ and $\mathcal{P}' = \{\text{psc}_k(D, D') : D \in \mathcal{D} \,\&\, 0 \leq k < \deg(D')\}$. Then $\mathcal{P} \cup \mathcal{P}'$, where $\mathcal{P}$ is the projection of $\mathcal{A}$, will be called the *augmented projection* of $\mathcal{A}$.

Theorem 6 *Let $\mathcal{A}$ be a set of real polynomials in r variables, $r \geq 2$. Let S be a connected subset of R^{r-1}. Let $\mathcal{P}^*$ be the augmented projection of $\mathcal{A}$. If every element of $\mathcal{P}^*$ is invariant on S then the roots of $\text{der}^j(A)$ are delineable on S for every $A \in \mathcal{A}$ and every $j \geq 0$.*

Proof. Let $A \in \mathcal{A}$, $j \geq 0$, $A^* = \text{der}^j(A)$, $\mathcal{B} = \{\text{red}^k(A^*) : k \geq 0 \,\&\, \deg(\text{red}^k(A^*)) \geq 1\}$, $\mathcal{L} = \{\text{ldcf}(B) : B \in \mathcal{B}\}$, $\mathcal{S} = \{\text{psc}_h(B, B') : 0 \leq h < \deg(B')\}$, and $\mathcal{P} = \mathcal{L} \cup \mathcal{S}$. By Theorem 4, it suffices to show that each element of $\mathcal{P}$ is invariant on S. If $k \geq 0$ and $\deg(\text{red}^k(A^*)) \geq 1$ then

$$\text{red}^k(A^*) = \text{red}^k(\text{der}^j(A)) = \text{der}^j(\text{red}^k A)$$

so

$$\text{ldcf}(\text{red}^k(A^*)) = \text{ldcf}(\text{der}^j(\text{red}^k(A)) = a \cdot \text{ldcf}(\text{red}^k(A))$$

for some positive integer a. Also $\deg(\text{red}^k(A)) \geq \deg(\text{red}^k(A^*)) \geq 1$ so $\text{ldcf}(\text{red}^k(A))$ is in the projection of $\mathcal{A}$. Hence every element of $\mathcal{L}$ is invariant on S. If $j = 0$ then the roots of $A = \text{der}^j(A)$ are delineable on S by

Theorem 5, so assume $j > 0$. If $j \geq \deg(\mathrm{red}^k(A))$ then $\mathrm{der}^j(\mathrm{red}^k(A))$ is an integer and hence invariant on S. Otherwise, $0 < j < \deg(\mathrm{red}^k(A))$ so if $B = \mathrm{red}^k(\mathrm{der}^j(A)) = \mathrm{der}^j(\mathrm{red}^k(A))$ then $\mathrm{psc}_h(B, B')$ belongs to the augmented projection of A and is invariant on S for $0 \leq h < \deg(B')$. Hence every element of S is invariant on S. ∎

We now complete this section with discussion and specification of the more important subalgorithms which will be needed for the quantifier elimination algorithm.

The quantifier elimination algorithm of the next section will require computation of the projection or augmented projection of A just in case A is finite and $\mathcal{P} = \mathbb{Z}[x_1, \ldots, x_{r-1}], r \geq 2$. Thus we assume the availability of an algorithm with the following specifications.

$$B = \mathrm{PROJ}(A)$$

Input: $A = (A_1, \ldots, A_x)$ is a list of distinct integral polynomials in r variables, $r \geq 2$.
Output: $\mathcal{B} = (B_1, \ldots, B_n)$ is a list of distinct integral polynomials in $r - 1$ variables, such that $(B_1, \ldots, B_n)$ is the projection of $(A_1, \ldots, A_n)$.

Another like algorithm, APROJ, is assumed for computing the augmented projection.

Now let $\mathcal{U}$ be a unique factorization domain, abbreviated u.f.d. If $a, b \in \mathcal{U}$ we say that a and b are *associates*, and write $a \sim B$, in case $a = ub$ for some unit u. An *ample set* for $\mathcal{U}$ (see Goldhaber and Ehrlich 1970) is a set $A \subseteq \mathcal{U}$ which contains exactly one element from each equivalence class of associates. Relative to A we can define a function gcd on $\mathcal{U} \times \mathcal{U}$ into $\mathcal{U}$ such that $\gcd(a, b) \in A$ and $\gcd(a, b)$ is a greatest common divisor of a and b for all $a, b \in \mathcal{U}$. We will assume, moreover, that A is *multiplicative*, i.e., closed under multiplication, from which $1 \in A$. Whenever $\mathcal{U}$ is a field we will have $A = \{0, 1\}$. For $\mathcal{U} = \mathbb{Z}$, we set $A = \{0, 1, 2, \ldots\}$. $\mathcal{U}[x]$ is also a u.f.d. and if A is an ample set for $\mathcal{U}$ we take $\{A : \mathrm{ldcf}(A) \in A\}$ as ample set for $\mathcal{U}[x]$ (see Musser 1971).

If $A(x) = \sum_{i=0}^n a_i x^i$ is a non-zero polynomial over $\mathcal{U}$, we set $\mathrm{cont}(A) = \gcd(a_n, a_{n-1}, \ldots, a_0)$, the content of A, and we set $\mathrm{cont}(0) = 0$. If $A \neq 0$ we define $\mathrm{pp}(A)$, the *primitive part* of A, to be the ample associate of $A/\mathrm{cont}(A)$, and we set $\mathrm{pp}(0) = 0$. The polynomial A is *primitive* in case $\mathrm{cont}(A) = 1$. Clearly $\mathrm{pp}(A)$ is primitive and $A \sim \mathrm{cont}(A) \cdot \mathrm{pp}(A)$ for all $A \neq 0$.

Let A be a set of primitive polynomials of positive degree over $\mathcal{U}$. A *basis* for A is a set $\mathcal{B}$ of ample primitive polynomials of positive degree over $\mathcal{U}$ satisfying the following three conditions:

(a) If $B_1, B_2 \in \mathcal{B}$ and $B_1 \neq B_2$ then $\gcd(B_1, B_2) = 1$.
(b) If $B \in \mathcal{B}$, then $B|A$ for some $A \in \mathcal{A}$.

·) If $A \in \mathcal{A}$, there exist $B_1, \ldots, B_n \in \mathcal{B}$ and positive integers $e_1, \ldots, e_n$ such that

$$A \sim \prod_{i=1}^{n} B_i^{e_i} \qquad \text{(with } n = 0 \text{ if } A \sim 1\text{)}.$$

If $\mathcal{A}$ is an arbitrary set of polynomials over $\mathcal{U}$, then a *basis* for $\mathcal{A}$ is a set $\mathcal{B} = \mathcal{B}_1 \bigcup \mathcal{B}_2$ where $\mathcal{B}_1 = \{\text{cont}(A) : A \in \mathcal{A} \,\& A \neq 0\}$ and $\mathcal{B}_2$ is a basis for $\{\text{pp}(A) : A \in \mathcal{A} \,\& \deg(A) > 0\}$.

If $\mathcal{A}$ is a set of primitive polynomials of positive degree then the set $\mathcal{P}$ of ample irreducible divisors of elements of $\mathcal{A}$ is clearly a basis for $\mathcal{A}$. If $\mathcal{B}_1$ and $\mathcal{B}_2$ are bases for $\mathcal{A}$, we say that $\mathcal{B}_1$ is a *refinement* of $\mathcal{B}_2$ in case every element of $\mathcal{B}_1$ is a divisor of some element of $\mathcal{B}_2$. $\mathcal{P}$ is the finest basis for $\mathcal{A}$ in the sense that it is a refinement of every other basis.

Every set $\mathcal{A}$ also has a *coarsest* basis, $\mathcal{C}$, in the sense that every basis for $\mathcal{A}$ is a refinement of $\mathcal{C}$, as we will now see. Let $\mathcal{P}$ be the set of all ample irreducible divisors of positive degree of elements of $\mathcal{A}$. For $P \in \mathcal{P}$, let $\sigma(P)$ be the set of all positive integers i such that, for some $A \in \mathcal{A}$, $P^i | A$ but not $P^{i+1} | A$. Let $e(P)$ be the greatest common divisor of the elements of $\sigma(P)$. For P, Q in $\mathcal{P}$, define $P \equiv Q$ in case, for every $A \in \mathcal{A}$, the orders of P and Q in A are identical. Let $\mathcal{C}$ be the set of all products $\{\prod_{Q \equiv P} Q^{e(P)}\}$ with $P \in \mathcal{P}$. Then it can be shown that $\mathcal{C}$ is a coarsest basis for $\mathcal{A}$.

If $\mathcal{A}$ is finite, its coarsest basis can be computed by g.c.d. calculation. Set $\mathcal{C} = \mathcal{A}$. If A and B are distinct elements of $\mathcal{C}$, set $C = \gcd(A, B)$, $\bar{A} = A/C$, $\bar{B} = B/C$. If $C \neq 1$, replace A and B in $\mathcal{C}$ by the non-units from among C, $\bar{A}$ and $\bar{B}$. Eventually the elements of $\mathcal{C}$ will be pairwise relatively prime and $\mathcal{C}$ will be a coarsest basis for $\mathcal{A}$.

A *squarefree basis* for $\mathcal{A}$ is a basis each of whose elements is squarefree. If A is any primitive element of $\mathcal{U}[x]$ of positive degree, there exist ample, square-free, relatively prime polynomials $A_1, \ldots, A_k$ and integers $e_1 < \ldots < e_k$ such that $A \sim \prod_{i=1}^{k} A_i^{e_i}$. $(A_1, \ldots, A_k)$ and $(e_1, \ldots, e_k)$ constitute the *squarefree factorization* of A. Musser (1971, 1975) discusses algorithms for squarefree factorization, which require, if $\mathcal{U}$ has characteristic zero, only differentiation, division and greatest common divisor calculations. We assume the availability of an algorithm for squarefree factorization in $\mathcal{U}[x]$ for the cases $\mathcal{U} = \mathbb{Z}[x_1, \ldots, x_{r-1}]$, $r \geq 1$, and $\mathcal{U} = \mathbb{Q}(\alpha)$, where $\mathbb{Q}(\alpha)$ is the real algebraic number field resulting from adjoining the real algebraic number α to the field $\mathbb{Q}$ of the rational numbers. For the case $\mathcal{U} = \mathbb{Q}(\alpha)$ we assume the following specifications.

$$\text{SQFREE}(\alpha, A, \mathcal{A}, e)$$

Inputs: α is a real algebraic number. A is a primitive element of $\mathbb{Q}(\alpha)[x]$ of positive degree.

Outputs: $\mathcal{A} = A_1, \ldots, A_k)$ and $e = (e_1, \ldots, e_k)$ constitute the squarefree factorization of A.

A similar algorithm for the case $\mathcal{U} = \mathbb{Z}[x_1, \ldots, x_{r-1}]$ is needed in order to compute a coarsest squarefree basis for integral polynomials, as follows.

If $A \sim \prod_{i=1}^{k} A_i^{e_i}$ is the squarefree factorization of A, then $\{A_1, \ldots, A_k\}$ is clearly a coarsest squarefree basis for $\{A\}$. Let $\bar{\mathcal{A}} = \{A_1, \ldots, A_m\}$ be a squarefree basis for $\mathcal{A}$, $\bar{\mathcal{B}} = \{B_1, \ldots, B_n\}$ a squarefree basis for $\mathcal{B}$. Consider the following algorithm proposed by R. Loos:

(1) For $j = 1, \ldots, n$ set $\bar{B}_j \leftarrow B_j$.
(2) For $i = 1, \ldots, m$ do $[\bar{A}_i \leftarrow A_i$; for $j = 1, \ldots, n$ do $(C_{i,j} \leftarrow \gcd(\bar{A}_i, \bar{B}_j);$ $\bar{A}_i \leftarrow \bar{A}_i/C_{i,j}$; $\bar{B}_j \leftarrow \bar{B}_j/C_{i,j})]$.
(3) Exit.

Upon termination, the distinct nonunits among the $\bar{A}_i$, the $\bar{B}_j$ and the $C_{i,j}$ constitute a squarefree basis $\bar{C}$ for $C = \mathcal{A} \cup \mathcal{B}$. Moreover, if $\bar{A}$ and $\bar{B}$ are coarsest squarefree bases, then so is C. Thus by squarefree factorization and application of Loos' algorithm we can successively obtain coarsest squarefree bases for $\{A_1\}$, $\{A_1, A_2\}$, $\ldots$, $\{A_1, A_2, \ldots, A_m\}$. Thus we assume the availability of the following basis algorithm.

$$\mathcal{B} = \text{BASIS}(\mathcal{A})$$

Input: $\mathcal{A} = (A_1, \ldots, A_m)$ is a list of distinct integral polynomials in r variables, $r \geq 1$.
Output: $\mathcal{B} = (B_1, \ldots, B_n)$ is a list of distinct integral polynomials in r variables such that $\{B_1, \ldots, B_n\}$ is a coarsest squarefree basis for $\{A_1, \ldots, A_m\}$.

A similar algorithm, ABASIS, with an additional input α, a real algebraic number, will be assumed for computing the coarsest squarefree basis when $\mathcal{A}$ is a finite list of univariate polynomials over $\mathbb{Q}(\alpha)$.

A recent Ph.D. thesis by Rubald (1974) provides algorithms for the arithmetic operations in the field $\mathbb{Q}(\alpha)$ and in the polynomial domain $\mathbb{Q}(\alpha)[x]$. Rubald also provides an efficient modular homomorphism algorithm for g.c.d. calculation in $\mathbb{Q}(\alpha)[x]$. An important feature of Rubald's work is that the minimal polynomial of α is not required. Instead, α is represented by any pair (A, I) such that A is a primitive squarefree integral polynomial of positive degree with $A(\alpha) = 0$, and $I = (r, s)$ is an open interval with rational number endpoints such that α is the unique zero of A in I. This feature is important because as yet (see Collins 1973a) no algorithm with polynomial-dominated maximum computing time is known for factoring a primitive univariate integral polynomial into its irreducible factors. A non-zero element β of $\mathbb{Q}(\alpha)$ is then represented by any polynomial $B(x) \in \mathbb{Q}[x]$ such that $\deg(B) < \deg(A)$ and $B(\alpha) = \beta$. Although this representation fails to be unique whenever A is reducible, no difficulties arise.

The next algorithm is easily obtained using Sturm's theorem, since Rubald's work provides an efficient algorithm for determining the sign of any element of $\mathbb{Q}(\alpha)$, and because his algorithm for g.c.d. calculation in $\mathbb{Q}(\alpha)[x]$ can be extended to the computation of Sturm sequences.

$$\text{ISOL}(\alpha, \mathcal{A}, I, \nu)$$

Inputs: α is a real algebraic number. $\mathcal{A} = (A_1, .., A_m)$ is a list of non-zero squarefree and pairwise relatively prime polynomials over $\mathbb{Q}(\alpha)$.
Outputs: $I = (I_1, \ldots, I_n)$ is a list of open intervals with rational endpoints with $I_1 < I_2 < \cdots < I_n$ such that each I_j contains exactly one real zero of $A = \prod_{i=1}^m A_i$, and every real zero of A belongs to some I_j. $\nu = (\nu_1, \ldots, \nu_n)$ is such that the zero of A in I_j is a zero of A_{ν_j}.

The Algorithm ISOL can be easily obtained by application of Sturm's theorem and repeated interval bisection. Heindel (1971) presents an algorithm of this type for the case of a single univariate integral polynomial. If the real zeros of each A_i are separately isolated, then the resulting intervals can be refined until they no longer overlap, while retaining the identity of the polynomials from which they came.

In the quantifier elimination algorithm, occasion will arise to reduce a multiple real algebraic extension of the rationals, $\mathbb{Q}(\alpha_1, \ldots, \alpha_m)$, to a simple extension $\mathbb{Q}(\alpha)$. This can be accomplished by iterating an algorithm of Loos (1975) based on resultant theory, with the following specifications.

$$\text{SIMPLE}(\alpha, \beta, \gamma, A, B)$$

Inputs: α and β are real algebraic numbers.
Outputs: γ is a real algebraic number. A and B are polynomials which represent α and β respectively as elements of $\mathbb{Q}(\gamma)$.

Finally, one additional subalgorithm, also provided in (Loos 1975), is the following.

$$\text{NORMAL}(\alpha, A, I, \bar{A}, \bar{I})$$

Inputs: α is a real algebraic number. A is a non-zero polynomial over $\mathbb{Q}(\alpha)$. $I = (I_1, \ldots, I_m)$ is a list of rational isolating intervals, $I_1 < I_2 < \cdots < I_m$, for the real zeros of A.
Outputs: $\bar{A}$ is a non-zero squarefree primitive integral polynomial such that every real zero of A is a real zero of $\bar{A}$. $\bar{I} = (\bar{I}_1, \ldots, \bar{I}_m)$ is a list of rational intervals with $\bar{I}_j \subseteq I_j$ such that if α_j is the zero of A in I_j then α_j is the unique zero of $\bar{A}$ in $\bar{I}_j$, $1 \leq j \leq m$.

3   The Main Algorithm

We define, by induction on r, a *cylindrical algebraic decomposition* of $\mathbb{R}^r$, abbreviated c.a.d. For $r = 1$, a c.a.d. of $\mathbb{R}$ is a sequence $(S_1, S_2, \ldots, S_{2\nu+1})$, where either $\nu = 0$ and $S_1 = \mathbb{R}$, or $\nu > 0$ and there exist ν real algebraic numbers $\alpha_1 < \alpha_2 < \ldots < \alpha_\nu$ such that $S_{2i} = \{\alpha_i\}$ for $1 \leq i \leq \nu$, S_{2i+1} is the open interval (α_i, α_{i+1}) for $1 \leq i < \nu$, $S_1 = (-\infty, \alpha_1)$ and $S_{2\nu+1} = (\alpha_\nu, \infty)$. Now let $r > 1$, and let $(S_1, \ldots, S_\mu)$ be any c.a.d. of $\mathbb{R}^{r-1}$. For $1 \leq i \leq \mu$, let $f_{i,1} < f_{i,2} < \cdots < f_{i,\nu_i}$ be continuous realvalued algebraic functions on S_i. If $\nu_i = 0$, set $S_{i,1} = S_i \times \mathbb{R}$. If $\nu_i > 0$ set $S_{i,2j} = f_{i,j}$, that is, $S_{i,2j} = \{(a,b) : a \in S_i \,\&\, b = f_{i,j}(a)\}$ for $1 \leq j \leq \nu_i$, set $S_{i,2j+1} = \{(a,b) : a \in S_i \,\&\, f_{i,j}(a) < b < f_{i,j+1}(a)\}$ for $1 \leq j < \nu_i$, set $S_{i,1} = \{(a,b) : a \in S_i \,\&\, b < f_{i,1}(a)\}$, and set $S_{i,2\nu_i+1} = \{(a,b) : a \in S_i \,\&\, f_{i,\nu_i}(a) < b\}$. A c.a.d. of $\mathbb{R}^r$ is any sequence $(S_{1,1}, \ldots, S_{i,2\nu_1+1}, \ldots, S_{\mu,1}, \ldots, S_{\mu,2\nu_\mu+1})$ which can be obtained by this construction from a c.a.d. of $\mathbb{R}^{r-1}$ and functions $f_{i,j}$ as just described.

It is important to observe that the cylinder $S_i \times \mathbb{R}$ is the disjoint union $\bigcup_{j=1}^{2\nu_i+1} S_{i,j}$ for $1 \leq i \leq \mu$. If $S = (S_1, \ldots, S_\mu)$ is any c.a.d. of $\mathbb{R}^r$, the S_i will be called the *cells* of S. Clearly every cell of a c.a.d. is a connected set. If $\mathcal{A}$ is a set of real polynomials in r variables, the c.a.d. S or $\mathbb{R}^r$ is $\mathcal{A}$-*invariant* in case each A in $\mathcal{A}$ is invariant on each cell of S.

A *sample* of the c.a.d. $S = (S_1, \ldots, S_\mu)$ is a tuple $\beta = (\beta_1, \ldots, \beta_\mu)$ such that $\beta_i \in S_i$ for $1 \leq i \leq \mu$. The sample β is *algebraic* in case each β_i is an algebraic point, i.e., each coordinate of β_i is an algebraic number. A *cylindrical sample* is defined by induction on r. For $r = 1$, any sample is cylindrical. For $(S_{1,1}, \ldots, S_{i,2\nu_1+1}, \ldots, S_{\mu,1}, \ldots, S_{\mu,2\nu_\mu+1})$ be a c.a.d. of $\mathbb{R}^r$ constructed from a c.a.d. $S^* = (S_1, \ldots, S_\mu)$ of $\mathbb{R}^{r-1}$, and let $\beta^* = (\beta_1, \ldots, \beta_\mu)$ be a sample of S^*. The sample $(\beta_{1,1}, \ldots, \beta_{i,2\nu_1+1}, \ldots, \beta_{\mu,1}, \ldots, \beta_{\mu,2\nu_\mu+1})$ of S is *cylindrical* if the first $r-1$ coordinates of $\beta_{i,j}$ are, respectively, the coordinates of β_i for all i and j, and β^* is cylindrical. Cylindrical algebraic sample will be abbreviated c.a.s.

Since a c.a.d. or $\mathbb{R}^r$ can be constructed from a unique c.a.d. of $\mathbb{R}^{r-1}$, any c.a.d. S of $\mathbb{R}^r$ determines, for $1 \leq k < r$, a c.a.d. S^* of $\mathbb{R}^k$, which will be called the c.a.d. of $\mathbb{R}^k$ *induced* by S. Similarly any c.a.s. β of S induces a unique c.a.s. β^* of S.

If S is an arbitrary subset of $\mathbb{R}^r$, the standard formula $\phi(x_1, \ldots, x_r)$ containing just $x_1, \ldots, x_r$ as free variables, *defines* S in case $S = \{(a_1, \ldots, a_r) : a_1, \ldots, a_r \in \mathbb{R} \,\&\, \phi(a_1, \ldots, a_r)\}$. A *standard definition* of the c.a.d. $S = (S_1, \ldots, S_\mu)$ is a sequence $(\phi_1, \ldots, \phi_\mu)$ such that, for $1 \leq i \leq \mu$, ϕ_i is a standard quantifier-free formula which defines S_i.

We are now prepared to describe a decomposition algorithm, DECOMP. The inputs to DECOMP are a finite set $\mathcal{A}$ of integral polynomials in r variables, $r \geq 1$, and an integer k, $0 \leq k \leq r$. The outputs of DECOMP are a c.a.s. β of some $\mathcal{A}$-invariant c.a.d. S of $\mathbb{R}^r$ and, if $k \geq 1$, a standard definition of ψ of the c.a.d. S^* of $\mathbb{R}^k$ induced by S.

Before proceeding to describe DECOMP we first explain its intended use in the quantifier elimination algorithm, ELIM, which will be described subsequently. ELIM has two distinct stages. Given as input a standard prenex formula ϕ, namely $(Q_{k+1}x_{k+1}) \ldots (Q_r x_r)\hat{\phi}(x_1, \ldots, x_r)$, ELIM applies DECOMP to the set $\mathcal{A}$ of all non-zero polynomials occurring in $\hat{\phi}$, and the integer k. The outputs β and ψ of DECOMP, together with the formula ϕ, are then input to an "evaluation" algorithm, EVAL, which produces a standard quantifier-free formula $\phi^*(x_1, \ldots, x_k)$ which is equivalent to ϕ. Thus, ELIM does little more than to successively invoke DECOMP and EVAL.

DECOMP uses a subalgorithm, DEFINE, for construction of the standard definition. The inputs to DEFINE are an integral polynomial $A(x_1, \ldots, x_r)$, $r \geq 2$, such that for some connected set $S \subset \mathbb{R}^{r-1}$ the real roots of A and of each derivative of A are delineable on S, and an algebraic point $\beta \in S$. The output of DEFINE is a sequence $(\phi_1, \ldots, \phi_{2m+1})$ of standard quantifier-free formulas ϕ_i such that if ϕ is any formula which defines S, then the conjunction $\phi \wedge \phi_i$ defines the i-th cell of the cylinder $S \times \mathbb{R}$ determined by the m real roots of A on S, as in the definition of a c.a.d. The description of DEFINE will be given following that of DECOMP.

DECOMP($\mathcal{A}, k, \beta, \psi$)

Inputs: $\mathcal{A} = (A_1, \ldots, A_m)$ is a list of distinct integral polynomials in r variables, $r \geq 1$. k is an integer such that $0 \leq k \leq r$.
Outputs: β is a c.a.s. for some $\mathcal{A}$-invariant c.a.d. S of $\mathbb{R}^r$. ψ is a standard definition of the c.a.d. S of $\mathbb{R}^k$ induced by S if $k > 0$, and ψ is the null list if $k = 0$.

Algorithm Description

(1) If $r > 1$, go to (4). Apply BASIS to $\mathcal{A}$, obtaining a coarsest squarefree basis $\mathcal{B} = (B_1, \ldots, B_h)$ for $\mathcal{A}$. Apply ISOL to $\mathcal{B}$, obtaining outputs $I = (I_1, \ldots, I_n)$ and $\nu = (\nu_1, \ldots, \nu_n)$. (Each I_j contains a unique zero, say α_j, of B_{ν_j}, and $\alpha_1 < \alpha_2 < \cdots < \alpha_n$ are all the real zeros of elements of $\mathcal{A}$. Thus the α_j determine an $\mathcal{A}$-invariant c.a.d. S of $\mathbb{R}$, and (B_{ν_j}, I_j) represents α_j).

(2) For $j = 1, \ldots, n$, where $I_j = (r_j, s_j)$, set $\beta_{2j-1} \leftarrow r_j$ and $B_{2j} \leftarrow \alpha_j$. If $n = 0$, set $\beta_{2n+1} \leftarrow 0$ and if $n > 0$, set $\beta_{2n+1} \leftarrow s_n$. Set $\beta \leftarrow (\beta_1, \ldots, \beta_{2n+1})$. ($\beta$ is now a c.a.s. of S.)

(3) If $k = 0$, set $\psi \leftarrow ()$ and exit. If $n = 0$, set $\psi_1 \leftarrow$ "$0 = 0$", $\psi \leftarrow (\psi_1)$ and exit. For $i = 1, \ldots, h$ do [$\sigma_{i,n} \leftarrow \text{sign}(\text{ldcf}(B_i))$; for $j = n - 1, \ldots, 0$ set $\sigma_{i,j} \leftarrow (-\sigma_{i,j+1}$ if $i = \nu_{j+1}$, $\sigma_{i,j+1}$ otherwise)]. (Now $\sigma_{i,j}$ is the sign of B_i in S_{2j+1}, where $S = (S_1, \ldots, S_{2n+1})$.) For $j = 1, \ldots, n$, where $r_j = a_j/b_j$ and $s_j = c_j/d_j$ with $b_j > 0$ and $d_j > 0$, set $\psi_{2j} \leftarrow$ "$B_{\nu_j} = 0 \ \& \ b_j x_1 - a_j > 0 \ \& \ d_j x_1 - c_j < 0$". (Now ψ_{2j} defines $S_j = \{\alpha_j\}$.) For $j = 1, \ldots, n - 1$, set $\psi_{2j+1} \leftarrow$ "$\sigma_{\nu_j,j} B_{\nu_j} > 0 \ \& \ \sigma_{\nu_{j+1},j} B_{\nu_{j+1}} > 0 \ \& \ b_j x_1 - a_j > 0 \ \& \ d_{j+1} x_1 - c_{j+1} < 0$". (If $\nu_j = \nu_{j+1}$ then the first two conjuncts are identical, so one can be omitted.) Set $\psi_1 \leftarrow$ "$\sigma_{\nu_1,0} B_{\nu_1} > 0 \ \& \ d_1 x_1 - c_1 < 0$" and $\psi_{2n+1} \leftarrow$ "$\sigma_{\nu_n,n} B_{\nu_n} > 0 \ \& \ b_n x_1 - a_n > 0$". Set $\psi \leftarrow (\psi_1, \ldots, \psi_{2n+1})$. ($\psi$ is now a standard definition of S.) Exit.

(4) Apply BASIS to $\mathcal{A}$, obtaining $\mathcal{B}$, a coarsest squarefree basis for $\mathcal{A}$. (This action is inessential; we could set $\mathcal{B} \leftarrow \mathcal{A}$. But the algorithm is likely more efficient if the coarsest squarefree basis is used, and it may be still more efficient, on the average, if the finest basis is computed here.) If $k < r$,

apply PROJ to $\mathcal{B}$, obtaining the projection, $\mathcal{P}$, of $\mathcal{B}$. If $k = r$, apply APROJ to $\mathcal{B}$, obtaining the augmented projection, $\mathcal{P}$ of $\mathcal{B}$.

(5) If $k = r$, set $k' \leftarrow k - 1$; otherwise, set $k' \leftarrow k$. Apply DECOMP (recursively) to $\mathcal{P}$ and k', obtaining outputs β' and ψ'. (For some $\mathcal{P}$-invariant c.a.d. S' of $\mathbb{R}^{r-1}$, β' is a c.a.s. of S' and ψ' is a standard definition of the c.a.d. S^* of $\mathbb{R}^{k'}$ induced by S', except that $\psi' = ()$ if $k' = 0$. Since $\mathcal{P}$ contains the projection of $\mathcal{B}$ and S' is $\mathcal{P}$-invariant the real zeros of $\mathcal{B}$ are delineable on each cell of S' by Theorem 5. Hence S', together with the real algebraic functions defined by elements of $\mathcal{B}$ on the cells of S', determines a c.a.d. S of $\mathbb{R}^r$. S is $\mathcal{B}$-invariant and therefore also $\mathcal{A}$-invariant since $\mathcal{B}$ is a basis for $\mathcal{A}$. Also, S^* is induced by S.)

(6) (This step extends the c.a.s. β' of S' to a c.a.s. β of S. Let $\beta' = (\beta'_1, \ldots, \beta'_l)$ and $\beta'_j = (\beta'_{j,1}, \ldots, \beta'_{j,r-1})$. We assume, inductively, that there is associated with each algebraic point β'_j an algebraic number α'_j such that $\mathbb{Q}(\beta'_{j,1}, \ldots, \beta'_{j,r-1}) = \mathbb{Q}(\alpha'_j)$ and polynomials $B'_{j,k}$ which represent the $\beta'_{j,k}$. The basis for this induction is trivial since the polynomial x represents $\beta'_{j,1} = \alpha'_j$ as an element of $\mathbb{Q}(\alpha'_j)$ if α'_j is irrational, and if α'_j is rational it represents itself as an element of $\mathbb{Q} = \mathbb{Q}(\alpha'_j)$.) Let $\mathcal{B} = (B_1, \ldots, B_h)$. For $j = 1, \ldots, l$ do [For $i = 1, \ldots, h$ set $B^*_{j,i}(x) \leftarrow B_i(\beta'_{j,1}, \ldots, \beta'_{j,r-1}, x)$. ($B^*_{j,i}$ is a polynomial over $\mathbb{Q}(\alpha'_j)$.) Apply ABASIS to α'_j and $(B^*_{j,1}, \ldots, B^*_{j,h})$, obtaining $\hat{B}_j = (\hat{B}_{j,1}, \ldots, \hat{B}_{j,m_j})$ a coarsest squarefree basis. Apply ISOL to α'_j and $\hat{B}_j$, obtaining outputs $I_j = (I_{j,1}, \ldots, I_{j,n_j})$ and $\nu_j = (\nu_{j,1}, \ldots, \nu_{j,n_j})$. $\hat{B}_{j,\nu_{j,k}}$ has a unique real zero $\gamma_{j,k}$ in $I_{j,k}$, and $\gamma_{j,1} < \cdots < \gamma_{j,n_j}$ are all the real zeros of elements of $\hat{B}_j$. For $k = 1, \ldots, m_j$ do [Set $\hat{I}_{j,k}$ to the subsequence of I_j consisting of those $I_{j,l}$ such that $\nu_{j,l} = k$. (Then $\hat{I}_{j,k}$ is a list of rational isolating intervals for the real roots of $\hat{B}_{j,k}$.) Apply NORMAL to α'_j, $\hat{B}_{j,k}$ and $\hat{I}_{j,k}$, obtaining as outputs $\bar{B}_{j,k}$ and $I^*_{j,k}$.] Merge the sequences $I^*_{j,k}$ into a single sequence $\bar{I}_j = (\bar{I}_{j,1}, \ldots, \bar{I}_{j,n_j})$ with $\bar{I}_{j,1} < \bar{I}_{j,2} < \cdots < \bar{I}_{j,n_j}$. (Now $\gamma_{j,k}$ is represented by $(\bar{B}_{j,k}, \bar{I}_{j,k})$.) If $n_j = 0$, set $\delta_{j,1} \leftarrow 0$. If $n_j > 0$, for $k = 1, \ldots, n_j$, where $\bar{I}_{j,k} = (r_{j,k}, s_{j,k})$, set $\delta_{j,2k-1} \leftarrow r_{j,k}$ and $\delta_{j,2k} \leftarrow \gamma_{j,k}$; also set $\delta_{j,2n_j+1} \leftarrow s_{j,n_j}$. For $k = 1, \ldots, 2n_j + 1$, set $\beta_{j,k} = (\beta'_{j,1}, \ldots, \beta'_{j,r-1}, \delta_{j,k})$. For $k = 1, \ldots, 2n_j + 1$ apply SIMPLE to α'_j and $\delta_{j,k}$, obtaining outputs $\alpha_{j,k}$, $A_{j,k}$ and $B_{j,k}$. (Now $\mathbb{Q}(\beta'_{j,1}, \ldots, \beta'_{j,r-1}, \delta_{j,k}) = \mathbb{Q}(\alpha'_j, \delta_{j,k}) = \mathbb{Q}(\alpha_{j,k})$, $A_{j,k}$ represents α'_j in $\mathbb{Q}(\alpha_{j,k})$, and $B_{j,k}$ represents $\delta_{j,k}$ in $\mathbb{Q}(\alpha_{j,k})$.) For $h = 1, \ldots, r - 1$ and $k = 1, \ldots, 2n_j + 1$, where $\alpha_{j,k}$ is represented by $(C_{j,k}, I^*_{j,k})$, set $D_{j,h,k}(x) \leftarrow B'_{j,h}(A_{j,k}(x))$ modulo $C_{j,k}(x)$. ($\alpha'_j = A_{j,k}(\alpha_{j,k})$, $\beta'_{j,h} = B'_{j,h}(\alpha'_j)$ and $C_{j,k}(\alpha_{j,k}) = 0$, so $D_{j,h,k}$ represents $\beta'_{j,h}$ in $\mathbb{Q}(\alpha_{j,k})$.)] Set $\beta \leftarrow (\beta_{1,1}, \ldots, \beta_{1,2n_1+1}, \ldots, \beta_{l,1}, \ldots, \beta_{l,2n_l+1})$. (Now β is a c.a.s. of S.)

(7) If $k < r$, set $\psi \leftarrow \psi'$ and exit. (If $k < r$, then $k' = k$ so ψ' is a standard definition for the c.a.d. S^* of $\mathbb{R}^k$ induced by S', and hence induced also by S. Otherwise $k = r$, $k' = r - 1$ and we next proceed to extend the standard definition ψ' of S' to a standard definition ψ of S. Since $k = r$, $\mathcal{P}$

is the augmented projection of $\mathcal{B}$ and, by Theorem 6, the real roots of every derivation of every element of $\mathcal{B}$ are delineable on every cell of S' because S' is $\mathcal{P}$-invariant.) For $j = 1, \ldots, l$ do [For $i = 1, \ldots, h$ apply DEFINE to B_i and β'_j, obtaining as output a sequence $\chi_{i,j} = (\chi_{i,j,1}, \ldots, \chi_{i,j,2n_{i,j}+1})$. ($\chi_{i,j,k}$ is a standard quantifier-free formula such that ψ'_j & $\chi_{i,j,k}$ defines the k-th cell of the cylinder $S'_j \times \mathbb{R}$ as determined by the real zeros of B_i on S'_j.) We next proceed to use the $\chi_{i,j,k}$ to define the cells of the cylinder $S'_j \times \mathbb{R}$ as determined by the real zeros of $B = \prod_{i=1}^{h} B_i$, that is, the cells of the j-th cylinder of S, using the results of step (6). Observe that B has n_j real zeros on S_j and that the k-th real zero is a zero of $\hat{B}_{j,\nu_{j,k}}$.) For $i = 1, \ldots, h$ and $k = 1, \ldots, n_j$, set $\delta_{i,j,k} = 1$ if $\hat{B}_{j,\mu_{j,k}}$ is a divisor of $B^*_{j,i}$, and $\delta_{i,j,k} = 0$ otherwise. (Now $\delta_{i,j,k} = 1$ just in case $B^*_{j,i}(\gamma_{j,k}) = 0$.) For $k = 1, \ldots, n_j$, set $\lambda_{j,k}$ to the least i such that $\delta_{i,j,k} = 1$. (Now $\gamma_{j,k}$ is a root of $B^*_{j,\lambda_{j,k}}$.) For $k = 1, \ldots, n_j$, set $\mu_{j,k} \leftarrow \sum_{i=1}^{\lambda_{j,k}} \delta_{i,j,k}$. (Now $\gamma_{j,k}$ is the $\mu_{j,k}$-th real root of $B^*_{j,\lambda_{j,k}}$. Hence the k-th real root of B on S'_j is the $\mu_{j,k}$-th real root of $B_{\lambda_{j,k}}$ on S'_j.) For $k = 1, \ldots, n_j$ set $\psi_{j,2k} \leftarrow \psi'_j$ & $\chi_{\lambda_{j,k},j,2\mu_{j,k}}$. For $k = 1, \ldots, n_j - 1$ set $\psi_{j,2k+1} \leftarrow \psi'_j$ & $\chi_{\lambda_{j,k},j,2\mu_{j,k}+1}$ & $\chi_{\lambda_{j,k+1},j,2\mu_{j,k+1}-1}$. (If $\nu_{j,k} = \nu_{j,k+1}$ then the last two conjuncts of $\psi_{j,2k+1}$ coincide so one may be omitted.) If $n_j > 0$, set $\psi_{j,1} \leftarrow \psi'_j$ & $\chi_{\lambda_{j,1},j,1}$ and $\psi_{j,2n_j+1} \leftarrow \psi'_j$ & $\chi_{\lambda_{j,n},j,2\mu_{j,n_j}+1}$. If $n_j = 0$, set $\psi_{j,1} \leftarrow \psi'_j$.] Set $\psi \leftarrow (\psi_{1,1}, \ldots, \psi_{1,2n_1+1}, \ldots, \psi_{1,2n_l+1})$. (Now ψ is a standard definition of S.) Exit.

$$\phi = \text{DEFINE}(B, \beta)$$

Inputs: B is an integral polynomial in r variables, $r \geq 2$, such that for some connected set $S \subseteq \mathbb{R}^{r-1}$ the real roots of B and of each derivative of B are delineable on S. β is an algebraic point of S.

Outputs: $\phi = (\phi_1, \ldots, \phi_{2m+1})$ is a sequence of standard quantifier-free formulas ϕ_i such that if ψ defines S then ψ & ϕ_i defines the i-th cell of the cylinder $S \times \mathbb{R}$ as determined by the m real roots of B on S.

Algorithm Description

(1) (We let $\beta = (\beta_1, \ldots, \beta_{r-1})$. As in DECOMP, we may assume that we are given an algebraic number α such that $\mathbb{Q}(\beta_1, \ldots, \beta_{r-1}) = \mathbb{Q}(\alpha)$, and polynomials B_i which represent β_i as elements of $\mathbb{Q}(\alpha)$.) Set $B^*(x) = B(\beta_1, \ldots, \beta_{r-1}, x)$. Apply SQFREE to α and B^*, obtaining the list $B^* = (B^*_1, \ldots, B^*_h)$ of squarefree factors of B^* and the list $(e_1, \ldots, e_h)$ of corresponding exponents. Apply ISOL to α and $\mathcal{B}^*$, obtaining as outputs the lists $(I_1, \ldots, I_m)$ and $(\nu_1, \ldots, \nu_m)$. (I_j isolates the j-th real zero, γ_j, of the elements of $\mathcal{B}^*$, and γ_j is a zero of B_{ν_j}.) If $m = 0$, set $\phi_1 \leftarrow$ "$0 = 0$", $\phi \leftarrow (\phi_1)$, and exit. For $i = 1, \ldots, m$ set $\mu_i \leftarrow e_{\nu_i}$. (Now γ_i is a zero of $B^*_{\nu_i}$ of multiplicity μ_i.) Set $\sigma_m \leftarrow \text{sign}(\text{ldcf}(B^*))$. For $j = m-1, \ldots, 0$ set $\sigma_j \leftarrow (-1)^{\mu_j+1}\sigma_{j+1}$. (Now σ_j is the sign of B in the $(2j+1)$-th cell of the B-invariant decomposition of the cylinder $S \times \mathbb{R}$.) If $m = 1$ and μ_1 is odd,

set $\phi_1 \leftarrow$ "$\sigma_0 B > 0$", $\phi_2 \leftarrow$ "$B = 0$", $\phi_3 \leftarrow \sigma_1 B > 0$", $\phi \leftarrow (\phi_1, \phi_2, \phi_3)$, and exit.

(2) Set $B^{*\prime} \leftarrow \mathrm{der}(B^*)$, $G \leftarrow \gcd(B^*, B^{*\prime})$ and $H \leftarrow B^{*\prime}/G$. (Now $H(\delta) = 0$ if and only if $B^{*\prime}(\delta) = 0$ and $B^*(\delta) \neq 0$.) Set $\bar{H} \leftarrow \gcd(H, H')$. ($\bar{H}$ is squarefree and has the same roots as H.) Apply ISOL to α and the list $(\bar{H})$, obtaining as output the list $I' = (I'_1, \ldots, I'_n)$ of isolating intervals for the roots of $B^{*\prime}$ which are not roots of B. For $j = 1, \ldots, m$ and $k = 1, \ldots, n$, if I_j and I'_k are non-disjoint, replace I_j by its left or right half, whichever contains a root of B^*, and replace I'_k by its left or right half, whichever contains a root of B', and repeat until I_j and I'_k are disjoint. Set n_0 to the number of intervals I'_k such that $I'_k < I_1$. For $j = 1, \ldots, m-1$, set n_j to the number of intervals I'_k such that $I_j < I'_k < I_{j+1}$. Set n_m to the number of intervals I'_k such that $I_m < I'_k$. Set $\lambda_1 \leftarrow n_0$. For $j = 1, \ldots, m$ set $\lambda_{2j} \leftarrow \{\lambda_{2j+1}$ if $\mu_j = 1$; $\lambda_{2j-1} + 1$ if $\mu_j > 1\}$, and $\lambda_{2j+1} \leftarrow \lambda_{2j} + n_j$. (Now λ_{2j-1} is the number of zeros of $B^{*\prime}$ less than γ_j, γ_{2j} is the number less than or equal to γ_j, and γ_{2m+1} is the number of all the zeros.)

(3) Set $B' \leftarrow \mathrm{der}(B)$. Apply DEFINE to B' and β, obtaining $(\phi'_1, \ldots, \phi'_l)$ as output. (Thus DEFINE is a recursive algorithm; its termination is assured because $\deg(B') < \deg(B)$.)

(4) (This step computes ϕ_{2i} for $1 \leq i \leq m$.) For $i = 1, \ldots, m$ if $\mu_i > 1$ set $\phi_{2i} \leftarrow \phi'_{2\lambda_{2i}}$. (If $\mu_i > 1$ then the i-th real zero of B is the λ_{2i}-th real zero of B'.) For $i = 1, \ldots, m$ if $\mu_i = 1$ set $\phi_{2i} \leftarrow$ "$B = 0$ & $\phi'_{2\lambda_{2i-1}+1}$". (There are λ_{2i-1} zeros of B less than the i-th zero of B, so the i-th zero of B is in the $\lambda_{2\lambda_{i-1}+1}$-th cell of the B' decomposition. By Rolle's theorem, any two real zeros of B are separated by a zero of B' so there is only one zero of B in this cell.)

(5) (This step defines ϕ_{2i+1} for $1 \leq i < m$. There are four cases.) For $i = 1, \ldots, m-1$ if $\mu_i > 1$ and μ_{i+1} set $\phi_{2i+1} \leftarrow \bigvee_{2\lambda_{2i}+1 \leq j \leq 2\lambda_{2i+2}-1} \phi'_j$. (In this case the i-th zero of B is the λ_{2i}-th zero of B' and the $(i+1)$-th zero of B is the λ_{2i+2}-th zero of B'.) For $i = 1, \ldots, m-1$ if $\mu_i = 1$ and $\mu_{i+1} > 1$ set $\psi_{2i+1} \leftarrow \{\sigma_i B > 0$ & $\psi'_{2\lambda_{2i}+1}\} \vee \{\bigvee_{2\lambda_{2i}+2 \leq j \leq 2\lambda_{2i+2}-1} \phi'_j\}$. (There are λ_{2i} zeros of B' less than the i-th zero of B. By Rolle's theorem the i-th zero of B is the only zero of B in the $(2\lambda_{2i}+1)$-th cell of the B' decomposition. Since $\mu_i = 1$, B changes sign from σ_{i-1} to σ_i at this zero.) For $i = 1, \ldots, m-1$ if $\mu_i > 1$ and $\mu_{i+1} = 1$ set $\psi_{2i+1} \leftarrow \{\sigma_i B > 0$ & $\phi'_{2\lambda_{2i+2}+1}\} \vee \{\bigvee_{2\lambda_{2i}+1 \leq j \leq 2\lambda_{2i+2}} \phi'_j\}$. (This case is similar to the preceding case.) For $i = 1, \ldots, m-1$ if $\mu_i = 1$ and $\mu_{i+1} = 1$ set $\phi_{2i+1} \leftarrow \{\sigma_i B > 0$ & $\sigma'_{2\lambda_{2i}+1}\} \vee \{\sigma_i B > 0$ & $\phi'_{2\lambda_{2i+2}+1}\} \vee \{\bigvee_{2\lambda_{2i}+2 \leq j \leq 2\lambda_{2i+2}} \phi'_j\}$.

(6) (This step defines ϕ_1 and ϕ_{2m+1}.) If $\mu_1 > 1$ set $\phi_1 \leftarrow \bigvee_{1 \leq j \leq 2\lambda_2 - 1} \phi'_j$. If $\mu_1 = 1$ set $\phi_1 \leftarrow \{\sigma_0 B > 0$ & $\phi'_{2\lambda_2+1}\} \vee \{\bigvee_{1 \leq j \leq 2\lambda_2} \phi'_j\}$. If $\mu > 1$ set $\phi_{2m+1} \leftarrow \bigvee_{2\lambda_{2m}+1 \leq j \leq 2\lambda_{2m+1}+1} \phi'_j$. If $\mu_m = 1$ set $\phi_{2m+1} \leftarrow \{\sigma_m B > 0$ & $\phi'_{2\lambda_{2m}+1}\} \vee \{\bigvee_{2\lambda_{2m}+2 \leq j \leq 2\lambda_{2m+1}+1} \phi'_j\}$. Set $\phi \leftarrow (\phi_1, \ldots, \phi_{2m+1})$ and exit.

Let ϕ be any formula in r free variables and let $S \subseteq \mathbb{R}^r$. ϕ is *invariant on S* in case either $(a_1, \ldots, a_r)$ is true for all $(a_1, \ldots, a_r) \in S$ or $(a_1, \ldots, a_r)$ is false for all $(a_1, \ldots, a_r) \in S$. If S is a c.a.d. of $\mathbb{R}^r$, we say that S is ϕ-*invariant* in case ϕ is invariant on each cell of S. If ϕ is a standard quantifier-free formula in r variables, $\mathcal{A}$ is the set of all non-zero polynomials which occur in ϕ, and S is an $\mathcal{A}$-invariant c.a.d. of $\mathbb{R}^r$, then clearly S is also ϕ-invariant.

If ϕ is a sentence, we will denote by $v(\phi)$ the truth value of ϕ, with "true" represented by 1, "false" by 0. Accordingly, if $(v_1, \ldots, v_n)$ is a vector of zeros and ones, then we define $\bigwedge_{i=1}^n v_i = 1$ if each $v_i = 1$ and $\bigwedge_{i=1}^n v_i = 0$ otherwise. Similarly, we define $\bigvee_{i=1}^n v_i = 0$ if each $v_i = 0$ and $\bigvee_{i=1}^n v_i = 1$ otherwise. If ϕ is a formula in r free variables and $a = (a_1, \ldots, a_r) \in \mathbb{R}^r$, we set $v(\phi, a) = v(\phi(a_1, \ldots, a_r))$. If ϕ is invariant on S, we set $v(\phi, S) = v(\phi, a)$ for any $a \in S$.

The following theorem is fundamental in the use of a c.a.d. for quantifier elimination.

Theorem 7 *Let $\phi(x_1, \ldots, x_r)$ be a formula in r free variables and let ϕ^* be $(\forall x_r)\phi$ or $(\exists x_r)\phi$. If $r > 1$, let S be a ϕ-invariant c.a.d. of $\mathbb{R}^r$, S^* the c.a.d. of $\mathbb{R}^{r-1}$ induced by S. Then S^* is ϕ-invariant. If $S^* = (S_1, \ldots, S_m)$ and $S = (S_{1,1}, \ldots, S_{1,n_1}, \ldots, S_{m,1}, \ldots, S_{m,n_m})$ where $(S_{i,1}, \ldots, S_{i,n_i})$ is the i-th cylinder of S, then $v((\forall x_r)\phi, S_i) = \bigwedge_{j=1}^{n_i} v(\phi, S_{ij})$ and $v((\exists x_r)\phi, S_i) = \bigvee_{j=1}^{n_i} v(\phi, S_{i,j})$. If $r = 1$ and $S = (S_1, \ldots, S_n)$ is a c.a.d. of $\mathbb{R}$, then $v((\forall x_1)\phi) = \bigwedge_{i=1}^n v(\phi, S_i)$ and $v((\exists x_1)\phi) = \bigvee_{i=1}^n (\phi, S_i)$.*

Proof. We will prove this theorem only for $r > 1$, and only for the case that ϕ^* is $(\forall x_r)\phi$. The omitted cases are similar. So let $S = (S_{1,1}, \ldots, S_{1,n_1}, \ldots, S_{m,1}, \ldots, S_{m,n_m})$ be a ϕ-invariant c.a.d. of $\mathbb{R}^r$, $S^*(S_1, \ldots, S_m)$ the c.a.d. of $\mathbb{R}^{r-1}$ induced by S. Let $1 \le i \le m$ and choose $(a_1, \ldots, a_{r-1}) \in S_i$. Assume $\phi^*(a_1, \ldots, a_{r-1})$ is true. Let $(b_1, \ldots, b_{r-1}) \in S_i$. Let $b_r \in \mathbb{R}$. Then for j, $(b_1, \ldots, b_r) \in S_{i,j}$. Choose a_r so that $(a_1, \ldots, a_r) \in S_{i,j}$. Since $\phi^*(a_1, \ldots, a_{r-1})$ is true, $\phi(a_1, \ldots, a_{r-1}, a)$ is true for all $a \in \mathbb{R}$. In particular, $\phi(a_1, \ldots, a_r)$ is true. Since S is ϕ-invariant, ϕ is invariant on $S_{i,j}$. So $\phi(b_1, \ldots, b_r)$ is true. Since b_r is an arbitrary element of $\mathbb{R}$, $\phi^*(b_1, \ldots, b_{r-1})$ is true. Since $(b_1, \ldots, b_{r-1})$ is an arbitrary element of S_i, ϕ^* is invariant on S_i. Since S_i is an arbitrary element of S, ϕ^* is S-invariant. This completes the proof of the first part.

Now assume $v(\phi^*, S_i) = 1$. Let $1 \le j \le n_i$. Choose $(a_1, \ldots, a_{r-1}) \in S_i$. By the first part, ϕ^* is S-invariant so $\phi^*(a_1, \ldots, a_{r-1})$ is true. Hence $\phi(a_1, \ldots, a_r)$ is true for all $a_r \in \mathbb{R}$. Choose a_r so that $(a_1, \ldots, a_r) \in S_{i,j}$. By the ϕ-invariance of S, $v(\phi, S_{i.j}) = 1$. Since j is arbitrary, $\bigwedge_{j=1}^{n_i} v(\phi.S_{i,j}) = 1$.

Next assume $v(\phi^*, S_i) = 0$. Choose $(a_1, \ldots, a_{r-1}) \in S_i$. Since ϕ^* is S-invariant, $\phi^*(a_1, \ldots, a_{r-1})$ is false. Hence for some $a_r \in \mathbb{R}$, $\phi(a_1, \ldots, a_r)$ is false. Let $(a_1, \ldots, a_r) \in S_{i,j}$. By the ϕ-invariance of S, $v(\phi, S_{i,j}) = 0$. Hence $\bigwedge_{j=1}^{n_j} v(\phi, S_{i,j}) = 0$. ∎

Let $a, b \in \mathbb{R}^r$ with $a = (a_1, \ldots, a_r)$ and $b = (b_1, \ldots, b_r)$. We define $a \sim_k b$ in case $a_i = b_i$ for $1 \le i \le k$. Note that $a \sim_r b$ if and only if $a = b$, while $a \sim_0 b$

for all $a, b \in \mathbb{R}^r$. We define $a < b$ in case $a \sim_k b$ and $a_{k+1} < b_{k+1}$ for some k, $0 \leq k < r$. The relation $a < b$ is a linear order on $\mathbb{R}^r$, which we recognize as the lexicographical order on $\mathbb{R}^r$ induced by the usual order on $\mathbb{R}$. We note that if $(\beta_1, \ldots, \beta_m)$ is a cylindrical sample of a c.a.d. S, then $\beta_1 < \beta_2 < \cdots < \beta_m$.

The cylindrical structure of a c.a.d. S is obtainable from any c.a.s. β of S. We define a grouping function g. Let $\beta = (\beta_1, \ldots, \beta_m)$ be any sequence of elements of $\mathbb{R}^r$. Then for $0 \leq k \leq r$, $g(k, \beta) = ((\beta_1, \ldots, \beta_{n_1}), (\beta_{n_1+1}, \ldots, \beta_{n_2}), \ldots, (\beta_{n_{l-1}+1}, \ldots, \beta_{n_l}))$ where $1 \leq n_1 < n_2 < \cdots < n_{l-1} < n_l = m, \beta_j \sim_k \beta_{j+1}$ for $n_i < j < n_{i+1}$, and $\beta_{n_i} \not\sim_k \beta_{n_i+1}$. Note that $g(0, \beta) = ((\beta_1, \ldots, \beta_m))$ and $g(r, \beta) = ((\beta_1), \ldots, (\beta_m))$. Also, if S is a c.a.d. of $\mathbb{R}^r$, $S^* = (S_1, \ldots, S_m)$ is the c.a.d. of $\mathbb{R}^k$ induced by S, and β is a c.a.s. of S, then $g(k, \beta) = (\beta_1^*, \ldots, \beta_m^*)$ where β_i^* is the list of those points in β which belong to $S_i^* \times \mathbb{R}^{r-k}$.

We define now an evaluation function e. Let $\phi(x_1, \ldots, x_r)$ be a standard quantifier-free formula, S a ϕ-invariant c.a.d. of $\mathbb{R}^r$, β a c.a.s. of S, and let $\phi^*(x_1, \ldots, x_k)$ be $(Q_{k+1}x_{k+1}) \ldots (Q_r x_r)\phi(x_1, \ldots, x_r)$, $0 \leq k \leq r$. Let $S^* = (S_1^*, \ldots, S_m^*)$ be the c.a.d. of $\mathbb{R}^k$ induced by S, $\beta^* = (\beta_1^*, \ldots, \beta_m^*) = g(k, \beta)$. Then we define $e(\phi^*, \beta_i^*)$ by induction on $r - k$, as follows. If $k = r$, then ϕ^* is ϕ, $\beta_i^* = (\beta_i)$, and we define $e(\phi^*, \beta_i^*) = v(\phi, \beta_i)$. If $k < r$, let $g(k + 1, \beta_i^*) = (\hat{\beta}_1, \ldots, \hat{\beta}_n) = \hat{\beta}$. Then each $\hat{\beta}_j$ is in the sequence $g(k + 1, \beta)$. Let $\hat{\phi}(x_1, \ldots, x_{k+1})$ be $(Q_{k+2}x_{k+2}) \ldots (Q_r x_r)\phi(x_1, \ldots, x_r)$. Then we define

$$e(\phi^*, \beta_i^*) = \bigwedge_{j=1}^n e(\hat{\phi}, \hat{\beta}_j), \quad \text{if } Q_{k+1} = \forall,$$
$$e(\phi^*, \beta_i^*) = \bigvee_{j=1}^n e(\hat{\phi}, \hat{\beta}_j), \quad \text{if } Q_{k+1} = \exists.$$

Theorem 8 *Let $\phi(x_1, \ldots, x_r)$ be a standard quantifier-free formula, S a ϕ-invariant c.a.d. of $\mathbb{R}^r$, β a cylindrical algebraic sample of S. Let*

$$\phi^*(x_1, \ldots, x_k) = (Q_{k+1}x_{k+1}) \ldots (Q_r x_r)\phi(x_1, \ldots, x_r), \ 0 \leq k \leq r.$$

If $k > 0$, let $S^* = (S_1^*, \ldots, S_m^*)$ be the c.a.d. of $\mathbb{R}^k$ induced by S and let $g(k, \beta) = \beta^* = (\beta_1^*, \ldots, \beta_m^*)$. Then $e(\phi^*, \beta_i^*) = v(\phi^*, S_i^*)$ for $1 \leq i \leq m$. If $k = 0$, then $e(\phi^*, \beta) = v(\phi^*)$.

Proof. By an induction on $r - k$, paralleling the definition of e and using Theorem 7. ∎

By Theorem 8, if $k = 0$, then $e(\phi^*, \beta)$ is the truth value of ϕ^*. If $k > 0$, let $\psi = (\psi_1, \ldots, \psi_m)$ be a standard definition of the c.a.d. S^*, as produced by DECOMP, and let ψ^* be the disjunction of those ψ_i such that $e(\phi^*, \beta_i^*) = 1$. Then ψ^* is a standard quantifier-free formula equivalent to ϕ^*.

The function e can be computed by an algorithm based directly on the definition of e. $e(\phi^*, \beta_i^*)$ is ultimately just some Boolean function of the truth values of ϕ at the sample points β_j in the list β_i^*, that is, of the $v(\phi, \beta)$. It is important to note, however, that usually not all $v(\phi, \beta_j)$ need be computed.

For example, if $Q_{k+1} = \forall$ then the computation of $e(\phi^*, \beta_i^*)$ can be terminated as soon as any j is found for which $e(\hat{\phi}, \beta_j) = 0$. Similarly, the computation of $v(\phi, \beta)$, β an algebraic point, is Boolean-reducible to the case in which ϕ is a standard atomic formula. This case itself amounts to determining the sign of $A(\beta_1, \ldots, \beta_r)$ where A is an integral polynomial and $\beta = (\beta_1, \ldots, \beta_r)$ is a real algebraic point. With β we are given an algebraic number α such that $Q(\beta_1, \ldots, \beta_r) = \mathbb{Q}(\alpha)$ and rational polynomials B_i such that $\beta_i = B_i(\alpha)$. We then obtain $\text{sign}(A(\beta_1, \ldots, \beta_r)) = \text{sign}(A(B_1(\alpha), \ldots, B_r(\alpha))) = \text{sign}(C(\alpha))$ using an algorithm of (Rubald 1974).

Since a standard formula ϕ may contain several occurrences of the same polynomial, we assume that the polynomials occurring in ϕ are stored uniquely in a list $\mathcal{A}$ inside the computer, and that the formula ϕ is stored so that the atomic formulas of ϕ contain references to this list in place of the polynomials themselves. Note also that the list $\mathcal{A}$ need not contain two different polynomials whose ratio is a nonzero rational number. In computing $v(\phi, \beta)$, a list σ should be maintained, containing $\text{sign}(A(\beta))$ for various polynomials $A \in \mathcal{A}$. Whenever the computation of $v(\psi, \beta)$ requires the computation of $\text{sign}(A(\beta))$ the list σ should be searched to determine whether $\text{sign}(A(\beta))$ was previously computed; if not, $\text{sign}(A(\beta))$ should be computed and placed on the list. Thus the computation of $v(\phi, \beta)$ will require at most one computation of $\text{sign}(A(\beta))$ for each $A \in \mathcal{A}$, and in some cases $\text{sign}(A(\beta))$ will not be computed for all $A \in \mathcal{A}$.

In terms of the functions g and e, the evaluation algorithm can now be described as follows.

$$\psi^* = \text{EVAL}(\phi^*, \beta, \psi)$$

Inputs: ϕ^* is a standard prenex formula $(Q_{k+1}x_{k+1}) \ldots (Q_r x_r)\phi(x_1, \ldots, x_r)$, where $0 \leq k \leq r$ and ϕ is quantifier-free. β is a c.a.s. of some ϕ-invariant c.a.d. S of $\mathbb{R}^r$. ψ is a standard definition of the c.a.d. S^* of $\mathbb{R}^k$ induced by S if $k > 0$, the null list if $k = 0$.
Output: $\psi^* = \psi(x_1, \ldots, x_k)$ is a standard quantifier-free formula equivalent to ϕ^*.

Algorithm Description

(1) If $k > 0$ go to (2). Set $v = e(\phi^*, \beta)$. If $v = 0$ set $\psi^* \leftarrow$ "$1 = 0$". If $v = 1$, set $\psi^* \leftarrow$ "$0 = 0$". Exit.

(2) Set $\beta^* \leftarrow g(k, \beta)$. Let $\beta^* = (\beta_1^*, \ldots, \beta_m^*)$ and $\psi = (\psi_1, \ldots, \psi_m)$. Set $\psi^* \leftarrow$ "$1 = 0$". For $i = 1, \ldots, m$ if $e(\phi^*, \beta_i^*) = 1$ set $\psi^* \leftarrow \psi_i \vee \psi$. Exit.

Finally we have the following quantifier elimination algorithm.

$$\psi^* = \text{ELIM}(\phi^*)$$

Input: ϕ^* is a standard prenex formula $(Q_{k+1}x_{k+1}) \ldots (Q_r x_r)\phi(x_1, \ldots, x_r)$, where $0 \leq k \leq r$ and ϕ is quantifier-free.
Output: ψ^* is a standard quantifier-free formula equivalent to ϕ^*.

Algorithm Description

(1) Determine k. Extract from ϕ the list $\mathcal{A} = (A_1, \ldots, A_m)$ of distinct non-zero polynomials occurring in ϕ.
(2) Apply DECOMP to $\mathcal{A}$ and k, obtaining β and ψ as outputs.
(3) Set $\psi^* \leftarrow \text{EVAL}(\phi^*, \beta, \psi)$ and exit.

4   Algorithm Analysis

Step (4) of the algorithm DECOMP provides for the optional computation of a basis $\mathcal{B}$ for a set $\mathcal{A}$ of integral polynomials. Experience with the algorithm provides a strong indication that this basis calculation is very important in reducing the total computing time of the algorithm. If the set $\mathcal{A}$ is the result of two or more projections, as in general it will be, then it appears that the polynomials in $\mathcal{A}$ have a considerable probability of having factors, common factors, and multiple factors. This will be discussed further in Section 5. But, as remarked in Section 3, the basis calculation of step (4) is not essential to the validity of the algorithm. In order to simplify the analysis of the algorithm, we will assume that this basis calculation is not performed. In general, the polynomials in the basis $\mathcal{B}$ will have smaller degrees than the polynomials of $\mathcal{A}$, but the number of polynomials in $\mathcal{B}$ may be either greater or less than the number in $\mathcal{A}$.

In Section 3 we gave conceptually simple definitions of projection and augmented projection, which definitions can be improved somewhat in order to reduce the sizes of these sets. It is easy to see that in the definition of the projection we can set $\mathcal{S}_2 = \{\text{psc}_k(\text{red}^i(A), \text{red}^j(B)) : A, B \in \text{ \& } A < B \text{ \& } i \geq 0 \text{ \& } j \geq 0 \text{ \& } 0 \leq k < \min(\deg(\text{red}^i(A)), \deg(\text{red}^j(B)))\}$, where "$<$" is an arbitrary linear ordering of the elements of $\mathcal{A}$. Also, in the definition of the augmented projection, we can set $\mathcal{P}' = \{\text{psc}_k(\text{der}^j(\text{red}^i(A)), \text{der}^{j+1}(\text{red}^i(A))) : A \in \mathcal{A} \text{ \& } i \geq 0 \text{ \& } j \geq 0 \text{ \& } 0 \leq k < \deg(\text{der}^{j+1}(\text{red}^i(A)))\}$. Then the set $\mathcal{S}_1$ in the definition of the projection is contained in $\mathcal{P}'$, and the augmented projection of $\mathcal{A}$ is $\mathcal{L} \cup \mathcal{P}_2 \cup \mathcal{P}'$.

Now suppose that the set $\mathcal{A}$ contains m polynomials, with the degree of each polynomial in each variable at most n. Assume $m \geq 1$ and $n \geq 1$. Then the set $\mathcal{L}$ contains at most mn elements. In the set $\mathcal{S}_2$, the pair (A, B) can be chosen in $\binom{m}{2}$ ways. Since $k < \min(\deg(\text{red}^i(A)), \deg(\text{red}^j(B))) \leq \min(n - i, n - j) = n - \max(i, j)$, we have $0 \leq i, j \leq n - k - 1$. Hence for given k, $0 \leq k \leq n - 1$, the pair (i, j) can be chosen in at most $(n - k)^2$ ways. Hence (i, j, k) can be chosen in $\sum_{h=0}^{n-1}(n - k)^2 = n(n + 1)(2n + 1)/6$ ways. So $\mathcal{S}_2$ has at most $\binom{m}{2}n(n + 1)(2n + 1)/6$ elements. In the definition of the set $\mathcal{P}'$, we must have $k < n - i - j - 1$. For given k, $0 \leq k \leq n - 2$, (i, j) can be chosen in $\sum_{h=0}^{n-k-2}(h + 1) = \binom{n-k}{2}$ ways. Hence (i, j, k) can be chosen in $\sum_{k=0}^{n-2}\binom{n-k}{2} = \binom{n+1}{3}$ ways. So $\mathcal{P}'$ has at most $m\binom{n+1}{3}$ elements. Altogether, the augmented projection has at most $\binom{m}{2}n(n + 1)(2n + 1)/6 + m\binom{n+1}{3} + mn$

elements. For $n = 1$, this reduces to $\binom{m}{2} + m = \binom{m+1}{2} \leq m^2$. For $n \geq 2$, $\binom{m}{2}n(n+1)(2n+1)/6 + m\binom{n+1}{3} + mn \leq (m^2/2)((15/4)n^3/6) + mn^3/6 + mn^3/4 < m^2n^3$. So in all cases the augmented projection of $\mathcal{A}$ has at most m^2n^3 elements.

By definition of a principal subresultant coefficient, each element of $\mathcal{S}_2$ or $\mathcal{P}'$ is the determinant of a matrix with at most $2n$ rows and columns, whose entries are coefficients of elements of $\mathcal{A}$. Hence the degree of any element of the augmented projection, in any variable, is at most $2n^2$.

In order to analyze the growth of coefficient length under the augmented projection operation, we need the concept of the norm of a polynomial. If A is any integral polynomial, the *norm* of A, denoted by $|A|_1$, is defined to be sum of the absolute values of the integer coefficients of A. This "norm" is actually just a semi-norm, having the important properties $|A + B|_1 \leq |A|_1 + |B|_1$ and $|A \cdot B|_1 \leq |A|_1 \cdot |B|_1$. Using these properties, it is easy to show (see Collins and Horowitz 1974) that if $\deg(A) = m$ and $\deg(B) = n$, then any square submatrix of the Sylvester matrix of A and B has a determinant whose norm is at most $|A|_1^n|B|_1^m$.

Let c be the maximum of the norms of the elements of $\mathcal{A}$. For any polynomial A with $\deg(A) = n$, $|A'|_1 \leq n|A|_1$. Hence it is easy to see that if $P \in \mathcal{P}'$ then $|P|_1 \leq (n^jc)^{n-j-1}(n^{j+1}c)^{n-j} \leq n^{n^2/2}c^{2n}$, while if $P \in \mathcal{L} \cup \mathcal{S}_2$ then $|P|_1 \leq c^{2n}$.

The *length* of any non-zero integer a, $L(a)$, is the number of bits in the binary representation of a, and $L(0) = 1$. It is easy to show that $L(ab) \leq L(a) + L(b)$ and hence $L(a^n) \leq nL(a)$ if $n > 0$. Also, $L(a) \leq a$ if $a > 0$. So if P is any element of the augmented projection of $\mathcal{A}$, then $L(|P|_1) \leq (1/2)n^2L(n) + 2nL(c) \leq (1/2)n^3 + 2nL(c)$.

The following theorem summarizes the several things we have proved.

Theorem 9 *Let $\mathcal{A}$ be a non-empty set of integral polynomials in r variables, $r \geq 2$. Let $\mathcal{A}^*$ be the augmented projection of $\mathcal{A}$. Let m be the number of elements of $\mathcal{A}$, n the maximum degree of any element of $\mathcal{A}$ in any variable, $n \geq 1$. Let d be the maximum of the lengths of the norms of the elements of $\mathcal{A}$. Let $m^*, n^*,$ and d^* be the same functions of $\mathcal{A}^*$. Then*

$$m^* \leq m^2n^3, \tag{2}$$

$$n^* \leq 2n^2, \tag{3}$$

$$d^* \leq (1/2)n^3 + 2nd. \tag{4}$$

When $\mathcal{A}$ is a set of polynomials in r variables, algorithm DECOMP computes a sequence of $r - 1$ projections or augmented projections. Using Theorem 9 we can now derive bounds for all such projections.

Theorem 10 *Let $\mathcal{A}, m, n$ and d be defined as in Theorem 9. Let $\mathcal{A}_1 = \mathcal{A}$ and let $\mathcal{A}_{i+1}$ be the augmented projection of $\mathcal{A}_i$ for $1 \leq i < r$. Let m_k be the number of elements of $\mathcal{A}_k$, n_k the degree maximum for $\mathcal{A}_k$, d_k the norm length* *maximum for $\mathcal{A}_k$.* Then

$$m_k \leq (2n)^{3^k} m^{2^{k-1}}, \tag{5}$$

$$n_k \leq (1/2)(2n)^{2^{k-1}}, \tag{6}$$

$$d_k \leq (2n)^{2^k} d. \tag{7}$$

Proof. One may first prove (6) by a simple induction on k, using (3). (5) obviously holds for $k = 1$. Assuming (5) holds for k, by (2) we have $m_{k+1} \leq \frac{1}{8}(2n)^a m^{2^k}$ where $a = 2 \cdot 3^k + 3 \cdot 2^{k-1} \leq 6 \cdot 3^{k-1} + 3 \cdot 3^{k-1} = 3^{k+1}$, proving (5). (7) obviously holds for $k = 1$. Assuming (7) holds for k, by (4) we have $d_{k+1} \leq \frac{1}{16}(2n)^a + (2n)^a d \leq 2(2n)^a d \leq (2n)^{a+1}d$ where $a = 3 \cdot 2^{k+1}$ so (7) is established. ∎

Using Theorem 10, we can now bound the time to compute all projections.

Theorem 11 *Let $\mathcal{A}, m, n, d$ and $\mathcal{A}_1, \ldots, \mathcal{A}_r$ be defined as in Theorem 10. Then there is an algorithm which computes $\mathcal{A}_2, \ldots, \mathcal{A}_r$ from $\mathcal{A}_1 = \mathcal{A}$ in time dominated by $(2n)^{3^{r+1}} m^{2^r} d^2$.*

Proof. Let A and B be integral polynomials in r variables, with degrees in each variable not exceeding $n \geq 2$, and with norms of lengths d or less. There is described in (Collins 1971a) an algorithm for computing the resultant of A and B, whose computing time is dominated by $n^{2r+2}d^2$. It is easy to see how to generalize this algorithm to compute $\mathrm{psc}_k(A, B)$, for any k, within the same time bound. By (6) and (7), any derivative of any element of $\mathcal{A}_k$ has a norm whose length is at most $d_k + L(n_k!) \leq d_k + n_k^2 \leq \frac{5}{4}(2n)^{2^k}d = d'_k$. Since the elements of $\mathcal{A}_k$ have $r - k - 1$ variables, each p.s.c. can be computed in time dominated by $n_k^{2(r-k+1)}d'^2_k$, and there are at most m_{k+1} such p.s.c.'s to be computed. Using the inequality $2(r - k + 1) \leq 2^{r-k+1}$, we thus find that the time to compute all p.s.c.'s of $\mathcal{A}_{k+1}$ is dominated by $(2n)^a m^{2^k} d^2$ where $a = 3^{k+1} + 2^r + 2^{k+1} \leq 3^r + 2^r + 2^r \leq 9 \cdot 3^{r-2} + 8 \cdot 3^{r-2} \leq 17 \cdot 3^{r-2} < 2 \cdot 3^r$. Hence the p.s.c.'s of $\mathcal{A}_{k+1}$ can be computed in time $(2n)^{2 \cdot 3^r} m^{2^r} d^2$. Multiplying by r and using $r \leq 2^{2^{r-1}}$, the p.s.c.'s of $\mathcal{A}_2, \ldots, \mathcal{A}_r$ can be computed in time $(2n)^{3^{r+1}} m^{2^r} d^2$. We have ignored the time required to compute reducta and derivatives, but this is relatively trivial. ∎

Let S be the c.a.d. of $\mathbb{R}^r$ computed by DECOMP and let S_k be the c.a.d. of $\mathbb{R}^k$ induced by S, for $1 \leq k \leq r$. Thus $S = S_r$. Let c_k be the number of cells in S_k. The cells of S_1 are determined by the real roots of m_r polynomials, each of degree n_r at most. There are at most $m_r n_r$ such roots and hence $c_1 \leq 2m_r n_r + 1$. For each value of k, $2 \leq k \leq r$, step (6) substitutes the $k - 1$ coordinates of each sample point of S_{k-1} for the first $k - 1$ variables of the k-variable polynomials in $\mathcal{A}_{r-k+1}$, thereby obtaining $u_{r-k+1} \leq c_{k-1} m_{r-k+1}$ univariate polynomials with real algebraic number coefficients, each of degree

n_{r-k+1} at most. These polynomials have at most $c_{k-1}m_{r-k+1}n_{r-k+1}$ real roots, and hence $c_k \leq 2m_{r-k+1}n_{r-k+1}c_{k-1} + 1$. For convenience, we set $u_r = m_r$. We can now prove the following theorem.

Theorem 12 *For* $1 \leq k \leq r$, *both* u_k *and* c_k *are less than* $(2n)^{3^{r+1}} m^{2^r}$.

Proof. We have shown above that $c_1 \leq 2m_r n_r + 1$; hence $c_1 \leq 4m_r n_r$. Similarly, from $c_k \leq 2m_{r-k+1}n_{r-k+1}c_{k-1} + 1$ it follows that $c_k \leq 4m_{r-k+1}$ $n_{r-k+1}c_{k-1}$. By induction on k, we then have $c_k \leq \prod_{i=r-k+1}^{r} 4m_i n_i$. Hence $c_k \leq 2^{2r} \prod_{i=1}^{r} m_i n_i$ for all k. In a similar manner it is easy to show that $u_k \leq 2^{2r} \prod_{i=1}^{r} m_i n_i$ for all k. By (5) and (6) we then deduce that $2^{2r} \prod_{i=1}^{r} m_i n_i \leq m^b(2n)^{a+b+2r}$ where $a = \sum_{k=1}^{r} 3^k < \frac{1}{2} \cdot 3^{r+1}$ and $b = \sum_{k=1}^{r} 2^{k-1} < 2^r$. Hence it suffices to show that $2^r + 2r \leq \frac{1}{2} \cdot 3^{r+1}$. But $2^r + 2r \leq 2 \cdot 2^r \leq 4 \cdot 2^{r-1} \leq 4 \cdot 3^{r-1} < \frac{1}{2} \cdot 3^{r+1}$. ∎

Our next goal is to bound the time for step (1) of DECOMP. We must first obtain bounds for the computing times of the subalgorithms BASIS and ISOL.

There exist (see Collins 1973a) polynomial greatest common divisor algorithms for univariate integral polynomials which, when applied to two polynomials of degree n or less and with norms of length d or less, have a maximum computing time dominated by $n^3 d^2$. Mignotte (1974) has recently shown that if A is a univariate integral polynomial with degree n and norm c, and if B is any divisor of A, then $|B|_1 \leq 2^n c$. From these two facts it easily follows that the squarefree factorization of A can be computed by the algorithm described in (Musser 1971) in time dominated by $n^6 + n^4 d^2$ where d is the length of $c = |A|_1$ and $n = \deg(A)$.

Now suppose the coarsest squarefree basis algorithm outlined in Section 3 is applied to a set of m univariate integral polynomials, with degrees and norm lengths bounded by n and d respectively. In each of the m applications of Loos' algorithm, each input basis set will contain at most mn polynomials, with degrees and norm lengths bounded by n and $n+d$ respectively. Hence the time for all applications of Loos' algorithm will be dominated by $m(mn)^2 n^3 (n+d)^2$, hence by $m^3 n^5 (n^2 + d^2) \leq m^3 n^7 d^2$. The time required for the m squarefree factorizations will be dominated by $mn^6 d^2$. Hence we arrive at a maximum computing time of $m^3 n^7 d^2$ for BASIS.

Now consider the computing time of ISOL when applied to a set $\mathcal{A}$ of m univariate integral squarefree and pairwise relatively prime polynomials, with a degree bound of n and a norm length bound of d. Collins has shown (Collins and Horowitz 1974) that if A is a univariate integral squarefree polynomial with $\deg(A) = n$ and $|A|_1 = c$, then the distance between any two roots of A is at least $\frac{1}{2}(e^{1/2}n^{3/2}c)^{-n}$. This theorem on root separation, together with the discussion of Heindel's algorithm in (Heindel 1971), implies that Heindel's algorithm will isolate the real roots of A in time dominated by $n^8 + n^7 d^3$. Hence the real roots of the m polynomials in $\mathcal{A}$ can be separately isolated in time dominated by $mn^8 + mn^7 d^3$. An isolating interval for a root of $A_i \in \mathcal{A}$

can be refined to length less than 2^{-h} in time dominated by $n^2h^3 + n^2d^2h$. By application of the root separation theorem to the product $A = \prod_{i=1}^{m} A_i$ of the elements of $\mathcal{A}$, the distance between any two roots of A is at least $\frac{1}{2}(e^{1/2}(mn)^{3/2}c^m)^{-mn} = \delta$. Hence if the isolating intervals for each A_i are refined to length 2^{-h} with $\delta/4 \leq 2^{-h} < \delta/2$ then all intervals for all A_i are disjoint. We then have h codominant with $mnL(mn) + m^2nd$, so the time to refine each interval is dominated by $m^3n^5L(mn)^3 + m^6n^5d^3$, hence by $m^4n^6 + m^6n^5d^3$. Since there are at most mn intervals to refine, the total time for ISOL is dominated by $(mn^8 + mn^7d^3) + mn(m^4n^6 + m^6n^5d^3)$, hence by $mn^8 + m^7n^7d^3$.

Theorem 13 *The computing time for step (1) of* DECOMP *is dominated by* $(2n)^{3^{r+3}}m^{2^{r+2}}d^3$.

Proof. The time to apply BASIS in step (1) is dominated by $m_r^3n_r^7d^2$. By Theorem 10, since $d + 1 \leq 2d$, $m_r^3n_r^7d^2$ is dominated by $(2n)^a m^b d^2$ where $b = 3 \cdot 2^{r-1} < 2^{r+1}$ and $a \leq 3 \cdot 3^r + 7 \cdot 2^{r-1} + 2 \cdot 2^r \leq 9 \cdot 3^{r-1} + 7 \cdot 2^{r-1} + 4 \cdot 2^{r-1} \leq 20 \cdot 3^{r-1} < 3^{r+2}$. The coarsest squarefree basis $\mathcal{B}$ obtained will have at most m_rn_r elements, with degrees bounded by n_r and norm length bounded by $n_r + d_r$ by Mignotte's theorem. Hence the time to apply ISOL will be dominated by $(m_rn_r)n_r^8 + (m_rn_r)^7n_r^7(n_r + d_r)^3$, which is dominated by $m_r^7n_r^{17}d^3$. By Theorem 10, $m_r^7n_r^{17}d^3 \leq (2n)^a m^b d^3$ where $b \leq 7 \cdot 2^{r-1} < 2^{r+2}$ and $a \leq 7 \cdot 3^r + 17 \cdot 2^{r-1} + 3 \cdot 2^r \leq 21 \cdot 3^{r-1} + 17 \cdot 3^{r-1} + 6 \cdot 3^{r-1} = 44 \cdot 3^{r-1} < 3^{r+3}$. ∎

Next we turn our attention to the "sizes" of the real algebraic numbers which arise in DECOMP. Two different representations are used, and hence there are two different definitions of "size". Regarded as an element of the field P of all real algebraic numbers, a real algebraic number α is represented by a primitive squarefree integral polynomial $A(x)$ such that $A(\alpha) = 0$ and an interval $I = (r, s)$ with rational endpoints r and s such that α is the unique root of A in I. We will assume moreover that r and s are *binary rationals*, that is, numbers of the form $a \cdot 2^{-k}$ where a and k are integers, $k \geq 0$, and a is odd if $k > 0$. Let λ be the minimum distance between two real roots of A. By the root separation theorem, $\lambda^{-1} \leq 2(e^{1/2}n^{3/2}c)^n$ where $n = \deg(A)$ and $c = |A|_1$. Hence $\log_2 \lambda^{-1} < 1 + n(1 + \frac{3}{2}L(n) + d)$ and $2^{-k} < \lambda$ if $k \geq 1 + n(1 + \frac{3}{2}L(n) + d)$, where $d = L(c)$. Hence we assume that $r = a \cdot 2^{-h}$ and $s = b \cdot 2^{-k}$ with h, $k \leq n(2 + 2L(n) + d)$. Since it follows from $A(\alpha) = 0$ that $|\alpha| < c$, we may also assume that $L(a), L(b) \leq 2n(1 + L(n) + d)$. Then the "size" of α will be characterized by $n = \deg(A)$ and $d = L(|A|_1)$.

Regarded as an element of the real algebraic number field $\mathbb{Q}(\alpha)$, the real algebraic number β is represented by a polynomial $B(x) \in \mathbb{Q}[x]$ with $\deg(B) < n = \deg(A)$. The rational polynomial $B(x)$ is itself represented in the form $B(x) = b^{-1} \cdot \bar{B}(x)$ where b is an integer, $\bar{B}(x)$ is an integral polynomial, and $\gcd(b, \bar{B}) = 1$. In this case the "size" of β is characterized by $L(b)$ and $L(|\bar{B}|_1)$.

Let P_k be the set of all points $\beta = (\beta_1, \ldots, \beta_k)$ belonging to the c.a.s. computed by DECOMP for the c.a.d. S_k. For each such point there is computed a real algebraic number α such that $\mathbb{Q}(\beta_1, \ldots, \beta_k) = \mathbb{Q}(\alpha)$, and a pair (A, I) which represents α. A is a squarefree univariate integral polynomial such that $A(\alpha) = 0$ and I is an isolating interval for α as a root of A. Let $\mathcal{A}_k^*$ be the set of all such polynomials A. Let n_k be the maximum degree of the elements of $\mathcal{A}_k^*$ and let d_k^* be the maximum norm length of the elements of $\mathcal{A}_k^*$.

For each coordinate β_i of a point $\beta \in P_k$, DECOMP computes a rational polynomial $B_i = b_i^{-1} \bar{B}_i$ which represents β_i as an element of $\mathbb{Q}(\alpha)$. Let $\mathcal{B}_k'$ be the set of all such rational polynomials B_i associated in this way with points β of P_k, and let d_k' be the maximum of $\max(L(b), L(|\bar{B}|_1))$ taken over all $B = b^{-1} B \in \mathcal{B}_k'$.

Our next goal is to obtain recurrence relations for n_k^*, d_k^* and d_k'. For $k = 1$, each algebraic number β_1 is a root of some element of $\mathcal{A}_r$, $\alpha = \beta_1$, and the polynomial A is an element of the basis for $\mathcal{A}_r$ which is computed in step (1). By Mignotte's theorem,

$$n_1^* \leq n_r, \tag{8}$$
$$d_1^* \leq n_r + d_r. \tag{9}$$

If β_1 is irrational, then $B(x) = x$ represents $\beta_1 = \alpha$ as an element of $\mathbb{Q}(\alpha)$. If β_1 is rational then $B(x) = \beta_1$ represents $\beta_1 = \alpha$ as an element of $\mathbb{Q}(\alpha)$. Referring to step (2) of DECOMP, β_1 arises as an endpoint of an isolating interval produced in step (1) by the application of ISOL to a basis $\mathcal{B}$ for $\mathcal{A}_r$. Let $\bar{B}$ be the product of the elements of $\mathcal{B}$. Then $\deg(\bar{B}) \leq m_r n_r$ and, since $\bar{B}$ is the product of at most $m_r n_r$ polynomials, each having a norm length of at most $n_r + d_r$, the norm length of $\bar{B}$ is at most $m_r n_r (n_r + d_r)$. In accordance with our previous discussion of the root separation theorem, we may therefore assume that the numerator and denominator of β_1 have lengths not exceeding $2 m_r n_r \{1 + L(m_r n_r) + m_r n_r (n_r + d_r)\} \leq 4 m_r^2 n_r^2 (n_r + d_r)$. Hence,

$$d_1' \leq 4 m_r^2 n_r^2 (n_r + d_r) \tag{10}$$

For each point $\beta = (\beta_1, \ldots, \beta_k)$ in P_k, and each polynomial $C(x_1, \ldots, x_{k+1})$ in $\mathcal{A}_{r-k}$, step (6) of DECOMP substitutes β_i for x_i, obtaining a polynomial $C^*(x) = C(\beta_1, \ldots, \beta_k, x)$ belonging to $\mathbb{Q}(\alpha)[x]$, $\mathbb{Q}(\alpha) = \mathbb{Q}(\beta_1, \ldots, \beta_k)$. This substitution may be performed in two stages. In the first stage we substitute $B_i(y)$ for x_i, where B_i represents β_i as an element of $\mathbb{Q}(\alpha)$, resulting in $\hat{C}(y, x) = C(B_1(y), \ldots, B_k(y), x)$, an element of $\mathbb{Q}[y, x]$. In the second stage, $\hat{C}(y, x)$ is reduced modulo $A(y)$, where A represents α, resulting in $C^*(y, x) \in \mathbb{Q}[y, x]$. $C^*(y, x)$ may be identified with $C^*(x)$ since the coefficients of $C^*(y, x)$ are elements of $\mathbb{Q}[y]$ which represent the coefficients of $C^*(x)$ as elements of $\mathbb{Q}(\alpha)$.

Instead of computing $\hat{C}(y, x)$ directly, we compute instead the integral polynomial $\bar{C}(y, x) = \{\prod_{i=1}^{k} b_i^{\nu_i}\} \hat{C}(y, x)$, where ν_i is the degree of C in x_i. To illustrate, suppose $k = 1$ and let $C(x_1, x_2) = \sum_{i=0}^{\nu_1} C_i(x_2) x_1^i$. Then $\bar{C}(y, x) = \sum_{i=0}^{\nu_1} C_i(x) \bar{B}_1(y)^i b_1^{\nu_1 - i}$. Since $|C|_1 = \sum_{i=0}^{\nu_1} |C_i|_1$, we see that $|\bar{C}|_1 \leq |C|_1 \cdot f^{\nu_1}$

where $f = \max(|b_1|, |B_1|_1)$. In general, recalling the definition of d'_k, we see that the length of the norm of $\bar{C}$ is at most $d_{r-k} + kn_{r-k}d'_k$. Also, $\hat{C}(y, x) = c_{-}^{-1}\bar{C}(y, x)$ with $L(c) \leq kn_{r-k}d'_k$. Furthermore, the degree of $\bar{C}(y, x)$ in y is at most $kn_{r-k}n^*_k$ since the degree of each B_i is less than n^*_k.

In reducing $\hat{C}(y, x)$ modulo $A(y)$, we compute the pseudo-remainder, $C'(y, x)$, of $\bar{C}(y, x)$ with respect to $A(y)$. $C'(y, x) \in \mathbb{Z}[y, x]$ and we have $c'\bar{C}(y, x) = A(y) \cdot Q(y, x) + C'(y, x)$ for some $c' \in \mathbb{Z}$ and some $Q(y, x) \in \mathbb{Z}[y, x]$, with the degree of C' in y less than $\deg(A)$. Hence

$$\hat{C}(y, x) = A(y)\{(cc')^{-1}Q(y, x)\} + (cc')^{-1}C'(y, x),$$

and $C^*(y, x) = (cc')^{-1}C'(y, x)$. Regarding the pseudo-remainder as a subresultant, we have $|C'|_1 \leq |\bar{C}|_1 \cdot |A|^n_1$ where n is the degree of $\bar{C}$ in y. Since $L(|\bar{C}|_1) \leq d_{r-k} + kn_{r-k}d'_k$, $L(|A|_1) \leq d^*_k$ and $n \leq kn_{r-k}n^*_k$, $L(|C'|_1) \leq d_{r-k} + kn_{r-k}d'_k + kn_{r-k}n^*_kd^*_k$. Also $c' = \{\text{ldcf}(A)\}^h$ with $h \leq n$, so $L(cc') \leq kn_{r-k}d'_k + kn_{r-k}n^*_kd^*_k$.

The polynomial $C^*(x)$ arises from a point $\beta \in P_k$ and a polynomial $C \in \mathcal{A}_{r-k}$. Keeping β fixed while C ranges over all elements of $\mathcal{A}_{r-k}$, we obtain a set $\mathcal{C}^*$ of univariate polynomials over $\mathbb{Q}(\alpha)$. Step (6) of DECOMP specifies the application of ABASIS to $\mathcal{C}^*$ to produce a coarsest squarefree basis $\mathcal{B}^*$. However, at present no theorem is known which provides a reasonable bound for the "sizes" of the coefficients of elements of $\mathcal{B}^*$. As an alternative we may therefore apply the algorithm NORMAL to α and $C^*(x)$, for each C^* in $\mathcal{C}^*$, producing an integral polynomial $D(x)$ such that every root of C^* is a root of D. Let $\mathcal{D}$ be the set of all polynomials D so obtained as C^* ranges over $\mathcal{C}^*$.

Given α, represented by the integral polynomial $A(y)$ and the isolating interval I, and the rational polynomial $C^*(y, x) = c^{-1}C'(y, x)$ representing $C^*(x)$, NORMAL proceeds as follows. Let $C'(y, x) = \sum_{i=0}^m C'_i(y)x^i$, where $C_m(\alpha) \neq 0$. The integral polynomial $C'(y, x)$ is divided by $\gcd(C_m(x), A(x))$, producing an integral polynomial $C''(y, x)$. Then $D(x)$ is the resultant, with respect to y, of $C''(y, x)$ and $A(y)$.

$\deg(A)$ and the degree in y of $C''(y, x)$ in y are both at most n^*_k. The degree of $C'(y, x)$ in x is at most n_{r-k}, the degree of $C(x_1, \ldots, x_{k+1})$ in x_{k+1}. Hence the degree of D is at most $n^*_kn_{r-k}$. Since the norm length of $C'(y, x)$ is at most $d_{r-k} + kn_{r-k}d'_k + kn_{r-k}n^*_kd^*_k$ and the norm length of A is at most d^*_k, the norm length of D is at most $n^*_kd^*_k + n^*_kd_{r-k} + kn_{r-k}n^*_kd'_k + kn_{r-k}n^{*2}_kd^*_k$.

Let $\bar{\mathcal{D}}$ be the coarsest squarefree basis of $\mathcal{D}$. Then every β_{k+1} with $(\beta_1, \ldots, \beta_k, \beta_{k+1}) \in P_{k+1}$ will be represented by a polynomial $\bar{D} \in \bar{\mathcal{D}}$ and, by Mignotte's theorem, the norm length of $\bar{D}$ is at most $n^*_kn_{r-k} + n^*_kd^*_k + n^*_kd_{r-k} + kn_{r-k}n^*_kd'_k + kn_{r-k}n^{*2}_kd^*_k$.

Next algorithm SIMPLE is applied to α and β_{k+1}, producing α' such that $\mathbb{Q}(\alpha, \beta_{k+1}) = \mathbb{Q}(\alpha')$. α and β_{k+1} are represented by the polynomials A and $\bar{D}$. α' is represented by a polynomial $A'(x)$, which is the resultant of $A(x - hy)$ and $\bar{D}(y)$ with respect to y, where h is some integer with $|h| < \deg(A) \cdot \deg(\bar{D})$. Since $|x - hy|_1 = |h| + 1$, we have $|A(x - hy)|_1 \leq$

$|A|_1 \cdot (|h| + 1)^{\deg(A)}$. Hence $L(|A(x - hy)|_1) \leq d_k^* + n_k^*(2L(n_k^*) + L(n_{r-k}))$. Since $\deg(\bar{D}) \leq n_k^* n_{r-k}$ and the degree of $A(x - hy)$ in y is $\deg(A) \leq n_k^*$, we have $L(|A|_1) \leq n_k^* n_{r-k}\{d_k^* + 2n_k^*L(n_k^*) + n_k^*L(n_{r-k})\} + n_k^*\{n_k^* n_{r-k} + n_k^* d_k^* + n_k^* d_{r-k} + kn_{r-k} n_k^{*2} d_k' + kn_{r-k} n_k^{*2} d_k\}$. Since the degree of $A(x - hy)$ in x is $\deg(A) \leq n_k^*$, we have $\deg(A') \leq n_k^{*2} n_{r-k}$. Thus we have proved

$$n_{k+1}^* \leq n_k^{*2} n_{r-k}. \tag{11}$$

Also, replacing $L(n_k^*)$ by n_k^*, $L(n_{r-k})$ by n_{r-k}, and making other simplifications in the inequality above for $L(|A'|_1)$, we have also

$$\begin{aligned}
d_{k+1}^* \leq{} & (d_k^* n_{r-k})n_k^* + (2n_{r-k}^2 + d_{r-k} + d_k^* + kn_{r-k}d_k')n_k^{*2} + \\
& (k + 2)n_{r-k}d_k^* n_k^{*3}.
\end{aligned} \tag{12}$$

It remains to obtain a relation for d_{k+1}'. Algorithm SIMPLE, when applied to α and β_{k+1}, produces, besides α' such that $\mathbb{Q}(\alpha, \beta_{k+1}) = \mathbb{Q}(\alpha')$, rational polynomials E and F which represent α and β_{k+1} as elements of $\mathbb{Q}(\alpha')$. The polynomial $A'(y)$ which represents α' is the resultant of $A(x - hy)$ and $\bar{D}(y)$. The monic greatest common divisor of $A(\alpha' - hx)$ and $\bar{D}(x)$ in $\mathbb{Q}(\alpha')$ is the polynomial $x - \beta_{k+1}$. This implies that if $G(y, x) = G_1(y)x + G_0(y)$ is the first subresultant of $A(y - hx)$ and $\bar{D}(x)$ then $\beta_{k+1} = -G_0(\alpha')/G_1(\alpha')$. Let $H = \gcd(A', G_1)$ and $\bar{A}' = A'/H$. Since $\bar{A}'$ is squarefree, $\bar{A}'$ and G_1 are relatively prime integral polynomials. Applying the extended Euclidean algorithm, we obtain integral polynomials U and V such that $\bar{A}'U + G_1V = c$, where $c \neq 0$ is the resultant of $\bar{A}'$ and G_1. Also, $G_1(\alpha') \neq 0$ so $H(\alpha') \neq 0$ and $\bar{A}'(\alpha') = 0$. Hence $G_1(\alpha')V(\alpha') = c$. Let $G_2 = -G_0V$. Then $c^{-1}G_2(\alpha') = -G_0(\alpha')c^{-1}V(\alpha') = -G_0(\alpha')/G_1(\alpha') = \beta_{k+1}$. Hence if $bG_2(y) = Q(y)A'(y) + G_3(y)$ where G_3 is the pseudo-remainder of G_2 and A', then $(bc)^{-1}G_3$ represents β_{k+1}. Also, $\alpha' = \alpha + h\beta_{k+1}$ so if $G_4(y) = bcy - hG_3(y)$ then $(bc)^{-1}G_4(y)$ represents α in $\mathbb{Q}(\alpha')$.

The same degree and norm length bounds, (11) and (12), which were derived for the resultant A' apply also to the subresultant coefficients G_0 and G_1. Since $\bar{A}'$ is a divisor of A', $\deg(\bar{A}') \leq n_{k+1}^*$ and, by Mignotte's theorem, $L(|\bar{A}'|_1) \cdot \leq n_{k+1}^* + d_{k+1}^*$. $\deg(V) \leq \deg(\bar{A}') \leq n_{k+1}^*$ and resultant bounds apply to c and $|V|_1$. Thus $L(c)$, $L(|V|_1) \leq n_{k+1}^*(2n_{k+1}^* + d_{k+1}^*)$. Hence $\deg(G_2) \leq 2n_{k+1}^*$ and $L(|G_2|_1) \leq 2n_{k+1}^{*2} + n_{k+1}^* d_{k+1}^* + d_{k+1}^*$. Therefore, $L(|G_3|_1) \leq (2n_{k+1}^{*2} + n_{k+1}^* d_{k+1}^* + d_{k+1}^*) + 2n_{k+1}^* d_{k+1}^* = 2n_{k+1}^{*2} + 3n_{k+1}^* d_{k+1}^* + d_{k+1}^*$, $L(b) \leq 2n_{k+1}^* d_{k+1}^*$ and $L(bc) \leq 2n_{k+1}^* + 3n_{k+1}^* d_{k+1}^*$. Since $L(|h| + 1) \leq L(n_{k+1}^*) \leq n_{k+1}^*$, $L(|G_4|_1) \leq 2n_{k+1}^{*2} + 3n_{k+1}^* d_{k+1}^* + n_{k+1}^* + d_{k+1}^* \leq 2n_{k+1}^{*2} + 4n_{k+1}^* d_{k+1}^*$.

Let $B_i = b_i^{-1}\bar{B}_i$ represent β_i as an element of $\mathbb{Q}(\alpha)$, $i \leq k$. Let $G' = g^{-1}\bar{G}'$ where $g = bc$ and $\bar{G}' = G_4$. Then G' represents α as an element of $\mathbb{Q}(\alpha')$. Hence $G'(\alpha') = \alpha$ and $B_i(\alpha) = \beta_i$ so $B_i(G'(\alpha')) = \beta_i$. Let $B_i^*(y) = q^{\nu_i}\bar{B}(G'(y))$ where $\nu_i = \deg(\bar{B}_i)$. B_i^* is an integral polynomial with $\deg(B_i^*) \leq \deg(\bar{B}_i) \cdot \deg(G') \leq n_k^* n_{k+1}^*$ and $L(|B_i^*|_1) \leq L(|\bar{B}_i|_1) + \deg(\bar{B}_i) \cdot (2n_{k+1}^{*2} + 4n_{k+1}^* d_{k+1}^*) \leq d_k' + n_k^*(2n_{k+1}^* + 4n_{k+1}^* d_{k+1}^*)$. Also, $L(b_i g^{\nu_i}) \leq d_k' + n_{k+1}^*(2n_{k+1}^{*2} + 4n_{k+1}^* d_{k+1}^*)$. Let

$\bar{B}'_i$ be the pseudo-remainder of B^*_i and A', $b^*_i B^*_i = A'Q + \bar{B}'_i$, and $b'_i = b^*_i b_i g^{\nu_i}$. Then $B'_i = b'^{-1}_i \bar{B}'_i$ represents β_i as an element of $\mathbb{Q}(\alpha')$. $L(|\bar{B}'_i|_1) \le L(|B^*_i|_1) + n^*_k n^*_{k+1} d^*_{k+1} \le d'_k + 2n^*_k n^{*~2}_{k+1} + 5n^*_k n^*_{k+1} d^*_{k+1}$, and $L(b'_i)$ satisfies the same bound.

Combining the last two paragraphs, $d'_k + 2n^*_k n^{*~2}_{k+1} + 5n^*_k n^*_{k+1} d^*_{k+1}$ is a norm length bound for the polynomials which represent $\beta_1, \ldots, \beta_{k+1}$ as elements of $\mathbb{Q}(\alpha')$ whenever β_{k+1} is a root of one of the polynomials which is obtained by substituting $\beta_1, \ldots, \beta_k$ for $x_1, \ldots, x_k$ in a polynomial of $\mathcal{A}_{r-k}$. But we must also consider the case that β_{k+1} is a rational endpoint of some isolating interval. We have seen that, for fixed $\beta_1, \ldots, \beta_k$, the isolated roots are all roots of the polynomials in the basis $\bar{\mathcal{D}}$ of $\mathcal{D}$. Let $\hat{D}$ be the product of the elements of $\mathcal{D}$. $\mathcal{D}$ has at most m_{r-k} elements, each of degree $n^*_k n_{r-k}$ at most. Hence $\bar{\mathcal{D}}$ has at most $m_{r-k} n^*_k n_{r-k}$ elements, each of degree $n^*_k n_{r-k}$ at most. We observed previously that $n^*_k d^*_k + n^*_k d_{r-k} + k n_{r-k} n^*_k d'_k + k n_{r-k} n^{*~2}_k d^*_k$ is a norm length bound for the elements of $\mathcal{D}$. Hence $\deg(\hat{D}) \le n^*_k m_{r-k} n_{r-k}$ and $L(|\hat{D}|_1) \le (n^*_k d^*_k + n^*_k d_{r-k} + k n_{r-k} n^*_k d'_k + k n_{r-k} n^{*~2}_k d^*_k) m_{r-k}$. If D^* is the greatest squarefree divisor of $\hat{D}$ then by Mignotte's theorem $L(|D^*|_1) \le (n^*_k n_{r-k} + n^*_k d^*_k + k n_{r-k} n^*_k d'_k + k n_{r-k} n^{*~2}_k d^*_k) m_{r-k} = L_k$, say. According to our earlier discussion, we may assume that the lengths of the numerators and denominators of the rational endpoints of isolating intervals for the roots of D^* do not exceed $2n^*_k m_{r-k} n_{r-k}(1 + L(2n^*_k m_{r-k} n_{r-k}) + L_k) \le 2n^*_k m_{r-k} n_{r-k} + 4n^{*~2}_k m^2_{r-k} n^2_{r-k} + 2n^*_k m_{r-k} n_{r-k} L_k \le 2\{k n_{r-k} d'_k + n_{r-k} d_{r-k} + (k+4) n^*_k n^2_{r-k} d^*_k\} n^{*~2}_k m^2_{r-k}$. Adding to this the bound $d'_k + 2n^*_k n^{*~2}_{k+1} + 5n^*_k n^*_{k+1} d^*_{k+1}$, we find that

$$
\begin{aligned}
d'_{k+1} \le{} & d'_k + 2n^*_k n^{*~2}_{k+1} + 5n^*_k n^*_{k+1} d^*_{k+1} + \\
& 2\{k n_{r-k} d'_k + n_{r-k} d_{r-k} + (k+4) n^*_k n^2_{r-k} d^*_k\} n^{*~2}_k m^2_{r-k} \quad (13)
\end{aligned}
$$

Now we use the recurrence relations (11), (12), and (13) to prove the following theorem.

Theorem 14 *With n^*_k, d^*_k and d'_k as defined above, we have*

$$
n^*_k \le (2n)^{2^{r+k-1}}, \tag{14}
$$

$$
d^*_k \le (2n)^{2^{r+k+2}} (2n)^{3^{r+1}} m^{2^{r+1}} d, \tag{15}
$$

$$
d'_k \le (2n)^{2^{r+k+2}} (2n)^{3^{r+1}} m^{2^{r+1}} d. \tag{16}
$$

Proof. To establish (14), we will show that $n^*_k \le (2n)^{s_k}$ where $s_k = \sum_{i=1}^k 2^{r+k-2i}$. For $k = 1$, by (8) and (6), $n^*_1 \le n_r (2n)^{s_1}$ since $s_1 = 2^{r-1}$. Assuming $n^*_k \le (2n)^{s_k}$, by (11) and (6), $n^*_{k+1} \le (2n)^a$ where $a \le 2s_k + 2^{r-k-1}$. But $2s_k + 2^{r-k-1} = \sum_{i=1}^k 2^{r+k+1-2i} + 2^{r+k+1-2(k+1)} = s_{k+1}$, completing the induction. And $s_k \le 2^{r+k-2}(1 + 2^{-2} + 2^{-4} + \cdots) < 2^{r+k-1}$, proving (14).

Using (14), we can now simplify the recurrence relation (12). We observe first that $k + 2 \le 2^{2^k}$, from which it follows by (6) that $(k+2) n_{r-k} \le (2n)^{2^r}$. It is then not difficult to derive from (12), using (6), (7), and (14), the inequality

$$d^*_{k+1} \leq (2n)^{r+k+1}(d + d^*_k + d'_k).$$ (17)

Similarly, we can simplify (13) using (5), (6), (7), and (14). We obtain

$$d'_{k+1} \leq (2n)^{2^{r+2}}(2n)^{2 \cdot 3^{r-k}} m^{2^{r-k}}(d + d^*_k + d^*_{k+1} + d'_k).$$ (18)

Let $\bar{D}_k$ be the common right-hand side of (15) and (16). Substituting from (17) for d^*_{k+1} into (18), we obtain

$$d_{k+1} \leq 2(2n)^{2^{r+k+1}+2^{r+2}}(2n)^{2 \cdot 3^{r-k}} m^{2^{r-k}}(d + d^*_k + d'_k).$$ (19)

Since, by (17), (19) also holds with d or d^*_{k+1} in place of d'_{k+1}, we have

$$D_{k+1} \leq 6\{(2n)^{2^{r+k+1}+2^{r+2}}(2n)^{2 \cdot 3^{r-k}} m^{2^{r-k}}\}D_k,$$ (20)

where $D_k = d + d^*_k + d'_k$. It suffices then to show that $D_k \leq \bar{D}_k$ for $1 \leq k \leq r$. We will prove instead the stronger inequality

$$D_k \leq (2n)^{2^{r+k+2}}(2n)^{u_k} m^{v_k} d,$$ (21)

where $u_k = 2 \sum_{i=1}^k 3^{r-i+1}$ and $v_k = \sum_{i=1}^k 2^{r-i+1}$. For $k \geq 2$, $6 \leq 2^3 \leq (2n)^3$ and $2^{r+k+2} + 2^{r+k+1} + 2^{r+2} + 3 \leq 2^{r+k+3}$. Since also $u_k + 2 \cdot 3^{r-k} = u_{k+1}$ and $v_k + 2^{r-k} = v_{k+1}$, (20) implies that (21) holds for $k+1$ if it holds for $k \geq 2$. It remains then to prove (21) for $k = 1$ and $k = 2$. By (9), (6), and (7),

$$d^*_1 \leq (2n)^{3 \cdot 2^{r-1}} d.$$ (22)

By (10), (5), (6), and (7),

$$d'_1 \leq (2n)^{5 \cdot 2^{r-1}}(2n)^{2 \cdot 3^r} m^{2^r} d.$$ (23)

If $r = 1$, then (22) and (23) imply (15) and (16). Otherwise, $r \geq 2$, $3 \leq (2n)^2$ and $5 \cdot 2^{r-1} + 2 \leq 2^{r+2}$, so (22) and (23) imply

$$D_1 \leq (2n)^{3 \cdot 2^r}(2n)^{u_1} m^{v_1} d,$$ (24)

which proves (21) for $k = 1$. Now (20) and (24) yield (21) for $k = 2$ since $6 \leq (2n)^3$ and $2^{r+3} + 2^{r+2} + 3 \cdot 2^r + 3 \leq 2^{r+4}$. ∎

As an easy corollary of Theorem 14, we obtain the following theorem.

Theorem 15 *For* $1 \leq k \leq r$,

$$n^*_k \leq (2n)^{2^{2r-1}},$$ (25)

$$d^*_k, d'_k \leq (2n)^{2^{2r+3}} m^{2^{r+1}} d.$$ (26)

Proof. This follows immediately from Theorem 14 by observing that $3^h \leq 2^{2h}$.
∎

We now have all of the information necessary to complete an analysis of the computing time of algorithm DECOMP. Theorem 12 bounds the number of cells in the decomposition as a function of m, n, and r. Theorems 10 and 15 together bound the degrees and coefficient lengths of all polynomials which arise in the mainstream of the calculation. From these bounds and from known computing time bounds for the various subalgorithms, it is straightforward, but tedious, to complete the analysis. We have given above computing time bounds for some, but not all of these subalgorithms. Those which we have not given may be found by the interested reader in the various references listed at the end of this paper. The critical property of these subalgorithms is that their computing times are all dominated by fixed powers of natural parameters such as degree products and maximum coefficient lengths or, as in the case of the BASIS algorithm, the number of polynomials in a list. Theorems 11 and 13 have illustrated the analysis of the computing time of certain parts of DE-COMP. Thus the completion of the analysis of DECOMP does not involve any novel techniques or subtleties. The seriously interested or skeptical reader may complete the analysis for himself. We therefore now state without additional proof the result of such analysis.

Theorem 16 *The computing time of* DECOMP *is dominated by*

$$(2n)^{2^{2r+8}} m^{2^{r+6}} d^3.$$

The exponents occurring in this theorem can likely be decreased in several ways. In the first place, the computing times used for the subalgorithms all presuppose that classical algorithms are used for integer multiplication and division. With the use of fast algorithms such as the Schönhage–Strassen algorithm (Schönhage and Strassen 1971) for integer arithmetic, it is evident that the exponent of d can be reduced from 3 to $2 + \epsilon$ for every $\epsilon > 0$. Secondly, it is probable that a tighter analysis without change of the subalgorithm would yield some improvement, for example perhaps the $2r + 8$ could be reduced to $2r + 4$. Thirdly, it is likely that improved mathematical knowledge would also improve the bound without change of the algorithms. For example, the analysis depends strongly on the root separation theorem, and it seems likely that this theorem is far from optimal. It should also be noted that the theorem bounds the maximum computing time, and the average computing time is likely much smaller.

By the remarks preceding the algorithm EVAL, if ϕ^* is a standard prenex formula with matrix ϕ, $\mathcal{A}$ is the set of polynomials occurring in ϕ, and β is a sample point, the truth value of each atomic formula in ϕ can be determined by at most one evaluation of $\text{sign}(A(\beta))$ for each $A \in \mathcal{A}$. Thus the application of EVAL to ϕ^*, a c.a.s. β, and a standard definition ψ involves mainly the evaluation of $\text{sign}(A(\beta_i))$ for each $A \in \mathcal{A}$ and each sample point $\beta_i \in \beta$. This is not essentially different from the calculations performed during the last phase of the application of DECOMP, and the bound of Theorem 16 again applies. However, one must not overlook the time required to compute the truth value of ϕ from the truth values of its atomic formulas, for each sample point β. The time required for this is obviously dominated, for each β, by the number, a, of occurrences of atomic formulas in ϕ. Thus the computing time bound of DECOMP, multiplied by a, is a bound for EVAL. Finally, consider step (1) of ELIM, which extracts from ϕ the set $\mathcal{A}$ of distinct polynomials occurring in ϕ. This involves, at most, ma polynomial comparisons, one comparison for each atomic formula with each element of the list of distinct polynomials already extracted. The time for each comparison is at most $(n+1)^r d \leq (2n)^{2^r} d$. We therefore obtain our final result:

Theorem 17 *The computing time of* ELIM *is dominated by*

$$(2n)^{2^{2r+8}} m^{2^{r+6}} d^3 a.$$

As a corollary of this analysis, we can obtain a computing time bound for ELIM as a function only of the length N of the formula ϕ. Obviously we must have m, r, d and a less than or equal to N and, assuming as is usual that x^k must be expressed as the product of k separate x's, we also have $n \leq N$. Since $r \geq 1$ and $N \geq 3$ for every formula, we have $2r + 8 \leq 5N$ and $r + 6 \leq 3N$. Hence by Theorem 17, the computing time is dominated by $(2N)^{2^{5N}} N^{2^{3N}} N^4$. But it is easy to prove by induction on h that $h^{2^k} \leq 2^{2^{h+k}}$. Hence $(2N)^{2^{5N}} N^{2^{3N}} N^4 \leq 2^{2^{7N}} 2^{2^{4N}} 2^{2^{2N}} \leq 2^{2^{8N}}$.

Theorem 18 *For a formula of length N, the computing time of* ELIM *is dominated by* $2^{2^{8N}}$.

In Theorem 17, d is the maximum norm length of the polynomials in ϕ, whereas in the Introduction the result of Theorem 17 was stated with coefficient d defined as the maximum coefficient length. But if d' is the maximum coefficient length then, since a polynomial in r variables with degrees bounded by n has at most $(n+1)^r$ integer coefficients, $d \leq L(n+1)^r + d' \leq rL(2n) + d' \leq 2rL(n) + d' \leq 2rn + d' \leq 2rnd'$, and $d^3 \leq (2n)^3 r^3 d'^3 \leq (2n)^{2^2} 2^{2^r} d'^3 \leq (2n)^{2^{2r+1}} d'^3$. Thus as a corollary of Theorem 17, the computing time of ELIM is dominated by $(2n)^{2^{2r+9}} m^{2^{r+6}} d'^3 a$. But in fact the exponent $2r + 9$ can be replaced by $2r + 8$ by converting from d to d' while carrying out the proof of Theorem 17.

5 Observations

For the sake of conceptual simplicity, and to facilitate the analysis, we have kept our quantifier elimination algorithm as simple as possible. But for practical

application many refinements and improvements are possible, some of which
will now be described.

It is unnecessary to form reducta when projecting a set of polynomials in
two variables. The leading coefficient of a non-zero bivariate polynomial A is
a non-zero univariate polynomial, which vanishes at only a finite number of
points. If the leading coefficient of A is invariant on a connected set $S \subseteq R$
then either S is a one-point set, on which the roots of A are trivially delineable,
or else $\mathrm{ldcf}(A) \neq 0$ on S.

We recall (van der Waerden 1970, chapter 4) that the discriminant of a
polynomial A, denoted by $\mathrm{discr}(A)$, satisfies $\mathrm{discr}(A) = \mathrm{res}(A, A')/\mathrm{ldcf}(A)$. If
A is a squarefree bivariate polynomial, then $\mathrm{discr}(A)$ is also a nonzero univariate
polynomial, which vanishes at only a finite number of points. Thus it is easy
to prove (confer Theorem 4) that if $A(x_1, x_2)$ is a squarefree polynomial, $\mathcal{L} =
\{\mathrm{ldcf}(A) : \deg(A) \geq 1\}$, $\mathcal{D} = \{\mathrm{discr}(A) : \deg(A) \geq 2\}$, S is a connected subset
of R, and every element of $\mathcal{L} \cup \mathcal{D}$ (there are at most two elements) is invariant
on S, then the roots of A are delineable on S.

If now $\mathcal{A}$ is a set of bivariate polynomials and $\mathcal{B}$ is a squarefree basis for
$\mathcal{A}$ then any two distinct elements of $\mathcal{B}$ have a non-zero resultant, which again
vanishes at only a finite number of points. Thus we see that in this case we can
define $\mathrm{proj}(\mathcal{A}) = \mathcal{P} = \mathcal{L} \cup \mathcal{D} \cup \mathcal{R}$ where $\mathcal{L} = \{\mathrm{ldcf}(B) : B \in \mathcal{B} \,\&\, \deg(B) \geq 1\}$,
$\mathcal{D} = \{\mathrm{discr}(B) : B \in \mathcal{B} \,\&\, \deg(B) \geq 2\}$ and $\mathcal{R} = \{\mathrm{res}(B_1, B_2) : B_1, B_2 \in
\mathcal{B} \,\&\, B_1 < B_2 \,\&\, \deg(B_1) \geq 1 \,\&\, \deg(B_2) \geq 1\}$, and Theorem 5 will still hold.

For the augmented projection of a set of bivariate polynomials, we must
be cautious; although A is squarefree, some derivative of A may fail to be
squarefree. But if A is any polynomial, $\mathcal{B}$ is a squarefree basis for $\{A\}$, and
the roots of $\mathcal{B}$ are delineable on a connected set S, then the roots of A are
delineable on S. So, with $\mathrm{proj}(\mathcal{A})$ defined as in the preceding paragraph, we
can define the augmented projection of $\mathcal{A}$ as follows. Let $\mathcal{D} = \{\mathrm{der}^j(A) : A \in
\mathcal{A} \,\&\, 1 \leq j \leq \deg(A) - 2\} = \{D_1, \ldots, D_k\}$. Let $\mathcal{B}_i$ be a squarefree basis for D_i
and $\mathcal{P}_i = \mathrm{proj}(\mathcal{B}_i)$. Then $\mathrm{proj}(\mathcal{A}) = \bigcup\{\bigcup_{i=1}^{k} \mathcal{P}_i\}$ will suffice for the augmented
projection of $\mathcal{A}$. Actually, this still gives us a little more than necessary; the
leading coefficients of elements of the $\mathcal{B}_i$ are superfluous. One might suppose
that these basis calculations would be time-consuming, but in most cases one
will quickly discover that D_i is squarefree so that $\mathcal{B}_i = \{D_i\}$ and for $\mathcal{P}_i$ we can
then use just $\{\mathrm{discr}(D_i)\}$.

Now let $\mathcal{A}$ be a set of polynomials in three variables. According to our
earlier definition, we begin the projection by forming the set $\mathcal{B}$ of all reducta
of elements of $\mathcal{A}$ such that the degree of the reductum is positive. In general,
this set $\mathcal{B}$ is much larger than necessary. The objective is just to ensure that
for each $A \in \mathcal{A}$, $\mathcal{B}$ contains some reductum of A whose leading coefficient is
invariantly non-zero on each cell S of the induced c.a.d. Indeed even this is
unnecessary for one-point cells S, because then the roots of A are trivially
delineable on S. So if the first i coefficients are simultaneously zero at only a
finite number of points of $\mathbb{R}^2$, then $\mathrm{red}^k(A)$ can be excluded from $\mathcal{B}$ for $k \geq i$.

Also, if the leading coefficient of A has no zeros in $\mathbb{R}^2$, as when its degree in both x_1 and x_2 is zero, then $\operatorname{red}^k(A)$ can be excluded from $\mathcal{B}$ for $k \geq 1$.

For the case $i = 2$, let $A_1 = \operatorname{ldcf}(A)$, $A_2 = \operatorname{ldcf}(\operatorname{red}(A))$. If $A_1(\alpha_1, \alpha_2) = A_2(\alpha_1, \alpha_2) = 0$ then $R_1(\alpha_1) = 0$ where $R_1(x_1)$ is the resultant of $A_1(x_1, x_2)$ and $A_2(x_1, x_2)$ with respect to x_2. Hence if $R_1 \neq 0$ there are only a finite number of α_1's. Similarly, if R_2 is the resultant with respect to x_1 and if $R_2 \neq 0$ then there are only finitely many α_2's. If $R_1 \neq 0$ and $R_2 \neq 0$ then there are only finitely many points (α_1, α_2).

Of course if either A_1 or A_2 has degree zero in either x_1 or x_2, then the resultants R_1 and R_2 cannot both be formed, but then there are alternatives. If A_1 is of degree zero in x_2 and A_2 is of degree zero in x_1 (or vice versa) then there are only finitely many solutions. Suppose A_1 is of degree zero in x_2 and A_2 is positive degree in both x_1 and x_2, say $A_2(x_1, x_2) = \sum_{i=0}^{n} A_{2,i}(x_1) \cdot x_2^i$. Then there are only finitely many solutions if $\gcd(A_1, A_{2,n}, \ldots, A_{2,1})$ is of degree zero. Also, there are only finitely many solutions if A_1 and A_2 are both of degree zero in x_2 (or x_1) and the degree of $\gcd(A_1, A_2)$ in x_1 (respectively x_2) is zero.

If the cases $i = 1$ and $i = 2$ fail, then we may try $i = 3$. Let $A_3 = \operatorname{ldcf}(\operatorname{red}^2(A))$. Then it suffices to show, as above, that A_1 and A_3, or A_2 and A_3, have only finitely many common solutions. Better still, we may compute $\operatorname{res}(\gcd(A_1, A_2), A_3)$.

There are obvious reasons to expect that in most cases it will be unnecessary to include in $\mathcal{B}$ $\operatorname{red}^k(A)$ for $k \geq 2$ when $\mathcal{A}$ is a set of polynomials in three variables. The reader can easily see for himself how to extend these methods to polynomials in $r \geq 4$ variables. For example, when $r = 4$ we can compute $\operatorname{res}(A_1, A_2)$, $\operatorname{res}(A_2, A_3)$ with respect to x_1, x_2 and x_3. It will usually be unnecessary to include in $\mathcal{B}$ $\operatorname{red}^k(A)$ for $k \geq r - 1$.

By a similar argument, for a given $B \in \mathcal{B}$, it is in general unnecessary to include in the set $\mathcal{S}_1$ $\operatorname{psc}_k(B, B')$ for all k such that $0 \leq k < \deg(B')$. If we can show that the equations $\operatorname{psc}_i(B, B') = 0$, for $0 \leq i \leq k$, have only a finite number of solutions then $\operatorname{psc}_i(B, B')$ may be omitted from $\mathcal{S}_1$ for $i > k$.

Thus, if A is a polynomial in r variables, it will usually suffice to include in $\mathcal{S}_1$ $\operatorname{psc}_j(\operatorname{red}^i(A), \operatorname{der}(\operatorname{red}^i(A)))$ only for $0 \leq i \leq r - 2$ and $0 \leq j \leq r - 2$, a total of $(r - 1)^2$ polynomials . One can even do better than this. For example, if the equations $\operatorname{ldcf}(A) = 0$ and $\operatorname{psc}_0(\operatorname{red}(A), \operatorname{der}(\operatorname{red}(A))) = 0$ have only finitely many common solutions, then we need not include $\operatorname{psc}_1(\operatorname{red}(A), \operatorname{der}(\operatorname{red}^i(A)))$. Thus we will usually need to include $\operatorname{psc}_j(\operatorname{red}^i(A), \operatorname{der}(\operatorname{red}^i(A)))$ only for $0 \leq i + j \leq r - 2$, a total of $\binom{r}{2}$ polynomials. Similar considerations apply to the set $\mathcal{S}_2$ of the projection and to the set of $\mathcal{P}'$ of the augmented projection. Whereas we derived an upper bound of $m^2 n^3$ for the number of polynomials in the augmented projection, the expected number will now be about $m^2 r^2$ when n is larger than r.

It is important to realize that the resultants used in testing for finitely many solutions need not be computed; it is only necessary to decide whether or not they are zero. If, in fact, a resultant is non-zero then this can usually be ascertained very quickly by computing the resultant modulo a large single-precision prime and modulo linear polynomials $x_i - a_i$.

As an experiment, we have computed the two successive projections of a set $\mathcal{A} = \{A_1(x_1, x_2, x_3), A_2(x_1, x_2, x_3)\}$ where A_1 and A_2 are polynomials of total degree two with random integers from the interval $[-9, +9]$ as coefficients. Each A_i is then a polynomial with 10 terms, of degree 2 in x_3, with constant leading coefficient. $\mathcal{A}_1$, the projection of $\mathcal{A}$, is then a set consisting of four elements, $B_1 = \mathrm{discr}(A_1)$, $B_2 = \mathrm{discr}(A_2)$, $B_3 = \mathrm{res}(A_1, A_2)$ and $B_4 = \mathrm{psc}_1(A_1, A_2)$. The leading coefficient of each B_i has degree zero in x_1 so the content of B_i is an integer. Dividing B_i by its content, we obtained the primitive part $\bar{B}_i$. We then found that each $\bar{B}_i$ was an irreducible polynomial in $\mathbb{Z}[x_1, x_2]$ so $\mathcal{B}_1 = \{\bar{B}_1, \bar{B}_2, \bar{B}_3, \bar{B}_4\}$ was a finest basis for $\mathcal{A}_1$. $\bar{B}_1$ and $\bar{B}_2$ are of degree 2 in each variable and have total degree 2. Their integer coefficients are two or three decimal digits in length. $\bar{B}_3$ is of degree 4, in each x_i separately and in total with 4-digit coefficients. $\bar{B}_4$ has degree 1 and 2-digit coefficients.

$\mathcal{A}_2$, the projection of $\mathcal{B}_1$, has as its elements $D_i = \mathrm{discr}(\bar{B}_i)$ for $i = 1, 2, 3$ and $R_{i,j} = \mathrm{res}(\bar{B}_i, \bar{B}_j)$ for $1 \leq i < j \leq 4$. Again we set $\bar{D}_i = \mathrm{pp}(D_i)$ (the primitive part of D_i) and $\bar{R}_{i,j} = \mathrm{pp}(R_{i,j})$. The contents of the D_i and $R_{i,j}$ are from 1 to 5 decimal digits in length. The largest of these primitive parts is $\bar{D}_3$, with degree 12 and coefficients about 22 to 24 decimal digits in length. Now we factor each $\bar{D}_i$ and each $\bar{R}_{i,j}$ to obtain a finest basis, $\mathcal{B}_2$, for $\mathcal{A}_2$. The results are as follows:

$$\bar{D}_1, \bar{D}_2, \bar{R}_{1,2}, \bar{R}_{1,4}, \bar{R}_{2,4} \quad \text{irreducible},$$

$$\bar{D}_3 = P_1 P_2^2,$$

$$\bar{R}_{3,4} = P_2^2,$$

$$\bar{R}_{1,3} = P_4^2,$$

$$\bar{R}_{2,3} = P_5^2. \tag{27}$$

Here each P_i is irreducible. Note that P_2 is a common factor $\bar{D}_3$ and $\bar{R}_{3,4}$.

Since random polynomials are almost always irreducible, the factorizations (27) strongly suggest some theorems. To help preclude chance events, the entire experiment was repeated using polynomials A_1 and A_2 with different random coefficients. With the new A_1 and A_2, the structure of the factorization (27) was exactly repeated; even the degrees of the irreducible factors remained the same.

Let us consider briefly what kinds of theorems are suggested by (27). Ignoring primitive parts, $\bar{D}_3 = \mathrm{discr}(\mathrm{res}(A_1, A_2))$, $R_{1,3} = \mathrm{res}(\mathrm{discr}(A_1), \mathrm{res}(A_1, A_2))$, $\bar{R}_{2,3} = \mathrm{res}(\mathrm{discr}(A_2), \mathrm{res}(A_1, A_2))$ and $\bar{R}_{3,4} = \mathrm{res}(\mathrm{psc}_0(A_1, A_2), \mathrm{psc}_1(A_1, A_2))$. Note that in each case the general form is a resultant of two psc's with "common ancestors". On the other hand, there are no "common ancestors" in the irreducible cases $\bar{D}_i = \mathrm{discr}(\mathrm{discr}(A_i))$ and $\bar{R}_{1,2} = \mathrm{res}(\mathrm{discr}(A_1), \mathrm{discr}(A_2))$. The cases $\bar{R}_{i,4} = \mathrm{res}(\mathrm{discr}(A_i), \mathrm{psc}_1(A_1, A_2))$ appear anomalous.

A number of other experiments have been performed which tend to substantiate the rather vague conclusions suggested by the above results. For example, we find that, "in general", $res(res(A_1, A_2), res(A_2, A_3))$ is reducible. These observations suggest very strongly the merit of performing basis calculations preceding each projection. However, it is not entirely clear whether a finest or coarsest squarefree basis should be used. Note that in the above example the two coincide.

We have seen now how to reduce substantially the number of polynomials which arise when the projections are performed, and we have seen empirical evidence, though not yet theorems, which indicates that the growth of degrees can be controlled somewhat by basis calculations. Also, the factorizations which reduce degrees tend to reduce coefficient lengths correspondingly. Hence there is some reason for optimism regarding the potential applicability of this method. A full implementation of the method within the SAC-1 computer algebra system (Collins 1973a) is still in progress. Nearly all of the necessary algebraic subalgorithms are already available, and some parts of the elimination algorithm itself exist in preliminary form. The completion of the implementation and its application to several non-trivial "real" problems within the next year or two is anticipated.

Ferrante and Rackoff (1975) have recently published a quantifier eliminating method for the first order theory of the additive ordered group of the real numbers, which they show to have a computing time dominated by $2^{2^{c^N}}$ for some unknown constant c. We obtain an alternative to their method as the special case of the method above in which every polynomial occurring in the formula ϕ has total degree 1. Setting $n = 1$ in Theorem 18, we obtain their result with $c = 8$. Their computing time bound is obtained as a function N, the length of ϕ, only. Setting $n = 1$ in Theorem 17, we obtain the more informative bound $2^{2^{2^{r+8}}} m^{2^{r+6}} d^3 a$.

We can easily improve this result. It is easy to see that in the special case we are considering, we may define $proj(\mathcal{A}) = \{res(A_i, A_j) : 1 \le i < j \le m\}$ where $\mathcal{A} = \{A_1, \ldots, A_m\}$. Thus $proj(\mathcal{A})$ has at most $\binom{m}{2} \le m^2/2$ members, and the augmented projection is never needed. With m_k and d_k defined as before, we then easily obtain

$$m_k \le 2(m/2)^{2^{k-1}}, \tag{28}$$
$$d_k \le 2^{k-1}d. \tag{29}$$

Instead of isolating roots, since they are rational, we now compute them exactly. If $r_1 < r_2 < \cdots < r_l$ are all the roots of a set of polynomials, then for the sample points s_i between roots we use the averages $(r_i + r_{i+1})/2$. If $r_1 < 0$ then we also use the sample point $s_0 = 2r_1 < r_1$; if $r \ge 0$ then we use instead the sample point $s_0 = -1$. Similarly we use $s_l = 2r_l$ or $+1$ as a greatest sample point. The rational number $r = a/b$ is represented by the linear integral polynomial $bx - a$ with norm $|a| + |b|$; we may call this the norm, $|r|_1$, of r. Note that $|s_i|_1 \le 2|r_i|_1 \cdot |r_{i+1}|_1$, $|s_0|_1 \le 2|r_1|_1$ and $|s_{l+1}|_1 \le 2|r_l|_1$.

The result of substituting $r = a/b$ for x_1 in a polynomial $C(x_1, \ldots, x_k) = c_1x_1 + \ldots + c_kx_k + c_0$ and then multiplying by b to obtain an integral polynomial is the same as the resultant of $bx_1 - a$ and $C(x_1, \ldots, x_k)$ with respect to x_1.

If c_k is the number of cells as before, then we have $c_1 \leq 2m_r + 1$ and $c_{k+1} \leq 2c_km_{r-k} < +1$, from which it follows that

$$c_k \quad \leq \quad 2^{k+1}(m/2)^{2^r}. \tag{30}$$

Let d_k' be the maximum norm length of the k-th coordinate of any sample point. Then $d_1' \leq 2d_r + 1$ and $d_{k+1}' \leq 2(d_{r-k} + d_1' + d_2' + \cdots + d_k') + 1$. It follows that

$$d_k' \quad \leq \quad 2^{r+k}d. \tag{31}$$

Using the bounds (28) to (31), it is not difficult to show that the computing time of the method is dominated by $m^{2^r}d^2a$ using classical arithmetic algorithms, $m^{2^r}d^{1+\epsilon}a$ using fast arithmetic algorithms.

Acknowledgments

I am indebted to numerous persons in various ways in connection with this paper. First to Frank Beckman, who first made it possible for me to pursue my interest in Tarski's method at IBM Corp. in 1955. Secondly, the late Abraham Robinson encouraged my interests in this subject at various times in the 1960s, as did also C. Elgot and D. A. Quarles, Jr. I am perhaps most indebted to R. Loos who, during several weeks at Stanford University in early 1973, through his keen appreciation and insightful discussions, persuaded me to persist to the final discovery of this method. Most recently I have become greatly indebted to H. Brakhage, who made it possible for me to spend a year at the University of Kaiserslautern in unencumbered pursuit of the further development of this method.

This research was partially supported by National Science Foundation grants GJ-30125X and DCR74-13278.

Super-Exponential Complexity of Presburger Arithmetic[1]

Michael J. Fischer and Michael O. Rabin

Lower bounds are established on the computational complexity of the decision problem and on the inherent lengths of proofs for two classical decidable theories of logic: the first-order theory of the real numbers under addition, and Presburger arithmetic — the first-order theory of addition on the natural numbers. There is a fixed constant $c > 0$ such that for every (nondeterministic) decision procedure for determining the truth of sentences of real addition and for all sufficiently large n, there is a sentence of length n for which the decision procedure runs for more than 2^{cn} steps. In the case of Presburger arithmetic, the corresponding bound is $2^{2^{cn}}$. These bounds apply also to the minimal lengths of proofs for any complete axiomatization in which the axioms are easily recognized.

1 Introduction and Main Theorems

We present some results obtained in the Fall of 1972 on the computational complexity of the decision problem for certain theories of addition. In particular we prove the following results.

Let L be the set of formulas of the first-order functional (predicate) calculus written using just $+$ and $=$. Thus, for example, $\sim[x+y = y+z] \vee x+x = x$ is a formula of L, and $\forall x \exists y[x+y = y]$ is a sentence of L. Even though this is not essential, we shall sometimes permit the use of the individual constants 0 and 1 in writing formulas of L. We assume a finite alphabet for expressing formulas of L, so a variable in general is not a single atomic symbol but is encoded by a sequence of basic symbols.

Let $\mathcal{N} = \langle \mathbb{N}, + \rangle$ be the structure consisting of the set $\mathbb{N} = \{0, 1, 2, \ldots\}$ of natural numbers with operation $+$ of addition. Let $\mathrm{Th}(\mathcal{N})$ be the first-order

---

[1]Reprinted from *SIAM-AMS Proceedings*, Volume VII, 1974, pp. 27–41, by permission of the American Mathematical Society.

theory of $\mathcal{N}$, i.e., the set of all sentences of L which are true in $\mathcal{N}$. For example, $\forall x \forall y[x + y = y + x]$ is in $\mathrm{Th}(\mathcal{N})$. Presburger (1930) has shown that $\mathrm{Th}(\mathcal{N})$ is decidable. For brevity's sake, we shall call $\mathrm{Th}(\mathcal{N})$ *Presburger arithmetic* and denote it by **PA**.

Theorem 1 *There exists a constant $c > 0$ such that for every decision procedure (algorithm)* **AL** *for* **PA**, *there exists an integer n_0 so that for every $n > n_0$ there exists a sentence F of L of length n for which* **AL** *requires more than $2^{2^{cn}}$ computational steps to decide whether $F \in$* **PA**.

The previous theorem applies also in the case of nondeterministic algorithms. This implies that not only algorithms require a super-exponential number of computational steps, but also proofs of true statements concerning addition of natural numbers are super-exponentially long. Let **AX** be a system of axioms in the language L (or in an extension of L) such that a sentence $F \in L$ is provable from **AX** (**AX** $\vdash F$) if and only if $F \in$ **PA**. Let **AX** satisfy the condition that to decide for a sentence F whether $F \in$ **AX**, i.e., whether F is an axiom, requires a number of computational steps which is polynomial in the length $|F|$ of F.

Theorem 2 *There exists a constant $c > 0$ so that for every axiomatization* **AX** *of Presburger arithmetic with the above properties there exists an integer n_0 so that for every $n > n_0$ there exists a sentence $F \in$* **PA** *such that the shortest proof of F from the axioms* **AX** *is longer than $2^{2^{cn}}$. By the length of a proof we mean the number of its symbols.*

With slight modifications, Theorem 2 holds for any (consistent) system **AX** of axioms in a language M in which the notion of integer and the operation $+$ on integers are definable by appropriate formulas so that under this interpretation, all the sentences of **PA** are provable from **AX**. The ordinary axioms **ZF** for set theory have this property.

The result concerning super-exponential length of proof applies, in this more general case, to the sentences of M which are encodings of sentences of **PA** under the interpretation, i.e., to sentences which express elementary properties of addition of natural numbers.

The previous results necessarily involve a cut-point $n_0(\mathbf{AL})$ or $n_0(\mathbf{AX})$ at which the super-exponential length of computation or proofs sets in. It is significant that a close examination of our proofs reveals that $n_0(\mathbf{AL}) = O(|\mathbf{AL}|)$ and $n_0(\mathbf{AX}) = O(|\mathbf{AX}|)$. Thus computations and proofs become very long quite early in the game.

The theory **PA** of addition of natural numbers is one of the simplest most basic imaginable mathematical theories. Unlike the theory of addition and multiplication of natural numbers **PA** is decidable. Yet any decision procedure for **PA** is inherently difficult.

Let us now consider the structure $\mathcal{R} = \langle \mathbb{R}, + \rangle$ of all *real* numbers $\mathbb{R}$ with addition. The theory $\mathrm{Th}(\mathcal{R})$ (in the same language L) is also decidable. In fact, to find a decision procedure for $\text{Th}(\mathcal{R})$ is even simpler than a procedure for **PA**; this is mainly because $\mathcal{R}$ is a divisible group without torsion. Yet the following holds.

Theorem 3 *There exists a constant $d > 0$ so that for the theory $\text{Th}(\mathcal{R})$ of addition of real numbers, the statement of Theorem 1 holds with the lower bound 2^{dn}.*

Similarly for the length of proofs of sentences in $\text{Th}(\mathcal{R})$.

Theorem 4 *There exists a constant $d > 0$ so that for every axiomatization* **AX** *for $\text{Th}(\mathcal{R})$ the statement of Theorem 2 holds with the lower bound 2^{dn}.*

Corollary 1 *The theory of addition and multiplication of reals (Tarski's algebra (Tarski 1951)) is exponentially complex in the sense of Theorems 3 and 4.*[1]

Ferrante and Rackoff (1973) strengthen results of Oppen (1973) to obtain decision procedures for $\text{Th}(\mathcal{R})$ and **PA** which run in deterministic space 2^{cn} and $2^{2^{dn}}$ (and hence in deterministic time $O(2^{2^{cn}})$ and $O(2^{2^{2^{dn}}})$) respectively, for certain constants c and d. That time $2^{2^{cn}}$ is sufficient even for Tarski's algebra has been announced by Collins (1974b) and also by Solovay (private communication in 1974) who extends a result of Monk (1974). Any substantial improvement in our lower bounds would settle some open questions on the relation between time and space. For example, a lower bound of time 2^{n^2} for the decision problem for $\text{Th}(\mathcal{R})$ would give an example of a problem solvable in space $S(n) = 2^{cn}$ but not in time bounded by a polynomial in $S(n)$ (cf. Stockmeyer 1974).

Variations of the methods employed in the proofs of Theorems 1–4 lead to complexity results for the (decidable) theories of multiplication of natural numbers, finite Abelian groups, and other classes of Abelian groups. Some of these results are stated in Section 7 and will be presented in full in a subsequent paper.

The fact that decision and proof procedures for such simple theories are exponentially complex is of significance to the program of theorem proving by machine on the one hand, and to the more general issue of what is knowable in mathematics on the other hand.

2   Algorithms

Since we intend to prove results concerning the complexity of algorithms, we must say what notion of algorithm we use. Actually our methods of proof and our results are strong enough to apply to any reasonable class of algorithms

---

[1]This result was obtained independently by V. Strassen.

or computing machines. However, for the sake of definiteness, we shall assume throughout this paper that our algorithms are the programs for Turing machines on the alphabet $\{0, 1\}$.

We proceed to give an informal description of these algorithms. The machine-tape is assumed to be one-way infinite extending to the right from an initial leftmost square. At any given time during the progress of a computation, all but a finite number of the squares of the tape contain 0. An *instruction* has the form:

"i: If 0 then print X_0, move M_0, go to one of $i_1, i_2, \cdots$;
if 1 then print X_1, move M_1, go to one of $j_1, j_2, \cdots,$."

Here $i, i_1, i_2, \ldots, j_1, j_2, \ldots$ are natural numbers, the so-called *instruction numbers*; X_0 and X_1 are either 0 or 1; and M_0 and M_1 are either R or L (for "move right" and "move left," respectively).

The possibility of going to one of several alternative instructions embodies the nondeterministic character of our algorithms. Another type of instruction is:

"i: Stop."

Instructions are abbreviated by dropping the verbal parts. Thus, "3: 0,1,L,72,5; 1,1,R,15,3." is an example of an instruction. A program **AL** is a sequence $I_1, \ldots, I_n$ of instructions. For the sake of definiteness we assume that the instruction number of I_i is i and that I_n is the instruction "n: Stop." Furthermore **AL** is assumed to be coded in the binary alphabet $\{0, 1\}$ in such a way that "Stop" also serves as an end-word indicating the end of the binary word **AL**.

Let $x \in \{0, 1\}^*$ be an input word. To describe the possible computations by the algorithm **AL** on x, we assume that x is placed in the left-most positions of the machine's tape and the scanning head is positioned on the left-most square of the tape. The computation starts with the first instruction I_1. A *halting computation* on x is a sequence $C = (I_{i_1}, \ldots, I_{i_m})$ of instructions of **AL** so that $i_1 = 1$ and $i_m = n$. At each step $1 \leq p \leq m$, the motion of the scanning head, the printing on the scanned square, and the transfer to the next instruction $I_{i_{p+1}}$ are according to the current instruction I_{i_p}. The length $l(C)$ of C is, by definition, m.

It is clear that a truly nondeterministic program may have several possible computations on a given input x.

3 Method for Complexity Proofs

Having settled on a definite notion of algorithm, we shall describe a general method for establishing lower bounds for theories of addition which are formalized in L. We do not develop our methods of proof in their fullest generality but rather utilize the fact that we deal with natural or real numbers to present the proofs in a more readily understandable and concrete form. The refinements and generalizations which are needed for other theories of addition will be introduced in a subsequent paper.

Theorem 5 *Let $f(n)$ be one of the two functions 2^n or 2^{2^n}. Assume for a complete theory T that there exists a polynomial $p(n)$ and a constant $d > 0$ so that for every program $\mathbf{AL}$ and binary word x, there exists a sentence $F_{\mathbf{AL},x}$ with the following properties:*

(a) *$F_{\mathbf{AL},x} \in T$ if and only if some halting computation C of $\mathbf{AL}$ on x satisfies $l(C) \leq f(|x|)$.*
(b) *$|F_{\mathbf{AL},x}| \leq d \cdot (|\mathbf{AL}| + |x|)$.*
(c) *$\sim F_{\mathbf{AL},x}$ is Turing machine calculable from $\mathbf{AL}$ and x in time less than $p(|\mathbf{AL}| + |x|)$.*

(We recall that all our objects such as F, $\mathbf{AL}$, etc. are binary words, and that $|w|$ denotes the length of w.)

Under these conditions, there exists a constant $c > 0$ so that for every decision algorithm $\mathbf{AL}$ for T there exists a number $n_0 = n_0(\mathbf{AL})$ so that for every $n > n_0$ there exists a sentence $\sigma \in T$ such that $|\sigma| = n$ and every computation by $\mathbf{AL}$ for deciding σ takes more than $f(cn)$ steps. Furthermore $n_0(\mathbf{AL}) = O(|\mathbf{AL}|)$.

Proof. There exists a number $c > 0$ and an m_0 so that for $m \geq m_0$ we have

$$p(2m) + f(c \cdot (2dm + 1)) \leq f(m). \tag{1}$$

Namely, let $c < 1/(2d)$ and recall that $p(n)$ is a polynomial, whereas $f(n)$ is 2^n or 2^{2^n}.

Let $\mathbf{AL}$ be a (nondeterministic) decision algorithm for T. We construct a new algorithm $\mathbf{AL}_0$ as follows. We do not care how $\mathbf{AL}_0$ behaves on an input word x which is not a program. If x is a program, then $\mathbf{AL}_0$ starts by constructing the sentence $F = \sim F_{x,x}$. The program $\mathbf{AL}_0$ then switches to $\mathbf{AL}$ which works on the input F. If $\mathbf{AL}$ stops on F and determines that $F \in T$, then $\mathbf{AL}_0$ halts; in all other cases $\mathbf{AL}_0$ does not halt. Thus, for a program x as input, $\mathbf{AL}_0$ halts if and only if the program x does not halt on the input x in fewer than $f(|x|)$ steps. Note that by possibly padding $\mathbf{AL}_0$ with irrelevant instructions, we may assume that $m_0 \leq |\mathbf{AL}_0| \leq |\mathbf{AL}_0| + k$, where k is independent of $\mathbf{AL}$.

Denote the binary word $\mathbf{AL}_0$ by z and let σ be the sentence $\sim F_{z,z}$. $F_{z,z}$ cannot be true, for if it were true, then $\sim F_{z,z}$ would be false and $\mathbf{AL}_0$ would not halt on z, whereas the truth of $F_{z,z}$ implies that $z(= \mathbf{AL}_0)$ does halt on the input z (even in at most $f(|z|)$ steps), a contradiction.

Thus, σ is true and hence $\mathbf{AL}_0(= z)$ halts on z. The truth of σ also implies that every halting computation of $\mathbf{AL}_0$ on z is longer than $f(|z|)$.

Let $m = |z|$. By (b), we have

$$n = |\sigma| \leq 2dm + 1. \tag{2}$$

Let t be the least number of steps that **AL** takes, by some halting computation, to decide σ. By the definition of $\mathbf{AL_0}$ and the fact that fewer then $p(2m)$ steps are required to find $\sigma = {\sim}F_{z,z}$ from z (this follows from (c) and $|z| = m$), there is a halting computation of the program $\mathbf{AL_0}$ on z requiring fewer than $p(2m) + t$ steps. By the truth of σ, $p(2m) + t > f(m)$. Using (1) and (2), $t > f(c \cdot (2dm + 1)) \geq f(cn)$.

Take n_0 to be $n = |\sigma|$. Then $n_0 \leq 2dm + 1 \leq 2d(|\mathbf{AL}| + k) + 1$, so $n_0 = O(|\mathbf{AL}|)$. The fact that the result holds for **AL** and every $n > n_0$ (with possibly a smaller constant c) is obtained by first padding $\mathbf{AL_0}$ by irrelevant instructions, and then padding the resulting σ by prefixing a quantifier $\exists x_j$ of an appropriate length, where $|\exists x_j| = 1 + |j|$. The details are left to the reader. ∎

For utilizing Theorem 5 we need a method for constructing sentences $F_{\mathbf{AL},w}$ with properties (a)–(c). One such method is provided by

Theorem 6 *Let $\mathcal{A} = \langle A, + \rangle$ be an additive structure such that $\mathbb{N} \subseteq A$, and on $\mathbb{N}$ the operation $+$ is ordinary addition. Let $f(n)$ again be one of the functions 2^n or 2^{2^n}. Assume that $T = \mathrm{Th}(\mathcal{A})$ is a theory of addition (formalized in the language L) for which there exists $c > 0$ such that, for every n and for every binary word w, $|w| = n$, there exist formulas $I_n(y)$, $J_n(y)$, $S_n(x, y)$ and $H_w(x)$ with the following properties:*

(i) $|S_n(x, y)| \leq cn$, $|I_n(y)| \leq cn$, $|J_n(y)| \leq cn$, *and* $|H_w(x)| \leq cn$.

(ii) $I_n(b)$ *is true in $\mathcal{A}$ for $b \in A$ if and only if $b \in \mathbb{N}$ and $b < f(n)^2$. $J_n(b)$ is true exactly for $b = f(n)$.*

(iii) S_n *codes all binary sequences of length $f(n)^2$. Namely, for every binary sequence $\beta \in \{0, 1\}^*$, $|\beta| = f(n)^2$, there exists an $\alpha \in A$ so that, for $i \in \mathbb{N}$, $0 \leq i < f(n)^2$, $S_n(\alpha, i)$ is true in $\mathcal{A}$ if $\beta(i) = 1$, and $S_n(\alpha, i)$ is false if $\beta(i) = 0$, where, for any sequence β, $\beta(i)$ denotes the $(i+1)$-st element of β, $0 \leq i < |\beta|$.*

(iv) $H_w(x)$ *is true for $\alpha \in A$ if and only if the first $f(n)$ symbols of the sequence coded by α in the sense of (iii) have the form $w0^p$, $p = f(n) - |w|$.*

(v) $S_n(x, y)$, $I_n(y)$, $J_n(y)$ *and $H_w(x)$ are Turing machine calculable from n and w in a polynomial number of steps.*

From such formulas S_n, I_n, J_n and H_w, a formula $F_{\mathbf{AL},w}$ with the properties (a)–(c) can be constructed, so that T satisfies the conclusion of Theorem 5.

Proof. We shall describe, by use of sequences of length $f(n)^2$, all possible halting computations of length at most $f(n)$ of a program **AL** on an input w. Let $C = (I_{i_1}, \ldots, I_{i_m})$ be such a computation. Assume that **AL** has k instructions; by our notational conventions every computation starts with the first instruction I_1 and the last instruction I_k of **AL** is "k: Stop." Thus, in C, $i_1 = 1$ and $i_m = k$.

Let us adopt the convention that after the stop instruction, the scanning head, the (stop) instruction, and the tape contents stay stationary and unchanged at all subsequent time instants. Since $m \leq f(n)$, the scanning head never moves beyond $f(n)$ squares from the initial left-most square of the tape. We assume also that the Turing machine never attempts to shift its head left off of the beginning of the tape.

The progress of the computation C on the input w will be described by stringing together $f(n)$ instantaneous descriptions of the computation in the following manner. Let W_j be the first (left-most) $f(n)$ symbols of the tape at time j, $1 \leq j \leq f(n)$. The the strings $W_1 W_2 \cdots W_{f(n)} = W \in \{0,1\}^*$ codes all the relevant information concerning the tape contents during the computation of C. We have $|W| = f(n)^2$. Also, $W_m = W_{m+1} = \cdots$.

To trace the motion of the scanning head and the sequence of instructions during the computation C, we define $U_j \in \{0, 1, \ldots, k\}^*$ to be $0^{p_j} i_j 0^{q_j}$ where $p_j + q_j + 1 = f(n)$ and p_j is the distance at time j of the scanning head from the start square, $1 \leq j \leq f(n)$. Recall that i_j is the instruction number of the j-th instruction executed in C. Also $i_m = i_{m+1} = \cdots = k$, the stop instruction. Put $U = U_1 U_2 \cdots U_{f(n)}$. We have $|U| = f(n)^2$.

The fact that the pair (W, U), where $W \in \{0,1\}^*$, $U \in \{0, 1, \ldots, k\}^*$, $|W| = |U| = f(n)^2$, describes a halting computation of **AL** on w, is equivalent to a number of statements which say, roughly, that the first $f(n)$ symbols are the initial configuration, that the transformation from a block of $f(n)$ symbols to the next block is by an instruction of **AL**, and that U contains k (the number of the halting instruction). More precisely, (W, U) codes a halting computation of length at most $f(n)$ of **AL** on w, where $|w| = n$, if and only if the following hold:

(α) $W(0) \cdots W(f(n) - 1) = w0^p$, $p = f(n) - |w|$.

(β) $U(0) \cdots U(f(n) - 1) = 10^{f(n)-1}$.

(γ) If $U(i) = 0$ and $i + f(n) < f(n)^2$, then $W(i + f(n)) = W(i)$.

(δ) If $U(i) = q$, $i + f(n) + 1 < f(n)^2$, $0 < q < k$, $W(i) = 0$, and I_q is, say, "$q : 0, 1, R, k_1, \ldots, k_t; 1, \ldots$," then $W(i+f(n)) = 1$, $U(i+f(n)+1) = k_1$ or $U(i+f(n)+1) = k_2$ or etc. (similarly for other instruction and tape-symbol combinations).

(ϵ) If, for $f(n) < i < f(n)^2$, $U(i) \neq 0$, then exactly one of $U(i - f(n)) \neq 0$, or $U(i - f(n) - 1) \neq 0$, or $U(i - f(n) + 1) \neq 0$ holds. Also, if $U(i) \neq 0$, then $U(i \pm 1) = U(i \pm 2) = 0$ (if they are defined).

(ζ) $U(i) = k$ for some i, $0 \leq i < f(n)^2$. If $U(i) = k$ and $i + f(n) < f(n)^2$, then $U(i + f(n)) = k$ and $W(i + f(n)) = W(i)$.

From the assumption that (W, U) satisfies (α)–(ζ), it can be proved by induction on $1 \leq j < f(n)$ that (W_{j+1}, U_{j+1}) is an instantaneous description which follows from (W_j, U_j) by an application of the instruction I_{i_j} whose number appears in U_j. Also, $(W_{f(n)}, U_{f(n)})$ is a halting instantaneous description.

Thus, the existence of a pair (W, U), $W \in \{0,1\}^*$, $U \in \{0, 1, \ldots, k\}^*$, $|W| = |U| = f(n)^2$, which satisfies (α)–(ζ) is a necessary and sufficient condition for the existence of a halting computation C on w with $l(C) \le f(n)$.

Conditions (i)–(v) provide means for making statements about arbitrary $(0, 1)$ sequences of length $f(n)^2$, about integers $0 \le i < f(n)^2$, and about the integer $f(n)$, all by use of formulas of L of size $O(n)$. Also, the ordinary ordering $\le$ on $\mathbb{N}$ restricted to integers of size less than $f(n)^2$ can be expressed by the length $O(n)$ formula

$$x \le_n y \leftrightarrow \exists z [I_n(x) \wedge I_n(y) \wedge I_n(z) \wedge x + z = y].$$

Hence, the existence of (W, U) satisfying (α)–(ζ) can be expressed by a sentence $F_{\mathbf{AL},w} = F$ with the desired properties (a)–(c). Namely, express $0, 1, \ldots, k$ in binary notation by words of equal length $p = |k|$. Then, via $S_n(x, y)$, a single element $a \in A$ exists which codes W, and elements $a_1, \ldots, a_p \in A$ code U. The sentence F will start with quantifiers and relativization:

$$F = \exists x \exists x_1 \cdots \exists x_p \forall y \forall z \left[I_n(y) \wedge J_n(z) \to E_\alpha \wedge E_\beta \wedge E_\gamma \wedge E_\delta \wedge E_\epsilon \wedge E_\zeta \right].$$

x codes the sequence W and $x_1, \ldots, x_p$ together code the sequence U. The clauses $E_\alpha, \ldots, E_\zeta$ express the corresponding conditions (α)–(ζ). Thus, for example, E_α is $H_w(x)$; E_β is $H_{u_1}(x_1) \wedge H_{u_2}(x_2) \wedge \cdots \wedge H_{u_p}(x_p)$, where $u_1 = 10^{n-1}$ and $u_j = 0^n$, $2 \le j \le p$; and E_γ is

$$[\sim S_n(x_1, y) \wedge \cdots \wedge \sim S_n(x_p, y) \wedge I_n(y + z) \to [S_n(x, y + z) \leftrightarrow S_n(x, y)]].$$

The reader can supply the details of the construction of the remaining expressions E_δ, E_ϵ, and E_ζ and verify that, altogether, the $F_{\mathbf{AL},x}$ thus formed satisfies (a)–(c) of Theorem 5. ∎

4 Proof of Theorem 3 (Real Addition)

We start by showing that for the theory $\mathrm{Th}(\mathcal{R})$ of real addition, there exist formulas $S_n(x, y)$, $I_n(y)$, etc. as postulated in Theorem 6 with $f(n) = 2^n$, thereby proving Theorem 3. Several of the results in this section will play a role later in the proof for **PA**.

Let $F(x, y)$ be any formula and consider the conjunction

$$G = F(x_1, y_1) \wedge F(x_2, y_2) \wedge F(x_3, y_3).$$

It is readily seen that $G \leftrightarrow G_1$ where

$$\begin{aligned} G_1 = \ & \forall x \forall y \left[((x = x_1 \wedge y = y_1) \vee (x = x_2 \wedge y = y_2) \vee \right. \\ & \left. (x = x_3) \wedge (y = y_3)) \to F(x, y) \right]. \end{aligned}$$

Note that $|G| = 3 \cdot |F(x, y)|$, whereas $|G_1| = |F(x, y)| + c$, where c is independent of $F(x, y)$. A similar rewriting exists for formulas F with more than two variables and for conjunctions of more than three instances of F. The above device, discovered independently by several people including V. Strassen, is a special case of a more general theorem due to M. Fischer and A. Meyer.

Theorem 7 *There exists a constant $c > 0$ so that for every n there is a formula $M_n(x, y, z)$ of L such that, for real numbers A, B, C,*

$$M_n(A, B, C) \text{ is true } \leftrightarrow A \in \mathbb{N} \wedge A < 2^{2^n} \wedge AB = C.$$

Also, $|M_n(x, y, z)| \leq c(n+1)$ and $M_n(x, y, z)$ is Turing machine computable from n in time polynomial in n.

Proof. The construction of $M_n(x, y, z)$ will be inductive on n. For $n = 0$ we have $2^{2^0} = 2$ and we define $M_0(x, y, z)$ as $[x = 0 \wedge z = 0] \vee [x = 1 \wedge z = y]$.

From M_k we get M_{k+1} by observing that $x \in \mathbb{N}$ and $x < 2^{2^{k+1}}$ if and only if there exist $x_1, x_2, x_3, x_4 \in \mathbb{N}$ all less than 2^{2^k} so that $x = x_1 x_2 + x_3 + x_4$. For this decomposition we have $z = xy = x_1(x_2 y) + x_3 y + x_4 y$. Hence, $M_{k+1}(x, y, z)$ is equivalent to

$$\exists u_1 u_2 \cdots u_5 x_1 \cdots x_4 \, [M_k(x_1, x_2, u_1) \wedge M_k(x_2, y, u_2) \wedge M_k(x_1, u_2, u_3)$$
$$\wedge M_k(x_3, y, u_4) \wedge M_k(x_4, y, u_5)$$
$$\wedge x = u_1 + x_3 + x_4 \wedge z = u_3 + u_4 + u_5].$$

(Strictly speaking, a triple sum such as $u_1 + x_3 + x_4$ should be written as a chain of sums of two variables, but we shall not do it here.) Now, $|M_{k+1}| \geq 5|M_k|$, which will not do. However, by using the device preceding the theorem, the five occurrences of M_k can be replaced by a single occurrence to yield M_{k+1}. Thus, $|M_{k+1}(x, y, z)| \leq |M_k(x, y, z)| + c$ for an appropriate $c > 0$. Hence, $|M_n(x, y, z)| \leq c(n+1)$. (We assume c is chosen large enough so $c \geq |M_0(x, y, z)|$.)

Actually, for the above bound to hold, it is necessary to show that the number of distinct variable names in M_n does not grow with n, for to encode one of v variables requires (on the average) a string of length $(\log v)$. In fact, 15 different variable names are sufficient to express M_n. This is because the new variables introduced in constructing M_{k+1} from M_k need only be distinct from each other and from the *free* variables of M_k; however, no difficulty arises if they coincide with variables *bound* inside M_k. A close look at the construction of M_{k+1} shows that 12 new variables are introduced, which must be distinct from the three free variables of M_k, giving a total of 15 distinct names needed.

∎

Corollary 2 *The formula $M_n(x, 0, 0)$ is true for a real number x if and only if $x \in \mathbb{N}$ and $x < 2^{2^n}$.*

The natural numbers $x < 2^{2^n}$ code all binary sequences of length 2^n. Namely write x in binary notation

$$x = x(0) + x(1) \cdot 2 + \cdots + x(2^n - 1) \cdot 2^{2^n - 1}.$$

We use the function 2^i to obtain the element $x(i)$ of x.

Theorem 8 *There exists a formula* $\mathrm{Pow}_n(x, y, z)$ *such that, for integers a, b, c for which $0 \le a, b^a, c < 2^{2^n}$, $\mathrm{Pow}_n(a, b, c)$ is true if and only if $b^a = c$. Also, $|\mathrm{Pow}_n(x, y, z)| \le d(n + 1)$ for an appropriate $d > 0$ and all n.*

Proof. Construct, by induction on k, a sequence, $E_k(x, y, z, u, v, w)$ of formulas with the property that for integers a, b, c for which $0 \le a < 2^{2^k}$, $0 \le b^a, c < 2^{2^n}$ and real numbers A, B, C, $E_k(a, b, c, A, B, C)$ is true in $\langle \mathbb{R}, + \rangle$ if and only if $A \in \mathbb{N}$, $A < 2^{2^n}$, $b^a = c$, and $AB = C$. Thus, E_k has M_n built into it since

$$E_k(0, 1, 1, A, B, C) \leftrightarrow M_n(A, B, C).$$

The case $k = 0$ is given by

$$[(x = 0 \wedge z = 1) \vee (x = 1 \wedge z = y)] \wedge M_n(u, v, w).$$

To obtain $E_{k+1}(x, y, z, u, v, w)$ from E_k, we again use the decomposition $x = x_1 x_2 + x_3 + x_4$ of every integer $0 \le x < 2^{2^{k+1}}$ in terms of integers $0 \le x_1, x_2, x_3, x_4 < 2^{2^k}$. Then we have $y^x = (y^{x_1})^{x_2} \cdot y^{x_3} \cdot y^{x_4}$. Now, y^{x_1} is expressed by a z_1 such that $E_k(x_1, y, z_1, 0, 0, 0)$; then $(y^{x_1})^{x_2}$ is a z_2 such that $E_k(x_2, z_1, z_2, 0, 0, 0)$, etc. Whenever we have to write a product such as $x_1 x_2$ or $(y^{x_1})^{x_2} \cdot y^{x_3}$, we use the formula $E_k(0, 1, 1, u, v, w)$. In this way we can write the formula $E_{k+1}(x, y, z, u, v, w)$. Using the usual device of contracting a conjunction of instances of E_k into one occurrence, we see that $|E_{k+1}| \le |E_k| + d$ for some $d > 0$, and hence $|E_n| \le d(n + 1) + c(n + 1)$, where $c(n + 1)$ is the bound on the length of M_n. As before, only a bounded number of variable names are needed.

Recalling the definition of $E_k(x, y, z, u, v, w)$, we see that

$$\mathrm{Pow}_n(x, y, z) \leftrightarrow E_n(x, y, z, 0, 0, 0)$$

has the desired properties. ∎

Theorem 9 *There exists a formula $S_n(x, y)$ of L which for $x, y \in \mathbb{R}$ is true in $\langle \mathbb{R}, + \rangle$ if and only if x and y are integers, $x < 2^{2^{2n}}$ and $y < 2^{2n}$, and the $(y+1)$-st digit $x(y)$ of x, counting from the low-order end of the binary representation of x, is 1. The formula $S_n(x, y)$ satisfies the conditions of Theorem 6 for $f(n) = 2^n$.*

Proof. That x and y are integers in the appropriate ranges is easily expressible by formulas of size $O(n)$. Recall that for the integers which satisfy $M_{2n}(x, 0, 0)$, i.e., $0 \le x < 2^{2^{2n}}$, the ordering $\le$ is expressible by a formula of length $O(n)$.

Now $x(y) = 1$ if and only if there exists an integer z, $2^y \le z < 2^{y+1}$ so that $x \ge z$ and 2^{y+1} divides $x - z$. This fact is easily expressible by a formula $S_n(x, y)$ of L using Pow_{2n} and M_{2n}. ∎

That formulas $I_n(y)$ and $J_n(y)$ with the properties listed in Theorem 6 exist is immediate. Thus to finish the proof of Theorem 3 we need the following.

Theorem 10 *For every binary word w, $|w| = n$, there exists a formula $H_w(x)$ of L which is true in $\langle \mathbb{R}, + \rangle$ for an integer $0 \le x < 2^{2^{2n}}$ if and only if $x(0) \cdots x(2^n - 1) = w0^p$, $p = 2^n - n$. The formula $H_w(x)$ satisfies the conditions of Theorem 6.*

Proof. Define for binary words u, by induction on $|u|$, formulas $K_u(z)$ as follows.

$$
\begin{aligned}
K_0(z) &\leftrightarrow z = 0, \\
K_1(z) &\leftrightarrow z = 1, \\
K_{u0}(z) &\leftrightarrow \exists y \, [K_u(y) \wedge z = y + y], \\
K_{u1}(z) &\leftrightarrow \exists y \, [K_u(y) \wedge z = y + y + 1].
\end{aligned}
$$

Clearly, if $K_w(z)$ is true, then, considered as a sequence, z satisfies $w(i) = z(i)$ for $0 \le i < |w|$, $z(i) = 0$ for $i \ge |w|$. Using this $K_w(z)$ and the formulas $S_n(x, y)$ and $J_n(y)$, we can write the formula $H_w(x)$ by formally expressing the statement that for z such that $K_w(z)$, $x(i) = z(i)$, $0 \le i < 2^n$. ∎

Thus we have proved, for $\text{Th}(\mathcal{R})$, the existence of formulas $S_n(x, y)$, $I_n(y)$, $J_n(y)$, and $H_w(x)$ which satisfy the conditions of Theorem 6 for $f(n) = 2^n$. This completes the proof of Theorem 3.

5   Proof of Theorem 4 (Lengths of Proofs for Real Addition)

We now show that for $\text{Th}(\mathcal{R})$ proofs are also exponentially long. This is an easy consequence of Theorem 3.

Let **AX** be a consistent system of axioms which is complete for $\text{Th}(\mathcal{R})$, i.e., every sentence $F \in \text{Th}(\mathcal{R})$ is provable from **AX** ($\mathbf{AX} \vdash F$). Furthermore, there exists an algorithm B which decides in polynomial time $p(|G|)$ for a sentence G of L whether $G \in \mathbf{AX}$.

Let $\dot{c}$ be the constant of Theorem 3. For every polynomial $q(x)$, there exists a constant $0 < d$ so that from a certain point on, $q(2^{dn}) < 2^{cn}$.

Construct a nondeterministic algorithm **AL** for $\text{Th}(\mathcal{R})$ as follows. Given a sentence F, **AL** writes down (nondeterministically) a binary sequence P. Then

AL checks whether P is a proof of F from **AX** or a proof of $\sim F$ from **AX**. The computation halts only if one of the two possibilities occurs. Because of the assumptions on **AX**, this check can be made in a polynomial number of steps $h(|P|)$. Thus the whole computation, if it halts, requires $|P| + h(|P|) = q(|P|)$ steps. If every true sentence F would have a proof P with $|P| < 2^{dn}$ where $n = |F|$, then for every such F there would be some halting computation of length less than $q(2^{dn})$, i.e., also less then 2^{cn} for all sufficiently large n, a contradiction.

6   Proof of Theorems 1 and 2 (Presburger Arithmetic)

The proof for Theorem 1 follows closely along the lines of the proof of Theorem 3 and utilizes our previous results. In particular we note that Theorems 7–8 apply, as they stand with the same proofs, to **PA**. Note also that the order $\leq$ on $\mathbb{N}$ is definable in **PA** using $+$. Throughout this section, let $f(n)$ be 2^{2^n}.

Theorem 11 *There exists a function* $g(n) \geq 2^{f(n)^2} = 2^{2^{2^{n+1}}}$ *so that for every* n *there exists a formula* $\mathrm{Prod}_n(x, y, z)$ *with the following properties. For integers* A, B, C,

$$\mathrm{Prod}_n(A, B, C) \text{ is true in } \mathcal{N} \leftrightarrow A, B, C < g(n) \text{ and } AB = C.$$

There exists a constant $c > 0$ *so that* $|\mathrm{Prod}_n| \leq c(n+1)$ *for all* n. *The formula* Prod_n *is Turing machine constructible from* n *in time polynomial in* n.

Proof. We shall use the Prime Number Theorem which says that the number of primes smaller than m is asymptotically equal to $m/\log_e(m)$; hence bigger than $m/\log_2(m)$ for all sufficiently large m. Thus, for $m = 2^{2^{n+2}}$, the number of primes $p < m$ exceeds $2^{2^{n+2}}/2^{n+2} > 2^{2^{n+1}} = f(n)^2$. Let $g(n) = \prod_{p<m} p$, where p runs over primes, $m = 2^{2^{n+2}}$; then $g(n) \geq 2^{f(n)^2}$ since $2 \leq p$ for all primes.

By use of the formula $M_{n+2}(x, y, z)$, we can write two formulas $\mathrm{Res}_{n+2}(x, y, z)$ and $P_{n+2}(x)$ of length $O(n)$ with the following meanings. Let $\mathrm{res}(x, y)$ denote the residue (remainder) of x when divided by y. Then

$$\mathrm{Res}_{n+2}(x, y, z) \quad \leftrightarrow \quad \left[y < 2^{2^{n+2}} \wedge \mathrm{res}(x, y) = z \right],$$

$$P_{n+2}(x) \quad \leftrightarrow \quad \left[x < 2^{2^{n+2}} \text{ and } x \text{ is prime} \right].$$

The formula Res_{n+2} is written in L as

$$z < y \wedge \exists q \exists w \left[M_{n+2}(y, q, w) \wedge x = w + z \right].$$

We recall that, for any q and w, $M_{n+2}(y, q, w)$ holds if and only if $y < 2^{2^{n+2}}$ and $yq = w$.

The formula $P_{n+2}(x)$ is simply,

$$M_{n+2}(x,0,0) \land \forall y \forall z \left[M_{n+2}(y,z,x) \to [y=1 \lor y=x]\right].$$

By formally saying that $x \geq 1$ is the smallest integer divisible by all primes $p < 2^{2^{n+2}}$, we can write a formula $G_{n+2}(x)$ which is true precisely for $x = g(n)$. Now $\mathrm{Prod}_n(x,y,z)$ is true if and only if

$$x,y,z < g(n) \land \forall u \left[u < 2^{2^{n+2}} \to \mathrm{res}(x,u) \cdot \mathrm{res}(y,u) = \mathrm{res}(z,u)\right]. \tag{3}$$

Namely, this implies that $xy = z \pmod{p}$ for all $p < 2^{2^{n+2}}$, which together with $x,y,z < g(n)$ is equivalent to $xy = z$. Now, by use of $G_{n+2}(x)$, $M_{n+2}(x,y,z)$ and $\mathrm{Res}_{n+2}(x,y,z)$, the above relation (3) can be expressed by a formula Prod_n with the desired properties. ∎

Exponentiation can be defined just as in the proof of Theorem 8 except that we now use $\mathrm{Prod}_n(x,y,z)$ instead of $M_n(x,y,z)$ to obtain a sequence of formulas $E'_k(x,y,z,u,v,w)$. For integers a, b, c, A, B, C for which $0 \leq a < 2^{2^k}$ and $0 \leq b^a, c < g(n)$, $E'_k(a,b,c,A,B,C)$ is true in $\mathcal{N}$ if and only if $A,B,C < g(n)$, $b^a = c$, and $AB = C$. Also $|E'_k| = O(n)$.

Having now multiplication up to $g(n)$ and exponentiation 2^i up to $i < 2^{2^{n+1}}$ expressed by formulas of length $O(n)$, we can code sequences of length $2^{2^{n+1}} = f(n)^2$ in exactly the same manner as in Section 4. This completes the proof of Theorem 1 by again appealing to Theorem 6.

The proof of Theorem 2 now follows exactly the lines of the proof of Theorem 4 given in Section 5.

7   Other Results

The techniques presented in this paper for proving lower bounds on logical theories may be extended in a number of directions to yield several other results. We outline some of them below without proof; they will be presented in full in a subsequent paper.

Theorem 12 *Let $\mathcal{U}$ be any class of additive structures, so if $\mathcal{A} = \langle A, + \rangle \in \mathcal{U}$, then $+$ is a binary associative operation on A. Let $\mathrm{Th}(\mathcal{U})$ be the set of sentences of L valid in every structure of $\mathcal{U}$. Assume $\mathcal{U}$ has the property that, for every $k \in \mathbb{N}$, there is a structure $\mathcal{A}_k = \langle A_k, + \rangle \in \mathcal{U}$ and an element $u \in A_k$ such that the elements $u, u+u, u+u+u, \ldots, k \cdot u$ are distinct. Then the statement of Theorem 1 holds for $\mathrm{Th}(\mathcal{U})$ with the lower bound 2^{dn} for some $d > 0$.*

Theorem 3 is an immediate corollary of this result, taking $\mathcal{U}$ to be the class of just one structure $\mathcal{R} = \langle \mathbb{R}, + \rangle$. Some other classes to which the result applies are the following:

1. the complex numbers under addition,

2. finite cyclic groups,
3. rings of characteristic p,
4. finite Abelian groups,
5. the natural numbers under multiplication.

The proof of Theorem 12 extends the ideas of Section 4. The element $n \cdot u$ is used as the representation of the integer n, and u itself is selected by existential quantification.

Special properties of certain theories permit us to obtain still larger lower bounds on the decision problem. For example, we get a lower bound of time $2^{2^{cn}}$ for (4), the theory of finite Abelian groups. This is obtained by encoding integer up to 2^{2^n} by formulas of length $O(n)$ just as in Theorem 12, but instead of representing a sequence by an integer, we let the structure itself encode the sequence. Let G be a finite Abelian group. Then the element $s(i)$ of the sequence s encoded by G is 1 if and only if G contains an element x of order p_i, where p_i is the $(i+1)$-st prime. The necessity of using primes as indices instead of integers considerably complicates the analog of Theorem 6.

Another example, where we get still larger bounds is (5), the theory of multiplication of the natural numbers (**MULT**). That **MULT** is at least as hard as **PA** is immediate, for the powers of 2 under multiplication are isomorphic to $\mathcal{N}$, and the property of being a power of 2 can be expressed in **MULT** (assuming we have the constant 2; otherwise we use an arbitrary prime). In fact, the bound can be increased yet another exponential to time $2^{2^{2^{cn}}}$ by using the encoding which associates a sequence s to a positive integer m, where $s(i) = 1$ if and only if q_i divides m, where q_i is the $(i+1)$-st prime in some fixed (but arbitrary) ordering of the primes. Again we are forced to use the primes as indices, and again the analog of Theorem 6 is considerably complicated. Rackoff (1974) shows a corresponding upper bound of deterministic space $2^{2^{2^{dn}}}$.

Acknowledgments

The first example of exponential and larger lower bounds on the complexity of logical theories and certain word problems from the theory of automata were obtained by A. R. Meyer (1973) and L. J. Stockmeyer (1974), and we gratefully acknowledge the influence of their ideas and techniques on our work. We are also indebted to C. Rackoff, R. Solovay, and V. Strassen for several helpful ideas and suggestions which led to and are incorporated in the present paper.

This research was supported in part by the National Science Foundation under research grant GJ–34671 to MIT Project MAC, and in part by the Artificial Intelligence Laboratory, an MIT research program sponsored by the Advanced Research Projects Agency, Department of Defense, under Office of Naval Research contract number N00014–70–A–0362–0003. The preparation of the manuscript was supported by the Hebrew University and the University of Toronto.

Cylindrical Algebraic Decomposition I: The Basic Algorithm[1]

Dennis S. Arnon, George E. Collins, and Scott McCallum

1   Introduction

Given a set of r-variate integral polynomials, a *cylindrical algebraic decomposition (cad)* of euclidean r-space E^r partitions E^r into connected subsets compatible with the zeros of the polynomials. By "compatible with the zeros of the polynomials" we mean that on each subset of E^r, each of the polynomials either vanishes everywhere or nowhere. For example, consider the bivariate polynomial

$$y^4 - 2y^3 + y^2 - 3x^2y + 2x^4.$$

Its zeros comprise the curve shown in Fig. 1.

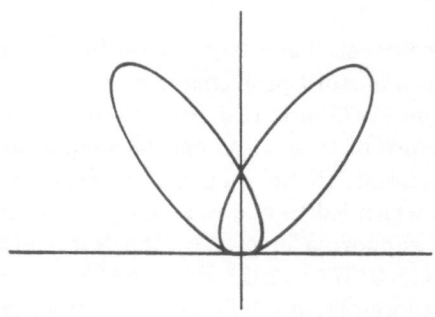

Figure 1

Figure 2 shows a cad of the plane compatible with its zeros. The cad consists of the distinct "dots," "arcs," and "patches of white space" of the figure (a rigorous definition of cad is given in Section 2).

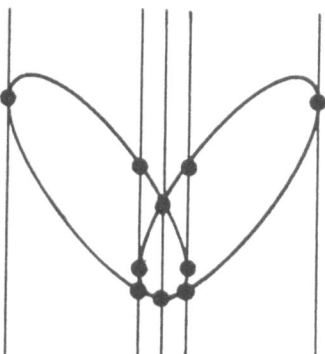

Figure 2

Cad's were introduced by Collins in 1973 (see Collins 1975, 1976) as part of a new quantifier elimination, and hence decision, method for elementary algebra and geometry. He gave an algorithm for cad construction, and proved that for any fixed number of variables, its computing time is a polynomial function of the remaining parameters of input size. As can be seen in the example above, cad's are closely related to the classical simplicial and CW-complexes of algebraic topology. In fact, the essential strategy of Collins' cad algorithm, induction on dimension, can be found in van der Waerden's 1929 argument (van der Waerden 1929, pp. 360–361) that real algebraic varieties are triangulable.

Collins' cad-based decision procedure for elementary algebra and geometry is the best known (see Ferrante and Rackoff 1979; very little besides a cad is needed for the decision procedure). J. Schwartz and M. Sharir (1983a, 1983b) used the cad algorithm to solve a motion planning problem. D. Lankford (1979) and N. Dershowitz (1979) pointed out that a decision procedure for elementary algebra and geometry could be used to test the termination of term-rewriting systems. P. Kahn (1979) used cad's to solve a problem on rigid frameworks in algebraic topology. Kahn (private communication in 1978) also observed that a cad algorithm provides a basis for a constructive proof that real algebraic varieties are triangulable, and thus for computing the homology groups of a real algebraic variety.

Implementation of Collins' cad algorithm began soon after its introduction, culminating in the first complete program in 1981 (Arnon 1981). The program has begun to find use; in May, 1982 the termination of the term-rewriting system for group theory in the Appendix of (Huet and Oppen 1980) was verified using it. It has also been utilized for display of algebraic curves (Arnon 1983).

In 1977, Müller (1978) implemented certain subalgorithms of the cad algorithm and used them to solve algebraic optimization problems.

We use a somewhat different (but equivalent) definition of cad than that in (Collins 1975); we devote Section 2 to it. We then take up the cad algorithm. Its intuitive strategy can be described by means of an example. Consider the curve of Figures 1 and 2. Given the bivariate polynomial which defines it, we will compute univariate polynomials whose roots constitute a "silhouette" of the curve. By this we mean that the roots of the univariate polynomials are the projections, onto the x-axis (E^1), of the "significant points" of the curve. The curve's "significant points" are its singularities (e.g., self-crossings, cusps, isolated points), and the points at which its tangent is vertical. Suppose that E^1 is decomposed into the points of the silhouette, and their complementary open intervals (this is done by finding the roots of the univariate polynomials). Then the portion of the curve "over" each of these points (intervals) consists of finitely many disjoint "dots" ("arcs"). Our cad of the plane is made by decomposing the line (strip) in the plane "over" each point (interval) in E^1 into the "dots" ("arcs") of the curve, and the "arcs" ("patches") of the complement of the curve, that it contains.

For our sample curve, we compute a single univariate polynomial (its discriminant):

$$2048x^{12} - 4608x^{10} + 37x^8 + 12x^6.$$

This polynomial has five roots, whose approximate values are -1.49, -0.23, 0.0, 0.23, and 1.49. All roots but the third are projections of points with vertical tangent. The third is the projection of the two singularities (self-crossings). Using the roots, we decompose the real line into points and open intervals (Fig. 3).

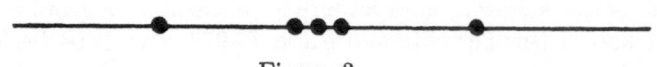

Figure 3

The Cartesian products of each of the eleven elements of this decomposition with a line, give us eleven vertical lines and strips. As we see in Fig. 2, each "significant point" of the curve lies on one of the vertical lines, and within each strip, the curve has finitely many disjoint "arcs." The "dots" and "arcs" which make up each line, and the "arcs" and "patches of white space" which make up each strip, give us the cad of Fig. 2.

The general algorithm consists of three phases: projection (computing successive sets of polynomials in $r - 1$, $r - 2, \ldots, 1$ variables; the zeros of each set contain a "silhouette" of the "significant points" of the zeros in the next higher dimensional space), base (constructing a decomposition of E^1), and extension (successive extension of the decomposition of E^1 to a decomposition of E^2, E^2 to $E^3, \ldots, E^{r-1}$ to E^r). In Sections 3, 4, and 5 we describe each of these phases in turn. In the interests of succinctness, we will at various times specify simple but inefficient methods of performing computations (for example, isolating the roots of a product of polynomials, rather than isolating the roots of each of the factors separately). In Section 6, we give a detailed example of the algorithm.

2   Definition of Cylindrical Algebraic Decomposition

Connectivity plays an important role in the theory of cad's. It is convenient to have a term for a nonempty connected subset of E^r; we will call such sets *regions*. For a region R, the *cylinder over R*, written $Z(R)$, is $R \times E^1$. A *section* of $Z(R)$ is a set s of points $\langle a, f(a) \rangle$, where a ranges over R, and f is a continuous, real-valued function on R. s, in other words, is the graph of f. We say such an s is the f-*section* of $Z(R)$. A *sector* of $Z(R)$ is a set $\hat{s}$ of all points $\langle \alpha, b \rangle$, where α ranges over R and $f_1(\alpha) < b < f_2(\alpha)$ for (continuous, real-valued) functions $f_1 < f_2$. The constant functions $f_1 = -\infty$, and $f_2 = +\infty$, are allowed. Such an $\hat{s}$ is the (f_1, f_2)-*sector* of $Z(R)$. Clearly sections and sectors of cylinders are regions. Note that if $r = 0$ and $R = E^0 = $ a point, then $Z(R) = E^1$, any point of E^1 is a section of $Z(R)$, and any open interval in E^1 is a sector of $Z(R)$.

For any subset X of E^r, a *decomposition* of X is a finite collection of disjoint regions whose union is X. Continuous, real-valued functions $f_1 < f_2 < \cdots < f_k, k \geq 0$, defined on R, naturally determine a decomposition of $Z(R)$ consisting of the following regions: (1) the (f_i, f_{i+1})-sectors of $Z(R)$ for $0 \leq i \leq k$, where $f_0 = -\infty$ and $f_{k+1} = +\infty$, and (2) the f_i-sections of $Z(R)$ for $1 \leq i \leq k$. We call such a decomposition a *stack over R*(determined by $f_1, \ldots, f_k$).

A decomposition D of E^r is *cylindrical* if either (1) $r = 1$ and D is a stack over E^0, or (2) $r > 1$, and there is a cylindrical decomposition D' of E^{r-1} such that for each region R of D', some subset of D is a stack over R. It is clear that D' is unique for D, and thus associated with any cylindrical decomposition D of E^r are unique *induced* cylindrical decompositions of E^i for $i = r-1, r-2, \ldots, 1$. Conversely, given a cad D' of E^i, $i < r$, a cad D of E^r is an *extension* of D' if D induces D'.

For $0 \leq i \leq r$, an i-*cell* in E^r is a subset of E^r which is homeomorphic to E^i. It is not difficult to see that if c is an i-cell, then any section of $Z(c)$ is an i-cell, and any sector of $Z(c)$ is an $(i + 1)$-cell (these observations are due to P. Kahn). It follows by induction that every element of a cylindrical decomposition is an i-cell for some i. Also, if c is an i-cell, we say that $Z(c)$ is an $(i + 1)$-*cylinder*, and that any stack over c is an $(i + 1)$-*stack*.

The decomposition of E^2 in Fig. 2 is cylindrical. Figure 3 shows the induced decomposition of E^1, consisting of five 0-cells and six 1-cells. The decomposition in Fig. 2 consists of eleven stacks. The first, or leftmost, stack consists of a single 2-dimensional sector; the next stack consists of two 1-dimensional sectors and one 0-dimensional section; and so forth.

A subset of E^r is *semi-algebraic* if it can be constructed by finitely many applications of the union, intersection, and complementation operations, starting from sets of the form

$$\{x \in E^r \mid F(x) \geq 0\},$$

where F is an element of $\mathbb{Z}[x_1, \ldots, x_r]$, the ring of integral polynomials in r variables. We write I_r to denote $\mathbb{Z}[x_1, \ldots, x_r]$. As we shall now see, a different (but equivalent) definition of semi-algebraic sets is possible, from which one obtains a useful characterization of them. By a *formula* we will mean a well-formed formula of the first order theory of real closed fields. (The "first order theory of real closed fields" is a precise name for what we referred to above as "elementary algebra and geometry"; see Kreisel and Krivine 1967). The formulas of the theory of real closed fields involve elements of I_r. A *definable set* in E^k is a set X such that for some formula $\Psi(x_1, \ldots, x_k)$, X is the set of points in E^k satisfying Ψ. Ψ is a *defining formula* for X. (We follow the convention that $\Psi(x_1, \ldots, x_k)$ denotes a formula Ψ in which all occurrences of $x_1, \ldots, x_k$ are free, each x_i may or may not occur in Ψ, and no variables besides $x_1, \ldots, x_k$ occur free in Ψ.) A definable set is *semi-algebraic* if it has a defining formula which is quantifier-free. The existence of a quantifier elimination method for real closed fields was established by Tarski (1951). Hence a subset of E^r is semi-algebraic if and only if it is definable.

A decomposition is *algebraic* if each of its regions is a semi-algebraic set. A *cylindrical algebraic decomposition* of E^r is a decomposition which is both cylindrical and algebraic.

Let X be a subset of E^r, and let F be an element of I_r. F is *invariant* on X (and X is *F-invariant*), if one of the following three conditions holds:

(1) $F(\alpha) > 0$ for all α in X. ("F has positive sign on X.")
(2) $F(\alpha) = 0$ for all α in X. ("F has zero sign on X.")
(3) $F(\alpha) < 0$ for all α in X. ("F has negative sign on X.")

Let $A = \{A_1, \ldots, A_n\}$, be a subset of I_r ("subset of I_r" will always mean "finite subset"). X is *A-invariant* if each A_i is invariant on X. A collection of subsets of E^r is *A-invariant* if each element of the collection is.

The decomposition in Fig. 2 is an A-invariant cad of E^2 for $A = \{y^4 - 2y^3 + y^2 - 3x^2 y + 2x^4\}$. Note that a set $A \subset I_r$ does not uniquely determine an A-invariant cad D of E^r. Since any subset of an A-invariant region is also A-invariant, we can subdivide one or more regions of D to obtain another, "finer," A-invariant cad.

3   The Cylindrical Algebraic Decomposition Algorithm: Projection Phase

Let us begin with a more precise version of the cad algorithm outline at the end of Section 1. Let $A \subset I_r$ denote the set of input polynomials, and suppose $r \geq 2$. The algorithm begins by computing a set $\text{PROJ}(A) \subset I_{r-1}$ ("PROJ" stands for "projection"), such that for any $\text{PROJ}(A)$-invariant cad D' of E^{r-1}, there is an A-invariant cad D of E^r which induces D'. Then the algorithm calls itself recursively on $\mathrm{PROJ}(A)$ to get such a D'. Finally D' is extended to D. If $r = 1$, an A-invariant cad of E^1 is constructed directly.

Thus for $r \geq 2$, if we trace the algorithm, we see it compute $\mathrm{PROJ}(A)$, then $\mathrm{PROJ}(\mathrm{PROJ}(A)) = \mathrm{PROJ}^2(A)$, and so on, until $\mathrm{PROJ}^{r-1}(A)$ has been computed. This is the projection phase. The construction of a $\mathrm{PROJ}^{r-1}(A)$-invariant cad of E^1 is the base phase. The successive extensions of the cad of E^1 to a cad of E^2, the cad of E^2 to a cad of E^3, and so on, until an A-invariant cad of E^r is obtained, are the extension phase. For the example of Section 1, where $A = \{y^4 - 2y^3 + y^2 - 3x^2y + 2x^4\}$, $\mathrm{PROJ}(A) = \{2048x^{12} - 4608x^{10} + 37x^8 + 12x^6\}$.

The key to the projection phase is to define the map PROJ (which takes a subset of I_r to a subset of I_{r-1}), and to prove that it has the desired property. We stated this property above as: any $\mathrm{PROJ}(A)$-invariant cad of E^{r-1} is induced by some A-invariant cad of E^r. To establish this, clearly it suffices to show that over any semi-algebraic, $\mathrm{PROJ}(A)$-invariant region in E^{r-1}, there exists an A-invariant algebraic stack. In this section, we define PROJ and outline the proof that it has this latter property.

Central to our definition of PROJ will be the notion of delineability. For $F \in I_r, r \geq 1$, let $V(F)$ denote the real variety of F, i.e., the zero set of F. Let R be a region in E^{r-1}. F is *delineable* on R if the portion of $V(F)$ lying in $Z(R)$ consists of k disjoint sections of $Z(R)$, for some $k \geq 0$. Clearly when F is delineable on R, it gives rise to a stack over R, namely the stack determined by the continuous functions whose graphs make up $V(F) \cap Z(R)$. We write $S(F, R)$ to denote this stack, and speak of the F-*sections* of $Z(R)$. One easily sees that $S(F, R)$ is F-invariant.

For example, consider again $F(x, y) = y^4 - 2y^3 + y^2 - 3x^2y + 2x^4$. F is delineable on each of the eleven cells shown in Fig. 3, and in fact the stacks which comprise the cad of Fig. 2 are just the stacks determined by F over these eleven cells.

Our tentative strategy for defining PROJ is: insure that for any $\mathrm{PROJ}(A)$-invariant region R, the following two conditions hold: (1) each $A_i \in A$ is delineable on R, and (2) the sections of $Z(R)$ belonging to different A_i and A_j are either disjoint or identical. If these conditions are met, then clearly we have an A-invariant stack over R, namely the stack determined by the functions whose graphs are the sections of the A_i's.

The lefthand drawing in Fig. 4 illustrates a region R and hypothetical bivariate polynomials A_1, A_2, and A_3 for which these conditions do not hold. A_1 and A_2 are delineable on R, but the A_1-section meets the A_2-section. A_3 is not delineable on R. The righthand drawing in the figure illustrates a partition of R into five regions, on each of which the conditions are satisfied.

The following example points out a difficulty with our tentative strategy. Let $A \subset I_2$ be the set $\{A_1(x, y), A_2(x, y)\} = \{x, y^2 + x^2 - 1\}$. $A_1(0, y)$ is the zero polynomial, hence A_1 vanishes everywhere on $Z(\{0\})$, hence A_1 is not delineable on the set $\{0\} \subset E^1$, nor on any superset of it. We resolve this difficulty as follows. We say $F \in I_r$ is *identically zero* on $X \subset E^{r-1}$ if $F(\alpha, x_r)$

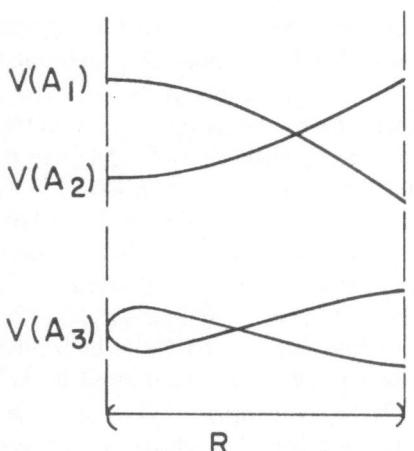
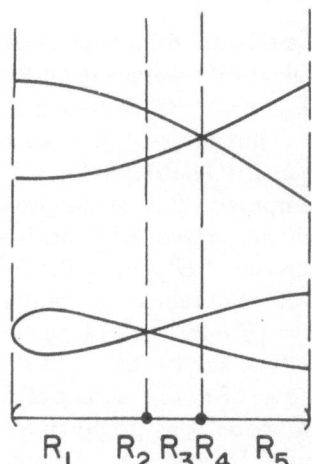

Figure 4

is the zero polynomial for every $\alpha \in X$. If F is identically zero on X, then any decomposition of $Z(X)$ will be F-invariant. Hence we may simply ignore F in decomposing $Z(X)$. In particular, in our example, we need only take account of the sections of A_2 in decomposing $Z(\{0\})$. Thus, we modify condition (1) above to read "(1') each $A_i \in A$ is either delineable or identically zero on R."

PROJ(A) will consist of two kinds of elements: those designed to attend to condition (1'), and those to attend to condition (2). Elements of both kinds are formed from the coefficients of the polynomials of A by addition, subtraction, and multiplication (remark: I_r consists of polynomials in x_r whose coefficients are elements of I_{r-1}). We now specify how this is done.

Let J be any unique factorization domain, and let F and G be nonzero elements of $J[x]$. We write $\deg(F)$ to denote the degree of F (the zero polynomial has degree $-\infty$). Let $n = \min(\deg(F), \deg(G))$. For $0 \leq j < n$, let $S_j(F, G)$ denote the j-th subresultant of F and G. $S_j(F, G)$ is an element of $J[x]$ of degree $\leq j$. (Each coefficient of $S_j(F, G)$ is the determinant of a certain matrix of F and G coefficients; see Loos 1982b, Brown and Traub 1971, or Collins 1975 for the exact definition.) For $0 \leq j < n$, the j-th *principal subresultant coefficient of F and G*, written $\mathrm{psc}_j(F, G)$, is the coefficient of x^j in $S_j(F, G)$. We define $\mathrm{psc}_n(F, G)$ to be $1 \in J$. Note that for $0 \leq j < n$, $\mathrm{psc}_j(F, G) = 0$ if and only if $\deg(S_j(F, G)) < j$.

The following theorem is the basis for definition of the first class of elements of PROJ(A). Some notation: suppose F is an element of I_r. The *derivative* of F, written F', is the partial derivative of F with respect to x_r. $\deg(F)$ is the degree of F in x_r. For $\alpha \in E^{r-1}$, we write $F_\alpha(x_r)$ or F_α to denote $F(\alpha, x_r)$.

Theorem 1 *Let $F \in I_r, r \geq 2$, and let R be a region in E^{r-1}. Suppose that $\deg(F_\alpha)$ is constant and nonnegative for $\alpha \in R$, and that if positive, then the* *least k such that $psc_k(F_\alpha, F'_\alpha) \neq 0$ is constant for $\alpha \in R$. Then F is delineable on R.*

A proof is given in (Arnon et al. 1982, theorem 3.6). The essential ideas are contained in the proof of theorem 4 of (Collins 1975). Collins (1975) uses a definition of delineability stronger than ours, but a polynomial delineable by that definition is delineable by ours.

Theorem 1 suggests that for a PROJ(A)-invariant region R, and for each $A_i \in A$, we should have $\deg((A_i)_\alpha)$ constant for $\alpha \in R$. This may be a nontrivial requirement. Suppose, for example, that $r = 3$ and A contains

$$F(x,y,z) = (y^2 + x^2 - 1)z^3 + (x-1)z^2 + (x-1)^2 + y^2.$$

If R is a region in the plane disjoint from the unit circle, then F_α has degree 3. If R is a subset of the unit circle which does not contain the point $\langle 1, 0 \rangle$, then F_α has degree 2. If R is the point $\langle 1, 0 \rangle$, then F_α is the zero polynomial. PROJ must separate these cases. Theorem 1 also suggests that for any PROJ(A)-invariant region R on which $\deg(F_\alpha)$ is constant and positive, we should insure that the least k such that $psc_k(F_\alpha, F'_\alpha) \neq 0$ is constant for $\alpha \in R$.

To achieve these goals, we introduce the notion of reductum of a polynomial. For any nonzero $F \in I_r = I_{r-1}[x_r]$, $ldcf(F)$ denotes the leading coefficient of F. The *leading term* of F, written $ldt(F)$, is

$$ldcf(F) \cdot x_r^{\deg(F)}.$$

The *reductum* of F, written $red(F)$, is $F - ldt(F)$. If $F = 0$, we define $red(F) = 0$. For any $k \geq 0$, the *k-th reductum of F*, written $red^k(F)$, is defined by induction on k:

$$red^0(F) = F.$$

$$red^{k+1}(F) = red(red^k(F))$$

For any $F \in I_r$, the *reducta set of F*, written REDI), is

$$\{red^k(F) | 0 \leq k \leq \deg(F) \ \& \ red^k(\ ) \neq 0\}.$$

Thus the reducta set of our sample $F(x,y,z)$ abov s

$$\{F(x,y,z), (x-1)z^2 + (x-1)^2 + y^2, \quad -1)^2 + y^2\}.$$

We now incorporate reducta into a specificatic of the (first) desired property of a PROJ(A)-invariant region R. For each $\in A$, there should exist an m such that $\deg(F_\alpha) = m$ for all $\alpha \in R$. Furthe iore, if m is positive, then where i is such that $\deg(red^i(F)) = m$, and $Q = 1$ $^i(F)$, the least k such that $psc_k(Q_\alpha, Q'_\alpha) \neq 0$ should be constant for $\alpha \in R$.

Let F and G be nonzero elements of $I_r[x]$. L $n = \min(\deg(F), \deg(G))$. The psc *set of F and G*, written PSC(F, G), is

$$\{psc_j(F,G) \mid 0 \leq j \leq n \ \& \ psc_j(\ G) \neq 0\}$$

If either $F = 0$ or $G = 0$, then $\mathrm{PSC}(F, G)$ is defined to be the empty set. Let $A = \{A_1, \ldots, A_n\}, n \geq 1$, be a set of polynomials in $I_r, r \geq 2$. $\mathrm{PROJ}_1(A) \subset I_{r-1}$, the first class of polynomials in $\mathrm{PROJ}(A)$, is defined as follows: For each i, $1 \leq i \leq n$, let $R_i = \mathrm{RED}(A_i)$. Then

$$\mathrm{PROJ}_1(A) = \bigcup_{i=1}^{n} \bigcup_{G_i \in R_i} (\{\mathrm{ldcf}(G_i)\} \cup \mathrm{PSC}(G_i, G_i')).$$

With the following simple observation, we can prove that PROJ_1 behaves as we want. Suppose F and G are nonzero elements of I_r, and suppose that for some $\alpha \in E^{r-1}$, $\deg(F) = \deg(F_\alpha) \geq 0$, and $\deg(G) = \deg(G_\alpha) \geq 0$. Let $n = \min(\deg(F), \deg(G))$. Then for every j, $0 \leq j \leq n$, it is the case that $(\mathrm{psc}_j(F, G))_\alpha = \mathrm{psc}_j(F_\alpha, G_\alpha)$. We see this as follows: For $j < n$, since $\deg(F) = \deg(F_\alpha)$ and $\deg(G) = \deg(G_\alpha)$, the matrix obtained by evaluating the entries of the Sylvester matrix of F and G at α is just the Sylvester matrix of F_α and G_α, hence if $j < n$ then $(S_j(F, G))_\alpha$ is equal to $S_j(F_\alpha, G_\alpha)$, and so $(\mathrm{psc}_j(F, G))_\alpha = \mathrm{psc}_j(F_\alpha, G_\alpha)$. If $j = n$, then $(\mathrm{psc}_j(F, G))_\alpha = \mathrm{psc}_j(F_\alpha, G_\alpha) = 1$.

Theorem 2 *For $A \subset I_r, r \geq 2$, if R is a $\mathrm{PROJ}_1(A)$-invariant region in E^{r-1}, then every element of A is either delineable or identically zero on R.*

Proof. Consider any $F \in A$. If $F = 0$, then F is identically zero on R. Suppose $F \neq 0$. By definition, $\mathrm{PROJ}_1(A)$ includes every nonzero coefficient of F, so each coefficient of F either vanishes everywhere or nowhere on R. Hence $\deg(F_\alpha)$ is constant for $\alpha \in R$. Let $\deg_R(F)$ denote this constant value. If $\deg_R(F) = -\infty$, then F is identically zero on R. If $\deg_R(F) = 0$, then obviously F is delineable on R. Suppose $\deg_R(F) \geq 1$. Then there is a unique reductum Q of F such that $\deg(Q) = \deg_R(Q) = \deg_R(F)$. Then $F_\alpha = Q_\alpha$ for all $\alpha \in R$, hence if Q is delineable on R, then F is delineable on R. Since $\mathrm{PSC}(Q, Q') \subset \mathrm{PROJ}(A)$, the least k such that $(\mathrm{psc}_k(Q, Q'))_\alpha \neq 0$ is constant for $\alpha \in R$. Hence by our observation above, the least k such that $\mathrm{psc}_k(Q_\alpha, Q'_\alpha) \neq 0$ is constant for $\alpha \in R$. Hence by Theorem 1, Q is delineable on R, hence F is delineable on R. Thus every element of A is either identically zero or delineable on R. ∎

The following theorem is the basis for definition of the second class of elements of $\mathrm{PROJ}(A)$.

Theorem 3 *Let $A \subset I_r$, $r \geq 2$, and let R be a region in E^{r-1}. Suppose that for every $F \in A$, the hypotheses of Theorem 1 are satisfied. Suppose also that for every $F, G \in A$, $F \neq G$, the least k such that $\mathrm{psc}_k(F_\alpha, G_\alpha) \neq 0$ is constant for $\alpha \in R$. Then every $F \in A$ is delineable on R, and for every $F, G \in A$, any F-section and any G-section of $Z(R)$ are either disjoint or identical.*

A proof is given in (Arnon et al. 1982, theorem 3.7). The essential ideas are contained in the proof of theorem 5 of (Collins 1975).

Let A and R_i be as in the definition of PROJ_1. Let

$$\mathrm{PROJ}_2(A) = \bigcup_{1 \le i < j \le n} \ \bigcup_{G_i \in R_i \ \& \ G_j \in R_j} \mathrm{PSC}(G_i, G_j).$$

We define $\mathrm{PROJ}(A)$ to be the union of $\mathrm{PROJ}_1(A)$ and $\mathrm{PROJ}_2(A)$. The following theorem establishes that PROJ works, i.e., that conditions $(1')$ and (2) are satisfied for a $\mathrm{PROJ}(A)$-invariant region:

Theorem 4 *For $A \subset I_r, r \ge 2$, if R is a $\mathrm{PROJ}(A)$-invariant region in E^{r-1}, then every element of A is either delineable or identically zero on R, and for every $F, G \in A$, any F-section and any G-section of $Z(R)$ are either disjoint or identical.*

Proof. By Theorem 2, every element of A is either delineable or identically zero on R. By an argument similar to that used in the proof of Theorem 2, but with Theorem 3 in place of Theorem 1, it follows that for every $F, G \in A$, any F-section and any G-section of $Z(R)$ are either disjoint or identical. ∎

This completes the proof that if R is a $\mathrm{PROJ}(A)$-invariant region in E^{r-1}, then there exists an A-invariant stack over R, namely the stack whose sections are the sections of those A_i's in A which are delineable on R. In general, when every element of $A \subset I_r$ is either delineable or identically zero on some $R \subset E^{r-1}$, we write $S(A, R)$ to denote this stack. Our agenda for this section will be completed by showing that if also R is semi-algebraic, then $S = S(A, R)$ is algebraic. By our remarks in Section 2, it suffices to show that each region of S is definable. Let x denote $\langle x_1, \ldots, x_{r-1} \rangle$ and y denote x_r. Any section of S is an F-section of $Z(R)$ for some $F \in A$ which is delineable on R; say that it is the j-th section of $S(F, R)$ (where sections are numbered from bottom to top). Then we can define it as the set of $\langle x, y \rangle$ satisfying a formula "$x \in R$ and y is the j-th real root of $F(x, y)$." If ϕ is a defining formula for R, then the following is such a formula:

$$\phi(x) \ \& \ (\exists y_1)(\exists y_2)\cdots(\exists y_{j-1}) \ [\ y_1 < y_2 < \cdots < y_{j-1} < y$$

$$\& \ F(x, y_1) = 0 \ \& \ F(x, y_2) = 0 \ \& \ \cdots \ \& \ F(x, y_{j-1}) = 0 \ \& \ F(x, y) = 0$$

$$\& \ (\forall y_{j+1}) \ \{(y_{j+1} \ne y_1 \ \& \ y_{j+1} \ne y_2 \ \& \ \cdots \ \& \ y_{j+1} \ne y_{j-1} \ \&$$

$$y_{j+1} \ne y \ \& \ F(x, y_{j+1}) = 0) \implies y_{j+1} > y \ \}].$$

The sectors of S can now be defined using the defining formulas for the sections: a sector is either the set of $\langle x, y \rangle$ between two sections of S, or the set of $\langle x, y \rangle$ above the topmost section of S, or the set of $\langle x, y \rangle$ below the bottommost section of S. This concludes the proof.

4 The Cylindrical Algebraic Decomposition Algorithm: Base Phase

Let us use the precise definition of cad given in Section 2 to give precise specifications for a cad algorithm. Its input is a set $A \subset I_r$, $r \geq 1$. Its output is a
description of an A-invariant cad D of E^r. This description should inform one
of the number of cells in the cad, how they are arranged into stacks, and the
sign of each element of A on each cell. We define in this section the index of a
cell in a cad; our cad algorithm meets the first two of the above requirements
by producing a list of indices of the cells of the cad of E^r that it constructs. We
also define in this section an exact representation for algebraic points in E^r,
that is, points whose coordinates are all real algebraic numbers. Our cad algorithm constructs, for each cell, an exact representation of a particular algebraic
point belonging to that cell (we call this a *sample point* for the cell). The sign of
$A_i \in A$ on a particular cell can then be determined by evaluating A_i (exactly)
at the cell's sample point, and in this way we meet the third requirement above.

Where $A \subset I_r$ is the input to the cad algorithm, in the projection phase
we computed $\text{PROJ}(A)$, $\text{PROJ}^2(A)$, and finally $K = \text{PROJ}^{r-1}(A) \subset I_1$. It is
the task of the base phase to construct a K-invariant cad D^* of E^1, that is, to
construct cell indices and sample points for the cells of such a cad. Let us now
define cell indices.

In a cylindrical decomposition of E^1, the index of the leftmost 1-cell (the 1-
cell with left endpoint $-\infty$), is (1). The index of the 0-cell (if any) immediately
to its right is (2), the index of the 1-cell to the right of that 0-cell (if any) is (3),
etc. Suppose that cell indices have been defined for cylindrical decompositions
of E^{r-1}; $r \geq 2$. Let D be a cylindrical decomposition of E^r. D induces a
cylindrical decomposition D' of E^{r-1}. Any cell d of D is an element of a stack
S over a cell c of D'. Let $(i_1, \ldots, i_{r-1})$ be the index of c. The cells of S may
be numbered from bottom to top, with the bottommost sector being called
cell 1, the section above it (if any) cell 2, the sector above that (if any) cell 3,
etc. If d is the j-th cell of the stack by this numbering, then its cell index is
$(i_1, \ldots, i_{r-1}, j)$.

The sum of the parities of the components of a cell index is the dimension
of the cell (even parity = 0, odd parity = 1). In a cylindrical decomposition of
E^2, for example, cell $(2, 4)$ would be a 0-cell, $(2, 5)$ would be a 1-cell.

We begin the base phase by constructing the set of all distinct (i.e., relatively
prime) irreducible factors of nonzero elements of K (see Kaltofen 1982, for
polynomial factorization algorithms). Let $M = \{M_1, \ldots, M_k\} \subset I_1$ be the set
of these factors. The real roots $\alpha_1 < \cdots < \alpha_n$, $n \geq 0$, of $\prod M$ will be the
0-cells of D^* (if $n = 0$ then D^* consists of the single 1-cell E^1). We determine
the α_j's by isolating the real roots of the individual M_i's (Collins and Loos
1982b). By their relative primeness, no two elements of M have a common
root. Hence by refining the isolating intervals for the α_j's, we obtain a collection
of disjoint left-open and right-closed intervals $(r_1, s_1], (r_2, s_2], \ldots, (r_n, s_n]$ with
rational endpoints, each containing exactly one α_j, and with $r_1 < s_1 \leq r_2 < \cdots$.

As soon as we know n, we can trivially write down the indices of the $2n + 1$ cells of D^*. Clearly each cell is definable, hence semi-algebraic. To describe sample point construction, we first define a representation for an algebraic point in E^i, $i \geq 1$. Loos (1982a, section 1) describes the representation of a real algebraic number γ by its minimal polynomial $M(x)$, and an isolating interval for a particular root of $M(x)$. With γ so represented, and letting $m = \deg(M)$, one can represent any element of $\mathbb{Q}(\gamma)$ as an element of $\mathbb{Q}[x]$ of degree $\leq m - 1$ (as Loos describes). For an algebraic point in E^i, there exists a real algebraic γ such that each coordinate of the point is in $\mathbb{Q}(\gamma)$; γ is a *primitive element* for the point. Our representation for the point is: a primitive element γ and an i-tuple of elements of $\mathbb{Q}(\gamma)$, all represented as described by Loos.

For the 1-cells of D^* we primarily use appropriately chosen (rational) endpoints from the isolating intervals above as sample points. However, if $s_i = r_{i+1}$ is a 0-cell, we find (by bisection) a positive rational ϵ, such that $(r_{i+1} + \epsilon, s_{i+1}]$ isolates α_{i+1}, and use $r_{i+1} + \epsilon$ as sample point for cell $(2i + 1)$. Also, we use $s_n + 1$ as a sample point for cell $(2n + 1)$. If $D^* = \{E^1\}$, we use an arbitrary rational number. Obviously the only point in a 0-cell is the cell itself. Its value is an algebraic number. Thus all our sample points for D^* are algebraic numbers, and hence can be trivially expressed in our just-defined algebraic point representation. Examples of sample points for a cad of E^1 are given in Section 6.

5   The Cylindrical Algebraic Decomposition Algorithm: Extension Phase

First, consider the extension of the cad D^* of E^1 to a cad of E^2. In the projection phase, we computed a set $J = \mathrm{PROJ}^{r-2}(A) \subset I_2$. Let c be a cell of D^*. We want to construct the stack $S(J, c)$ (as defined following Theorem 4). Let α be the sample point for c, and let $J_c(y)$ be the product of all nonzero $G(\alpha, y)$, $G \in J$ (we construct $J_c \in \mathbb{Q}(\alpha)[y]$ using algorithms for exact arithmetic in $\mathbb{Q}(\alpha)$; Loos 1982a). We isolate the real roots of $J_c(y)$ (Loos 1982a, section 2). This determines $S(J, c)$: β is a root of $J_c(y)$ if and only if $\langle \alpha, \beta \rangle$ lies on a section of $S(J, c)$. For each such β, we use the representation for α, the isolating interval for β, and the algorithms NORMAL and SIMPLE of (Loos 1982b) to construct a primitive element γ for $\mathbb{Q}(\alpha, \beta)$; we use γ to construct a representation of the form we require for $\langle \alpha, \beta \rangle$. We get sector sample points for $S(J, c)$ from α and the (rational) endpoints of the isolating intervals for the roots of J_c, much as was done in Section 4 for E^1. Thus sector sample points are of the form $\langle \alpha, r \rangle$, r rational, so we can take $\gamma = \alpha$ for them. Given the cell index for c, and the isolated roots of J_c, we can trivially write down the indices for the cells of $S(J, c)$ (as for E^1 in Section 4).

After processing each cell c of D^* in this fashion, we have determined a cad of E^2 and constructed a sample point for each cell.

Extension from E^{i-1} to E^i for $3 \leq i \leq r$ is essentially the same as from E^1 to

E^2. The only difference is that a sample point in E^{i-1} has $i-1$, instead of just one, coordinates. But where α is the primitive element of an E^{i-1} sample point, and $F = F(x_1, \ldots, x_i)$ an element of I_i, arithmetic in $\mathbb{Q}(\alpha)$ still suffices for constructing the univariate polynomial over $\mathbb{Q}(\alpha)$ that results from substituting the coordinates $\langle \alpha_1, \ldots, \alpha_{i-1} \rangle$ of the sample point for $\langle x_1, \ldots, x_{i-1} \rangle$ in F.

The following abstract algorithm summarizes our discussion of the cad algorithm.

$$\text{CAD}(r, A; I, S)$$

Inputs: r is a positive integer. A is a list of $n \geq 0$ integral polynomials in r variables.

Outputs: I is a list of the indices of the cells comprising an A-invariant cad D of E^r. S is a list of sample points for D.

(1) $[r = 1]$. If $r > 1$ then go to 2. Set $I \leftarrow$ the empty list. Set $S \leftarrow$ the empty list. Isolate the real roots of the irreducible factors of the nonzero elements of A. Construct the indices of the cells of D and add them to I. Construct sample points for the cells of D and add them to S. Exit.

(2) $[r > 1]$. Set $P \leftarrow \text{PROJ}(A)$. Call CAD recursively with inputs $r - 1$ and P to obtain outputs I' and S' that specify a cad D' of E^{r-1}. Set $I \leftarrow$ the empty list. Set $S \leftarrow$ the empty list. For each cell c of D', let i denote the index of c, let α denote the sample point for c, and carry out the following four steps: first, set $A_c(x_r) \leftarrow \prod \{ A_j(\alpha, x_r) \,|\, A_j \in A \ \& \ A_j(\alpha, x_r) \neq 0 \}$; second, isolate the real roots of $A_c(x_r)$; third, use i, α, and the isolating intervals for the roots of A_c to construct cell indices and sample points for the sections and sectors of $S(A, c)$; fourth, add the new indices to I and the new sample points to S. Exit.

6 An Example

We now show what algorithm CAD does for a particular example in E^2. Let

$$A_1(x, y) = 144y^2 + 96x^2 y + 9x^4 + 105x^2 + 70x - 98,$$

$$A_2(x, y) = xy^2 + 6xy + x^3 + 9x,$$

and $A = \{A_1, A_2\}$. CAD is called with input A. We compute $\text{PROJ}(A)$:
ldcf$(A_1) = 144$,
$\text{psc}_0(A_1, A_1') = -580608(x^4 - 15x^2 - 10x + 14)$,
$\text{psc}_1(A_1, A_1') = 1$,
ldcf$(\text{red}(A_1)) = 96x^2$,
$\text{psc}_0(\text{red}(A_1), [\text{red}(A_1)]') = 1$,
ldcf$(\text{red}^2(A_1)) = 9x^4 + 105x^2 + 70x - 98$,
ldcf$(A_2) = x$,
$\text{psc}_0(A_2, A_2') = 4x^5$,

$\text{psc}_1(A_2, A_2') = 1$,
$\text{ldcf}(\text{red}(A_2)) = 6x$,
$\text{psc}_0(\text{red}(A_2), [\text{red}(A_2)]') = 1$,
$\text{ldcf}(\text{red}^2(A_2)) = x(x^2 + 9)$,
$\text{psc}_0(A_1, A_2) = x^2(81x^8 + 3330x^6 + 1260x^5 - 37395x^4 - 45780x^3 - 32096x^2 + 167720x + 1435204)$,
$\text{psc}_1(A_1, A_2) = 96x(x^2 - 9)$,
$\text{psc}_2(A_1, A_2) = 1$,
$\text{psc}_0(\text{red}(A_1), A_2) = x(81x^8 + 5922x^6 + 1260x^5 + 31725x^4 - 25620x^3 + 40768x^2 - 13720x + 9604)$,
$\text{psc}_1(\text{red}(A_1), A_2) = 1$,
$\text{psc}_0(A_1, \text{red}(A_2)) = -36x(3x^4 - 33x^2 - 70x - 226)$,
$\text{psc}_1(A_1, \text{red}(A_2)) = 1$,
$\text{psc}_0(\text{red}(A_1), \text{red}(A_2)) = 1$.

It turns out that the roots of $p_1(x) = x^4 - 15x^2 - 10x + 14$ and $p_2(x) = x$ give us a "silhouette" of $V(A_1) \cup V(A_2)$, hence for simplicity in this example, let us set $\text{PROJ}(A) = \{p_1(x), p_2(x)\}$ (in general, $\text{PROJ}(A)$ may contain superfluous elements; Collins (1975) and Arnon (1981) describe techniques for detecting and eliminating such elements).

p_1 and p_2 are both irreducible, so we have $M_1 = p_1$ and $M_2 = p_2$ in the notation of Section 5. M_1 has four real roots with approximate values -3.26, -1.51, 0.7, and 4.08; M_2 has the unique root $x = 0$. The following collection of isolating intervals for these roots satisfies the conditions set out in Section 5:

$$(-4, -3], (-2, -1], (-1, 0], (1/2, 1], (4, 8].$$

Since there are five 0-cells, the cell indices for the cad are (1), (2), ..., (11).

We now construct representations for the sample points of the induced cad of E^1. Each 1-cell will have a rational sample point, hence any rational γ will be a primitive element. We arbitrarily choose $\gamma = 0$. $(-1, 0]$ is an isolating interval for γ as a root of its minimal polynomial x. We may take the 1-cell sample points to be -4, -2, -1, $1/2$, 4, and 9. The four irrational 0-cells have as their primitive elements the four roots of $M_1(x)$. The representation for the leftmost 0-cell, for example, consists of $M_1(x)$, the isolating interval $(-4, 3]$ for the leftmost root γ of M_1, and the 1-tuple $\langle x \rangle$, where x corresponds to the element γ of $\mathbb{Q}(\gamma)$. The 0-cell $x = 0$ is represented in the same fashion as the rational 1-cell sample points.

We now come to the extension phase of the algorithm. Let c be the leftmost 1-cell of the cad D' of E^1. $A_1(-4, y) \neq 0$ and $A_2(-4, y) \neq 0$, hence

$$A_c(y) = A_1 A_2(-4, y) = 24(y^2 + 6y + 25)(24y^2 + 256y + 601).$$

$y^2 + 6y + 25$ has no real roots, but $24y^2 + 256y + 601$ has two real roots, which can be isolated by the intervals $(-8, -7]$ and $(-4, -2]$. Thus the stack $S(A, c)$ has two sections and three sectors; the indices for these cells are $(1, 1)$, $(1, 2)$, ..., $(1, 5)$. From the endpoints of the isolating intervals we obtain sector sample points of $\langle -4, -8 \rangle$, $\langle -4, -4 \rangle$, and $\langle -4, -1 \rangle$ (which will be represented in the customary fashion). The two roots γ_1 and γ_2 of $24y^2 + 256y + 601$ are not only y-coordinates for the section sample points, but also primitive elements for these sample points. Thus the (representations for the) section sample points are

$$\{24y^2 + 256y + 601, (-8, -7], \langle -4, y \rangle\}$$

and

$$\{24y^2 + 256y + 601, (-4, -2], \langle -4, y \rangle\}.$$

Now let c be the leftmost 0-cell of D'; let α also denote this point. $A_1(\alpha, y) \neq 0$ and $A_2(\alpha, y) \neq 0$; we have

$$A_c(y) = A_1 A_2(\alpha, y) = (y^2 + 6y + \alpha^2 + 9)(y + (1/3)\alpha^2)^2.$$

$y^2 + 6y + \alpha^2 + 9 \in \mathbb{Q}(\alpha)[y]$ has no real roots, but obviously $y + \frac{1}{3}\alpha^2$ has exactly one; $(-8, 8]$ is an isolating interval for it. Hence $S(A, c)$ has one section and two sectors; the indices of these cells are $(2, 1)$, $(2, 2)$, and $(2, 3)$. The appropriate representations for $\langle -\alpha, -8 \rangle$ and $\langle -\alpha, 9 \rangle$ are the sector sample points. Since $y + \frac{1}{3}\alpha^2$ is linear in y, its root is an element of $\mathbb{Q}(\alpha)$. Hence

$$\{M_1(x), (-4, 3], \langle x, -(1/3)x^2 \rangle\}$$

is the representation of the section sample point. Thus in this particular case it was not necessary to apply the NORMAL and SIMPLE algorithms of (Loos 1982a) to find primitive elements for the sections of $S(A, c)$. In general, however, for a 0-cell $c = \alpha$ of D', $A_c(y)$ will have nonlinear factors with real roots, and it will be necessary to apply NORMAL and SIMPLE. Saying this another way, where α is a 0-cell of D' and $\langle \alpha, \beta \rangle$ is a section sample point of D, we had in our example above $\mathbb{Q}(\alpha, \beta) = \mathbb{Q}(\alpha)$, but in general, $\mathbb{Q}(\alpha)$ will be a proper subfield of $\mathbb{Q}(\alpha, \beta)$.

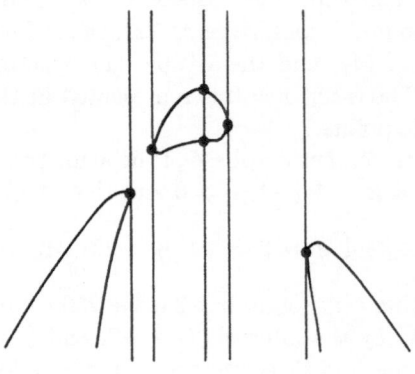

Figure 5

The steps we have gone through above for a 1-cell and a 0-cell are carried out for the remaining cells of D' to complete the determination of the A-invariant cad D of E^2.

Although information of the sort we have described is all that would actually be produced by CAD, it may be useful to show a picture of the decomposition of the plane to which the information corresponds. The curve defined by $A_1(x, y) = 0$ has three connected components which are easily identified in Fig. 5. The curve defined by $A_2(x, y) = 0$ is just the y-axis, i.e., the same curve as defined by $x = 0$, which cuts through the middle of the second component of $V(A_1)$. Figure 5 shows the A-invariant cad that CAD constructs.

We remark that the curve $A_1(x, y)$ is from (Hilton 1932, p. 329).

Cylindrical Algebraic Decomposition II: An Adjacency Algorithm for the Plane[1]

Dennis S. Arnon, George E. Collins, and Scott McCallum

1   Introduction

In Part I of the present paper we defined cylindrical algebraic decompositions (cad's), and described an algorithm for cad construction. In Part II we give an algorithm that provides information about the topological structure of a cad of the plane. Informally, two disjoint cells in E^r, $r \geq 1$, are *adjacent* if they touch each other; formally, they are adjacent if their union is connected. In a picture of a cad, e.g., Fig. 2 of Part I, it is obvious to the eye which pairs of cells are adjacent. However, the cad algorithm of Part I does not actually produce this information, nor did the original version of that algorithm in (Collins 1975).

Adjacency information has already been an essential part of certain applications of cad's. For example, for $r = 2$ and $r = 3$, a cad construction algorithm has been developed which uses adjacency information to circumvent certain time-consuming steps of the original algorithm (Arnon 1981). Schwartz and Sharir (1983b) use cell adjacency in their solution to the mover's problem, as do Arnon and McCallum (1984) in their recently developed algorithm to determine the topological type of a real algebraic curve. We remark that adjacency of cells is a slight generalization of the notion of incidence of cells in algebraic topology (for which one may consult Massey 1978).

We present here an algorithm which, given a set of polynomials $A \subset I_2$, constructs an A-invariant cad of E^2, and determines all pairs of adjacent cells in that cad. For certain inputs A, the cad constructed by this algorithm (which we call a "proper" cad) is different from the cad constructed by algorithm CAD of Part I. In a proper cad, knowing only some of the pairs of adjacent cells

[1]Reprinted with permission from the *SIAM Journal on Computing*, volume 13, number 4, pp. 878–889. Copyright 1984 by the Society for Industrial and Applied Mathematics, Philadelphia, Pennsylvania. All rights reserved.

(those in which both cells are sections) suffices for determining the rest, and the algorithm for constructing the set of sufficient pairs is attractively simple.

The following definitions and observations are our starting point for adjacency determination. If R_1 and R_2 are adjacent regions, we call the set $\{R_1, R_2\}$ an *adjacency*. If both R_1 and R_2 are sections, it is a *section-section* adjacency. Recall that a cad of E^r, $r \geq 2$, is the union of certain stacks over the cells of some induced cad of E^{r-1}. Clearly the adjacencies of any cad can be divided into those in which both cells are in the same stack, and those in which the two cells involved are in different stacks. The first kind may be called *intrastack* adjacencies, and the second kind *interstack* adjacencies. To determine the intrastack adjacencies of a cad it suffices to know the number of sections in each of its stacks. This is because in any stack, each section is adjacent to the sector immediately below it, and to the sector immediately above it, and every adjacency involving two cells of that stack is of one of these two forms. Any algorithm for cad construction must determine how many sections are in each stack (cf. the specifications for cad algorithms in Section 4 of Part I). Hence, the only hard part of determining the adjacencies of a cad is determining the interstack adjacencies.

The contents of the paper is as follows. Section 2 defines the notion of a proper cad, and shows that in a proper cad of the plane, if one knows the section-section interstack adjacencies, then one can infer from them all other interstack adjacencies. Section 3 presents an algorithm (called SSADJ2) to determine the section-section interstack adjacencies between a pair of stacks in E^2 satisfying certain hypotheses (which will be satisfied by the stacks in a proper cad). In Section 4 we develop an algorithm (called CADA2) which, given $A \subset I_2$, constructs a proper A-invariant cad of E^2, and its adjacencies. Section 5 traces CADA2 on the example of Section 6 of Part I.

Algorithm CADA2 has been implemented (Arnon 1981). The basic idea of algorithm SSADJ2 is due to McCallum (1979). We remark that Schwartz and Sharir (1983b) propose adjacency algorithms for cad's which are quite different from ours, and whose feasibilities have not yet been determined. The notion of proper cad (called "basis-determined cad") was first used in (Arnon 1981). A similar notion ("well-based cad") was used in (Schwartz and Sharir 1983b).

2   Adjacencies in Proper Cylindrical Algebraic Decompositions

We say that a cad D of E^r is *proper* if (1) there exists some $F \in I_r$ such that $V(F)$ is equal to the union of the sections of D, and (2) if $r > 1$, then the induced cad D' of E^{r-1} is proper. When such an F exists for a cad D, we say it is a *defining polynomial* for D. Such polynomials are not unique. For example, given a defining polynomial $F(x, y)$ for a cad D of E^2, $(y^2 + 1)F$ is also one.

The following theorem provides a useful characterization of proper cad's.

Recall that if D is a cad of E^r, then $D = \bigcup_{c \in D'} S(c)$, where for each $c \in D'$, $S(c)$ denotes the unique stack over c which is a subset of D.

Theorem 1 *If D is a proper cad of E^r, $r \geq 2$, then (1) D' is proper, (2) any defining polynomial $F \in I_r$ for D is delineable on every $c \in D'$, and (3) for any defining polynomial F for D, $D = \bigcup_{c \in D'} S(F, c)$. Conversely, if D' is proper, if there exists $F \in I_r$ which is delineable on every $c \in D'$, and if $D = \bigcup_{c \in D'} S(F, c)$, then D is proper.*

Proof. Suppose D is proper, and suppose F is a defining polynomial for it. By definition of proper cad, D' is proper, and for any $c \in D'$, every section of $S(c)$ is in $V(F)$, and no sector of $S(c)$ meets $V(F)$. Hence F is delineable on c, and $S(c) = S(F, c)$. Thus $D = \bigcup_{c \in D'} S(F, c)$. Conversely, if D' is proper, if there exists $F \in I_r$ which is delineable on every $c \in D'$, and if $D = \bigcup_{c \in D'} S(F, c)$, then clearly $V(F)$ is equal to the union of the sections of D, and so D is proper. ∎

Let us illustrate how a cad constructed by algorithm CAD of Part I may not be proper. Given input $\{xy\}$, CAD would construct the $\{xy\}$-invariant cad of E^2 shown in Fig. 1. A superscript attached to the name of a cell will denote its dimension, e.g., c^0 is a 0-cell, c^1 is a 1-cell. Assume $F(x, y)$ is a defining polynomial for this cad. By definition, $c_1^1 \subset V(F)$. Hence it follows that $F(x, 0) = 0$ for all x, so $\langle 0, 0 \rangle \in V(F)$. But $\langle 0, 0 \rangle$ does not belong to any section of the cad, a contradiction.

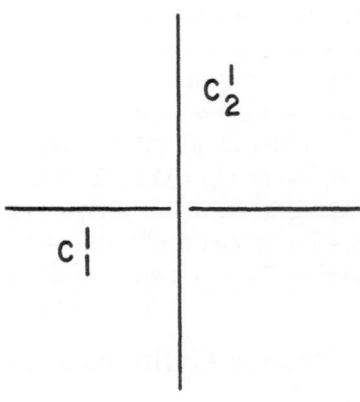

Figure 1

Figure 2 shows a proper $\{xy\}$-invariant cad with defining polynomial y.

In dealing with interstack adjacencies, it will be convenient not to have to treat points at infinity specially. We therefore introduce the following notation and terminology. We write E^* for $E \cup \{-\infty, +\infty\}$, the usual two-point compactification of the real line. For a subset X of any E^i, we write $Z^*(X)$ to denote $X \times E^*$, which we call the *extended cylinder* over X. If R is a region in

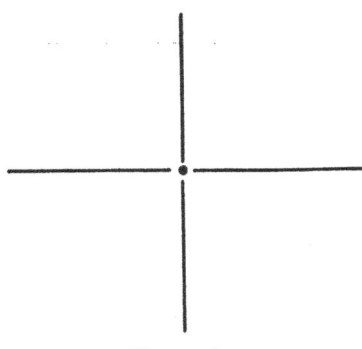

Figure 2

E^i, a *section of* $Z^*(R)$ is either a section of $Z(R)$ or $R \times \{-\infty\}$ or $R \times \{+\infty\}$. $R \times \{-\infty\}$ and $R \times \{+\infty\}$ are respectively the $-\infty$-*section* and the $+\infty$-*section*, or collectively the *infinite sections*, of $Z^*(R)$. If S is a stack over R, then S^* is S plus the infinite sections of $Z^*(R)$. S^* is the *extension* of S, and we say also that it is an *extended stack* over R.

Given a cad $D = \bigcup_{c \in D'} S(c)$, for any $c \in D'$ we write $S^*(c)$ to denote the extension of $S(c)$. If c has cell index (i), and $S(c)$ has j sections, then the cell indices of the $-\infty$ and $+\infty$-sections of $S^*(c)$ are defined to be $(i,0)$ and $(i, 2j+2)$, respectively.

Theorem 2 *Let* $c^0 = \alpha$ *be a 0-cell in* E^1, *let* R *be a region in* E^1 *which is adjacent to* c^0, *and let* $F(x,y) \in I_2$ *be such that* $F(\alpha, y) \neq 0$ *(thus* F *is delineable on* c^0*), and* F *is delineable on* R. *If* s *is a section of* $S^*(F, R)$, *then* s *has a unique limit point* p *in* $Z^*(c^0)$, *and* p *is a section of* $S^*(F, c^0)$.

Proof. If s is an infinite section of $S^*(F, c^0)$, then the assertion is obvious. Suppose s is the graph of a continuous function $f : R \to E$. Then $F(x, f(x)) = 0$ for all $x \in R$, hence f is an algebraic function. Hence sufficiently close to c^0, f is monotone as $x \in R$ approaches c^0. Then where $\{a_1, a_2, \ldots\}$ is a sequence of points in R converging to c^0, the sequence $\{f(a_1), f(a_2), \ldots, \}$ converges to a limit γ in E^*, and $p = \langle \alpha, \gamma \rangle$ is the unique limit point of s in $Z^*(c^0)$.

If $p = \langle \alpha, -\infty \rangle$ or $p = \langle \alpha, +\infty \rangle$ we are done. Suppose p is neither of these. It is a standard fact that the variety of a polynomial is a closed set. Hence $p \in V(F)$. By hypotheses, $F(\alpha, y) \neq 0$, hence it has finitely many real roots, hence since $S(F, c^0)$ is F-invariant, $F \neq 0$ at every point of every sector of $S(F, c^0)$. Hence p is a section of $S(c^0)$. ∎

We call p the *boundary section* of s in $S^*(F, c^0)$.

Our next theorem requires a general notion of boundary. For a subset X of a topological space T, the *boundary of* X, written ∂X, is $\overline{X} - X$ ($\overline{X}$ denotes the closure of X). One can easily show that ∂X is the set of all limit points of X which do not belong to X. We introduce some notation: for a region R in E^i, suppose that sections s_1 and s_2 of $Z^*(R)$ are respectively the f_1-section and f_2-section of $Z^*(R)$, and that $f_1 < f_2$. We write (s_1, s_2) to denote the (f_1, f_2)-sector of $Z(R)$, and $[s_1, s_2]$ to denote $s_1 \cup s_2 \cup (s_1, s_2)$ (see Part I, Section 2, for the notation f_1-section, f_2-section, and (f_1, f_2)-sector).

Theorem 3 *Let $c^0 = \alpha$ be a 0-cell in E^1, let R be a region in E^1 which is adjacent to c^0, and let $F(x, y) \in I_2$ be such that $F(\alpha, y) \neq 0$, and F is delineable on R. Let $s = (s_1^1, s_2^1)$ be a sector of $S(F, R)$; let t_1^0 and t_2^0 be the respective boundary sections of s_1^1 and s_2^1 in $S^*(F, c^0)$. Then the portion of ∂s contained in $Z^*(c^0)$ is $[t_1^0, t_2^0]$.*

Proof. Suppose s_1^1 is the f_1-section, and s_2^1 the f_2-section, of $Z^*(R)$. The following argument will assume that both t_1^0 and t_2^0 are finite, but can easily be modified for the cases where either or both is infinite. Let p be a point of $[t_1^0, t_2^0]$. p can be written in the form

$$\gamma \, t_1^0 + (1 - \gamma)t_2^0, \quad 0 \leq \gamma \leq 1.$$

Let $\{a_1, a_2, \ldots, \}$ be a sequence of points in R converging to c^0. Then the sequence

$$\{\langle a_1, \gamma f_1(a_1) + (1 - \gamma)f_2(a_1)\rangle, \ldots\}$$

in s converges to p, hence p is a limit point of s. Suppose p is a limit point of s in $Z^*(c^0)$. Then there is some sequence of points $\{\langle x_i, y_i \rangle\}$ in s converging to p. we have $f_1(x_i) < y_i < f_2(x_i)$, hence $p = \lim\{\langle x_i, y_i \rangle\}$ is an element of $[t_1^0, t_2^0]$. Hence $\partial s \cap Z^*(c^0) = [t_1^0, t_2^0]$. ∎

   Given two disjoint regions, it is not difficult to see that they are adjacent if and only if one contains a limit point of the other. Using this fact, we now show that if c^0, R, and $F(x, y)$ are as in the hypotheses of Theorems 2 and 3, then all interstack adjacencies between $S(F, c^0)$ and $S(F, R)$ can be determined from their section-section interstack adjacencies. Let s^1 be a section of $S^*(F, R)$. By Theorem 2 and the fact that regions are adjacent if and only if one contains a limit point of the other, if s^1 is adjacent to a cell of $S^*(F, c^0)$, then that cell is a section of $S^*(F, c^0)$. Hence any interstack adjacency involving a section of $S^*(F, R)$ is a section-section adjacency. Let $s^2 = (s_1^1, s_2^1)$ be a sector of $S^*(F, R)$. Where t_1^0 and t_2^0 are the respective boundary sections of s_1^1 and s_2^1 in $S^*(F, c^0)$, by Theorem 3, s^2 is adjacent to all cells of $S^*(F, c^0)$ between t_1^0 and t_2^0 inclusive, and to only those cells of $S^*(F, c^0)$. Hence knowledge of the section-section adjacencies between $S(F, c^0)$ and $S(F, R)$ suffices for determining all interstack adjacencies between them.

   We now relate our general development to proper cad's. We first establish that we can make use of Theorems 2 and 3.

Theorem 4 *Let D be a proper cad of E^2, let $c^0 = \alpha$ and c^1 be adjacent cells of D', and let $F(x, y) \in I_2$ be a defining polynomial for D. Then $F(\alpha, y) \neq 0$, F is delineable on c^1, $S(F, c^0) = S(c^0)$, and $S(F, c^1) = S(c^1)$.*

Proof. By Theorem 1, F is delineable on every $c \in D'$, and $D = \bigcup_{c \in D'} S(F, c)$. The conclusions of the theorem now follow directly. ∎

Theorems 2–4 give us:

Corollary 1 *Let D be a proper cad of E^2, and let $c^0 = \alpha$ and c^1 be adjacent cells of D'. If s is a section of $S^*(c^1)$, then s has a unique limit point p in $Z^*(c^0)$, and p is a section of $S^*(c^0)$. If $s = (s_1^1, s_2^1)$ is a sector of $S(c^1)$, and t_1^0 and t_2^0 are the respective boundary sections of s_1^1 and s_2^1 in $S^*(c^0)$, then the portion of ∂s contained in $Z^*(c^0)$ is $[t_1^0, t_2^0]$.*

The same argument we used in the more general setting can be applied to show that if D is a proper cad of E^2, and if c^0 and c^1 are adjacent cells of D', then all interstack adjacencies between $S(c^0)$ and $S(c^1)$ can be inferred from their section-section interstack adjacencies. It follows that all interstack adjacencies of D can be inferred from knowledge of its section-section interstack adjacencies.

3   Determination of Section-Section Adjacencies

To find the section-section interstack adjacencies of a proper cad D of E^2, we use the following strategy: for each 0-cell c^0 of D', and for the two 1-cells c_1^1 and c_2^1 of D' adjacent to c^0, we find all adjacencies between sections of $S^*(c^0)$ and sections of $S^*(c_1^1)$, and all adjacencies between sections of $S^*(c^0)$ and sections of $S^*(c_2^1)$. Algorithm SSADJ2, which we develop in this section, handles a particular triple c^0, c_1^1, c_2^1. The application of SSADJ2 to each triple c^0, c_1^1, c_2^1 of D' is done by algorithm CADA2 of Section 4.

Our results in this section apply to a setting similar to that which we had for Theorems 2 and 3. That is, we have a 0-cell $c^0 = \alpha$ in E^1, a region R in E^1 which is adjacent to c^0, and an $F(x, y) \in I_2$ such that $F(\alpha, y) \neq 0$, and F is delineable on R. The application of these general results to the context of proper cad's is straightforward.

Suppose e^0 is a section of $S(F, c^0)$. Theorems 5 and 6 below provide a method of determining how many sections of $S(F, R)$ are adjacent to e^0. This method can be described as: draw a suitable "box" centered at e^0; then the number of sections of $S(F, R)$ adjacent to e^0 is equal to the number of intersections of $V(F)$ with the left (right) side of the box (R is to the left (right) of c^0).

But knowing how many sections of $S(F, R)$ are adjacent to e^0 is not enough; we must determine *which* sections are adjacent to it. This we accomplish by processing the sections of $S(F, c^0)$ in consecutive order from bottom to top, after an initial step suggested by Theorem 7: determine how many sections of $S(F, R)$ are adjacent to the $-\infty$-section of $S^*(F, c^0)$, i.e. how many sections of $S(F, R)$ tend to $-\infty$ as $x \to \alpha$ in R. If there are $n \geq 0$ such sections, then by Theorem 2 they are the bottom n sections of $S(F, R)$. Next, we apply the "box" method of Theorems 5 and 6 to the bottommost section e^0 of $S(F, c^0)$. If we determine that there are $m \geq 0$ sections of $S(F, R)$ adjacent to e^0, then by Theorem 2 these are the $(n+1)$-st,...,$(n+m)$-th sections (counting upwards) of $S(F, R)$. Next, let e^0 be the second (from the bottom) section of $S(c^0)$, and apply the box method to find out how many sections of $S(F, R)$ are adjacent to it. If there are $q \geq 0$ such sections, then these are the $(n+m+1)$-st,...,$(n+m+q)$-th sections of $S(F, R)$. We continue in this fashion until we have processed all sections of $S(F, c^0)$. If there remain sections "at the top" of $S(F, R)$ which are not adjacent to the top section of $S(F, c^0)$, then they are adjacent to the $+\infty$-section of $S^*(F, c^0)$, i.e., they tend to $+\infty$ as $x \to \alpha$ in R.

The above steps can be described as analyzing (either the left or the right) "sides" of a collection of "boxes" stacked on top of each other. This is indeed what SSADJ2 does, but with a refinement: a complete set of suitable boxes is determined at the beginning of the algorithm, all of which have the same width (they may have different heights). Thus one might picture a single "ladder", whose "compartments" are the "boxes" for the different sections of $S(F, c^0)$ (cf. Fig. 8 in Section 5).

We now give Theorems 5–7, followed by SSADJ2.

Theorem 5 *Let $c^0 = \alpha$ be a 0-cell in E^1, suppose for some $b \in E$, $\alpha < b$, that $F(\dot{x}, y) \in I_2$ is delineable on $R = (\alpha, b]$, and suppose for $s, t, u \in E$, $s < t < u$, that t is the unique real root of $F(\alpha, y)$ in $[s, u]$, and neither $F(x, s)$ nor $F(x, u)$ has real roots in $[\alpha, b]$. Then the number of sections of $S(F, R)$ which are adjacent to the section $e^0 = \langle \alpha, t \rangle$ of $S(F, c^0)$, is equal to the number of real roots of $F(b, y)$ in (s, u).*

Proof. Let σ be a section of $S(F, R)$ which is adjacent to e^0. Then e^0 is a limit point of σ, so clearly there is a point of σ in $(\alpha, b) \times (s, u)$. For some $y \in E$, $\langle b, y \rangle \in \sigma$. If y is not in (s, u), then by the Intermediate Value Theorem, either $F(x, s)$ or $F(x, u)$ has a real root in $[\alpha, b]$, contrary to hypothesis. Hence $y \in (s, u)$, and σ gives rise to a real root of $F(x, b)$ in (s, u).

Consider any real root $y \in (s, u)$ of $F(x, b)$. By definition of $S(F, R)$, $\langle b, y \rangle$ lies in a section σ of $S(F, R)$. If σ contains any point outside of $R \times (s, u)$, then by the Intermediate Value Theorem, we violate our hypotheses. By Theorem 2, σ has a limit point $\langle \alpha, z \rangle$ in $Z^*(c^0)$. We must have $z \in [s, u]$, since $\sigma \subset R \times (s, u)$. Since $V(F)$ is closed, $\langle \alpha, z \rangle \in V(F)$, hence by hypothesis, $z = t$, hence σ is adjacent to $\langle \alpha, t \rangle$. ∎

Let us see an example of how we will use Theorem 5. Let $F = y^2 - x^3$, $\alpha = 0$, $b = 1$, $s = -2$, $t = 0$, and $u = 2$. Figure 3 shows the curve defined by $F = 0$, and Fig. 4 illustrates our complete situation.

Clearly we satisfy the hypotheses of Theorem 5. From the fact that $F(1, y)$ has two real roots in $(-2, 2)$, we learn that there are two sections of $S(F, R)$ adjacent to the section $\langle 0, 0 \rangle$ of $S(F, c^0)$.

There is a companion to Theorem 5 which differs from it only in that $b < \alpha$ rather than $\alpha < b$.

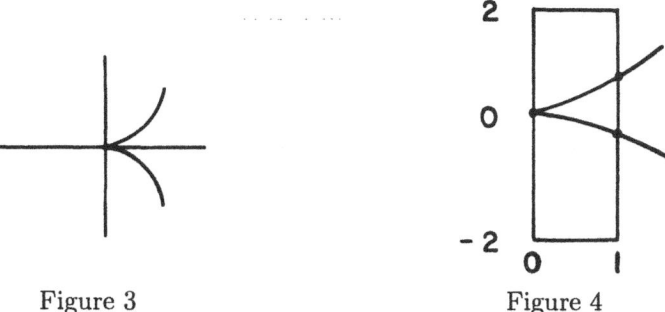

Figure 3   Figure 4

We must deal with the following problem before we can apply Theorem 5 to proper cad's. Suppose, for a proper cad D of E^2 with defining polynomial $F(x,y)$, that we have constructed the induced cad D' of E^1. Consider a 0-cell $c^0 = \alpha$ of D' and the adjacent 1-cell d^1 of D' immediately to its right. Let t be a particular real root of $F(\alpha, y)$. By isolating the real roots of $F(\alpha, y)$ we can obtain s and u such that $s < t < u$ and t is the unique real root of $F(\alpha, y)$ in $[s, u]$. Let b be the sample point for d^1; F is delineable on $(\alpha, b]$. Thus we have found $\alpha, b, s, t,$ and u which nearly satisfy the hypotheses of Theorem 5, except that $F(x, s)$ or $F(x, u)$ might have a real root in $[\alpha, b]$. This can indeed occur, as we show in Fig. 5 for $F = y^2 - x^3$, $\alpha = 0$, $b = 1$, $s = -1/2$, $t = 0$, and $u = 1/2$.

Theorem 6 establishes that we can "shrink" the width of the box, i.e., move b closer to α, so that neither $F(x, s)$ nor $F(x, u)$ has a real root in $[\alpha, b]$.

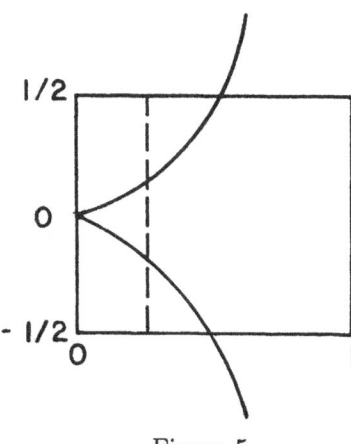

Figure 5

Theorem 6 *Suppose that the hypotheses of Theorem 5 are fulfilled, except that possibly either $F(x,s)$ or $F(x,u)$ has a real root in $[\alpha,b]$. Then there exists b^* in E, $\alpha < b^* \leq b$, such that neither $F(x,s)$ nor $F(x,u)$ has a real root in $[\alpha,b^*]$.*

Proof. By hypothesis, $F(x,s) \neq 0$ and $F(x,u) \neq 0$. Hence each has finitely many real roots, hence since $F(\alpha,s) \neq 0$ and $F(\alpha,u) \neq 0$, there exists b^*, $\alpha < b^* \leq b$, such that neither $F(x,s)$ nor $F(x,u)$ has real roots in $[\alpha,b^*]$. ∎

The proof of the next theorem is similar to the proof of Theorem 5.

Theorem 7 *Let $c^0 = \alpha$ be a 0-cell in E^1, suppose for some $b \in E$, $\alpha < b$, that $F(x,y) \in I_2$ is delineable on $R = (\alpha,b]$, and suppose for $s \in E$ that $F(\alpha,y)$ has no real roots in $(-\infty,s]$, and $F(x,s)$ has no real roots in $[\alpha,b]$. Then the number of sections of $S(F,R)$ which are adjacent to the $-\infty$-section of $S^*(F,c^0)$ is equal to the number of real roots of $F(b,y)$ in $(-\infty,s)$.*

Consider an example. Let $F = xy + 1$, $\alpha = 0$, $b = 1$, and $s = 0$. Figure 6 shows the curve $F = 0$, and Fig. 7 illustrates the overall situation. From the fact that $F(1,y)$ has one real root in $(-\infty,0)$, we learn that there is one section of $S(F,R)$ adjacent to the $-\infty$-section of $S^*(F,c^0)$.

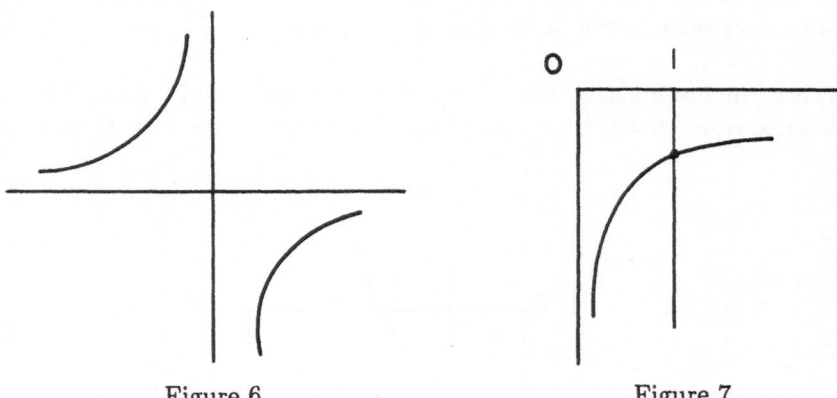

Figure 6   Figure 7

There are companions to Theorem 7 with $b < a$ rather than $a < b$, and with $+\infty$ in place of $-\infty$. Also, a result similar to Theorem 6, but tailored for 7 instead of 5, exists.

We now give algorithm SSADJ2. We adopt the convention that the sections of a stack are numbered consecutively from bottom to top, starting with section 1 (the lowest finite section), $2, \ldots, n$ (the highest finite section). The sections of the stack's extension are numbered starting with section 0 (the $-\infty$-section), 1 (the lowest finite section), $2, \ldots, n$ (the highest finite section), $n + 1$ (the $+\infty$-section).

$$\text{SSADJ2}(F(x,y),\alpha,b_1,b_2;L_1,L_2)$$

Inputs: $F(x,y)$ is an element of I_2. α is a real algebraic number such that $F(\alpha,y) \neq 0$ (we view α as also being the 0-cell c^0 in the real line). b_1 and b_2 are rational numbers such that $b_1 < \alpha < b_2$, F is delineable on $R_1 = [b_1,\alpha)$, and F is delineable on $R_2 = (\alpha,b_2]$.

Outputs: L_1 is a list of all section-section interstack adjacencies between $S^*(F,c^0)$ and $S^*(F,R_1)$. L_2 is a list of all section-section interstack adjacencies between $S^*(F,c^0)$ and $S^*(F,R_2)$.

(1) [Construct tops and bottoms of "ladder compartments".] Set $L_1 \leftarrow$ the empty list and $L_2 \leftarrow$ the empty list. Isolate the real roots $t_1,\ldots,t_m$, ($m \geq 0$), of $F(\alpha,y)$, obtaining rational $s_0,\ldots,s_m$, such that $s_0 < t_1 < s_1 < \cdots < t_m < s_m$. (If $m = 0$, set $s_0 \leftarrow$ an arbitrary rational number).

(2) [Construct left and right "sides" of "ladder".] Set $u \leftarrow b_1$ and $v \leftarrow b_2$. While there is an s_j, $0 \geq j \geq m$, such that $F(x,s_j)$ has a real root in $[u,v]$, set b^* to a rational approximate midpoint of (u,v) different from α. Set u,v to whichever of b_1,b^* and b^*,b_2 yields the property that (u,v) contains α.

(3) [Adjacencies of the $-\infty$-section of $S^*(F,c^0)$ and sections of $S^*(F,R_1)$.] Set $n \leftarrow$ the number of real roots of $F(u,y)$ in $(-\infty,s_0)$. Record in L_1 that section 0 of $S^*(F,c^0)$ is adjacent to sections $0,1,\ldots,n$ of $S^*(F,R_1)$.

(4) [Adjacencies of finite sections of $S^*(F,c^0)$ and finite sections of $S^*(F,R_1)$.] For $j = 1,\ldots,m$ do the following three things: First, set $n_j \leftarrow$ the number of real roots of $F(u,y)$ in (s_{j-1},s_j). Second, record in L_1 that section j of $S^*(F,c^0)$ is adjacent to sections $n+1,\ldots,n+n_j$ of $S^*(F,R_1)$. Third, set $n \leftarrow n + n_j$.

(5) [Adjacencies of the $+\infty$-section of $S^*(F,c^0)$ and sections of $S^*(F,R_1)$.] Set $n_{m+1} \leftarrow$ the number of real roots of $F(u,y)$ in $(s_m,+\infty)$. Record in L_1 that section $m+1$ of $S^*(F,c^0)$ is adjacent to sections $n+1,\ldots,n+n_{m+1},n+n_{m+1}+1$ of $S^*(F,R_1)$.

(6) [Adjacencies of sections of $S^*(F,c^0)$ and sections of $S^*(F,R_2)$.] Repeat Steps (3), (4), and (5) with v in place of u, and L_2 in place of L_1. Exit.

4 Construction of Proper Cylindrical Algebraic Decompositions

Theorem 8 tells us how, given $A \subset I_2$, to construct a proper A-invariant cad of E^2. For nonzero $F \in I_r = I_{r-1}[x_r]$, the *content* of F is the greatest common divisor of its coefficients. If $F = 0$, its content is 1. F is *primitive* if its content is 1. The *primitive part* of F, written $pp(F)$, is F divided its content. The reader may wish to refer back to Section 2 of Part I for other terms and notation used in the theorem.

Theorem 8 *Let $A \subset I_2$, let A^* be the primitive part of the product of the nonzero elements of A, and let D' be the proper cad of E^1 for which $P(x) = \prod \mathrm{PROJ}(A)$ is a defining polynomial. Then*

(1) A^* is delineable on every $c \in D'$.

(2) $\bigcup_{c \in D'} S(A^*, c)$ is a proper A-invariant cad of E^2, and A^* is a defining polynomial for it.

Proof. (1) For any nonzero $F \in I_2$, and any $\alpha \in E$, one can easily see that $F(\alpha, y) = 0$ if and only if α is a root of content(F). Hence since A^* is primitive, $A^*(\alpha, y) \neq 0$ for all $\alpha \in E$, and so A^* is delineable on every 0-cell of D'. Let c be a 1-cell of D'. Let A_i be any nonzero element of A. content(A_i) divides ldcf(A_i), hence since ldcf$(A_i) \in \mathrm{PROJ}(A)$, and D' is $\mathrm{PROJ}(A)$-invariant, content(A_i) is nonvanishing at all points of c, hence $A_i(\alpha, y) \neq 0$ for all $\alpha \in c$, hence by Theorem 4 of Part I, A_i is delineable on c. By another application of the same theorem, it is clear that the product F of the nonzero elements of A is delineable on c. Since the content of F is the product of the contents of the nonzero A_i's, $V(F) \cap Z(c) = V(\mathrm{pp}(F)) \cap Z(c)$, and so since $A^* = \mathrm{pp}(F)$, A^* is delineable on c.

(2) Since A^* is delineable on every $c \in D'$, $\bigcup_{c \in D'} S(A^*, c)$ is a cad of E^2. By Theorem 1 it is proper, and clearly A^* is a defining polynomial for it. Let A_i be any nonzero element of A. If c is a 1-cell of D', then as just shown, A_i is delineable on c, and every A_i-section of $Z(c)$ is an A^*-section of $Z(c)$, hence $S(A^*, c)$ is A_i-invariant. If c is a 0-cell with sample point α, then either content(A_i) vanishes at α and A_i is identically zero on c, or content(A_i) does not vanish at α, A_i is delineable on c, and every A_i-section of $Z(c)$ is an A^*-section of $Z(c)$. In either case, $S(A^*, c)$ is A_i-invariant. Hence $S(A^*, c)$ is A-invariant. ∎

Here now is our algorithm for construction of a proper A-invariant cad of E^2, and its adjacencies. It is not difficult to see that the arguments we pass to SSADJ2 each time we call it in Step (3) satisfy the hypotheses on its inputs. Theorem 4 implies that the adjacencies we get back from SSADJ2 are indeed adjacencies of our cad of E^2.

$$\mathrm{CADA2}(A; I, L, S)$$

Input: A is a subset of I_2.

Outputs: I is a list of the indices of the cells of a proper A-invariant cad D of E^2. L is a list of all adjacencies of D, plus the adjacencies involving infinite sections. S is a list of sample points for D.

(1) [Construct sample points for induced cad D' of E^1.] Set $P \leftarrow \mathrm{PROJ}(A)$. Isolate the real roots of the irreducible factors of the nonzero elements of P to determine the 0-cells of D'. Construct a sample point for each cell of D'. Set $A^* \leftarrow$ primitive part of the product of the nonzero elements of A.

(2) [Construct cell indices, determine intrastack adjacencies for D.] Let $a_1 < a_2 < \cdots < a_{2n} < a_{2n+1}$, $n \geq 0$, be the sample points for D' (each a_{2i+1} is a rational sample point for a 1-cell; each a_{2i} is an algebraic sample point for a 0-cell). Set $I \leftarrow$ the empty list. Set $L \leftarrow$ the empty list. For $i = 1, \ldots, 2n + 1$ do the following three things: first, isolate the real roots of $A^*(a_i, y)$ to determine the sections of a stack T in E^2, second, construct cell indices for the cells of T and add to I, third, record the intrastack adjacencies for T in L.

(3) [Section-section adjacency determination.] For $i = 1, \ldots, n$, call SSADJ2 with inputs A^*, a_{2i}, a_{2i-1}, and a_{2i+1}, and add the contents of its outputs L_1 and L_2 to L. (The section numbers which occur in the adjacencies returned by SSADJ2 must first be converted into the indices of the corresponding cells of D. For example, if the list L_1 returned by the i-th call to SSADJ2 contains the adjacency $\{3, 2\}$, it must be converted to $\{(2i, 6), (2i - 1, 4)\}$ before being added to L.)

(4) [Inference of remaining interstack adjacencies.] Use the current contents of L to infer the remaining interstack adjacencies of D, as described at the end of Section 2. Add them to L.

(5) [Construct sample points.] Set $S \leftarrow$ the empty list. Use the sample points for D' and the isolating intervals constructed in Step (2) to construct sample points for the cells of D, adding them to S. Exit.

5   An Example

We now trace CADA2 for the sample cad of E^2 discussed in Section 6 of Part I. The input polynomials are:

$$A_1(x, y) = 144y^2 + 96x^2y + 9x^4 + 105x^2 + 70x - 98,$$

$$A_2(x, y) = xy^2 + 6xy + x^3 + 9x,$$

As in Part I, we may take PROJ(A) to consist of the two elements $p_1(x) = x^4 - 15x^2 - 10x + 14$ and $p_2(x) = x$. The four real roots of p_1, which are -3.26, -1.51, 0.7, 4.08 (the presence of a decimal point in a number will indicate that its value is approximate), and the unique root $x = 0$ of p_2 become the 0-cells of D'. We take the 1-cell sample points to be -4, -2, -1, $1/2$, 4, and 9. We set A^* to be the primitive part of $A_1 A_2$. Since A_1 has content 1 and A_2 has content x, $A^* = (A_1 A_2)/x$, which is

$$144y^4 + 96x^2y^3 + 864y^3 + 9x^4y^2 + 825x^2y^2 + 70xy^2 + 1198y^2 + 150x^4y +$$
$$1494x^2y + 420xy - 588y + 9x^6 + 186x^4 + 70x^3 + 847x^2 + 630x - 882.$$

After constructing cell indices and determining intrastack adjacencies, CADA2 makes a total of five calls to SSADJ2, moving from left to right along the cells of the induced cad of 1-space. Let us look at the third of these calls. The inputs to SSADJ2 are $F = A^*$, $\alpha = 0$, $a_1 = -1$, and $a_2 = 1/2$. Let c^0, c_1^1, and c_2^1 denote the cells of D' whose respective sample points are α, a_1, and a_2. SSADJ2 begins by isolating the real roots of $A^*(0, y)$, which are -3.0, -0.82, and 0.82. We obtain $s_0, \ldots, s_3 = -4, -1, 0, 1$. At the start of Step 2, we set $u \leftarrow -1$ and $v \leftarrow 1/2$. We check to see if $F(x, s_0) = F(x, -4)$ has a root in $[u, v]$. The real roots of $F(x, -4)$ are -4.58 and -3.30, so this does not happen. However, the real roots of $F(x, s_1) = F(x, -1)$ are -1.39 and -0.79, so since $u < -0.79 < v$, we shrink our "ladder". We set $u \leftarrow -1/4$, and continue checking for the new u and v. We find that $F(x, 0)$ has real roots of -1.25 and 0.68, and $F(x, 1)$ has no real roots, so we have completed Step 2.

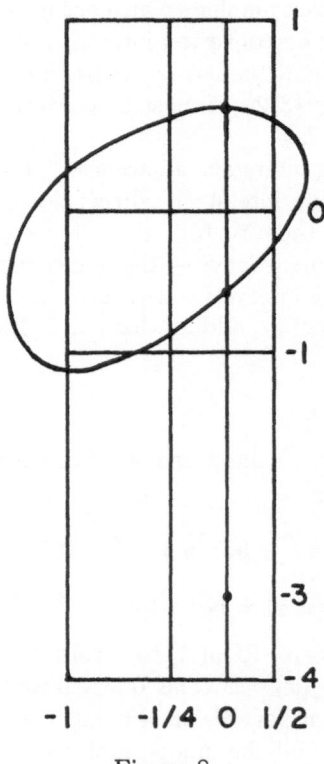

Figure 8

Figure 8 illustrates what occurred in Step 2. We see that with the wide "ladder" we started with, the curve crosses one of the horizontal "rungs" (i.e. $y = s_1 = -1$), but when we shrink the "ladder", this no longer occurs. Thus for the narrower "ladder", the intersections of the curve with the vertical sides of a "compartment" are in 1–1 correspondence with the adjacencies in that compartment.

In Step 3, we find that $F(-1/4, y)$ has no real roots in $(-\infty, -4)$, and so we record only that section 0 of $S^*(c^0)$ is adjacent to section 0 of $S^*(c_1^1)$. In Step 4, we have $m = 3$, so the loop will be executed three times. The real roots of

$F(-1/4, y)$ are -0.89 and 0.85; comparing these with $s_0, \ldots, s_3 = -4, -1, 0, 1$, we can see what will happen as we do the loop. The first time through we record no adjacencies (section 1 of $S^*(c^0)$ is an isolated point). The second time we record that section 2 of $S^*(c^0)$ is adjacent to section 1 of $S^*(c_1^1)$. The third time we record that section 3 of $S^*(c^0)$ is adjacent to section 2 of $S^*(c_1^1)$. In Step 5 we record only that section 4 of $S^*(c^0)$ (its $+\infty$-section) is adjacent to section 3 of $S^*(c_1^1)$ (its $+\infty$-section). The events of Step 6 are similar.

We saw above that $A^*(0, y)$ has three real roots, and thus the cad D constructed by CADA2 has three sections over the 0-cell $c^0 = 0$ of D'. Yet the cad constructed by algorithm CAD of Part I for the same input, shown in Fig. 5 of Part I, has only two sections over $c^0 = 0$. We have another example of how CAD and CADA2 may construct different cad's for the same input. Since content$(A_2) = x$, $A_2(0, y) = 0$, and CAD ignores the fact that the point $\langle 0, -3 \rangle$ is in $V(\mathrm{pp}(A_2))$, whereas CADA2 makes that point a section of $S(c^0)$.

An Improvement of the Projection Operator in Cylindrical Algebraic Decomposition[1]

Hoon Hong

1 Introduction

The cylindrical algebraic decomposition (CAD) of Collins (1975) provides a potentially powerful method for solving many important mathematical problems, provided that the required amount of computation can be sufficiently reduced. An important component of the CAD method is the projection operation. Given a set A of r-variate polynomials, the projection operation produces a certain set P of $(r-1)$-variate polynomials such that a CAD of r-dimensional space for A can be constructed from a CAD of $(r-1)$-dimensional space for P. The CAD algorithm begins by applying the projection operation repeatedly, beginning with the input polynomials, until univariate polynomials are obtained. This process is called the projection phase.

McCallum (1984) made an important improvement to the original projection operation. He showed, using a theorem from real algebraic geometry, that the original projection set can be substantially reduced in size, provided that input polynomials are well-oriented.

In this paper, we present another improvement to the original projection operation. However our improvement does not impose any restrictions on input polynomials. This improvement is essentially obtained by generalizing a lemma used in Collins' original proof.

Let m be the number of polynomials contained in A, and let n be a bound for the degree of each polynomial in A in the projection variable. The number of polynomials produced by the original projection operation is dominated by $m^2 n^3$ whereas the number of polynomials produced by our projection operation is dominated by $m^2 n^2$.

Preliminary experiments show that our projection operation can sometimes significantly speed up the projection phase of the CAD method. In fact, we present an experimental result that shows 85 times speedup.

The organization of the paper is as follow. In Section 2 we present our projection operation and prove that it satisfies the requirements that any projection operation must satisfy. In Section 3 we analyze and compare the original projection operation and ours in terms of the number of polynomials produced by the projection operations. In Section 4 we present some empirical comparisons between the original projection operation and ours.

2   Idea

In this section, we present our projection operation and prove its validity. We assume that the reader is familiar with the basic terminology of (Arnon et al. 1984a; Collins 1975), including reducta, principal subresultant coefficients, sign-invariance, delineability, regions, sections, sectors, cylinders, and stacks.

Let I_r be the set of all r-variate polynomials with integral coefficients, $r \geq 2$. A *projection operator* is a mapping $\text{PROJ} : 2^{I_r} \longrightarrow 2^{I_{r-1}}$ such that for any finite subset A of I_r and any $\text{PROJ}(A)$-invariant region R in E^{r-1} the following two conditions hold: (1) every element of A is either delineable or identically zero on R, and (2) the sections of $Z(R)$ belonging to different $F, G \in A$ are either disjoint or identical.

2.1   Original Projection Operator

Collins (1975) proposed the following mapping PROJC as a projection operator in his pioneering paper on cylindrical algebraic decomposition:

$$
\begin{aligned}
\text{PROJC}(A) &= \text{PROJ}_1(A) \cup \text{PROJ}_2(A), \\
\text{PROJ}_1(A) &= \bigcup_{\substack{F \in A \\ F^* \in \text{RED}(F)}} (\{\text{ldcf}(F^*)\} \cup \text{PSC}(F^*, F^{*\prime})), \\
\text{PROJ}_2(A) &= \bigcup_{\substack{F, G \in A \\ F < G}} \bigcup_{\substack{F^* \in \text{RED}(F) \\ G^* \in \text{RED}(G)}} \text{PSC}(F^*, G^*),
\end{aligned}
$$

where "<" denotes an arbitrary linear ordering of the elements of A, and for any $F, G \in I_r$, $\mathrm{PSC}(F, G)$ denotes the set

$$\{\mathrm{psc}_k(F, G) | 0 \leq k < \min(\deg(F), \deg(G)), \mathrm{psc}_k(F, G) \neq 0\},$$

and $\mathrm{RED}(F)$ denotes the set

$$\{\mathrm{red}^i(F) | 0 \leq i \leq \deg(F), \mathrm{red}^i(F) \neq 0\}.$$

2.2 Improved Projection Operator

Now we propose another mapping PROJH as a projection operator:

$$\begin{aligned}
\mathrm{PROJH}(A) &= \mathrm{PROJ}_1(A) \cup \mathrm{PROJ}_2^*(A), \\
\mathrm{PROJ}_2^*(A) &= \bigcup_{\substack{F, G \in A \\ F < G}} \bigcup_{F^* \in \mathrm{RED}(F)} \mathrm{PSC}(F^*, G),
\end{aligned}$$

where $\mathrm{PROJ}_1(A)$ is defined as above.

A proof that PROJH is a projection operator is almost the same as the proof for PROJC. Therefore we will show only the differences between these two proofs. First we recall the following. Let F, G be two polynomials in I_r, $r \geq 2$, such that $\deg(F) = m \geq 1$ and $\deg(G) = n \geq 1$. Let $\alpha \in E^{r-1}$ such that $\mathrm{ldcf}(F)(\alpha) \neq 0$ and $\mathrm{ldcf}(G)(\alpha) \neq 0$. Then

$$\mathrm{psc}_k(F, G)(\alpha) = \mathrm{psc}_k(F(\alpha, x_r), G(\alpha, x_r)),$$

for $0 \leq k < \min(m, n)$. This obvious equality is used in proving PROJC to be a projection operator. Now we give a lemma which generalizes this observation. In fact, this lemma contains the key idea leading to the new projection operator PROJH.

Lemma 1 *Let F, G be two polynomials in I_r, $r \geq 2$, such that $\deg(F) = m \geq 1$ and $\deg(G) = n \geq 1$. Let $\alpha \in E^{r-1}$ such that $\mathrm{ldcf}(F)(\alpha) \neq 0$ and $\deg(G(\alpha, x_r)) = \ell \geq 1$. Then we have*

$$\mathrm{psc}_k(F, G)(\alpha) = [\mathrm{ldcf}(F)(\alpha)]^{n-\ell} \, \mathrm{psc}_k(F(\alpha, x_r), G(\alpha, x_r))$$

for $0 \leq k < \min(m, \ell)$.

The lemma follows immediately from the definition of principal subresultant coefficients (psc).

Now we recall two more lemmas which were used in proving PROJC to be a projection operator and which will also be used in proving PROJH to be a projection operator.

Lemma 2 *Let A be a finite subset of I_r, $r \geq 2$, and let R be a $\mathrm{PROJ}_1(A)$-invariant region in E^{r-1}. Then every element of A is either delineable or identically zero on R.*

The original proof is given in (Collins 1975, theorem 4). A slightly different proof is also given in (Arnon et al. 1982).

Lemma 3 *Let A be a finite subset of I_r, $r \geq 2$, and let R be a $\mathrm{PROJ}_1(A)$-invariant region in E^{r-1}. Let F and G be any two different polynomials in A. If the least integer k such that $\mathrm{psc}_k(F(\alpha, x_r), G(\alpha, x_r)) \neq 0$ is constant for $\alpha \in R$, then the sections of $Z(R)$ belonging to F and G are either disjoint or identical.*

The original proof is given in (Collins 1975, theorem 5). A slightly different proof is also given in (Arnon et al. 1982, theorem 3.7).

Now we are ready to prove that PROJH is a projection operator.

Theorem 1 PROJH *is a projection operator. To be specific, let A be a finite subset of I_r, $r \geq 2$. Then for any $\mathrm{PROJH}(A)$-invariant region R in E^{r-1} the following two conditions hold: (1) every element of A is either delineable or identically zero on R, and (2) the sections of $Z(R)$ belonging to different F, $G \in A$ are either disjoint or identical.*

Proof. Let R be a $\mathrm{PROJH}(A)$-invariant region R in E^{r-1}. Since the set $\mathrm{PROJH}(A)$ contains the set $\mathrm{PROJ}_1(A)$, clearly R is also $\mathrm{PROJ}_1(A)$-invariant. Then, by Lemma 2, every element of A is either delineable or identically zero on R. Therefore PROJH satisfies the first condition. So we continue to prove that PROJH satisfies the second condition also.

Let F, G be two polynomials in A such that $F < G$. Since $\mathrm{PROJH}(A)$ contains every nonzero coefficient of F and G, each coefficient of F and G either vanishes everywhere or nowhere on R. Hence $\deg(F(\alpha, x_r))$ and $\deg(G(\alpha, x_r))$ are constant for $\alpha \in R$. Let m denote the constant $\deg(F(\alpha, x_r))$ and let ℓ denote the constant $\deg(G(\alpha, x_r))$. Let n denote $\deg(G)$.

If $m \leq 0$ or $\ell \leq 0$, then the second condition is trivially satisfied. So from now on assume that $m \geq 1$ and $\ell \geq 1$. Let F^* be the unique reductum of F such that $m = \deg(F^*) = \deg(F^*(\alpha, x_r)) = \deg(F(\alpha, x_r))$ and let k be any integer such that $0 \leq k < \min(m, \ell)$. Then $\mathrm{psc}_k(F^*, G)$ is contained in $\mathrm{PROJH}(A)$ and thus the sign of $\mathrm{psc}_k(F^*, G)(\alpha)$ is constant for $\alpha \in R$. Then, by Lemma 1, the sign of $[\mathrm{ldcf}(F^*)(\alpha)]^{n-\ell} \, \mathrm{psc}_k(F^*(\alpha, x_r), G(\alpha, x_r))$ is constant for $\alpha \in R$.

Since $\mathrm{ldcf}(F^*)$ is contained in $\mathrm{PROJH}(A)$, the sign of $\mathrm{ldcf}(F^*)(\alpha)$ is constant for $\alpha \in R$. Also from the way F^* is defined, $\mathrm{ldcf}(F^*)(\alpha)$ is nonzero for $\alpha \in R$. Thus the sign of $[\mathrm{ldcf}(F^*)(\alpha)]^{n-\ell}$ is nonzero and constant for $\alpha \in R$. Hence the sign of $\mathrm{psc}_k(F^*(\alpha, x_r), G(\alpha, x_r))$ is constant for $\alpha \in R$. Then, since $F^*(\alpha, x_r) = F(\alpha, x_r)$ for $\alpha \in R$, the sign of $\mathrm{psc}_k(F(\alpha, x_r), G(\alpha, x_r))$ is constant for $\alpha \in R$. Hence the least k such that $\mathrm{psc}_k(F(\alpha, x_r), G(\alpha, x_r)) \neq 0$ is also constant for $\alpha \in R$. Then, by Lemma 3, the sections of $Z(R)$ belonging to F and G are either disjoint or identical. Therefore the second condition is satisfied. ∎

3   Analysis

In this section, we analyze and compare Collins' projection operator and our projection operator in terms of the number of polynomials produced by the projection operators. In the following discussion, let A be a finite subset of I_r, let m be the number of polynomials contained in A, and let n be a bound for the degree of each polynomial in A in the projection variable.

Theorem 2 (PROJC) *The number of polynomials contained in* PROJC(A) *is dominated by* $m^2 n^3$.

A proof is given in (Collins 1975, pp. 161, 163).

Theorem 3 (PROJH) *The number of polynomials contained in* PROJH(A) *is dominated by* $m^2 n^2$

Proof. The set $\text{PROJ}_1(A)$ contains at most $m(n + 1) + m(n - 1) = 2mn$ polynomials. So let us now continue with the set $\text{PROJ}_2^*(A)$.

In the set $\text{PROJ}_2^*(A)$, a pair (F, G) can be chosen in $\binom{m}{2}$ ways. So let us count the number of psc's such as $\text{psc}_k(\text{red}^i(F), \text{red}(G))$ for a fixed pair (F, G).

Since $0 \leq k < \min(\deg(\text{red}^i(F)), \deg(B)) \leq n - i$, k can be chosen in $n - i$ ways for a given i, $0 \leq i \leq n - 1$. Hence (i, k) can be chosen in $\sum_{i=0}^{n-1}(n - i) = \frac{n(n+1)}{2}$ ways. So the set $\text{PROJ}_2^*(A)$ contains at most $\binom{m}{2}\frac{n(n+1)}{2}$ polynomials. Therefore PROJH(A) contains at most $2mn + \binom{m}{2}\frac{n(n+1)}{2}$ polynomials. Hence the number of polynomials in PROJH(A) is dominated by $m^2 n^2$. ∎

4   Empirical Results

In this section we present empirical comparisons between the original projection operator and our projection operator on several problems.

We implemented both projection operators as a part of the **qepcad** system (Collins and Hong 1991) which carries out quantifier elimination with a partially built CAD. The **qepcad** system is implemented in the ALDES/SAC-2 computer algebra system (Collins and Loos 1990).

The **qepcad** system, given a set A of r-variate integral polynomials, applies a projection operator repeatedly, until univariate projection polynomials are obtained. Following the suggestions made in section 5 of (Collins 1975), it computes a squarefree basis of the polynomials each time before applying the projection operation. Explicitly, the following algorithm is used, in which finest squarefree bases are computed.

Projection Phase

(1) Set $J_r \leftarrow A$. Set $k \leftarrow r$.
(2) Compute the contents C_k and the primitive parts P_k of J_k.
(3) Compute the finest squarefree basis B_k of the primitive parts P_k.
(4) If $k = 1$, then return. Set $R_{k-1} \leftarrow \mathrm{PROJ}(B_k)$. Set $J_{k-1} \leftarrow C_k \cup R_{k-1}$.
    Remove from J_{k-1} all the constant polynomials if there are any. Set $k \leftarrow k - 1$. Go to Step 2.

To both projection operators, we also made additional refinements as described in (Collins 1975; Hong 1989).

We have carried out experiments on the following sets of input polynomials.

Input Set 1 The following three polynomials in $\mathbb{Z}[a, b, r, s, t]$ are obtained from a condition that a cubic polynomial has three real roots, counting multiplicities.

$$r + s + t$$
$$rs + st + tr - a$$
$$rst - b$$

Input Set 2 The following three polynomials in $\mathbb{Z}[a, b, c, x, y]$ are obtained from the x-axis ellipse problem (Arnon and Mignotte 1988).

$$b^2 x^2 - 2b^2 xc + b^2 c^2 + a^2 y^2 - a^2 b^2$$
$$x^2 + y^2 - 1$$
$$ab$$

Input Set 3 The following five polynomials in $\mathbb{Z}[d, c, b, a]$ are from a formula used by Davenport and Heintz (1988) in order to show the time complexity of the quantifier elimination in elementary algebra and geometry.

$$a - d$$
$$b - c$$
$$a - c$$
$$b - 1$$
$$a^2 - b$$

Tables 1, 2, and 3 show the performance comparisons on Input Sets 1, 2, and 3, respectively. In these tables, T_2, T_3, and T_4 are the times (in seconds) taken to carry out Step 2, 3, and 4 of the above algorithm Projection Phase respectively. The numbers N_B and N_J are the numbers of polynomials in the sets B_k and J_{k-1} respectively. N_R is the size of the set R_{k-1} considered as a multiset; that is, it counts the number of times each polynomial in R_{k-1} is produced by the projection operator. The last rows show the total times taken, including the times taken for garbage collection. All the experiments were done on a SUN3/50 running Unix using 4 megabytes of memory for lists.

For Input Set 1, our projection operator speeds up the projection phase $21620.2/256.2 \approx 85$ times.

Table 1: Comparison on Input Set 1

k	Proj.	T_2	T_3	T_4	N_B	N_R	N_J
5	Orig.	0.4	1.6	14.2	3	8	7
	Impr.	0.4	1.6	10.6	3	8	7
4	Orig.	0.5	9.2	60.3	6	42	30
	Impr.	0.5	9.2	33.2	6	32	21
3	Orig.	1.6	26.2	1513.1	14	324	228
	Impr.	1.0	20.1	50.9	8	64	44
2	Orig.	4.3	415.5	16167.4	77	3079	2031
	Impr.	0.5	17.0	73.4	16	151	106
1	Orig.	3.3	998.9	N/A	1	N/A	N/A
	Impr.	0.2	17.6	N/A	1	N/A	N/A
Total	Orig.			21620.2		N/A	
time	Impr.			256.2		N/A	

Table 2: Comparison on Input Set 2

k	Proj.	T_2	T_3	T_4	N_B	N_R	N_J
5	Orig.	0.3	26.6	37.4	2	6	6
	Impr.	0.3	26.6	37.2	2	6	6
4	Orig.	1.3	35.0	124.0	5	22	22
	Impr.	1.3	34.7	27.3	5	18	18
3	Orig.	1.6	18.2	177.8	13	135	74
	Impr.	1.1	5.8	17.8	7	32	25
2	Orig.	1.6	58.2	53.7	30	482	280
	Impr.	0.3	12.5	2.2	9	49	32
1	Orig.	0.9	125.1	N/A	111	N/A	N/A
	Impr.	0.1	17.6	N/A	7	N/A	N/A
Total	Orig.			727.0		N/A	
time	Impr.			179.5		N/A	

Table 3: Comparison on Input Set 3

k	Proj.	T_2	T_3	T_4	N_B	N_R	N_J
4	Orig.	0.2	1.7	0.9	3	7	6
	Impr.	0.2	1.6	0.8	3	7	6
3	Orig.	0.2	0.7	3.3	5	15	10
	Impr.	0.2	0.7	3.1	5	15	10
2	Orig.	0.1	2.2	0.6	6	21	11
	Impr.	0.1	2.2	0.6	6	21	11
1	Orig.	0.1	1.5	N/A	4	N/A	N/A
	Impr.	0.1	1.5	N/A	4	N/A	N/A
Total	Orig.			12.6		N/A	
time	Impr.			12.5		N/A	

For Input Set 2, our projection operator speeds up the projection phase $727.0/179.5 \approx 4$ times. Note also that the number of univariate basis polynomials (N_B when $k = 1$) is reduced from 111 to 7. A similar improvement was obtained by Arnon and Mignotte (1988); however their approach, unlike ours, is based on problem specific information.

Reducing the size of the projection set is important because it causes additional speedup in the following phases of CAD algorithm, namely the base phase and the extension phase (Arnon et al. 1984a).

For Input Set 3, our projection operator does not show any improvement either on the time taken for the projection phase or the number of basis polynomials produced at each iteration. It is mainly because the input polynomials are very simple.

Acknowledgments

This research was done as a part of the author's PhD dissertation research. The author is grateful for all the guidance given by his PhD adviser, Professor G. E. Collins, who also made several valuable suggestions on both the structure and the content of this paper.

Partial Cylindrical Algebraic Decomposition for Quantifier Elimination[1]

George E. Collins and Hoon Hong

1 Introduction

Cylindrical Algebraic Decomposition (CAD) by Collins (1975) provides a potentially powerful method for solution of many important mathematical problems by means of quantifier elimination, provided that the required amount of computation can be sufficiently reduced. Arnon (1981) introduced the important method of clustering for reducing the required computation and McCallum (1984) introduced an improved projection operation which is also very effective in reducing the amount of computation. In this paper we introduce yet another method for reducing the amount of computation which we will call *partial CAD construction*.

The method *partial CAD construction* is based on the simple observation that we can *very often* complete quantifier elimination by a *partially* built CAD if we utilize more information contained in the input formula. Our method utilizes three aspects of the input formula: the quantifiers, the Boolean connectives, and the absence of some variables from some polynomials occurring in the input formula. Preliminary observations show that the new method is always more efficient than the original, and often significantly more efficient.

A simple example can illustrate how we utilize the quantifier information. Let us consider a sentence in two variables $(\exists x)(\exists y)F(x, y)$. The original CAD method computes a certain decomposition D_1 of $\mathbb{R}$ and then lifts this to a decomposition D_2 of $\mathbb{R}^2$ by constructing a stack of cells in the cylinder over each cell of D_1. Then the quantifier elimination proceeds by determining the set of all cells of D_1 in which $(\exists y)F(x, y)$ is true. Finally, it computes the truth value of $(\exists x)(\exists y)F(x, y)$ by checking whether the set is empty. In contrast, our

---

[1]Reprinted from *Journal of Symbolic Computation*, Volume 12, Number 3, 1991, pp. 299–328, by permission of Academic Press.

method constructs only one stack at a time, aborting the CAD construction as soon as a cell of D_1 is found which satisfies $(\exists y)F(x, y)$, if any such cell exists. If, instead, the given sentence were $(\forall x)(\exists y)F(x, y)$, our method would stop as soon as any cell is found in which $(\exists y)F(x, y)$ is false. The quantifier $(\exists y)$ could be changed to $(\forall y)$ without effect. The method illustrated above for two variables extends in an obvious way to more variables, with even greater effectiveness because the CAD construction can be partial in each dimension. This idea applies equally to formulas in which some variables are free, and the CAD construction can be partial again as in the above example.

Another simple example can illustrate how we utilize the Boolean connectives and the absence of some variables from some polynomials. Let $(Qx)(Qy)$ $F(x, y)$ be a sentence in two variables such that $F(x, y) = F_1(x) \wedge F_2(x, y)$, where F_1 and F_2 are quantifier-free formulas and Q's are arbitrary quantifiers. Note that the variable y is absent from the formula F_1 and thus from the polynomials contained in F_1. The original CAD method determines the truth value of $(Qy)F(x, y)$ in a cell c of D_1 by building a stack over it. In contrast, our method first determines the truth value of $F_1(x)$ by evaluating the associated polynomials on the sample point of c. If it is false, $(Qy)F(x, y)$ is clearly false in the cell c, so we do not need to build a stack over it. Likewise, if $F(x, y) = F_1(x) \vee F_2(x, y)$ and $F_1(x)$ is true in a cell c, $(Qy)F(x, y)$ is clearly true in the cell c, so we do not build a stack over it either. This method can be generalized in an obvious way to arbitrary number of variables and an arbitrary formula $F(x_1, \ldots, x_r)$.

Remember that our method builds only one stack at a time. But at a certain time there might be many candidate cells on which we can build stacks, and so we need to choose one of them. We have dealt with this problem as follows. We have designed an interactive program which allows the user to choose among candidate cells, or to choose one of a small number of selection algorithms, called cell-choice strategies, which the program will use repeatedly.

Some of these strategies were designed to be efficient for each of several sub-problem classes, namely collision problems in robot motion planning (Buchberger et al. 1988), termination proof of term rewrite systems based on polynomial interpretation (Lankford 1979; Huet and Oppen 1980), and consistency of polynomial strict inequalities (McCallum 1987). For these problem classes we achieved very large reductions in required computation time.

The plan of the paper is as following: In Section 2 we elaborate the main ideas underlying our method. In Section 3 we present the partial CAD construction algorithm. This algorithm is generic in the sense that it does not impose any particular cell-choice strategy. In Section 4 we describe various cell-choice strategies. In Section 5 we illustrate our method by showing a terminal session on a simple but "real" example. In Section 6, we present some empirical comparisons for several problems from diverse application areas.

2 Main Idea

In this section, we elaborate the main ideas underlying our method by showing how our main algorithm evolved from the original one. We assume that the reader is familiar with the basic terminology of (Arnon et al. 1984a), including that of CAD, induced CAD, cells, cell indices, cylinders, stacks, sample points, defining·formulas, delineability, sign-invariance, and truth-invariance.

We also assume that the reader is familiar with the basic terminology of graph theory and trees, including that of roots, leaves, children, parents, ancestors, descendants, and levels. (The level of the root node is defined to be 0.)

2.1 Original Algorithm

Let $F^* = (Q_{f+1}x_{f+1})\cdots(Q_r x_r)F(x_1,\ldots,x_r)$, $0 \leq f < r$, be any formula of elementary algebra, where the formula $F(x_1,\ldots,x_r)$ is quantifier-free. Let D_r be any truth-invariant CAD of $\mathbb{R}^r$ for $F(x_1,\ldots,x_r)$. For $f \leq k < r$, D_r induces a CAD D_k of $\mathbb{R}^k$ in which the formula $F_k^* = (Q_{k+1}x_{k+1})\cdots(Q_r x_r)F(x_1,\ldots,x_r)$ is truth-invariant. Thus if c is any cell of D_k, the formula F_k^* has a constant truth value throughout c. This observation leads to the following definition.

Definition 1 (Truth value) *Let c be a cell of D_k, $f \leq k \leq r$, and let F_k^* be the formula $(Q_{k+1}x_{k+1})\cdots(Q_r x_r)F(x_1,\ldots,x_r)$. We define $v(c)$, called the truth value of the cell c, to be the constant truth value of F_k^* throughout c.*

In order to treat the case $f = k = 0$ uniformly we hypothesize a decomposition D_0 of $\mathbb{R}^0$ having a single cell, whose truth value is assumed to be the truth value of the sentence $(Q_1 x_1)\cdots(Q_r x_r)F(x_1,\ldots,x_r)$. From Definition 1 the following theorem is immediate.

Theorem 1 (Evaluation) *Let c be a cell of D_r and let $s = (s_1,\ldots,s_r)$ be a sample point of c, where each s_i is the i-th coordinate of the sample point. Then $v(c) =$ the truth value of $F(s_1,\ldots,s_r)$.*

This theorem provides a way to evaluate the truth values of cells of D_r from their sample points. From theorem 7 of (Collins 1975), the following theorem is immediate.

Theorem 2 (Propagation) *Let c be a cell of D_k, $f \leq k < r$, and let $c_1,\ldots,c_n$ be the cells in the stack over c. If $Q_{k+1} = \exists$, $v(c) = \bigvee_{i=1}^n v(c_i)$. If $Q_{k+1} = \forall$, $v(c) = \bigwedge_{i=1}^n v(c_i)$.*

This theorem provides a way to propagate the truth values of the cells of D_{k+1} to the truth values of the cells of D_k. From Definition 1, the following theorem is immediate.

Theorem 3 (Solution) *Let $S = \{c \in D_f | v(c) = true\}$ and let $W = \bigcup_{c \in S} c$.*
Then $(Q_{f+1} x_{f+1}) \cdots (Q_r x_r) F(x_1, \ldots, x_r) \iff (x_1, \ldots, x_f) \in W$.

This theorem provides a way to compute the solution set of an input formula. From Theorems 1, 2, and 3, Collins' original quantifier elimination algorithm follows as:

Original Algorithm

(1) [CAD construction.] Build a CAD D_r of $\mathbb{R}^r$.
(2) [Evaluation.] Evaluate the truth values of the cells of D_r by using Theorem 1.
(3) [Propagation.] For $k = r-1, r-2, \ldots, f$, determine the truth values of the cells of D_k from the truth values of the cells of D_{k+1} by using Theorem 2.
(4) [Solution.] Get the solution set W by using Theorem 3. (Now a defining formula for the set W is equivalent to the input formula.)

2.2 First Improvement

Our first improvement of the original algorithm is based on a simple observation on Theorem 2 that the truth value of a cell c can *often* be determined from the truth values of *only some* of the cells in the stack over c. To be specific, let c be a cell of D_k, $f \le k < r$. In case $Q_{k+1} = \exists$, $v(c)$ can be determined to be *true* as soon as the truth value of one of the cells in the stack over c is known to be *true*. Likewise, in case $Q_{k+1} = \forall$, $v(c)$ can be determined to be *false* as soon as the truth value of one of the cells in the stack over c is known to be *false*.

This observation was already made in (Collins 1975, p. 158) and used to reduce the number of truth evaluations in Step 2. We now, however, carry the observation further to reduce the number of cells constructed in Step 1. This can be done by intermingling CAD construction with truth evaluation so that parts of the CAD are constructed only as needed to further truth evaluation and aborts CAD construction as soon as no more truth evaluation is needed. By doing so, we can often complete quantifier elimination by a *partially built* CAD, thus reducing the amount of required computation.

In order to take the full advantage of this idea of *partial* CAD *construction*, we should also do our best to avoid any computations which will not be used later on in the process. This can be done by postponing all the computations which are not needed at a given moment, and carry them out only when it becomes clear that they are needed.

One such computation is the conversion of representation of a sample point. Let c be a cell of D_k, $1 \le k \le r$. As Arnon (1988b) points out, during CAD construction, the sample point of the cell c is represented in either one of the two representations: (a) *primitive*, consisting of a real algebraic number α and a k-tuple of elements of $Q(\alpha)$. (b) *extended*; consisting of a real algebraic number

α, and a $(k-1)$-tuple of elements of $Q(\alpha)$, a non-zero squarefree polynomial $g(x) \in Q(\alpha)[x]$, and an isolating interval for a real root of $g(x)$. (This root is the k-th coordinate of the sample point.) From now on we will call a sample point *an extended sample point* if it is in an extended representation, and *a primitive sample point* if it is in a primitive representation.

During CAD construction, we first get an extended sample point of a cell while building a stack to which the cell belongs (except when the cell is a sector or it belongs to D_1; in this case we trivially get a primitive sample point). Then the original CAD algorithm converts it into a primitive one immediately by using the NORMAL and SIMPLE algorithms of (Loos 1982a). This conversion process, however, is known to be often very expensive. So we would like to postpone the conversion process and carry it out only when it is needed. A primitive sample point of a cell is needed only when we want to build a stack over the cell or to evaluate its truth value[1]. Therefore we should carry out the conversion process just before building a stack or evaluating truth value. In this way, we do not waste time converting the representation of sample points which will not be used in the remaining process of the *partial* CAD construction.

Let us now devise an algorithm which utilizes these ideas. First, we define some terminology. A *partial* CAD is a tree of cells such that the root node is the hypothetical single cell of D_0, the children of the root node are the cells of D_1, and the children of a non-root node c are the cells in the stack over the cell (node) c. We denote the level of a cell c in a *partial* CAD tree as $l(c)$. A cell c is a leaf if either $l(c) = r$ or there is no stack built on it yet. Each cell in a partial CAD may have the following two data structures associated with it: *sample point* and *truth value*. The truth value associated with a cell c at any time can be one of the three: *true*, *false*, and *undetermined*. The value associated is *undetermined* if either $l(c) < f$ or it is not determined yet. A *candidate* cell is a leaf whose truth value is *undetermined*. Now here is an algorithm that utilizes the ideas discussed above:

First Improvement

(1) [Projection.] Compute the projection polynomials of all orders.
(2) [Initialization.] Initialize the partial CAD D as the single cell of D_0 whose truth value is set to be *undetermined*.
(3) [Choice.] If there is no candidate cell in the current partial CAD D, go to Step (8). Choose a candidate cell c from the current partial CAD D.
(4) [Conversion.] If $l(c) > 1$ and c is a section, convert the extended sample point of c into a primitive one by using the NORMAL and SIMPLE algorithms. If $l(c) = r$, go to Step (6).
(5) [Augmentation.] Construct a stack over c and form a sample point for each child cell. (If $l(c) = 0$, construct a primitive sample point for every

---

[1]Actually, as you will see later in this section, we can carry out truth evaluation with an extended sample point.

child. Otherwise, construct a primitive sample point for each sector and an extended sample point for each section.) Go to step (3).

(6) [Evaluation.] Evaluate the truth value of the cell c by using Theorem 1.

(7) [Propagation.] Determine the truth values of as many ancestors of c as possible by Theorem 2, removing from the tree the subtrees of each cell whose truth value is thus determined. Go to Step (3).

(8) [Solution.] (At this point, the truth value of every cell of D_f has been determined.) Get the solution set W by using Theorem 3. (Now a defining formula for the set W is equivalent to the input formula.)

It is important to note that the efficiency of this algorithm depends on which candidate cell is chosen at each iteration. It is also important to note that this algorithm is generic in the sense that it does not impose any particular strategy for choosing candidate cells. Thus it naturally raises a question: *Which strategy for candidate cell choice is best?* We will discuss this problem in Section 4.

2.3 Second Improvement

Further improvement of the generic algorithm is based on the observation that we can *often* carry out truth evaluation on a cell c such that $l(c) < r$, if some variables are absent from some polynomials occurring in the input formula. This is important because if we succeed in evaluating the truth value of the cell c, we do not have to build a partial CAD over c, thus reducing the amount of computation.

In this section, we also make another minor improvement by allowing propagation below free variable space which often results in further tree pruning.

These improvements are essentially generalization of Definition 1 and Theorems 1, 2, and 3. So we will describe the improvements by showing how we generalize the definition and the theorems.

Definition 1 defined the *truth value* for a cell c such that $f \leq l(c) \leq r$ (i.e., on or above free variable space). Now we would like to define the *truth value* for a cell c such that $0 \leq l(c) < f$ (i.e. below free variable space). The following definition is found to be useful.

Definition 2 (Truth value of a cell below free variable space) *Let c be a cell of D_k, $0 \leq k < f$. Let $F^* = (Q_{f+1}x_{f+1}) \cdots (Q_r x_r)F(x_1, \ldots, x_r)$. If the formula F^* has a constant truth value, say w, throughout $c \times \mathbb{R}^{f-k}$, then we define $v(c)$ to be the constant truth value w. Otherwise, $v(c)$ is undefined.*

Let us now generalize Theorem 2. Theorem 2 provided a way to propagate truth values, but only *on* or *above* free variable space, and now we would like to propagate truth values even *below* free variable space.

Theorem 4 (Propagation below free variable space) *Let c be a cell of D_k, $0 \leq k < f$, and let $c_1, \ldots, c_n$ be the cells in the stack over c. If $v(c_i) = w$ for all i, $1 \leq i \leq n$, then $v(c) = w$.*

Proof. Suppose that $v(c_i) = w$ for all i. Then from Definition 2 it is clear that $F^* = (Q_{f+1}x_{f+1}) \cdots (Q_r x_r) F(x_1, \ldots, x_r)$ has the constant truth value w throughout $c_i \times \mathbb{R}^{f-(k+1)}$ for all i. Therefore F^* has the constant truth value w throughout $(\bigcup_{i=1}^n c_i) \times \mathbb{R}^{f-(k+1)}$. Then, since $\bigcup_{i=1}^n c_i = c \times \mathbb{R}$, F^* has the constant truth value w throughout $c \times \mathbb{R}^{f-k}$. Thus $v(c)$ is defined and $v(c) = w$. ∎

Let us now generalize Theorem 1. Theorem 1 provided a way to carry out truth evaluation on a cell c such that $l(c) = r$. Now we would like to generalize it so that we can also carry out truth evaluation on a cell c such that $0 < l(c) < r$, if possible.

Let $f(x_1, \ldots, x_r)$ be a non-constant r-variate integral polynomial. We define $n(f)$, called *the effective number of variables in the polynomial f*, as the biggest integer k, $1 \le k \le r$, such that $\deg_k(f) > 0$, where $\deg_k(f)$ is the degree of the polynomial f with respect to the variable x_k. Let $A = \{A_1, \ldots, A_n\}$, $n \ge 1$, be the set of the non-constant r-variate integral polynomials occurring in an input formula $F(x_1, \ldots, x_r)$. Let $A^{[k]}$, $1 \le k \le r$, be the subset of A such that $A^{[k]} = \{f \in A | n(f) = k\}$. Let $A^{(k)}$, $1 \le k \le r$, be the subset of A such that $A^{(k)} = \{f \in A | n(f) \le k\}$.

In quantifier elimination problems arising in applications, $A^{(k)}$ is often a non-empty set for some $k < r$. In this case, we can determine the signs of the polynomials in $A^{(k)}$ on a cell c of D_k by evaluating the polynomials on its sample point, and thereby the truth values of the atomic formulas in which the polynomials occur. Then it may happen that the truth value of $F(x_1, \ldots, x_r)$ is determined by the truth values of these atomic constituents. In other words, $F(s_1, \ldots, s_k, x_{k+1}, \ldots, x_r)$ may have a constant truth value throughout $\mathbb{R}^{r-k}$, where $(s_1, \ldots, s_k)$ is the sample point of c.

For example, consider the formula $F(x, y) = A_1(x) > 0 \land A_2(x, y) < 0$. Let c be a cell of D_1, and let $s = (s_1)$ be its sample point. We can determine the sign of $A_1(x)$ by evaluating $A_1(x)$ on s_1. Suppose that $A_1(s_1) < 0$. Then we know right away $F(s_1, y)$ is *false* no matter what y may be.

In such cases we can evaluate $v(c)$ by using the following generalized evaluation theorem.

Theorem 5 (Generalized Evaluation) *Let c be a cell of D_k, $1 \le k \le r$, and let $s = (s_1, \ldots, s_k)$ be its sample point. If $F(s_1, \ldots, s_k, x_{k+1}, \ldots, x_r)$ has a constant truth value, say w, throughout $\mathbb{R}^{r-k}$, then $v(c) = w$.*

Proof. We will use mathematical induction on k from $k = r$ down to $k = 1$. In case $k = r$, the theorem is reduced to Theorem 1, and so we are done. Assume that the theorem is true for k, $r \ge k \ge 2$. Now we only need to prove that the theorem is true for $k - 1$. Let c be a cell of D_{k-1}, and let $c_1, \ldots, c_n$ be the cells in the stack over c. Let $s_1, \ldots, s_{k-1}$ be a sample point of c, and let $s_1, \ldots, s_{k-1}, s_k^{(i)}$ be a sample point of c_i ($1 \le i \le n$). Suppose that $F(s_1, \ldots, s_{k-1}, x_k, \ldots, x_r)$ has a constant truth value, say w, throughout

$\mathbb{R}^{r-(k-1)}$. Then $F(s_1, \ldots, s_{k-1}, s_k^{(i)}, x_{k+1}, \ldots, x_r)$ has the constant truth value w throughout $\mathbb{R}^{r-k}$ for all i. So, by the induction hypothesis, $v(c_i) = w$ for all i. If $k - 1 \geq f$, from Theorem 2 we have $v(c) = w$. Otherwise, from Theorem 4 we have $v(c) = w$. ∎

Now we need to generalize Theorem 3 in order to facilitate Theorems 4 and 5. From Definition 2 the following theorem is immediate.

Theorem 6 (Generalized Solution) *Let $Z = \{c_1, \ldots, c_n\}$ be a set of cells such that $\mathbb{R}^f = \bigcup_{c \in Z} c \times \mathbb{R}^{f-l(c)}$ and that $v(c_i)$ is defined for all i. Let S be a subset of Z such that $S = \{c \in Z | v(c) = true\}$ and let $W = \biguplus_{c \in S} c \times \mathbb{R}^{f-l(c)}$. Then $(x_1, \ldots, x_f) \in W \iff (Q_{f+1}x_{f+1}) \cdots (Q_r x_r) F(x_1, \ldots, x_r)$.*

The following algorithm utilizes Theorems 5, 2, 4, and 6.

Second Improvement

(1) [Projection.] Compute the projection polynomials of all orders.
(2) [Initialization.] Initialize the partial CAD D as the single cell of D_0 whose truth value is set to be *undetermined*.
(3) [Choice.] If there is no candidate cell in the current partial CAD D, go to Step (8), otherwise, choose a candidate cell c from the current partial CAD D.
(4) [Conversion.] If $l(c) > 1$ and c is a section, convert the extended sample point of c into a primitive one by using the NORMAL and SIMPLE algorithms. If $A^{[l(c)]} \neq \emptyset$, go to Step (6).
(5) [Augmentation.] Construct a stack over c and form a sample point for each child cell. (If $l(c) = 0$, construct a primitive sample point for every child. Otherwise, construct a primitive sample point for each sector and an extended sample point for each section.) Go to Step (3).
(6) [Trial Evaluation.] Try to evaluate the truth value of the cell c by using the Boolean connectives and Theorem 5. If the evaluation fails, go to Step (5).
(7) [Propagation.] Determine the truth values of as many ancestors of c as possible by Theorems 2 and 4, removing from the tree the subtrees of each cell whose truth value is thus determined. Go to Step (3).
(8) [Solution.] Set $Z \leftarrow$ the set of all leaves of the partial CAD D. (At this point, the set Z satisfies the hypothesis of Theorem 6.) Get the solution set W by using Theorem 6. (Now a defining formula for the set W is equivalent to the input formula.)

2.4 Final Improvement

We now make our final improvement based on the observation that we can carry out the trial truth evaluation with extended sample points. This is important because, as mentioned earlier, the conversion process is often very expensive.

Let $A = (A_1, \ldots, A_n)$, $n \geq 1$, be a set of r-variate integral polynomials. $r \geq 1$. Let f be the number of free variables in the input formula. Let $C = (C_1, \ldots, C_n)$ be the set of $(r-1)$-variate integral polynomials where C_j is the content of A_j for $1 \leq j \leq n$. Let $P = (P_1, \ldots, P_n)$ be the set of r-variate integral polynomials where P_j is the primitive part of A_j for $1 \leq j \leq n$. We include the signs into the contents so that $A_j = C_j P_j$ for $1 \leq j \leq n$. Let $B = (B_1, \ldots, B_l)$ be a finest squarefree basis of P. Let $\mathrm{PROJ}(B)$ denote any projection set for B sufficient to ensure delineability.[2] Then we define $\mathrm{Proj}(A)$ to be $C \cup \mathrm{PROJ}(B)$. We further define $\mathrm{Proj}^0(A)$ as A itself and $\mathrm{Proj}^k(A)$ as $\mathrm{Proj}(\mathrm{Proj}^{k-1}(A))$ for $1 \leq k \leq r-1$.

Let c be a cell of D_k, $1 \leq k < r$, and let $c_1, \ldots, c_n$ be the children of c. Arnon (1988b) presents an algorithm, InputSignaturesOverCell, that evaluates the signs of the polynomials in $\mathrm{Proj}^{r-(k+1)}(A)$ on $c_1, \ldots, c_n$, given the extended representations for the cells' sample points. In Section 3 we present another algorithm SIGNJ which does the same thing more efficiently. This algorithm together with the following theorem provides a way to determine the signs of the polynomials in $A^{(k+1)}$ on $c_1, \ldots, c_n$ even though the sample points of $c_1, \ldots, c_n$ exist in extended representations.

Theorem 7 *Let $A = \{A_1, \ldots, A_n\}$, $n \geq 1$, be a set of the non-constant r-variate integral polynomials. Then $A^{(k)} \subseteq \mathrm{Proj}^{r-k}(A)$ for $1 \leq k \leq r$.*

Proof. We will use mathematical induction on k from $k = r$ down to $k = 1$. In case $k = r$, we have $A^{(r)} = A = \mathrm{Proj}^0(A)$. So we are done. Assume that the theorem is true for k, $r \geq k \geq 2$. Now we only need to prove that the theorem is true for $k-1$. From the induction hypothesis and the obvious fact $A^{(k-1)} \subseteq A^{(k)}$, it is clear that $A^{(k-1)} \subseteq \mathrm{Proj}^{r-k}(A)$. Let C be the set of the contents of the polynomials in $\mathrm{Proj}^{r-k}(A)$. Then $A^{(k-1)} \subseteq C$. Since $C \subseteq \mathrm{Proj}(\mathrm{Proj}^{r-k}(A))$, we have $A^{(k-1)} \subseteq \mathrm{Proj}(\mathrm{Proj}^{r-k}(A)) = \mathrm{Proj}^{r-(k-1)}$. ∎

In summary, let c be a cell of D_k, $1 \leq k < r$, and let $c_1, \ldots, c_n$ be the children of c. We try to evaluate the truth values of $c_1, \ldots, c_n$ in three steps: (1) Use SIGNJ to compute the signs of the polynomials in $\mathrm{Proj}^{r-(k+1)}(A)$ on $c_1, \ldots, c_n$. (2) Use Theorem 7 in order to determine the signs of the polynomials in $A^{(k+1)}$ on $c_1, \ldots, c_n$. In fact, this amounts to searching through the set $\mathrm{Proj}^{r-(k+1)}(A)$ to find an element of $A^{(k+1)}$.[3] (3) Try to evaluate the truth values of $F(x_1, \ldots, x_r)$ on $c_1, \ldots, c_n$ from the signs of the polynomials in $A^{(k+1)}$ on $c_1, \ldots, c_n$. According to Theorem 5, $v(c_i) = $ the evaluated truth value if the evaluation was successful for the cell c_i.

The following algorithm utilizes this idea.

---

[2] For such projection sets, (Collins 1975; McCallum 1984; Hong 1990a).

[3] Actually this search is not necessary if we compute the signs of only those elements of $\mathrm{Proj}^{r-(k+1)}$ that are also in $A^{(k+1)}$. The details of this idea are discussed in (Hong 1990b). But the statistics given in Section 6 of this paper are obtained without this further improvement.

Final Improvement

(1) [Projection.] Compute the projection polynomials of all orders.

(2) [Initialization.] Initialize the partial CAD D as the single cell of D_0 whose truth value is set to be *undetermined.*

(3) [Choice.] If there is no candidate cell in the current partial CAD D, go to Step (8), otherwise, choose a candidate cell c from the current partial CAD D.

(4) [Conversion.] If $l(c) > 1$ and c is a section, convert the extended sample point of c into a primitive one by using the NORMAL and SIMPLE algorithms.

(5) [Augmentation.] Construct a stack over c and form a sample point for each child cell. (If $l(c) = 0$, construct a primitive sample point for every child. Otherwise, construct a primitive sample point for each sector and an extended sample point for each section.) If $A^{[l(c)+1]} = \emptyset$, go to Step (3).

(6) [Trial Evaluation.] Try to determine the truth values of the children of c by using SIGNJ, Theorem 7, the Boolean connectives, and Theorem 5.

(7) [Propagation.] Determine the truth value of c, if possible and if it is, the truth values of as many ancestors of c as possible by Theorems 2 and 4, removing from the tree the subtrees of each cell whose truth value is thus determined. Go to Step (3).

(8) [Solution.] Set $Z \leftarrow$ the set of all leaves of the partial CAD D. (At this point, the set Z satisfies the hypothesis of Theorem 6.) Get the solution set W by using Theorem 6. (Now a defining formula for the set W is equivalent to the input formula.)

3 Partial CAD Construction Algorithm

The partial CAD construction algorithm consists of a main algorithm QEPCAD and seven subalgorithms: PROJM, CHOOSE, CCHILD, EVALTV, SIGNJ, SIGNB, and PRPTV.

The main algorithm QEPCAD is a slightly more formal rendering of the algorithm developed in Section 2.

The subalgorithm PROJM essentially computes the projections J of all orders. It, however, in preparation for the algorithm SIGNJ, keeps also several intermediate results, namely the contents C and primitive parts P of the projections J (see Equation 1), and the finest squarefree bases B of the primitive parts P (see Equation 2). For each squarefree basis it also computes a matrix E of the multiplicities of the basis elements in the primitive parts (see Equation 2).

$$J_i = C_i P_i \tag{1}$$

$$P_i = \prod_{j=1}^{m} B_j^{E_{i,j}} \tag{2}$$

The subalgorithm CHOOSE chooses a candidate cell c from the current partial CAD. Discussion of CHOOSE is deferred to Section 4

The subalgorithm CCHILD builds a stack over the chosen cell c. Note that this subalgorithm, like PROJM, also keeps several intermediate results, namely substituted algebraic polynomials A^* (see Equation 3), their leading coefficients L^* and monic associates P^* (see Equation 4), the coarsest squarefree basis B^* of the monic associates P^* (see Equation 5), and the real roots I^* of the squarefree basis B^*. Furthermore, it computes a matrix E^* of multiplicities of the squarefree basis B^* in the monic associates P^* (see Equation 5).[4]

$$B_j(s, x_{k+1}) \;=\; A_j^* \tag{3}$$

$$A_j^* \;=\; L_j^* P_j^* \tag{4}$$

$$P_j^* \;=\; \prod_{g=1}^{l} B_g^{* \, E_{j,g}^*} \tag{5}$$

In the following discussions, let $\sigma(X)$ stand for the sign of the polynomial X on the chosen cell c, and $\sigma_i(X)$ stand for the sign of X on the i-th child of c.

The subalgorithm SIGNB, given the real roots I^* and the squarefree basis B or B^*, computes the signs of the squarefree basis element on each child, that is $\sigma_h(B_j)$ or $\sigma_h(B_g^*)$. This algorithm does the same thing as Arnon's BasisSignaturesOverCell (Arnon 1988b), but it avoids some unnecessary computation.

The subalgorithm SIGNJ given the signs of the squarefree basis elements on the children $(\sigma_h(B_j)$ or $\sigma_h(B_g^*))$, computes the signs of the projections polynomials on each children, that is $\sigma_h(J_i)$. First, note the following equations resulting immediately from Equations 1, 2, 3, 4, and 5. (In case $l(c) = 0$, Equations 6, 7, and 8 are irrelevant.)

$$\sigma_h(P_j^*) \;=\; \prod_{g=1}^{l} \sigma_h(B_g^*)^{E_{j,g}^*} \tag{6}$$

$$\sigma_h(A_j^*) \;=\; \sigma(L_j^*)\sigma_h(P_j^*) \tag{7}$$

$$\sigma_h(B_j) \;=\; \sigma_h(A_j^*) \tag{8}$$

$$\sigma_h(P_i) \;=\; \prod_{j=1}^{m} \sigma_h(B_j)^{E_{i,j}} \tag{9}$$

$$\sigma_h(J_i) \;=\; \sigma(C_i)\sigma_h(P_i) \tag{10}$$

In case $l(c) > 0$, we first compute $\sigma_h(B_g^*)$ by calling SIGNB, and then $\sigma(L_j^*)$ and $\sigma(C_i)$ by evaluating them on the primitive sample point of the cell c, then

---

[4]For the details on how to compute multiplicity matrices along with squarefree bases, see (Collins and Hong 1990).

finally $\sigma_h(J_i)$ from $\sigma_h(B_g^*)$ and $\sigma(L_j^*)$ and $\sigma(C_i)$ by applying Equations 6, 7, 8, 9, and 10 successively. In case $l(c) = 0$, we first compute $\sigma_h(B_j)$ by calling SIGNB, and then $\sigma(C_i)$ (C_i are integers), then finally $\sigma_h(J_i)$ from $\sigma_h(B_j)$ and $\sigma(C_i)$ by applying Equations 9 and 10 successively.

The subalgorithm EVALTV, given the signs of the projection polynomials on the children ($\sigma_h(J_i)$), tries to evaluate the truth values of the children as discussed in Section 2.4.

The subalgorithm PRPTV simply implements Theorems 2 and 4.

$$F' \leftarrow \text{QEPCAD}(F^*)$$

Algorithm QEPCAD (Quantifier Elimination by Partial Cylindrical Algebraic Decomposition).
Inputs: F^* is a quantified formula $(Q_{f+1}x_{f+1}) \cdots (Q_r x_r) F(x_1, \ldots, x_r)$, where $0 \le f \le r$ and F is a quantifier-free formula.
Outputs: $F' = F'(x_1, \ldots, x_f)$ is a quantifier-free formula equivalent to F^*.

(1) [Projection.] Extract the integral polynomials $A = (A_1, \ldots, A_d)$ from the quantifier-free formula F. Compute the projections of A of all orders by calling $\hat{J} \leftarrow \text{PROJM}(A)$.
(2) [Initialization.] Initialize the partial CAD D as the single cell of D_0 whose truth value is set to be *undetermined*.
(3) [Choice.] If there is no candidate cell in the partial CAD D, go to Step 8. Otherwise choose a candidate cell c from the partial CAD D by calling $c \leftarrow \text{CHOOSE}(D)$.
(4) [Conversion.] If $l(c) > 1$ and c is a section, convert the extended sample point of c into a primitive one by using the NORMAL and SIMPLE algorithms.
(5) [Augmentation.] Augment the partial CAD D through constructing the children of the cell c by calling $T \leftarrow \text{CCHILD}(D, c, \hat{J})$. If $A^{[l(c)+1]} = \emptyset$, go to Step 3.
(6) [Trial Evaluation.] Try to determine the truth values of the children of the cell c by calling $\text{EVALTV}(D, c, F, A, T, \hat{J})$.
(7) [Propagation.] Determine the truth value of c, if possible and if it is, the truth values of as many ancestors of c as possible, removing from the partial CAD D the subtrees of each cell whose truth value is thus determined, by calling $\text{PRPTV}(D, c, Q, f)$. Go to Step 3.
(8) [Solution.] Let S be the set of all leaves whose truth values are *true*. Set $W \leftarrow \biguplus_{c \in S} c \times \mathbb{R}^{f - l(c)}$. (Now a defining formula F' for the set W is equivalent to the input formula F^*.)

$$\hat{J} \leftarrow \text{PROJM}(A)$$

Algorithm PROJM (Projection with Multiplicity information).
Input: $A = (A_1, \ldots, A_d)$, $d \ge 1$, is a list of r-variate integral polynomials of positive degree. $r \ge 1$.

Output: $\hat{J} = (\hat{J}_1, \ldots, \hat{J}_r)$ where each $\hat{J}_k$ is a list of 5-tuples (J, C, P, B, E); $J = (J_1, \ldots, J_n)$ is $\mathrm{Proj}^{r-k}(A)$. $C = (C_1, \ldots, C_n)$ where C_i is the content of J_i. $P = (P_1, \ldots, P_n)$ where P_i is the primitive part of J_i. Note that the contents are signed so that $J_i = C_i P_i$. $B = (B_1, \ldots, B_m)$ is a finest squarefree basis of P. $E = (E_{i,j})$ is the matrix of the multiplicities of B in P such that $P_i = \prod_{j=1}^{m} B_j^{E_{i,j}}$ for $1 \le i \le n$.

(1) [Initialize.] Set $J \leftarrow A$. Set $k \leftarrow r$.
(2) [Compute C, P, B and E from J.] Compute the contents C and the primitive parts P of J. Compute the finest squarefree basis B of the primitive parts P, obtaining the multiplicity matrix E also. Set $\hat{J}_k \leftarrow (J, C, P, B, E)$.
(3) [Are we done?] If $k = 1$, return.
(4) [Compute $\mathrm{Proj}(J)$ from B and C.] Set $J \leftarrow C \cup \mathrm{PROJ}(B)$. Set $k \leftarrow k - 1$. Go to Step 2.

$$T \leftarrow \mathrm{CCHILD}(D, c, \hat{J})$$

Algorithm CCHILD (Construct the Children of a given cell).
Inputs: D is a partial CAD. c is a candidate cell of D. $\hat{J}$ is a structure produced by PROJM.
Output: The partial CAD D is augmented with the children of the cell c. Let k be the level of the cell c. Let (J, C, P, B, E) be the $(k+1)$-th element of the list $\hat{J}$. Let $B = (B_1, \ldots, B_m)$. Then T is a 6-tuple $(A^*, L^*, P^*, B^*, E^*, I^*)$; Case $k = 0$. A^*, L^*, P^*, B^* and E^* are not computed. $I^* = (I_1, b_1, \ldots, I_u, b_u)$ where $I_1 < I_2 < \cdots < I_u$ are disjoint isolating intervals for all the real roots of $\prod_{j=1}^{m} B_j$ and each b_i is the unique B_j which has a root in I_i. Case $k > 0$. Let s be the primitive sample point of the cell c. $A^* = (A_1^*, \ldots, A_m^*)$ where each $A_j^* = B_j(s, x_{k+1})$. $L^* = (L_1^*, \ldots, L_m^*)$ where each L_j^* is the leading coefficient of A_j^*. $P^* = (P_1^*, \ldots, P_m^*)$ where each P_j^* is the monic associate of A_j^*. $B^* = (B_1^*, \ldots, B_l^*)$ is a squarefree basis of P^*. $E^* = (E_{j,g}^*)$ is the matrix of the multiplicities of B^* in P^* such that $P_j^* = \prod_{g=1}^{l} B_g^{*E_{j,g}^*}$ for $1 \le j \le m$. $I^* = (I_1, b_1, \ldots, I_u, b_u)$ where $I_1 < I_2 < \cdots < I_u$ are disjoint isolating intervals for all the real roots of $\prod_{g=1}^{l} B_g^*$ and each b_i is the unique B_g^* which has a root in I_i.

(1) [Setup.] Let k be the level of the cell c. Let (J, C, P, B, E) be the $(k+1)$-th element of the list $\hat{J}$. If $k > 0$, go to Step 3.
(2) [If c is the root cell.] Isolate the real roots of the finest squarefree basis B, obtaining I^*. Construct the children cells and the primitive sample point for each child. (A^*, L^*, P^*, B^* and E^* are not computed.) Set $T \leftarrow (A^*, L^*, P^*, B^*, E^*, I^*)$. Return.
(3) [If c is not the root cell.] Substitute the primitive sample point of the cell c into the squarefree basis B, obtaining A^*. Compute the leading coefficients L^* and the monic associates P^* of A^*. Compute the squarefree basis B^* of the monic polynomials P^*, also obtaining the multiplicity matrix E^*. Isolate the real roots of the squarefree basis B^*, obtaining I^*. Construct

the children cells and the sample point for each child. (For sectors construct primitive sample points and for sections construct extended sample points.) Set $T \leftarrow (A^*, L^*, P^*, B^*, E^*, I^*)$. Return.

$$\Sigma \leftarrow \text{SIGNB}(I, B)$$

Algorithm SIGNB (Sign matrix of a squarefree Basis).
Inputs: $B = (B_1, \ldots, B_v)$, $v \geq 0$, is a squarefree basis. $I = (I_1, b_1, \ldots, I_u, b_u)$, $u \geq 0$, where $I_1 < I_2 < \cdots < I_u$ are disjoint isolating intervals for all the real roots of $\prod_{j=1}^{v} B_j$ and each b_i is the unique B_k which has a root in I_i.
Output: Σ is a $(2u + 1)$ by v matrix $(\Sigma_{i,j})$ where $\Sigma_{i,j}$ is the sign of B_j in the i-th cell.

(1) [The rightmost cell (i.e., the $(2u+1)$-th cell.)] Set $\Sigma_{2u+1,j} \leftarrow 1$ for $1 \leq j \leq v$. Set $i \leftarrow u$.
(2) [Are we done?] If $i = 0$, return.
(3) [The $(2i)$-th and $(2i - 1)$-th cells.] Let k be such that $B_k = b_i$. For $j = 1, \ldots, v$ do { If $j = k$, set $\Sigma_{2i,j} \leftarrow 0$ and set $\Sigma_{2i-1,j} \leftarrow -\Sigma_{2i+1,j}$. Otherwise, set $\Sigma_{2i,j} \leftarrow \Sigma_{2i+1,j}$ and set $\Sigma_{2i-1,j} \leftarrow \Sigma_{2i+1,j}$. }. Set $i \leftarrow i-1$. Go to Step 2.

$$\Sigma \leftarrow \text{SIGNJ}(c, T, \hat{J})$$

Algorithm SIGNJ (Sign matrix of projection polynomials).
Inputs: c is a cell in D. $\hat{J}$ is a structure produced by PROJM. T is a structure produced by CCHILD.
Output: Σ is a w by n matrix $(\Sigma_{h,i})$ where $\Sigma_{h,i}$ is the sign J_i in the h-th child of c. (For w,n, and J see Step 1.) In this algorithm, $\sigma(X)$ stands for the sign of X on the cell c, and $\sigma_i(X)$ stands for the sign of X on the i-th child of the cell c.

(1) [Setup.] Let k be the level of the cell c. Let the $(k + 1)$-th element of $\hat{J}$ be (J, C, P, B, E) where $J = (J_1, \ldots, J_n)$, $C = (C_1, \ldots, C_n)$, $P = (P_1, \ldots, P_n)$, $B = (B_1, \ldots, B_m)$, and E is an n by m matrix. Let T be $(A^*, L^*, P^*, B^*, E^*, I^*)$ where $A^* = (A_1^*, \ldots, A_m^*)$, $L^* = (L_1^*, \ldots, L_m^*)$, $P^* = (P_1^*, \ldots, P_m^*)$, $B^* = (B_1^*, \ldots, B_l^*)$, and E is an m by l matrix. Let $(c_1, \ldots, c_w)$ be the children of the cell c. If $k > 0$, go to Step 3.
(2) [If c is the root cell.] Compute $\sigma_h(B_j)$ for $1 \leq j \leq m$ and $1 \leq h \leq w$ by calling $\Sigma \leftarrow \text{SIGNB}(I^*, B)$, where $\Sigma_{h,j} = \sigma_h(B_j)$. Compute $\sigma(C_i)$ for $1 \leq i \leq n$. For $h = 1, 2, \ldots, w$ do { Set $\sigma_h(P_i) \leftarrow \prod_{j=1}^{m} \sigma_h(B_j)^{E_{i,j}}$ for $1 \leq i \leq n$. Set $\sigma_h(J_i) \leftarrow \sigma(C_i)\sigma_h(P_i)$ for $1 \leq i \leq n$. }. Return.
(3) [If c is not a root cell.] Compute $\sigma_h(B_g^*)$ for $1 \leq g \leq l$ and $1 \leq h \leq w$ by calling $\Sigma^* \leftarrow \text{SIGNB}(I^*, B^*)$, where $\Sigma_{h,g}^* = \sigma_h(B_g^*)$. Compute $\sigma(C_i)$ for $1 \leq i \leq n$ by evaluating C_i on the primitive sample point of the cell c. Compute $\sigma(L_j^*)$ for $1 \leq j \leq m$ by evaluating L_j^* on the primitive sample point of the cell c. For $h = 1, 2, \ldots, w$ do { Set $\sigma_h(P_j^*) \leftarrow \prod_{g=1}^{l} \sigma_h(B_g^*)^{E_{j,g}^*}$

for $1 \leq j \leq m$. Set $\sigma_h(A_j^*) \leftarrow \sigma(L_j^*)\sigma_h(P_j^*)$ for $1 \leq j \leq m$. Set $\sigma_h(B_j) \leftarrow$ $\sigma_h(A_j^*)$ for $1 \leq j \leq m$. Set $\sigma_h(P_i) \leftarrow \prod_{j=1}^{m} \sigma_h(B_j)^{E_{i,j}}$ for $1 \leq i \leq n$. Set $\sigma_h(J_i) \leftarrow \sigma(C_i)\sigma_h(P_i)$ for $1 \leq i \leq n$. }. Return.

EVALTV$(D, c, F, A, T, \hat{J})$

Algorithm EVALTV (Evaluate Truth Values).
Inputs: D is a partial CAD. c is a cell in D. $F = F(x_1, \ldots, x_r)$ is a quantifier-free formula. $A = (A_1, \ldots, A_d)$ is the list of polynomials occurring in F. $\hat{J}$ is a structure produced by PROJM. T is a structure produced by CCHILD.
Outputs: The truth values of some children of c might be determined, resulting in modification of D.

In this algorithm, $\sigma_i(X)$ stands for the sign of X on the i-th child of the cell c.

(1) [Setup.] Let k be the level of the cell c. Let the $(k+1)$-th element of $\hat{J}$ be (J, C, P, B, E), where $J = (J_1, \ldots, J_n)$. Let $(c_1, \ldots, c_w)$ be the children of the cell c.
(2) [Compute the signs of the projection J.] Compute $\sigma_h(J_i)$ for $1 \leq h \leq w$ and $1 \leq i \leq n$ by calling $\Sigma \leftarrow \text{SIGNJ}(c, T, \hat{J})$, where $\Sigma_{h,i} = \sigma_h(J_i)$.
(3) [Get the signs of the input polynomials A.] For $h = 1, 2, \ldots, w$ and $p = 1, 2, \ldots, d$ do { If there exists i such that $A_p = J_i$, set $\sigma_h(A_p) \leftarrow \sigma_h(J_i)$. }.
(4) [Try to evaluate the truth values of the children of c.] For $h = 1, 2, \ldots, w$ do { Try to determine $v(c_h)$ by trying to evaluate the truth value of the quantifier-free formula F on the cell c_h from $\sigma_h(A_1), \ldots, \sigma_h(A_d)$. }. Return.

PRPTV(D, c, Q, f)

Algorithm PRPTV(Propagate Truth Value).
Inputs: D is a partial CAD. c is a cell in D. $Q = (Q_{f+1}, \ldots, Q_r)$ is the list of the quantifiers occurring in the quantified formula. f is the number of free variables.
Outputs: The truth values of the cell c and some of its ancestors might be determined. The descendants of each cell whose truth value has been determined through the propagation are removed from the partial CAD D.

(1) [Initialize.] Set $k \leftarrow$ the level of the cell c. Set $c' \leftarrow c$.
(2) [Are we done with propagating on or above the free variable space?] If $k < f$, go to Step 5.
(3) [Propagate on or above the free variable space.] **If** the truth value of every child of c' is *true* **then** set $v(c') \leftarrow$ *true* **else if** the truth value of every child of c' is *false* **then** set $v(c') \leftarrow$ *false* **else if** there exists a *true* child of c' and $Q_{i+1} = \exists$ **then** set $v(c') \leftarrow$ *true* **else if** there exists a *false* child of c' and $Q_{i+1} = \forall$ **then** set $v(c') \leftarrow$ *false* **else** return.

(4) [Remove the descendants of c' and loop.] Remove the descendants of the cell c' from the partial CAD D. Set $k \leftarrow k - 1$. Set $c' \leftarrow$ the parent of c'. Go to Step 2.

(5) [Are we done with propagating below the free variable space?] If $k < 0$, return.

(6) [Propagate below the free variable space.] **If** the truth value of every child of c' is *true* **then** set $v(c') \leftarrow$ *true* **else if** the truth value of every child of c' is *false* **then** set $v(c') \leftarrow$ *false* **else** return.

(7) [Remove the descendants of c' and loop.] Remove the descendants of the cell c' from the partial CAD D. Set $k \leftarrow k - 1$. Set $c' \leftarrow$ the parent of c'. Go to Step 5.

4 Strategy for Cell Choice

In this section, as promised, we discuss the algorithm CHOOSE, whose task is to choose a candidate cell from a partial CAD. We begin by recalling that the main algorithm QEPCAD repeatedly chooses a candidate cell – over which stack construction, trial evaluation and propagation are carried out – until there are no more candidate cells. Let c_i be the i-th cell chosen by the algorithm CHOOSE. We will call the list $(c_1, \ldots, c_n)$ a *choice path*. Note that for a given input formula many different choice paths will be possible and that each may require a different amount of time.

We implemented our partial CAD construction system as an interactive environment where a user can manually choose a candidate cell or invoke an already built-in strategy algorithm. When cells are chosen manually, a user can query for diverse information, such as the current partial CAD, the sample points, the signs of basis polynomials in a specified cell, and so on.[5]

In Fig. 1 we present several cell-choice strategies devised for several problem-classes: the strategy HL-LI for collision problems from robot motion planning (Buchberger et al. 1988), the strategy SR-HL-LI for consistency of a system of polynomial strict inequalities (McCallum 1987), the strategy TC-LD-HL-GI for termination proof of term rewrite systems (Lankford 1979; Huet and Oppen 1980), and the strategy TC-LD-HL-LI for any other unclassified problems, where HL stands for higher level first, LI for lesser index first, GI for greater index first, SR for sector first, TC for trivial conversion first, and LD for lesser degree of minimal polynomial first.

These strategies were based on a limited amount of experience and study, but the experiments showed that they significantly reduced the amount of required computation time.

The algorithm ORDER defines a total ordering among cells, in the way that it compares two input cells according to the ordering and returns the greater cell. Each strategy uses its own version of the algorithm ORDER

---

[5]The user's manual for the partial CAD system will be available to any interested researchers in the very near future.

$$c \leftarrow \text{ORDER}(c_1, c_2) \qquad \text{(Strategy: TC-LD-HL-LI)}$$

(1) [By the complexity of conversion. Trivial conversion first. (TC)]
    if $p(c_1) > p(c_2)$ then { set $c \leftarrow c_1$. return. }
    if $p(c_1) < p(c_2)$ then { set $c \leftarrow c_2$. return. }

(2) [By the degree of the minimal polynomial. Less degree first. (LD)]
    if $d(c_1) < d(c_2)$ then { set $c \leftarrow c_1$. return. }
    if $d(c_1) > d(c_2)$ then { set $c \leftarrow c_2$. return. }

(3) [By the level. Higher Level first. (HL)]
    if $l(c_1) > l(c_2)$ then { set $c \leftarrow c_1$. return. }
    if $l(c_1) < l(c_2)$ then { set $c \leftarrow c_2$. return. }

(4) [By the index. Less Index first. (LI)]
    if $i(c_1) < i(c_2)$ then { set $c \leftarrow c_1$. return. }
    if $i(c_1) > i(c_2)$ then { set $c \leftarrow c_2$. return. }

$$c \leftarrow \text{ORDER}(c_1, c_2) \qquad \text{(Strategy: TC-LD-HL-GI)}$$

The same as TC-LD-HL-LI except replacing Step 4 with the following.

(4) [By the index. Greatest Index first. (GI)]
    if $i(c_1) > i(c_2)$ then { set $c \leftarrow c_1$. return. }
    if $i(c_1) < i(c_2)$ then { set $c \leftarrow c_2$. return. }

$$c \leftarrow \text{ORDER}(c_1, c_2) \qquad \text{(Strategy: SR-HL-LI)}$$

The same as TC-LD-HL-LI except replacing Steps 1 and 2 with the following.

(1) [By the cell type. Sector first. (SR)]
    if $s(c_1) < s(c_2)$ then { set $c \leftarrow c_1$. return.}
    if $s(c_1) > s(c_2)$ then { set $c \leftarrow c_2$. return.}

$$c \leftarrow \text{ORDER}(c_1, c_2) \qquad \text{(Strategy: HL-LI)}$$

The same as TC-LD-HL-LI except omitting Steps 1 and 2.

$$c \leftarrow \text{CHOOSE}(D)$$

(1) Let $c_1, \ldots, c_n$ be the candidate cells in D. Set $c \leftarrow c_1$.
    For $i = 2, \ldots, n$ do set $c \leftarrow \text{ORDER}(c, c_i)$.

Figure 1: Cell Choice Strategies

(i.e., a definition of ordering among cells). The algorithm CHOOSE finds the greatest candidate cell by utilizing the algorithm ORDERin an obvious way.

In the algorithm ORDER, the following functions are used.

$l(c)$ denotes the level of the cell c.

$s(c)$ denotes the cell type of the cell c. It is 0 if c is a sector, and 1 if c is a section.

$i(c)$ denotes the index of the cell c. $i(c_1) > i(c_2)$ if $i(c_1)$ is after $i(c_2)$ lexicographically.

$d(c)$ denotes the presumed degree of the minimal polynomial of the primitive
sample point of the cell c. More specifically, if the cell c already has a
primitive sample point, $d(c)$ is m where m is the degree of the corresponding minimal polynomial. Otherwise, $d(c)$ is mm' where m is the degree of
the minimal polynomial for the non-last coordinates and m' is the degree
of the algebraic polynomial for the last coordinate.

$p(c)$ denotes the triviality of the conversion of the sample point of the cell c.
It is 1 if the conversion is definitely trivial or not needed, and 0 if the
conversion *can* be nontrivial. More specifically, $p(c)$ is 0 if the cell c has
an extended sample point such that both the minimal polynomial for the
non-last coordinates and the algebraic polynomial for the last coordinate
are nonlinear. Otherwise, $p(c)$ is 1.

Now we describe the motivation that led to the strategies presented here.
Let us begin by TC-LD-HL-LI, the strategy for any unclassified problems. The
motive for TC-LD is to choose a cell on which stack construction might be
cheapest. The motive for HL is to choose a cell that might lead to truth
evaluations earliest. LI is not significant, it could be replaced with a random
choice.

In termination proof of term rewrite systems, GI is superior to LI for reasons
explained in Hong and Küchlin (1990). This is the motive for using GI in the
strategy TC-LD-HL-GI.[6]

In consistency problems of a system of polynomial strict inequalities, as
McCallum (1987) points out, we only need to consider sectors. This is the
motive for using SR in the strategy SR-HL-LI.[7]

In collision problems, we are interested in whether several moving objects
would collide, and if so, we may also be interested in finding the time of the
earliest collision. The strategy HL-LI chooses candidate cells in such an order
that this time can be easily detected.

5   Illustration

In this section we illustrate our algorithm by a simple collision problem from
robot motion planning.

Consider two semi-algebraic objects: a circle and a square (see Fig. 2).
The circle has diameter 2 and is initially centered at $(0,0)$ and is moving with
the velocity $v_x = 1$ and $v_y = 0$. The square has side-length 2 and is initially
centered at $(0,-8)$ and is moving with the velocity $v_x = 17/16$ and $v_y = 17/16$.
Now we want to decide if these two objects would collide.

---

[6]In order to fully utilize this strategy, we also need to modify the propagation
step slightly as discussed in Section 7.

[7]In order to fully utilize this strategy, we also need to modify the propagation

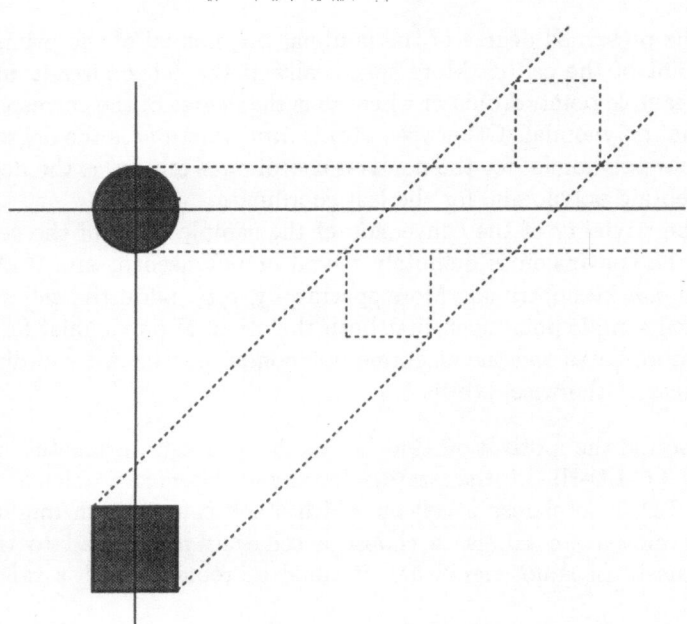

Figure 2: Collision Problem

The moving circle can be described algebraically as:

$$(x - t)^2 + y^2 \leq 1$$

The moving square can be described algebraically as:

$$-1 \leq x - (17/16)t \leq 1 \quad \wedge \quad -9 \leq y - (17/16)t \leq -7$$

From these we can express the collision problem as a decision problem of a sentence in elementary algebra and geometry:

$$
\begin{aligned}
(\exists t)(\exists x)(\exists y)(\quad & t > 0 \\
\wedge \quad & x - (17/16)t \geq -1 \\
\wedge \quad & x - (17/16)t \leq 1 \\
\wedge \quad & y - (17/16)t \geq -9 \\
\wedge \quad & y - (17/16)t \leq -7 \\
\wedge \quad & (x - t)^2 + y^2 \leq 1 \ )
\end{aligned}
$$

We can refine this sentence by observing that collision can occur only when the top side of the square is above or on the line $y = -1$ and the bottom side

_____

step slightly as discussed in Section 7.

of the square is below or on the line $y = 1$. Translating this observation into a restriction on the range of t, we get the following sentence:

$$(\exists t)(\exists x)(\exists y)(\quad (17/16)t \geq 6$$
$$\wedge \quad (17/16)t \leq 10$$
$$\wedge \quad x - (17/16)t \geq -1$$
$$\wedge \quad x - (17/16)t \leq 1$$
$$\wedge \quad y - (17/16)t \geq -9$$
$$\wedge \quad y - (17/16)t \leq -7$$
$$\wedge \quad (x - t)^2 + y^2 \leq 1 \ )$$

We entered this sentence into an implementation (Arnon 1981) of the original CAD algorithm, and it reported collision, after constructing 25 cells in 1-space, 263 cells in 2-space, and 1795 cells in 3-space, taking 1 hour and 44 minutes on a Sun3/50 running Unix.

We tried our partial CAD algorithm on the same sentence using the cell choice strategy HL-LI. It reported collision, after constructing 25 cells in 1-space, 11 cells in 2-space, and 25 cells in 3-space, taking 20.5 seconds (7.8 seconds for projection, 12.7 seconds for stack construction).

Below we present the trace of the actual computation carried out by the partial CAD system. In order to save space the trace output has been compressed. In this trace, a partial CAD is represented as a tree of cells, where each cell is represented by its cell-index along with one character indicating its truth value (T,F,? respectively for *true*, *false*, *undetermined*). The text in italics consists of comments inserted afterwards.

```
Enter a prenex formula:
(Et)(Ex)(Ey) (   17/16 t >= 6
             & 17/16 t <= 10
             & x - 17/16 t >= -1
             & x - 17/16 t <=  1
             & y - 17/16 t >= -9
             & y - 17/16 t <= -7
             & y^2 + x^2 - 2 t x + t^2 <= 1 )
```

```
The initial partial CAD:
() ?
```

This indicates that the initial partial CAD consists of a root cell whose truth value is undetermined.

```
The cell chosen: ()
Constructing children of the cell ()......
Trying to evaluate truth values of the children of ()......

The current partial CAD over ():
()---(1)  F
  ---(2)  F
  ---(3)  F
  ---(4)  F
  ---(5)  F
  ---(6)  ?
  ---(7)  ?
  ---(8)  ?
  ---(9)  ?
  ---(10) ?
  ---(11) ?
  ---(12) ?
  ---(13) ?
  ---(14) ?
  ---(15) ?
  ---(16) ?
  ---(17) ?
  ---(18) ?
  ---(19) ?
  ---(20) ?
  ---(21) ?
  ---(22) ?
  ---(23) F
  ---(24) F
  ---(25) F
```

The truth values of the cells (1), (2), (3), (4), *and* (5) *are already determined to be false because* $\frac{17}{16}t \not\geq 6$ *for these cells. Likewise the truth values of the cells* (23), (24), *and* (25) *are already determined to be false because* $\frac{17}{16}t \not\leq 10$ *for these cells. Therefore we do not need to build stacks over them, resulting in reduction of the computation.*

```
-----------------------------------------------------------
The cell chosen: (6)
Constructing children of the cell (6)......
Trying to evaluate truth values of the children of (6)......

The current partial CAD over (6):
(6)---(6,1) F
   ---(6,2) F
```

```
---(6,3) F
---(6,4) ?
---(6,5) ?
---(6,6) ?
---(6,7) ?
---(6,8) ?
---(6,9) ?
---(6,10) ?
---(6,11) F
```

The truth values of the cells $(6,1)$, $(6,2)$, *and* $(6,3)$ *are already determined to be false because* $x - \frac{17}{16}t \not\geq -1$ *for these cells. Likewise the truth value of the cell* $(6,11)$ *is already determined to be false because* $x - \frac{17}{16}t \not\leq 1$ *for this cell.*

```
------------------------------------------------------------
The cell chosen: (6,4)
Constructing children of the cell (6,4)......
Trying to evaluate truth values of the children of (6,4)......

The current partial CAD over (6,4):
(6,4)--(6,4,1) F
     --(6,4,2) F
     --(6,4,3) F
     --(6,4,4) F
     --(6,4,5) F
     --(6,4,6) F
     --(6,4,7) F
     --(6,4,8) F
     --(6,4,9) F


Propagating the truth values....

The current partial CAD over (6):
(6)---(6,1) F
   ---(6,2) F
   ---(6,3) F
   ---(6,4) F
   ---(6,5) ?
   ---(6,6) ?
   ---(6,7) ?
   ---(6,8) ?
   ---(6,9) ?
   ---(6,10) ?
   ---(6,11) F
```

The truth value of the cell $(6,4)$ *has been determined to be false because the truth values of all the children of* $(6,4)$ *are false.*

The cell chosen: (6,5)

    Actually it is unnecessary to construct a stack over the cell (6,5). One can give an argument that because the objects in this problem are closed sets and because the truth value of the adjacent cell (6,4) is false, the truth value of the cell (6,5) must also be false.

Constructing children of the cell (6,5)......
Trying to evaluate truth values of the children of (6,5)......

The current partial CAD over (6,5):
```
(6,5)--(6,5,1) F
      --(6,5,2) F
      --(6,5,3) F
      --(6,5,4) F
      --(6,5,5) F
      --(6,5,6) F
      --(6,5,7) F
      --(6,5,8) F
      --(6,5,9) F
```

Propagating the truth values....

The current partial CAD over (6):
```
(6)---(6,1) F
   ---(6,2) F
   ---(6,3) F
   ---(6,4) F
   ---(6,5) F
   ---(6,6) ?
   ---(6,7) ?
   ---(6,8) ?
   ---(6,9) ?
   ---(6,10) ?
   ---(6,11) F
```

    The truth value of the cell $(6,5)$ *has been determined to be false because the truth values of all the children of* $(6,5)$ *are false.*

The cell chosen: (6,6)
Constructing children of the cell (6,6)......
Trying to evaluate truth values of the children of (6,6)......

```
The current partial CAD over (6,6):
(6,6)--(6,6,1) F
     --(6,6,2) F
     --(6,6,3) F
     --(6,6,4) T
     --(6,6,5) F
     --(6,6,6) F
     --(6,6,7) F
```

```
Propagating the truth values....
```

```
The current partial CAD over ():
() T
```

Now the truth value of the the root cell () is determined to be true and so the input sentence is also true, which tells us that the two objects will collide. The truth value of the root cell () is determined to be true since the truth value of the cell $(6, 6, 4)$ is true and the quantifiers Q_3, Q_2, and Q_1 are all existential. It is important to note that we did not have to build stacks over the cells $(6, 7)$, $(6, 8)$, $(6, 9)$, $(6, 10)$, and (7) through (22).

```
---------------------- Statistics ----------------------
                  Level-1    Level-2    Level-3
Number of stacks:    1          1          3
Number of cells:    25         11         25
```

```
Time taken for projection: 7.816 seconds
Time taken for choosing candidate cells: 0.250 seconds
Total time taken: 20.516 seconds
-----------------------------------------------------------
```

6   Empirical Results

In this section we present empirical comparisons between the partial CAD method and the original method on several problems from diverse application areas.[8]

The original method has been implemented by Arnon (1981) in the ALDES/SAC-2 computer algebra system (Collins and Loos 1990). We used

---

[8] Added during proof: More recent experimental results for these and various other problems are found in Hong (1990b, 1991a). For example, in Hong (1990b) the x-axis ellipse problem and the quartic problem (Arnon and Mignotte 1988) are shown to be solved completely mechanically in less than a minute.

this implementation for our experiments, but with a slight modification as follows. Arnon's implementation avoids some redundant computations in CAD construction by partitioning the set of cells of a CAD into disjoint subsets called *conjugacy equivalence classes* and by carrying out those computations only once for each class. This partitioning is done for CAD's of all levels. However, this process is not needed for a CAD of the last level since we do not need to build stacks over it. Experience also shows that this process can be very time consuming. Therefore we modified Arnon's implementation so that it does not carry out partitioning at the last level.

We implemented the partial CAD method also in ALDES/SAC-2. But we incorporated two more improvements into the implementation. First, NORMAL and SIMPLE operations are bypassed in certain trivial conversions of sample points. Let c be a cell of D_k which has an extended sample point consisting of: (1) a real algebraic number α and a $(k-1)$-tuple $(b_1, \ldots, b_{k-1})$ of elements of $Q(\alpha)$, and (2) a non-zero monic squarefree polynomial $g(x) \in Q(\alpha)[x]$, and an isolating interval for a real root β of $g(x)$. Now if $\deg(g) = 1$ (i.e., $g(x) = x + a_0$), we trivially have $\beta = -a_0 \in Q(\alpha)$. In this case, therefore, we can get a primitive sample point of the cell c without carrying out the NORMAL and SIMPLE operation, consisting of the real algebraic number α and the k-tuple $(b_1, \ldots, b_{k-1}, -a_0)$ of elements of $Q(\alpha)$.

Second, several databases are used in order to avoid some redundant computations. As mentioned earlier Arnon's implementation uses conjugacy equivalence classes for this purpose. This method, however, is not compatible with our partial CAD method for obvious reasons, and therefore we chose instead a database approach. When we need to compute a certain result, we first search an appropriate database for an earlier instance of the same computation. If the search is successful, the result is retrieved and used; otherwise we compute the result and enter it into the database. The details will be discussed in a subsequent paper.

Table 1 shows the performance comparisons on several problems from diverse application areas. In this table, T is the total computation time in seconds and C_i is the number of cells constructed in i-space. For each problem three rows are given, the top one for the original method with the slight modification as mentioned above, the bottom one for our method, and the middle one for our method as modified to construct a full CAD.

Therefore the differences between the top one and the middle one are due to (1) the avoidance of NORMAL and SIMPLE operations at the last level, (2) the use of databases instead of conjugacy equivalence class computation, and (3) the bypassing of NORMAL and SIMPLE operation in trivial conversions of sample points. The differences between the middle and the bottom one are due to the construction of only a partial CAD through the use of propagation and trial truth evaluation.

All the experiments were carried out on a Sun3/50 running Unix using 4 megabytes of memory for lists.

Table 1: Comparisons

Problem	Method	T	C_1	C_2	C_3	C_4	Strategy
Collision	original	5,255	25	263	1,795		
	full	585	25	263	1,795		
	partial	22	25	11	33		HL-LI
Consistency	original	4,961	11	57	365		
	full	179	11	57	365		
	partial	26	11	15	43		SR-HL-LI
Termination	original	333	17	177	1,099		
	full	254	17	177	1,099		
	partial	7	17	13	7		TC-LD-HL-GI
Collins	original	1,759	19	269	2,149		
Johnson	full	641	19	269	2,149		
	partial	228	19	142	524		TC-LD-HL-LI
Davenport	original	1,007	7	73	667	4,949	
Heintz	full	1,464	7	73	667	4,949	
	partial	217	7	73	649	486	TC-LD-HL-LI

Collision Problem This problem is same as the one used in Section 5, except that the square moves with the velocity $v_x = 15/16$, $v_y = 15/16$.

$$(\exists t)(\exists x)(\exists y)(\quad (15/16)t \geq 6 \ \wedge \ (15/16)t \leq 10$$
$$\wedge \quad x - (15/16)t \geq -1 \ \wedge \ x - (15/16)t \leq 1$$
$$\wedge \quad y - (15/16)t \geq -9 \ \wedge \ y - (15/16)t \leq -7$$
$$\wedge \quad (x - t)^2 + y^2 \leq 1 \ )$$

Consistency in Strict Inequalities (McCallum 1987) This problem decides whether the intersection of the open ball with radius 1 centered at the origin and the open circular cylinder with radius 1 and axis the line $x = 0$, $y + z = 2$ is nonempty.

$$(\exists x)(\exists y)(\exists z)(x^2 + y^2 + z^2 < 1 \ \wedge \ x^2 + (y + z - 2)^2 < 1)$$

Termination of Term Rewrite System (Lankford 1979; Huet and Oppen 1980) This problem decides whether we should orient the equation $(xy)^{-1} = y^{-1}x^{-1}$ into $(xy)^{-1} \rightarrow y^{-1}x^{-1}$ in order to get a terminating rewrite system for group theory. It uses a polynomial interpretation: $xy \Rightarrow x + 2xy$, $x^{-1} \Rightarrow x^2$, and $1 \Rightarrow 2$.

$$(\exists r)(\forall x)(\forall y)(x > r \wedge y > r \implies x^2(1 + 2y)^2 > y^2(1 + 2x^2))$$

Collins and Johnson (Collins and Johnson 1989b) The formula below gives necessary and sufficient conditions on the complex conjugate roots $a \pm bi$ so that there exists a cubic polynomial with a single real r root in $(0, 1)$ yet more than one variation is obtained.

$$(\exists r)(\qquad 3a^2r + 3b^2r - 2ar - a^2 - b^2 < 0$$
$$\wedge \quad 3a^2r + 3b^2r - 4ar + r - 2a^2 - 2b^2 + 2a > 0$$
$$\wedge \quad a \geq 1/2 \wedge b > 0 \wedge r > 0 \wedge r < 1 \ )$$

Davenport and Heintz (Davenport and Heintz 1988) This formula is a special case of a more general formula used by Davenport and Heintz in order to show the time complexity of the quantifier elimination in elementary algebra and geometry.

$$(\exists c)(\forall b)(\forall a)((a = d \wedge b = c) \ \vee \ (a = c \wedge b = 1) \ \implies \ a^2 = b)$$

7 Conclusion

In this paper we developed a method called *partial CAD construction*, that completes quantifier elimination by a *partially* built CAD, utilizing several aspects of the input formula such as: the quantifiers, the Boolean connectives, and the absence of some variables from some polynomials occurring in the input formula.

The resulting algorithm was nondeterministic in nature, and thus we also presented several strategies that heuristically try to guide the algorithm along a cheap computational path. These strategies were based on very limited amount of experience and study, but even these crude strategies significantly reduce the amount of computation as the above experiments show. We hope that these strategies can be refined more in order to achieve further reduction of the required computation.

We can also sometimes make further improvement of our method by customizing the propagation algorithm for each problem-class. For example, in termination proof of term rewrite systems, we can determine the truth value of the root cell as soon as the truth value of the cell with the greatest index in a CAD of 1-space is determined (Hong and Küchlin 1990). Similarly, in consistency problems of polynomial strict inequalities, the truth value of a cell can be determined to be *false* as soon as the truth values of all the sector-children of the cell are known to be *false*.

Finally, we hope that our method and Arnon's clustering method (Arnon 1981, 1988b) may be combined in the near future in order to take advantage of both methods.

Acknowledgments

The authors would like to thank a referee who made some constructive criticisms.

Simple Solution Formula Construction in Cylindrical Algebraic Decomposition Based Quantifier Elimination[1]

Hoon Hong

1 Introduction

Since Tarski (1951) gave the first quantifier elimination algorithm for real closed fields, various improvements and new methods have been devised and analyzed (Arnon 1981, 1988b; Ben-Or et al. 1986; Böge 1980; Buchberger and Hong 1991; Canny 1988; Cohen 1969; Collins 1975; Collins and Hong 1991; Fitchas et al. 1990a; Grigor'ev 1988; Grigor'ev and Vorobjov 1988; Heintz et al. 1989a; Holthusen 1974; Hong 1989, 1990a, 1990b, 1991a, 1991b, 1991c; Johnson 1991; Langemyr 1990; Lazard 1990; McCallum 1984; Renegar 1992a, 1992b, 1992c; Seidenberg 1954).

In this paper, we present several improvements to the last step (solution formula construction step) of Collins' cylindrical algebraic decomposition (CAD) based quantifier elimination algorithm (Collins 1975).

Collins' original algorithm constructs a solution formula by forming a disjunction of defining formulas of solution cells. This method produces complex solution formulas and often requires a great amount of computation because of the augmented projection which is needed.

Arnon and Mignotte (1988) and Collins and Johnson (1989b) succeeded in obtaining simple solution formulas without using augmented projection for several nontrivial quantifier elimination problems by plotting the solution cells and by using problem-specific geometric intuition.

---

[1]Reprinted with permission from *Proceedings of the International Symposium on Symbolic & Algebraic Computation*, edited by P. S. Wang, ACM Press, 1992, pp. 177–188. Copyright 1992, Association for Computing Machinery, Inc.

In this paper we present a new method for constructing simple solution formulas, which does not require any human interaction. The key features of this method are as follows:

- It uses the *improved* CAD algorithm of (Hong 1990b) (see also Collins and Hong 1991), which is designed so that quantifier elimination is often completed with a partially built CAD. (From now on, the improved CAD algorithm will be called *the partial* CAD *algorithm.*)
- It does *not* use the expensive augmented projection, but instead tries to construct solution formulas using only projection polynomials. Thus it can fail to produce a solution formula, but our experiments with many quantifier elimination problems from diverse application areas suggest that it will *rarely* fail.

  In case it does fail, it will however produce both a necessary condition and a sufficient condition to the input formula, which are fairly "close" to the input formula so that they together give a narrow "approximation" to a solution formula.
- It *simplifies* solution formulas by using three valued logic minimization. The simplification is carried out not only on the logical connectives but also on the relational operators.

The resulting algorithm has been successfully used in constructing simple solution formulas for various quantifier elimination problems (Hong 1990b, 1991a; Buchberger and Hong 1991). For example, for the quartic problem studied by Arnon and Mignotte (1988) and Lazard (1988), the solution formula produced by the original algorithm consists of 401 atomic formulas, but that by the improved algorithm consists of 5 atomic formulas. (See the end of Section 5 for the solution formulas.)

The plan of the paper is as follows: In Section 2 we precisely state the problem tackled in the subsequent sections. In Section 3 we show how to construct a solution formula from the output of the partial CAD algorithm, and in Section 4, we show how to simplify this solution formula by using three valued logic minimization. Finally in Section 5, we report experimental data comparing the solution formulas produced by the original algorithm and the improved algorithm.

We assume that the reader is familiar with the basic terminology of (Arnon et al. 1984a; Hong 1990b; Collins and Hong 1991), including that of projection, augmented projection, CAD, partial CAD, stacks, cells, truth-invariance, sign-invariance, defining formulas, propagation, and trial evaluation.

2   Problem Statement

Let $F^* = (Q_{f+1}x_{f+1}) \cdots (Q_r x_r) F(x_1, \ldots, x_r)$ be a quantified prenex formula in the first order theory of real closed fields. Here is the problem which we will tackle in the following sections.

Given: the output of the partial CAD algorithm for the formula F^*.
Task: construct a simple quantifier-free formulawhich is equivalent to F^* (we will call any such formula a solution formula).

Now we give a brief review of the outputs of the partial CAD algorithm (for the details, see (Hong 1990b; Collins and Hong 1991)). Given a formula F^*, the partial CAD algorithm produces:

- A list of (projection) polynomials $(P_1, \ldots, P_n)$ where $P_i \in \mathbb{Z}[x_1, \ldots, x_f]$.
- A decomposition of f-dimensional Euclidean space where the formula F^* has an invariant truth value in each cell. In particular for each cell the following information is produced:

  - the constant truth value of F^*,
  - the mechanism used for determining the truth value: propagation or trial evaluation.
  - the projection (partial) signature: $(\sigma_1, \ldots, \sigma_n)$ where σ_i is the sign of P_i in the cell if it is invariant, ? otherwise.

Note that the decomposition produced by the partial CAD algorithm is *not necessarily* sign-invariant with respect to all of the projection polynomials, while this is the case for the decomposition produced by the original algorithm. In general the decomposition produced by the original CAD algorithm is a refinement of that produced by the partial CAD algorithm.

3   (Complex) Solution Formula Construction

In this section, we show how to construct solution formulas from the outputs of the partial CAD algorithm. The resulting solution formulas are in general complex, and in the next section we show how to simplify them. We begin by introducing several definitions.

Definition 1 (Cell types) *A* solution (non-solution) cell *is a cell whose truth value is true (false). A* propagation (trial evaluation) *cell is a cell whose truth value is determined by propagation (trial evaluation).*

Definition 2 (Signature compatibility) *Let s and s' be two signs. We say s and s' are* compatible *if $s = s'$ or $s = $ '?' or $s' = $ '?'. Let $\Sigma = (s_1, \ldots, s_n)$ and $\Sigma' = (s'_1, \ldots, s'_n)$ be two projection signatures. We say that the two projection signatures are* compatible *if s_i and s'_i are compatible for every i.*

Definition 3 (Signatures-based formula) *Let W be a set of projection signatures. Then the W-based formula is the quantifier-free formula:*

$$\bigvee_{(\sigma_1, \ldots, \sigma_n) \in W} \bigwedge_{\sigma_i \neq ?} P_i \; p(\sigma_i) \; 0$$

where $p(+), p(0), p(-)$ are $>, =, <$ respectively.

Now the following theorem gives a way to construct a necessary condition and a sufficient condition for the input formula F^*.[1]

Theorem 1 *Let $W_t(W_f)$ be the set of all projection signatures of solution (non-solution) cells. Let W_c be the set $\{\Sigma \in W_t | \Sigma$ is compatible with some signature in $W_f.\}$. Let $W_t^* = W_t - W_c$. Let F_n be the W_t-based formula, and let F_s be the W_t^*-based formula. Then F_n is a necessary condition of F^*, and F_s is a sufficient condition of F^*.*

If $W_c = \emptyset$, we have $F_n = F_s$ which is a solution formula for F^*. Extensive experiments with many quantifier elimination problems arising from real applications show that almost always $W_c = \emptyset$ (Hong 1990b, 1991a, Buchberger and Hong 1991), suggesting that this method is an effective alternative to the original method, especially because it does not involve the enormous cost of augmented projection and related operations (in the algorithm DEFINE of (Collins 1975)). It is important to note that if $W_c \neq \emptyset$, then it is impossible to construct a solution formula from projection polynomials only. Therefore in order to obtain a solution formula, it is necessary to add some other polynomials which distinguish solution cells and non-solution cells. In fact this is the essential idea behind Collins' augmented projection. A strategy for adding polynomials is under development, but in the interim our method just produces a necessary and a sufficient conditions. The following theorem gives a cheaper way to compute W_c.

Theorem 2 *Let $T_t(T_f)$ be the set of the projection signatures of all propagation solution (non-solution) cells. Then we have $W_c = T_t \cap T_f$.*

From now on for the sake of simple presentation, we assume that $W_c = \emptyset$. One can easily generalize the following discussions to the case $W_c \neq \emptyset$ (for such generalizations, see (Hong 1990b)). Now we make an important improvement to Theorem 1.

Definition 4 (Normalized formula) *A formula is said to be* normalized *if (1) all the polynomials occurring in the formula are positive, irreducible, and of positive degree in at least one variable, and (2) it uses only $\wedge$ and $\vee$ as logical connectives.*

Definition 5 (Level) *Let $f \in \mathbb{Z}[x_1, \ldots, x_r]$. The level of f is the largest integer k such that $\deg_{x_k}(f) > 0$. An atomic formula is said to be of level k if the polynomial appearing in it is of level k.*

Theorem 3 *Let $F^* = (Q_{f+1}x_{f+1}) \cdots (Q_r x_r) F(x_1, \ldots, x_r)$ be a normalized formula. Let T_t be the set of the signatures of all propagation solution cells. Let F' be the T_t-based formula. Let $\bar{F}$ be the quantifier-free formula obtained from F by replacing every atomic formula of level greater than f with F'. Then $\bar{F}$ is equivalent to F^*.*

---

[1]Due to page limits, we do not include proofs for the theorems in this paper. The proofs can be found in (Hong 1990b).

$$\text{SOLUTION}(D, F, P, f; t, F_e, F_n, F_s)$$

Algorithm SOLUTION (Solution Formula Construction).
Inputs: D is a truth-invariant CAD produced by the partial CAD algorithm.
F is the normalized quantifier-free part of the input formula. P is the list of
projection polynomials. f is the number of free variables in the input formula.
Outputs: t is either **equ** or **inequ**. F_e is a quantifier-free formula equivalent
to the input formula if t is **equ**, otherwise F_e is undefined. F_n is a quantifier-
free formula necessary for the input formula if t is **inequ**, otherwise F_n is
undefined. F_s is a quantifier-free formula sufficient for the input formula if t is
inequ, otherwise F_s is undefined.

(1) [Decision Problem.] If $f > 0$, go to Step 2. Set $t \leftarrow$ **equ**. If the truth value
    of the root cell D is **true**, then set $F_e \leftarrow$ "$0 = 0$" and return, otherwise set
    $F_e \leftarrow$ "$0 \neq 0$" and return.
(2) [Signature tables.] Set T_t (T_f) to the set of the projection signatures of all
    propagation solution (non-solution) cells.
(3) [Any indistinguishable cell?] Set $T_c \leftarrow T_t \cap T_f$. Set $T_t^* \leftarrow T_t - T_c$. If
    $T_c = \emptyset$, then set $t \leftarrow$ **equ**, else set $t \leftarrow$ **inequ** and go to Step 5.
(4) [F_e.] Let F' be a T_t-based formula. Obtain F_e from F by replacing every
    atomic formula of level greater than f with F'. Return.
(5) [F_n.] Let F' be a T_t-based formula. Obtain F_n from F by replacing every
    atomic formula of level greater than f with F'.
(6) [F_s.] Let F' be a T_t^*-based formula. Obtain F_s from F by replacing every
    atomic formula of level greater than f with F'. Return.

Since any formula can be easily converted into an equivalent normalized
form, the above theorem tells us that one needs to consider *only propagation
cells* while forming signature-based formulas. The algorithm SOLUTION implements the ideas contained in the above theorems.

4   Simplification of Solution Formulas

The solution formulas produced by the algorithm SOLUTION is complex in
general, and now we show how to simplify them. Let us begin by making
general observations on various ways to simplify formulas as illustrated by the
examples below:

- *Logical connectives*: Consider the quantifier-free formula $(\phi_1 \wedge \phi_2) \vee \phi_1$, where
  ϕ_1 and ϕ_2 are atomic formulas. It can be simplified to the formula ϕ_1 since
  $\phi_1 \wedge \phi_2$ implies ϕ_1.
- *Relational operators*: The formula $x > 0 \vee x = 0$ can be simplified to the
  formula $x \geq 0$ due to the obvious reason.
- *Polynomials*: The formula $x > 0 \wedge x + 1 > 0$ can be simplified to the formula
  $x > 0$ since $x > 0$ implies $x + 1 > 0$.

- *Unsatisfiable formulas*: Suppose that $F_t \equiv \phi_1 \wedge \phi_2$ is a defining formula for the solution set (i.e., a solution formula), and that $F_f \equiv (\neg\phi_1) \wedge \phi_2$ is a defining formula for the non-solution set. Then the formula $\phi_1 \wedge (\neg\phi_2)$ is unsatisfiable and therefore the formula F_t is equivalent to $[\phi_1 \wedge \phi_2] \vee [\phi_1 \wedge (\neg\phi_2)]$, which can be simplified to ϕ_1.

In the following subsections, we show that we can algorithmically carry out simplifications based on logical connectives, relational operators, and unsatisfiable formulas, by using multiple-valued logic minimization. The simplification based on the properties of polynomials is left for the future research.

4.1   Multiple-Valued Logic Minimization Problem

We begin with a review of the multiple-valued logic minimization. The notation and terminology are from (Rudell 1986; Brayton et al. 1984).

An *n-variate m-valued logic function* L is a mapping

$$L : M^n \longrightarrow \{\texttt{true}, \texttt{false}, \texttt{dc}\}$$

where $\texttt{dc}$ stands for "don't care" and M is a set of m logic values. A *minterm* is an element of the domain of the function, namely M^n. The *TRUE-set* is the set of minterms for which the function values are $\texttt{true}$. The *FALSE-set* and the *DC-set* are defined in the similar way.

Let x be a variable[2] taking a value from the set M, and let S be a subset of M. A *literal*, x^S, is the Boolean function

$$x^S(x) = \begin{cases} \texttt{true} & \text{if } x \in S, \\ \texttt{false} & \text{if } x \notin S. \end{cases}$$

A *product* is a conjunction of literals. A *sum-of-products* is a disjunction of products. A *cover* of a logic function L is a sum-of-products which evaluates to $\texttt{true}$ for all minterms of the TRUE-set, $\texttt{false}$ for all minterms of the FALSE-set, and either $\texttt{true}$ or $\texttt{false}$ for all the minterms of the DC-set.

A logic function is uniquely defined by giving its TRUE-set, FALSE-set. A TRUE-set can be given by a sum-of-products H_t which evaluates to $\texttt{true}$ for all minterms of the TRUE-set and to $\texttt{false}$ for all other minterms. A FALSE-set can be given in the similar way. A *complexity function* is a mapping from the set of sum-of-products to the set of the real numbers. Now we are ready to give a precise definition of the multiple-valued logic minimization problem:

Given: two sum-of-products H_t and H_f which together define a multiple-valued logic function L, and a complexity function,

Task: find a minimum-complexity cover of L.

This problem has been extensively studied for the applications in circuit design, and recently many significant advances have been made in connection

---

[2]It is a propositional variable. Not to be confused with polynomial variables.

with VLSI synthesis (Bartholomeus and Man 1985; Brayton et al. 1984; Hong et al. 1974; McCluskey 1956; Quine 1952, 1955, 1959; Rudell 1986; Sasao 1982; Simanyi 1983) producing several efficient algorithms such as MINI, POP, Presto, ESPRESSO, ESPRESSO-MV, etc. A worst case time complexity of the m-valued logic minimization problem is known to be 2^{2^m} (Garey and Johnson 1979).

4.2 Simplification by Three-Valued Logic Minimization

Now we show how to obtain simple solution formulas by using multiple-valued logic minimization. The main idea is to view each projection polynomial P_i as a propositional variable x_i taking one of the three logic values $(-, 0, +)$. Then the output of the partial CAD algorithm defines an n-variate 3-valued logic function L where n is the number of projection polynomials. Then simplification of formulas turns into simplification of 3-valued logic functions. Now we elaborate this idea.

Definition 6 (Conversion function) *The conversion function is the invertible mapping from the set of atomic formulas to the set of literals such that $0 \neq 0$, $P_i > 0$, $P_i = 0$, $P_i < 0$, $P_i \geq 0$, $P_i \leq 0$, $P_i \neq 0$, $0 = 0$ are mapped respectively to $x_i^{\{\}}$, $x_i^{\{+\}}$, $x_i^{\{0\}}$, $x_i^{\{-\}}$, $x_i^{\{0,+\}}$, $x_i^{\{-,0\}}$, $x_i^{\{-,+\}}$, $x_i^{\{-,0,+\}}$. We extend this mapping naturally to the mapping from the set of disjunctions of conjunctions of atomic formulas to the set of sum-of-products.*

Definition 7 (Cover function) *A cover function Ψ is a mapping from the set of logic functions to the set of sum-of-products such that $\Psi(H_t, H_f)$ is a cover of the logic function defined by H_t and H_f.*

Theorem 4 *Let T_t (T_f) be the set of the projection signatures of all propagation solution (non-solution) cells, and let F_t (F_f) be the T_t (T_f)-based formula. Let Ω be the conversion function and Ψ be a cover function. Let*

$$F' = \Omega^{-1}\Psi(\Omega(F_t), \Omega(F_f)).$$

Then F' is true on all propagation solution cells and false on all propagation non-solution cells.

If a logic minimization algorithm is used as a cover function, the above theorem shows us how to obtain a simple quantifier-free formula which is true on all propagation solution cells and false on all propagation non-solution cells. The following theorem is a generalization of Theorem 3.

Theorem 5 *Let $F^* = (Q_{f+1}x_{f+1}) \cdots (Q_r x_r) F(x_1, \ldots, x_r)$ be a normalized formula. Let $F'(x_1, \ldots; x_f)$ be any quantifier-free formula which is true on all propagation solution cells and false on all propagation non-solution cells. Let $\bar{F}$ be the quantifier-free formula obtained from F by replacing every atomic formula of level greater than f with F'. Then $\bar{F}$ is equivalent to F^*.*

The above two theorems show us how to obtain a simple solution formula. In order to utilize these theorems, we need to make a slight modification of the algorithm SOLUTION. For instance, in Step (4) of the algorithm F' needs to be set to the formula

$$\Omega^{-1}\Psi(\Omega(F_t), \Omega(F_f))$$

instead of the T_t-based formula F_t. Similar modifications need to be done on Step (5) and (6).

4.3 Multiple-Valued Logic Minimization Algorithms

In this subsection, we give a brief review of the existing multiple-valued logic minimization algorithms and show why they do not fit to our application. Then we present a new algorithm tailored for our application. We begin by introducing several terminology.

Definition 8 *A product is said to* capture *a minterm iff it evaluates to* true *for the minterm. An* implicant *of a logic function L is a product which does not capture any minterm in the FALSE-set of L but captures at least one minterm in the TRUE-set of L. A* prime implicant *of a logic function L is an implicant of L such that no shorter implicant of L captures all the minterms captured by it.*

Definition 9 *Let M be the set of logic values and let S be a subset of M. A literal x^S is said to be* proper *if $1 \le |S| < |M|$. A proper literal x^S is said to be* simple *if $|S| = 1$. A proper literal x^S is said to be* compound *if $|S| > 1$.*

Definition 10 *The* length *of a product is the number of proper literals in it. The* length *of a sum-of-products is the sum of the lengths of the products in it.*

Currently there exist several multiple-valued logic minimization algorithms such as MINI, POP, Presto, ESPRESSO, ESPRESSO-MV, etc. They, though differing greatly from each other in the details, share the following general structure developed by Quine (1955) and McCluskey (1956):

1. Generate a set P of prime implicants which together capture all the minterms in the TRUE-set of the given logic function.
2. Select an "optimal" subset P' of the set P, which together still capture all the minterms in the TRUE-set of the logic function.

Further they also agree in the following two aspects:

- They generate prime implicants by *gradually shortening* the products in the input sum-of-products defining the TRUE-set.
- They are designed for VLSI synthesis where one is most interested in reducing the physical size of a chip. As a result they measure the complexity of a cover by the *number of products* in it.

However these do not seem to fit our application for the reasons:

- In our application, our experience to date suggests that most prime implicants selected for a cover are very short. Thus if they are generated by gradually shortening the input products, much computation is required. For example, it takes about 480 seconds for ESPRESSO-MV to simplify a solution formula of the x-axis ellipse problem (Arnon and Mignotte 1988) on Sun SPARC (Hong 1990b). This suggests that we need to devise another method which generates short prime implicants more efficiently.
- In our application, we are interested in reducing not only the number of products in a cover but also the lengths of products. In other words, we would like to obtain a short cover (a cover with small length).

A multiple-valued logic minimization algorithm, MVLMASF, motivated by the above observations is presented. The cover produced by this algorithm is not necessarily shortest, but "sufficiently" short.

$$\text{MVLMASF}(H_t, H_f)$$

Algorithm MVLMASF (Multiple-Valued Logic Minimization Algorithm, Shorter prime implicant First).
Inputs: H_t and H_f are sum-of-products which together define a logic function L.
Output: H is a cover of L, hopefully short, but not necessarily shortest.

(1) [Generate.] Call $P \leftarrow \text{GENPIMP}(H_t, H_f)$, which generates prime implicants of L, shorter ones first, until they together capture all the minterms in the TRUE-set of L. P be the set of all the generated prime implicants.
(2) [Select.] Select an optimal subset P' of the set P, which together still capture all the minterms in the TRUE-set of L, using Quine's method of essential prime implicants and cover. Set $H \leftarrow$ the disjunction of the prime implicants in the set P'. Return.

In Step 1, we generate prime implicants of L, shorter ones first, until they together capture all the minterms in the TRUE-set of L. Among the prime implicants of the same length, those with more compound literals are generated first. The algorithm GENPIMP which does this is presented below.

$$P \leftarrow \text{GENPIMP}(H_t, H_f)$$

Algorithm GENPIMP (Generate Prime Implicants).
Inputs: H_t and H_f are sum-of-products which together define a logic function L.
Output: P is a set of prime implicants of L.
Comment: At any point during the execution of this algorithm, the set P contains the prime implicants found until that point (not necessarily all the prime implicants). The set Z contains the minterms of the TRUE-set of L which are not yet captured by any prime implicant in P.

(1) [Setup.] Set P to the empty set. Set Z to the TRUE-set of L. Set M_t to the TRUE-set of L and M_f to the FALSE-set of L.

(2) [Generate the next product p.] Generate the next product p. (Shorter ones first, among these, those with more compound literals first.)

(3) [Does p capture any minterm in M_f?] If the product p captures any minterm in M_f, then go to Step 2.

(4) [Does p capture any minterm in M_t?] Set T to the set of the minterms of M_t captured by p. If T is empty, then go to Step 2.

(5) [Is p prime?] If there is an implicant in P which captures all the minterms in T, then go to Step 2.

(6) [Register p as a prime implicant.] At this point we know that the product p is a prime implicant, so insert p into the set P. Set Z to $Z - T$. If Z is not empty, then go to Step 2.

(7) [Done.] At this point we know that the prime implicants in P together capture all the minterms in M_t, and so return.

Step 5 of this algorithm needs some explanation. At the beginning of this step, the product p is necessarily an implicant since it passed the tests in Step 3 and 4. Now we would like to know whether it is also a prime implicant. Recall that by definition an implicant is prime iff there is no shorter implicant that captures all the minterms captured by it. The brute-force check based on this definition will require much computation, but we can do better based on the following theorem.

Theorem 6 *If there is no implicant in P capturing every minterm captured by p, then p is prime.*

In Step 2 of the algorithm MVLMASF, we select an optimal subset P' of the set P, which together still captures all the minterms in the TRUE-set of L, by using Quine's method of essential prime implicants and cover (Quine 1959). Quine's method is chosen because, although it is an algorithm for an NP-hard problem (Garey and Johnson 1979), it seems to have a very good average-time behavior for our application problems and because it ensures that the resulting cover is shortest among the covers which can be constructed from the prime implicants in the set P. For a detailed exposition of Quine's method of essential prime implicants and cover, see (Hohn 1966, pp. 201–220).

5 Experiments

5.1 Test Environment

We have used the following two programs for the experiments: (1) Arnon's program (Arnon 1981) implementing the original CAD algorithm along with the original solution formula construction algorithm. (2) The author's program implementing the improved CAD algorithm along with the solution formula construction algorithm presented in this paper. Both programs were written in SAC2/ALDES (Collins and Loos 1990; Loos 1976), and translated into C.

All the experiments were carried out on a disk-less SUN SPARC Station SLC with a 12 MIPS processor and 8 megabyte main memory running Unix. Out of 8 megabytes, 6 megabytes were used for list cells.

5.2   Test Problems

We carried out experiments on the following four quantifier elimination problems:

- *Quartic* (Arnon and Mignotte 1988; Delzell 1984)   Find a necessary and sufficient condition on the coefficients of a quartic polynomial such that it is non-negative for all real x.

$$(\forall x)\ x^4 + px^2 + qx + r \geq 0$$

- *Collins and Johnson* (Collins and Johnson 1989b)   Find a necessary and sufficient condition on real number a and b ($a \geq 1/2$ and $b > 0$ without loss of generality) such that a cubic polynomial with $a \pm bi$ as roots and a real root r between 0 and 1 has more than one coefficient sign variation.

$$\begin{aligned}
(\exists r)(\quad & 3a^2r + 3b^2r - 2ar - a^2 - b^2 < 0 \\
\wedge\ & 3a^2r + 3b^2r - 4ar + r - 2a^2 - 2b^2 + 2a > 0 \\
\wedge\ & a \geq 1/2 \\
\wedge\ & b > 0 \\
\wedge\ & r > 0 \\
\wedge\ & r < 1\ ).
\end{aligned}$$

- *X-axis Ellipse* (Arnon 1988b; Arnon and Mignotte 1988; Arnon and Smith 1983; Kahan 1975; Mignotte 1986; Lauer 1977; Lazard 1988)   Consider two objects: an ellipse of semi-axes a and b centered at $(c, 0)$ and a circle of radius 1 centered at $(0, 0)$. Find a necessary and sufficient condition for a, b, and c such that the ellipse is inside the circle (including touching).

$$(\forall x)(\forall y)[E(x, y) = 0 \Rightarrow C(x, y) \leq 0],$$

where $E(x, y) = (x - c)^2/a^2 + y^2/b^2 - 1$ and $C(x, y) = x^2 + y^2 - 1$. We can refine this formula based on some obvious observations, like symmetries:

$$(\forall x)(\forall y)[0 < a \leq 1 \wedge 0 < b \leq 1 \wedge 0 \leq c \leq 1 - a \wedge$$
$$[[c - a < x < c + a \wedge E(x, y) = 0] \Rightarrow C(x, y) \leq 0]].$$

- *Solotareff* (Achieser 1956)   The formula below arises as a sub-case of Solotareff's first problem for cubic polynomials.

$$\begin{aligned}
(\exists x)(\exists y)[& 1 \leq 4a \wedge 4a \leq 7 \wedge -3 \leq 4b \wedge 4b \leq 3 \\
\wedge\ & {-1} \leq x \wedge x \leq 0 \wedge 0 \leq y \wedge y \leq 1 \\
\wedge\ & 3x^2 - 2x - a = 0 \wedge x^3 - x^2 - ax = 2b - a + 2 \\
\wedge\ & 3y^2 - 2y - a = 0 \wedge y^3 - y^2 - ay = a - 2].
\end{aligned}$$

Table 1: Experiment Results

Problem	Method	L	T	T_{CAD}
Collins/Johnson	original	649	20.3	82.3
	improved	4	0.2	8.1
Quartic	original	401	10.7	9706.9
	improved	5	2.3	18.5
X-axis ellipse	original	?	?	$\gg$55454.3
	improved	8	0.7	39.5
Solotareff	original	?	?	$\gg$59267.3
	improved	6	8.5	10.6

Problem	n	N_t	N_f	N_p	N_{pi}	N_s
Collins/Johnson	15	44	19	57	2	2
Quartic	8	48	54	76	2	2
X-axis ellipse	23	92	20	36	5	2
Solotareff	19	1	64	4844	1	1

5.3   Test Results and Discussion

The solution formulas produced by both programs are shown at the end of this subsection. In Table 1 we show various quantities measured from the experiments, where

L is the length of the solution formula; i.e, the number of atomic formulas in it,

T is the time (in seconds) taken for constructing the solution formula from the CAD,

T_{CAD} is the time (in seconds) taken for constructing the CAD,

n is the number of the projection polynomials,

N_t is the number of the propagation solution cells,

N_f is the number of the propagation non-solution cells,

N_p is the number of the products generated by the algorithm GENPIMP,

N_{pi} is the number of the prime implicants among the generated products, and

N_s is the number of the prime implicants selected by Quine's procedure.

Arnon's program (the original algorithm) failed on X-axis ellipse and Solotareff problems because it ran out of memory. For these problems, T_{CAD} is the time taken until the failure took place. More detailed statistics in (Hong 1990b) suggest that the length of the solution formula produced by the original algorithm for X-ellipse problem would be at least several hundreds of thousands.

Solotareff's problem needs some explanation. As the table shows, *many* products ($N_p = 4844$) were generated before *one* prime implicant was found.

This phenomenon happened because (1) the prime implicant is a conjunction of two polynomial equalities (see the solution formula at the end of this subsection), but (2) the algorithm GENPIMP generates *shorter* products first and among the products of the same length the products with more *compound* literals first. If one chooses different ordering for product generation, one might be able to improve the situation. Currently we are experimenting with various other orderings.

Note also that the solution formula for Solotareff's problem (see the solution formula at the end of this subsection) can be further simplified, by taking advantage of linear polynomials, to the formula: $a = 1 \wedge b = -11/27$. This kind of simplification needs to be studied systematically. For example, one could use Buchberger's Gröbner basis algorithm (Buchberger 1965) for this purpose. See (Buchberger and Hong 1991) for certain initial attempts.

Here are the solution formulas produced by the original algorithm and the improved algorithm for the four quantifier elimination problems. As mentioned earlier, the original algorithm failed for X-axis ellipse problem and Solotareff problem, thus we could not obtain their solution formulas.

Collins/Johnson

Original Algorithm

```
[ 2 a - 1 = 0 /\ [ b > 0 /\ [ [ b < 0 /\ 3 b^2 + 3 a^2 - 2 a < 0
] \/ b = 0 \/ [ b > 0 /\ 3 b^2 + 3 a^2 - 2 a < 0 ] ] ] ] /\ [ 2
a - 1 = 0 /\ [ b > 0 /\ 3 b^2 + 3 a^2 - 2 a = 0 ] ] /\ [ 2 a - 1
= 0 /\ [ [ b > 0 /\ 3 b^2 + 3 a^2 - 2 a > 0 ] /\ [ [ b < 0 /\ b^
2 + a^2 - a < 0 ] \/ b = 0 \/ [ b > 0 /\ b^2 + a^2 - a < 0 ] ] ]
] /\ [ 2 a - 1 = 0 /\ [ b > 0 /\ b^2 + a^2 - a = 0 ] ] /\ [ 2 a
- 1 = 0 /\ [ [ b > 0 /\ b^2 + a^2 - a > 0 ] /\ [ [ [ [ [ b < 0
/\ 18 b^2 + 6 a^2 - 6 a - 1 < 0 ] \/ b = 0 \/ [ b > 0 /\ 18 b^2
+ 6 a^2 - 6 a - 1 < 0 ] ] /\ 6 b^3 + 6 a^2 b - 6 a b - b < 0 ] \
/ [ b > 0 /\ 18 b^2 + 6 a^2 - 6 a - 1 = 0 ] \/ [ [ b > 0 /\ 18 b
^2 + 6 a^2 - 6 a - 1 > 0 ] /\ 6 b^3 + 6 a^2 b - 6 a b - b < 0 ]
] /\ 3 b^4 + 6 a^2 b^2 - 6 a b^2 - b^2 + 3 a^4 - 6 a^3 + 3 a^2 <
0 ] \/ [ [ b > 0 /\ 18 b^2 + 6 a^2 - 6 a - 1 > 0 ] /\ 6 b^3 + 6
a^2 b - 6 a b - b = 0 ] \/ [ [ [ b > 0 /\ 18 b^2 + 6 a^2 - 6 a -
1 > 0 ] /\ 6 b^3 + 6 a^2 b - 6 a b - b > 0 ] /\ 3 b^4 + 6 a^2 b^
2 - 6 a b^2 - b^2 + 3 a^4 - 6 a^3 + 3 a^2 < 0 ] ] ] ] /\ [ [ 2 a
- 1 > 0 /\ 3 a - 2 < 0 ] /\ [ b > 0 /\ [ [ b < 0 /\ 3 b^2 + 3 a^
2 - 2 a < 0 ] \/ b = 0 \/ [ b > 0 /\ 3 b^2 + 3 a^2 - 2 a < 0 ] ]
] ] /\ [ [ 2 a - 1> 0 /\ 3 a - 2 < 0 ] /\ [ b > 0 /\ 3 b^2 + 3 a
^2 - 2 a = 0 ] ] /\ [ [ 2 a - 1 > 0 /\ 3 a - 2 < 0 ] /\ [ [ b >
0 /\ 3 b^2 + 3 a^2 - 2 a > 0 ] /\ [ [ [ [ b < 0 /\ 18 b^2 + 6
a^2 - 6a - 1 > 0 ] /\ 6 b^3 + 6 a^2 b - 6 a b - b > 0 ] \/ [ b <
```

0 /\ 18 b^2 + 6 a^2 - 6 a - 1 = 0] \/ [[[b < 0 /\ 18 b^2 + 6
a^2 - 6 a - 1 < 0] \/ b = 0 \/ [b > 0 /\ 18 b^2 + 6 a^2- 6 a -
1 < 0]] /\ 6 b^3 + 6 a^2 b - 6 a b - b > 0]] /\ 3 b^4 + 6 a^
2 b^2 - 6 a b^2 - b^2 + 3 a^4 - 6 a^3 + 3 a^2 > 0] \/ [[[b <
0 /\ 18 b^2 + 6 a^2 - 6 a - 1 < 0] \/ b = 0 \/ [b > 0 /\ 18 b^
2 + 6 a^2 - 6 a - 1 < 0]] /\ 6 b^3 + 6 a^2 b - 6 a b - b = 0]
\/ [[[[[b < 0 /\ 18 b^2 + 6 a^2 - 6 a - 1 < 0] \/ b = 0 \/
[b > 0 /\ 18 b^2 + 6 a^2 - 6 a - 1 < 0]] /\ 6 b^3 + 6 a^2 b -
6 a b - b < 0] \/ [b > 0 /\ 18 b^2 + 6 a^2 - 6 a - 1 = 0] \/
[[b > 0 /\ 18 b^2 + 6 a^2 - 6 a - 1 > 0] /\ 6 b^3 + 6 a^2 b -
6 a b - b < 0]] /\ 3 b^4 + 6 a^2 b^2 - 6 a b^2 - b^2 + 3 a^4 -
6 a^3 + 3 a^2 > 0]]]] /\ [[2 a - 1 > 0 /\ 3 a - 2 < 0] /\
[[[[b < 0 /\ 18 b^2 + 6 a^2 - 6 a - 1 < 0] \/ b = 0 \/ [
b > 0 /\ 18 b^2 + 6 a^2 - 6 a - 1 < 0]] /\ 6 b^3 + 6 a^2 b - 6
a b - b < 0] \/ [b > 0 /\ 18 b^2 + 6 a^2 - 6 a - 1 = 0] \/ [
[b > 0 /\ 18 b^2 + 6 a^2 - 6 a - 1 > 0] /\ 6 b^3 + 6 a^2 b - 6
a b - b < 0]] /\ 3 b^4 + 6 a^2 b^2 - 6 a b^2 - b^2 + 3 a^4 - 6
a^3 + 3 a^2 = 0]] /\ [[2 a - 1 > 0 /\ 3 a - 2 < 0] /\ [[
[[[b < 0 /\ 18 b^2 + 6 a^2 - 6 a - 1 < 0] \/ b = 0 \/ [b
> 0 /\ 18 b^2 + 6 a^2 - 6 a - 1 < 0]] /\ 6 b^3 + 6 a^2 b - 6 a
b - b < 0] \/ [b > 0 /\ 18 b^2 + 6 a^2 - 6 a - 1 = 0] \/ [[
b > 0 /\ 18 b^2 + 6 a^2 - 6 a - 1 > 0] /\ 6 b^3 + 6 a^2 b - 6 a
b - b < 0]] /\ 3 b^4 + 6 a^2 b^2 - 6 a b^2 - b^2 + 3 a^4 - 6 a
^3 + 3 a^2 < 0] \/ [[b > 0 /\ 18 b^2 + 6 a^2 - 6 a - 1 > 0]
/\ 6 b^3 + 6 a^2 b - 6 a b - b = 0] \/ [[[b > 0 /\ 18 b^2 +
6 a^2 - 6 a - 1 > 0] /\ 6 b^3 + 6 a^2 b - 6 a b - b > 0] /\ 3
b^4 + 6 a^2 b^2 - 6 a b^2 - b^2 + 3 a^4 - 6 a^3 + 3 a^2 < 0]]
/\ [[b < 0 /\ 3 b^2 + 3 a^2 - 4 a + 1 < 0] \/ b = 0 \/ [b >
0 /\ 3 b^2 + 3 a^2 - 4 a + 1 < 0]]]] /\ [[2 a - 1 > 0 /\ 3
a - 2 < 0] /\ [b > 0 /\ 3 b^2 + 3 a^2 - 4 a + 1 = 0]] /\ [[
2 a - 1 > 0 /\ 3 a - 2 < 0] /\ [[b > 0 /\ 3 b^2 + 3 a^2 - 4 a
+ 1 > 0] /\ [[b< 0 /\ b^2 + a^2 - a < 0] \/ b = 0 \/ [b > 0
/\ b^2 + a^2 - a < 0]]]] /\ [[2 a - 1 > 0 /\ 3 a - 2 < 0]
/\ [b > 0 /\ b^2 + a^2 - a = 0]] /\ [[2 a - 1 > 0 /\ 3 a -
2 < 0] /\ [[b > 0 /\ b^2 + a^2 - a > 0] /\ [[[[[b < 0
/\ 18 b^2 + 6 a^2 - 6 a - 1 < 0] \/ b = 0 \/ [b > 0 /\ 18 b^2 +
6 a^2 - 6 a - 1 < 0]] /\ 6 b^3 + 6 a^2 b - 6 a b - b < 0] \/
[b > 0 /\ 18 b^2 + 6 a^2 - 6 a - 1 = 0] \/ [[b > 0 /\ 18 b^2
+ 6 a^2 - 6 a - 1 > 0] /\ 6 b^3 + 6 a^2 b - 6 a b - b < 0]] /
\ 3 b^4 + 6 a^2 b^2 - 6 a b^2 - b^2 + 3 a^4 - 6 a^3 + 3 a^2 < 0
] \/ [[b > 0 /\ 18 b^2 + 6 a^2 - 6 a - 1 > 0] /\ 6 b^3 + 6 a^
2 b - 6 a b - b = 0] \/ [[[b > 0 /\ 18 b^2 + 6 a^2 - 6 a - 1
> 0] /\ 6 b^3 + 6 a^2 b - 6 a b - b > 0] /\ 3 b^4 + 6 a^2 b^2
- 6 a b^2 - b^2 + 3 a^4 - 6 a^3 + 3 a^2 < 0]]] /\ [3a - 2
= 0 /\ [b > 0 /\ [[[[b < 0 /\ 18 b^2 + 6 a^2 - 6 a - 1 >

0] /\ 6 b^3 + 6 a^2b - 6 a b - b > 0] \/ [b < 0 /\ 18 b^2 + 6
a^2 - 6 a - 1 = 0] \/ [[[b < 0 /\ 18 b^2 + 6 a^2 - 6 a - 1 <

....

Slightly more then three pages of output have been deleted from here.

....

> 0 /\ 18 b^2 + 6 a^2 - 6 a - 1 > 0] /\ 6 b^3 + 6 a^2 b - 6 a b
- b = 0] \/ [[[b > 0 /\ 18 b^2 + 6 a^2 - 6 a - 1 > 0] /\ 6
b^3 + 6 a^2 b - 6 a b - b > 0] /\ 3 b^4 + 6 a^2 b^2 - 6 a b^2 -
b^2 + 3 a^4 - 6 a^3 + 3 a^2 < 0]]]] /\ [[a - 1 > 0 /\ 12 a
^2 - 12 a - 1 < 0 /\ 32 a - 35 < 0] /\ [[[[[b < 0 /\ 18
b^2 + 6 a^2 - 6 a - 1 < 0] \/ b = 0 \/ [b > 0 /\ 18 b^2 + 6 a^
2 - 6 a - 1 < 0]] /\ 6 b^3 + 6 a^2 b - 6 a b - b < 0] \/ [b
> 0 /\ 18 b^2 + 6 a^2 - 6 a - 1 = 0] \/ [[b > 0 /\ 18 b^2 + 6
a^2 - 6 a - 1 > 0] /\ 6 b^3 + 6 a^2 b - 6 a b - b < 0]] /\ 3
b^4 + 6 a^2 b^2 - 6 a b^2 - b^2 + 3 a^4 - 6 a^3 + 3 a^2 < 0] \/
[[b > 0 /\ 18 b^2 + 6 a^2 - 6 a - 1 > 0] /\ 6 b^3 + 6 a^2 b -
6 a b - b = 0] \/ [[[b > 0 /\ 18 b^2 + 6 a^2 - 6 a - 1 > 0]
/\ 6 b^3 + 6 a^2 b - 6 a b - b > 0] /\ 3 b^4 + 6 a^2 b^2 - 6 a
b^2 - b^2 + 3 a^4 - 6 a^3 + 3 a^2 < 0]]]

Improved Algorithm

[b^2 + a^2 - a <= 0 \/ 3 b^4 + 6 a^2 b^2 - 6 a b^2 - b^2 + 3 a^
4 - 6 a^3 + 3 a^2 < 0] /\ 2 a - 1 >= 0 /\ b > 0

Quartic

Original Algorithm

[p < 0 /\ [q < 0 /\ 27 q^2 + 8 p^3 > 0] /\ 256 r^3 - 128 p^2
r^2 + 144 p q^2 r + 16 p^4r - 27 q^4 - 4 p^3 q^2 = 0] /\ [p <
0 /\ [q < 0 /\ 27 q^2 + 8 p^3 > 0] /\ 256 r^3 - 128p^2 r^2 + 1
44 p q^2 r + 16 p^4 r - 27 q^4 - 4 p^3 q^2 > 0] /\ [p < 0 /\ [
q < 0 /\ 27 q^2+ 8 p^3 = 0] /\ [[6 r - p^2 > 0 /\ 48 r^2 - 16
p^2 r + 9 p q^2 + p^4 > 0] /\ 256 r^3 - 128 p^2 r^2 + 144 p q^2
r + 16 p^4 r - 27 q^4 - 4 p^3 q^2 = 0]] /\ [p < 0 /\ [q < 0
/\ 27 q^2 + 8 p^3 = 0] /\ [[6 r - p^2 > 0 /\ 48 r^2 - 16 p^2
r + 9 p q^2 + p^4 > 0] /\ 256 r^3 - 128 p^2 r^2 + 144 p q^2 r +
16 p^4 r - 27 q^4 - 4 p^3 q^2 > 0]] /\ [p < 0 /\ [[[q < 0 /
\ 27 q^2 + 8 p^3 < 0] \/ q = 0 \/ [q > 0 /\ 27 q^2 + 8 p^3 < 0
]] /\ [q < 0 /\ 9 q^2 + 2 p^3 > 0]] /\ [[6 r - p^2 > 0 /\
48 r^2 - 16 p^2 r + 9 p q^2 + p^4 > 0] /\ 256 r^3 - 128 p^2 r^2
+ 144 p q^2 r + 16 p^4 r - 27 q^4 - 4 p^3 q^2 = 0]] /\ [p < 0
/\ [[[q < 0 /\ 27 q^2 + 8 p^3 < 0] \/ q = 0 \/ [q > 0 /\ 27

q^2 + 8 p^3 < 0]] /\ [q < 0/\ 9 q^2 + 2 p^3 > 0]] /\ [[6
r - p^2 > 0 /\ 48 r^2 - 16 p^2 r + 9 p q^2 + p^4 > 0] /\ 256 r^
3 - 128 p^2 r^2 + 144 p q^2 r + 16 p^4 r - 27 q^4 - 4 p^3 q^2 >
0]] /\ [p < 0 /\ [q < 0 /\ 9 q^2 + 2 p^3 = 0] /\ [[6 r -
p^2 > 0 /\ 48 r^2 - 16 p^2 r + 9 p q^2 + p^4 > 0] /\ 256 r^3 -
128 p^2 r^2 + 144 p q^2 r + 16 p^4 r - 27 q^4 - 4 p^3 q^2 = 0]
] /\ [p < 0 /\ [q < 0 /\ 9 q^2 + 2 p^3 = 0] /\ [[6 r - p^2
> 0 /\ 48 r^2 - 16 p^2 r + 9 p q^2 + p^4 > 0] /\ 256 r^3 - 128
p^2 r^2 + 144 p q^2 r + 16 p^4 r - 27 q^4 - 4 p^3 q^2 > 0]] /\
[p < 0 /\ [[[q < 0 /\ 9 q^2 + 2 p^3 < 0] \/ q = 0 \/ [q >
0 /\ 9 q^2 + 2 p^3 < 0]] /\ [q < 0 /\ 27 q^2 + 4 p^3 > 0]]
/\ [[6 r - p^2 > 0 /\ 48 r^2 - 16 p^2 r + 9 p q^2 + p^4 > 0]
/\ 256 r^3 - 128 p^2 r^2 + 144 p q^2 r + 16 p^4 r - 27 q^4 - 4 p
^3 q^2 = 0]] /\ [p < 0 /\ [[[q < 0 /\ 9 q^2 + 2 p^3 < 0]
\/ q = 0 \/ [q > 0 /\ 9 q^2 + 2 p^3 < 0]] /\ [q < 0 /\ 27 q^
2 + 4 p^3 > 0]] /\ [[6 r - p^2 > 0 /\ 48 r^2 - 16 p^2 r + 9
p q^2 + p^4 > 0] /\ 256 r^3 - 128 p^2 r^2 + 144 p q^2 r + 16 p^
4 r - 27 q^4 - 4 p^3 q^2 > 0]] /\ [p < 0 /\ [q < 0 /\ 27 q^2
+ 4 p^3 = 0] /\ [[6 r - p^2 > 0 /\ 48 r^2 - 16 p^2 r + 9 pq^2
+ p^4 > 0] /\ 256 r^3 - 128 p^2 r^2 + 144 p q^2 r + 16 p^4 r -
27 q^4 - 4 p^3 q^2 = 0]] /\ [p < 0 /\ [q < 0 /\ 27 q^2 + 4 p^
3 = 0] /\ [[6 r - p^2 > 0 /\ 48 r^2 - 16 p^2 r+ 9 p q^2 + p^4
> 0] /\ 256 r^3 - 128 p^2 r^2 + 144 p q^2 r + 16 p^4 r - 27 q^4
- 4 p^3 q^2 > 0]] /\ [p < 0 /\ [[[q < 0 /\ 27 q^2 + 4 p^3
< 0] \/ q = 0 \/ [q > 0 /\ 27 q^2 + 4 p^3 < 0]] /\ q < 0] /
\ [[6 r - p^2 > 0 /\ 48 r^2 - 16 p^2 r + 9 p q^2 + p^4 > 0] /
\ 256 r^3 - 128 p^2 r^2 + 144 p q^2 r + 16 p^4 r - 27 q^4 - 4 p^
3 q^2 = 0]] /\ [p < 0 /\ [[[q < 0 /\ 27 q^2 + 4 p^3 < 0]
\/ q = 0 \/ [q > 0 /\ 27 q^2 + 4 p^3 < 0]] /\ q < 0] /\ [[
6 r - p^2 > 0 /\ 48 r^2 - 16 p^2 r + 9 p q^2 + p^4 > 0] /\ 256
r^3 - 128 p^2 r^2 + 144 p q^2 r + 16 p^4 r - 27 q^4 - 4 p^3 q^2
> 0]] /\ [p < 0 /\ q = 0 /\ [6 r - p^2> 0 /\ 48 r^2 - 16 p^2
r + 9 p q^2 + p^4 = 0]] /\ [p < 0 /\ q = 0 /\ [6 r - p^2 > 0
/\ 48 r^2 - 16 p^2 r + 9 p q^2 + p^4 > 0]] /\ [p < 0 /\ [q >
0 /\ [[q < 0 /\ 27 q^2 + 4 p^3 < 0] \/ q = 0 \/ [q > 0 /\ 27
q^2 + 4 p^3 < 0]]] /\ [[6 r - p^2 > 0 /\ 48 r^2 - 16 p^2 r
+ 9 p q^2 + p^4 > 0] /\ 256 r^3 - 128 p^2 r^2 + 144 p q^2 r + 1
6 p^4 r - 27 q^4 - 4 p^3 q^2 = 0]] /\ [p < 0 /\ [q > 0 /\ [
[q < 0 /\ 27 q^2 + 4 p^3 < 0] \/ q = 0 \/ [q > 0 /\ 27 q^2 +
4 p^3 < 0]]] /\ [[6 r - p^2 > 0 /\ 48 r^2 - 16 p^2 r + 9 p
q^2 + p^4 > 0] /\ 256 r^3 - 128 p^2 r^2 + 144 p q^2 r + 16 p^4
r - 27 q^4 - 4 p^3 q^2 > 0]] /\ [p < 0 /\ [q > 0 /\ 27 q^2 +
4 p^3 = 0] /\ [[6 r - p^2 > 0 /\ 48 r^2 - 16 p^2 r + 9 pq^2 +
p^4 > 0] /\ 256 r^3 - 128 p^2 r^2 + 144 p q^2 r + 16 p^4 r - 27
q^4 - 4 p^3 q^2 = 0]] /\ [p < 0 /\ [q > 0 /\ 27 q^2 + 4 p^3 =

0] /\ [[6 r - p^2 > 0 /\ 48 r^2 - 16 p^2 r+ 9 p q^2 + p^4 > 0
] /\ 256 r^3 - 128 p^2 r^2 + 144 p q^2 r + 16 p^4 r - 27 q^4 - 4
p^3 q^2 > 0]] /\ [p < 0 /\ [[q > 0 /\ 27 q^2 + 4 p^3 > 0]
/\ [[q < 0 /\ 9 q^2 + 2 p^3 < 0] \/ q = 0 \/ [q > 0 /\ 9 q^2
+ 2 p^3 < 0]]] /\ [[6 r - p^2 > 0 /\ 48 r^2 - 16 p^2r + 9 p
q^2 + p^4 > 0] /\ 256 r^3 - 128 p^2 r^2 + 144 p q^2 r + 16 p^4
r - 27 q^4 - 4 p^3q^2 = 0]] /\ [p < 0 /\ [[q > 0 /\ 27 q^2
+ 4 p^3 > 0] /\ [[q < 0 /\ 9 q^2 + 2 p^3 < 0] \/ q = 0 \/ [
q > 0 /\ 9 q^2 + 2 p^3 < 0]]] /\ [[6 r - p^2 > 0 /\ 48 r^2
- 16 p^2r + 9 p q^2 + p^4 > 0] /\ 256 r^3 - 128 p^2 r^2 + 144 p
q^2 r + 16 p^4 r - 27 q^4 - 4 p^3q^2 > 0]] /\ [p < 0 /\ [q >
0 /\ 9 q^2 + 2 p^3 = 0] /\ [[6 r - p^2 > 0 /\ 48 r^2 - 16 p^2
r + 9 p q^2 + p^4 > 0] /\ 256 r^3 - 128 p^2 r^2 + 144 p q^2 r +
16 p^4 r - 27 q^4 - 4 p^3 q^2 = 0]] /\ [p < 0 /\ [q > 0 /\ 9
q^2 + 2 p^3 = 0] /\ [[6 r - p^2 > 0 /\ 48 r^2 - 16 p^2 r + 9
p q^2 + p^4 > 0] /\ 256 r^3 - 128 p^2 r^2 + 144 p q^2 r + 16 p^
4 r - 27q^4 - 4 p^3 q^2 > 0]] /\ [p < 0 /\ [[q > 0 /\ 9 q^2
+ 2 p^3 > 0] /\ [[q < 0 /\ 27 q^2 + 8 p^3 < 0] \/ q = 0 \/ [
q > 0 /\ 27 q^2 + 8 p^3 < 0]]] /\ [[6 r - p^2 > 0 /\ 48 r^2
- 16 p^2 r + 9 p q^2 + p^4 > 0] /\ 256 r^3 - 128 p^2 r^2 + 144
p q^2 r + 16 p^4 r - 27 q^4 - 4 p^3 q^2 = 0]] /\ [p < 0 /\ [
[q > 0 /\ 9 q^2 + 2 p^3 > 0] /\ [[q < 0 /\ 27 q^2 + 8 p^3 <
0] \/ q = 0 \/ [q > 0 /\ 27 q^2 + 8 p^3 < 0]]] /\ [[6 r -
p^2 > 0 /\ 48 r^2 - 16 p^2 r + 9 p q^2 + p^4 > 0] /\ 256 r^3 -
128 p^2 r^2 + 144 p q^2 r + 16 p^4 r - 27 q^4 - 4 p^3 q^2 > 0]
] /\ [p < 0 /\ [q > 0 /\ 27 q^2 + 8 p^3 = 0] /\ [[6 r - p^2
> 0 /\ 48 r^2 - 16 p^2 r + 9 p q^2 + p^4 > 0] /\ 256 r^3 - 128
p^2 r^2 + 144 p q^2 r + 16 p^4 r - 27 q^4 - 4 p^3 q^2 = 0]] /\
[p < 0 /\ [q > 0 /\ 27 q^2 + 8 p^3 = 0] /\ [[6 r - p^2 > 0
/\ 48 r^2 - 16 p^2 r + 9 p q^2 + p^4 > 0] /\ 256 r^3 - 128 p^2
r^2 + 144 p q^2 r + 16 p^4 r - 27 q^4 - 4 p^3 q^2 > 0]] /\ [p
< 0 /\ [q > 0 /\ 27 q^2 + 8 p^3 > 0] /\ 256 r^3 - 128 p^2 r^2
+ 144 p q^2 r + 16 p^4 r - 27 q^4 - 4 p^3 q^2 = 0] /\ [p < 0 /
\ [q > 0 /\ 27 q^2 + 8 p^3 > 0] /\ 256 r^3 - 128 p^2 r^2 + 144
p q^2 r + 16 p^4 r - 27 q^4- 4 p^3 q^2 > 0] /\ [p = 0 /\ q < 0
/\ 256 r^3 - 128 p^2 r^2 + 144 p q^2 r + 16 p^4 r - 27 q^4 - 4 p
^3 q^2 = 0] /\ [p = 0 /\ q < 0 /\ 256 r^3 - 128 p^2 r^2 + 144
p q^2 r + 16 p^4r - 27 q^4 - 4 p^3 q^2 > 0] /\ [p = 0 /\ q = 0
/\ 256 r^3 - 128 p^2 r^2 + 144 p q^2 r + 16 p^4 r - 27 q^4 - 4 p
^3 q^2 = 0] /\ [p = 0 /\ q = 0 /\ 256 r^3 - 128 p^2 r^2 + 144
p q^2r + 16 p^4 r - 27 q^4 - 4 p^3 q^2 > 0] /\ [p = 0 /\ q > 0
/\ 256 r^3 - 128 p^2 r^2 + 144 p q^2 r + 16 p^4 r - 27 q^4 - 4 p
^3 q^2 = 0] /\ [p = 0 /\ q > 0 /\ 256 r^3 - 128 p^2 r^2 + 144
p q^2 r + 16 p^4 r - 27 q^4 - 4 p^3 q^2 > 0] /\ [p > 0 /\ [q
< 0 /\ 27 q^2 - p^3 > 0] /\ 256 r^3 - 128 p^2 r^2 + 144 p q^2 r

+ 16 p^4 r - 27 q^4 - 4 p^3 q^2 = 0] /\ [p > 0/\ [q < 0 /\ 27
q^2 - p^3 > 0] /\ [256 r^3 - 128 p^2 r^2 + 144 p q^2 r + 16 p^
4 r - 27 q^4 - 4 p^3 q^2 > 0 /\ 8 p r - 9 q^2 - 2 p^3 < 0]] /\
[p > 0 /\ [q < 0 /\ 27 q^2 - p^3 > 0] /\ 8 p r - 9 q^2 - 2 p^
3 = 0] /\ [p > 0 /\ [q < 0 /\ 27 q^2 - p^3 > 0] /\ 8 p r - 9
q^2 - 2 p^3 > 0] /\ [p > 0 /\ [q < 0 /\ 27 q^2 - p^3 = 0] /\
256 r^3 - 128 p^2 r^2 + 144 p q^2 r + 16 p^4 r - 27 q^4 - 4 p^3
q^2 = 0] /\ [p > 0 /\ [q < 0 /\ 27 q^2 - p^3 = 0] /\ [256 r^
3 - 128 p^2 r^2 + 144 p q^2 r + 16 p^4 r - 27 q^4 - 4 p^3 q^2 >
0 /\ 8 p r - 9 q^2 - 2 p^3 < 0]] /\ [p > 0 /\ [q < 0 /\ 27 q
^2 - p^3 = 0] /\ 8 p r - 9 q^2 - 2 p^3 = 0] /\ [p > 0 /\ [q
< 0 /\ 27 q^2 - p^3 = 0] /\ 8 p r - 9 q^2 - 2 p^3 > 0] /\ [p
> 0 /\ [[[q < 0 /\ 27 q^2 - p^3 < 0] \/ q = 0 \/ [q > 0 /\
27 q^2 - p^3 < 0]] /\ q < 0] /\ 256 r^3 - 128 p^2 r^2 + 144 p
q^2 r + 16 p^4 r - 27 q^4 - 4 p^3 q^2 = 0] /\ [p > 0 /\ [[[
q < 0 /\ 27 q^2 - p^3 < 0] \/ q = 0 \/ [q > 0 /\ 27 q^2 - p^3
< 0]] /\ q < 0] /\ [256 r^3 - 128 p^2 r^2 + 144 p q^2 r + 16
p^4 r - 27 q^4 - 4 p^3 q^2 > 0 /\ 8 p r - 9 q^2 - 2 p^3 < 0]]
/\ [p > 0 /\ [[[q < 0 /\ 27 q^2 - p^3 < 0] \/ q = 0 \/ [q
> 0 /\ 27 q^2 - p^3 < 0]] /\ q < 0] /\ 8 p r - 9 q^2 - 2 p^3
= 0] /\ [p > 0 /\ [[[q < 0 /\ 27 q^2 - p^3 < 0] \/ q = 0 \
/ [q > 0 /\ 27 q^2 - p^3 < 0]] /\ q < 0] /\ 8 p r - 9 q^2- 2
p^3 > 0] /\ [p > 0 /\ q = 0 /\ [[6 r - p^2 < 0 /\ 48 r^2 - 1
6 p^2 r + 9 p q^2 + p^4> 0] /\ 256 r^3 - 128 p^2 r^2 + 144 p q^
2 r + 16 p^4 r - 27 q^4 - 4 p^3 q^2 = 0]] /\ [p> 0 /\ q = 0 /
\ [[[6 r - p^2 < 0 /\ 48 r^2 - 16 p^2 r + 9 p q^2 + p^4 > 0]
/\ 256 r^3 - 128 p^2 r^2 + 144 p q^2 r + 16 p^4 r - 27 q^4 - 4 p
^3 q^2 > 0] \/ [6 r - p^2 < 0 /\ 48r^2 - 16 p^2 r + 9 p q^2 +
p^4 = 0] \/ [[6 r - p^2 < 0 /\ 48 r^2 - 16 p^2 r + 9 p q^2 +
p^4 < 0] \/ 6 r - p^2 = 0 \/ [6 r - p^2 > 0 /\ 48 r^2 - 16 p^2
r + 9 p q^2 + p^4 < 0]]]] /\ [p > 0 /\ q = 0 /\ [6 r - p^2
> 0 /\ 48 r^2 - 16 p^2 r + 9 p q^2 + p^4 = 0]] /\ [p > 0 /\ q
= 0 /\ [6 r - p^2 > 0 /\ 48 r^2 - 16 p^2 r + 9 p q^2 + p^4 > 0
]] /\ [p > 0 /\ [q > 0 /\ [[q < 0 /\ 27 q^2 - p^3 < 0] \/
q = 0 \/ [q > 0 /\ 27 q^2 - p^3 < 0]]] /\ 256 r^3 - 128 p^2
r^2 + 144 p q^2 r + 16 p^4 r - 27 q^4 - 4 p^3 q^2 = 0] /\ [p >
0 /\ [q > 0 /\ [[q < 0 /\ 27 q^2 - p^3 < 0] \/ q = 0 \/ [q
> 0 /\ 27 q^2 - p^3 < 0]]] /\ [256 r^3 - 128 p^2 r^2 + 144 p
q^2 r + 16 p^4 r - 27 q^4 - 4 p^3 q^2 > 0 /\ 8 p r - 9 q^2 - 2 p
^3 < 0]] /\ [p > 0 /\ [q > 0 /\ [[q < 0 /\ 27 q^2 - p^3 <
0] \/ q = 0 \/ [q> 0 /\ 27 q^2 - p^3 < 0]]] /\ 8 p r - 9 q^
2 - 2 p^3 = 0] /\ [p > 0 /\ [q > 0 /\ [[q< 0 /\ 27 q^2 - p^
3 < 0] \/ q = 0 \/ [q > 0 /\ 27 q^2 - p^3 < 0]]] /\ 8 p r -
9 q^2 - 2 p^3 > 0] /\ [p > 0 /\ [q > 0 /\ 27 q^2 - p^3 = 0]
/\ 256 r^3 - 128 p^2 r^2 + 144 p q^2 r + 16 p^4 r - 27 q^4 - 4 p

^3 q^2 = 0] /\ [p > 0 /\ [q > 0 /\ 27 q^2 - p^3 = 0] /\ [25
6 r^3 - 128 p^2 r^2 + 144 p q^2 r + 16 p^4 r - 27 q^4 - 4 p^3 q^
2 > 0 /\ 8 p r - 9 q^2 - 2 p^3 < 0]] /\ [p > 0 /\ [q > 0 /\
27 q^2 - p^3 = 0] /\ 8 p r - 9 q^2 - 2 p^3 = 0] /\ [p > 0 /\
[q > 0 /\ 27 q^2 - p^3 = 0] /\ 8 p r - 9 q^2 - 2 p^3 > 0] /\
[p > 0 /\ [q > 0 /\ 27 q^2 - p^3 > 0] /\ 256 r^3 - 128 p^2 r^
2 + 144 p q^2 r + 16 p^4 r - 27 q^4 - 4 p^3q^2 = 0] /\ [p > 0
/\ [q > 0 /\ 27 q^2 - p^3 > 0] /\ [256 r^3 - 128 p^2 r^2 + 14
4 p q^2 r + 16 p^4 r - 27 q^4 - 4 p^3 q^2 > 0 /\ 8 p r - 9 q^2 -
2 p^3 < 0]] /\ [p > 0 /\ [q > 0 /\ 27 q^2 - p^3 > 0] /\ 8 p
r - 9 q^2 - 2 p^3 = 0] /\ [p > 0 /\ [q > 0 /\ 27 q^2 - p^3 >
0] /\ 8 p r - 9 q^2 - 2 p^3 > 0]

Improved Algorithm

[[256 r^3 - 128 p^2 r^2 + 144 p q^2 r + 16 p^4 r - 27 q^4 - 4
p^3 q^2 >= 0 /\ 8 p r - 9 q^2 - 2 p^3 <= 0] \/ [27 q^2 + 8 p^3
>= 0 /\ 8 p r - 9 q^2 - 2 p^3 >= 0]] /\ r >= 0

X-axis ellipse

Original Algorithm

Failed

Improved Algorithm

[b^2 - a <= 0 \/ b^2 c^2 + b^4 - a^2 b^2 - b^2 + a^2 <= 0] /\
c + a - 1 <= 0 /\ c >= 0 /\ a - 1 <= 0 /\ a > 0 /\ b - 1 <= 0 /
\ b > 0

Solotareff

Original Algorithm

Failed

Improved Algorithm

a - 1 = 0 /\ 27 b^2 - 18 a b + 56 b - a^3 + 2 a^2 - 19 a + 29 = 0
/\ 4 a - 1 >= 0 /\ 4 a - 7 <= 0 /\ 4 b + 3 >= 0 /\ 4 b - 3 <= 0

Acknowledgments

The author would like to thank George Collins and Bruno Buchberger for helpful suggestions on the draft of this paper.

Recent Progress on the Complexity of the Decision Problem for the Reals [1]

James Renegar

This paper concerns recent progress on the computational complexity of decision methods and quantifier elimination methods for the first order theory of the reals. The paper begins with a quick introduction of terminology followed by a short survey of some complexity highlights. We then discuss ideas leading to the most (theoretically) efficient algorithms known. The discussion is necessarily simplistic as a rigorous development of the algorithms forces one to consider a myriad of details.

This paper is similar to a talk the author gave at the DIMACS workshop on Algebraic Methods in Geometric Computations, held in May, 1990. A complete development can be found in (Renegar 1992a, 1992b, 1992c).

1   Some Terminology

A *sentence* is an expression composed from certain ingredients. Letting $\mathbb{R}$ denote a real closed field, the following is an example of a sentence:

$$(\exists x_1 \in \mathbb{R}^{n_1})(\forall x_2 \in \mathbb{R}^{n_2})[(g_1(x_1, x_2) > 0) \vee (g_2(x_1, x_2) = 0)] \wedge (g_3(x_1, x_2) \neq 0). \tag{1}$$

The ingredients are: vectors of variables (x_1 and x_2); the quantifiers $\exists$ and $\forall$; atomic predicates (e.g., $g_1(x_1, x_2) > 0$) which are real polynomial inequalities ($>, \geq, =, \neq, \leq, <$); and a Boolean function holding the atomic predicates ($[B_1 \vee B_2] \wedge B_3$).

A sentence asserts something. The above sentence asserts that there exists $x_1 \in \mathbb{R}^{n_1}$ such that for all $x_2 \in \mathbb{R}^{n_2}$, (i) either $g_1(x_1, x_2) > 0$ or $g_2(x_1, x_2) = 0$,

---

[1]Reprinted from *DIMACS Series*, Volume 6, 1991, pp. 287–308, by permission of the American Mathematical Society.

and (ii) $g_3(x_1, x_2) \neq 0$. Depending on the specific coefficients of the atomic predicate polynomials this assertion is either true or false.

The set of all true sentences constitutes the first order theory of the reals. A *decision method* for the first order theory of the reals is an algorithm that, given any sentence, correctly determines if the sentence is true. Decision methods for the reals were first proven to exist by Tarski (1951), who constructed one.

A sentence is a special case of a more general expression, called a *formula*. Here is an example of a formula:

$$(\exists x_1 \in \mathbb{R}^{n_1})(\forall x_2 \in \mathbb{R}^{n_2})[(g_1(z, x_1, x_2) > 0) \\ \vee (g_2(z, x_1, x_2) = 0)] \wedge (g_3(z, x_1, x_2) \neq 0). \tag{2}$$

A formula has one thing that a sentence does not, namely, a vector $z \in \mathbb{R}^{n_0}$ of *free variables*. When specific values are substituted for the free variables, the formula becomes a sentence.

A vector $\bar{z} \in \mathbb{R}^{n_0}$ is a *solution* for the formula if the sentence obtained by substituting $\bar{z}$ is true.

Two formulae are *equivalent* if they have the same solutions.

A *quantifier elimination method* is an algorithm that, given any formula, computes an equivalent quantifier-free formula, i.e., for the above formula $(\exists x_1 \in \mathbb{R}^{n_1})(\forall x_2 \in \mathbb{R}^{n_2})P(z, x_1, x_2)$ such a method would compute an equivalent formula $Q(z)$ containing no quantified variables.

When a quantifier elimination method is applied to a sentence, it becomes a decision method. Thus, a quantifier elimination method is in some sense more general than a decision method.

Tarski (1951) actually constructed a quantifier elimination method.

Both (1) and (2) are said to be in *prenex* form, i.e., all quantifiers occur in front. More generally, a formula can be constructed from other formulae just as (1) was constructed from the atomic predicates, i.e., we can construct Boolean combinations of formulae, then quantify any or all of the free variables. If we quantify all free variables then we obtain a sentence.

2   Some Complexity Highlights

We now present a brief survey of some complexity highlights for quantifier elimination methods, considering only formulae in prenex form. General bounds follow inductively. (If a formula is constructed from other formulae, first apply quantifier elimination to the innermost formulae, then to the innermost formulae of the resulting formula, etc.)

We consider the general formula

$$(Q_1 x_1 \in \mathbb{R}^{n_1}) \cdots (Q_\omega x_\omega \in \mathbb{R}^{n_\omega}) P(z, x_1, \ldots, x_\omega) \tag{3}$$

where $Q_1, \ldots, Q_\omega$ are quantifiers, assumed without loss of generality to alternate, i.e., Q_i is not the same as Q_{i+1}. Let m denote the number of distinct polynomials occurring among the atomic predicates and let $d \geq 2$ be an upper bound on their degrees.

In the case of Turing machine computations where all polynomial coefficients are restricted to be integers, we let L denote the maximal bit length of the coefficients. In this context we refer to the number of *bit operations* required by a quantifier elimination method. In the general and idealized case that the coefficients are real numbers we rely on the computational model of Blum et al. (1989), and refer to *arithmetic operations*, these essentially being field operations, including comparisons.

The sequential bit operation bounds that have appeared in the literature are all basically of the form

$$(md)^E[L^{O(1)} + \text{Cost}] \tag{4}$$

where E is some exponent and *Cost* is the worst-case cost of evaluating the Boolean function holding the atomic predicates, i.e., worst-case over 0-1 vectors. The Cost term is generally relatively negligible compared to the factor $(md)^E$. Recent complexity bound improvements concern the exponent E.

The first reasonable upper bound for a quantifier elimination method was proven around 1973 by Collins (1975). He obtained $E = 2^{O(n)}$ where $n :=$ $n_0 + \cdots + n_\omega$. Collins' bound is thus *doubly exponential* in the number of variables. His method requires the formula coefficients to be integers, the number of arithmetic operations (not just bit operations) growing with the size of the integers. Collins' algorithm was not shown to parallelize, although enough is now known from work of Neff (1994) that a parallel version probably could be developed. Collins' work has been enormously influential in the area.

The next major complexity breakthrough was made about 1985 by Grigor'ev (1988) who developed a decision method for which $E \approx [O(n)]^{4\omega}$. Grigor'ev's bound is doubly exponential only in the number of quantifier alternations. Many interesting problems can be cast as sentences with only a few quantifier alternations. For these, Grigor'ev's result is obviously significant. Like Collins' quantifier elimination method, Grigor'ev's decision method requires integer coefficients and was not proven to completely parallelize.

Slightly incomplete ideas of Ben-Or et al. (1986) (approximately 1986) were completed by Fitchas et al. (1990a) (approximately 1987) to construct a quantifier elimination method with arithmetic operation bound

$$(md)^E \text{Cost} \tag{5}$$

where $E = 2^{O(n)}$. This provides an arithmetic operation analog of Collins' bit operation bound. When restricted to integer coefficients, the method also yields the Collins' bound if the arithmetic operations are carried out bit by bit. Moreover, the algorithm parallelizes. Assuming each arithmetic operation requires one time unit, the resulting time bound is

$$[E \log(md)]^{O(1)} + \text{Time}(N) \tag{6}$$

if $(md)^E N$ parallel processors are used, where Time(N) is the worst-case time required to evaluate the Boolean function holding the atomic predicates using N parallel processors. The analogous time bound for bit operations is also valid, namely, $[E \log(Lmd)]^{O(1)} +$ Time(N) if $(md)^E [L^{O(1)} + N]$ parallel processors are used.

In 1989 (Renegar 1992a, 1992b, 1992c), the author introduced a new quantifier elimination method for which $E = \prod_{k=0}^{\omega} O(n_k)$. This was established for arithmetic operations and bit operations, i.e., (5) and (4). Regarding bit operations, the dependence of the bounds on L was shown to be very low, $L^{O(1)}$ in (4) being replaced by $L(\log L)(\log \log L)$, i.e., the best bound known for multiplying two L-bit integers. Moreover, the method was shown to parallelize, resulting in the arithmetic operation time bound (6) if $(md)^E N$ parallel processors are used, and the bit operation time bound $\log(L)[E \log(md)]^{O(1)} +$ Time(N) if $(md)^E [L^2 + N]$ parallel processors are used.

Independently and simultaneously, Heintz, Roy, and Solernó (1990) developed a quantifier elimination method for which $E = O(n)^{O(\omega)}$, both for arithmetic and bit operations. Their method also completely parallelizes.

The various bounds are best understood by realizing that quantifier elimination methods typically work by passing through a formula from back to front. First the vector x_ω is focused on, then the vector $x_{\omega-1}$, and so on. The work arising from each vector results in a factor for E. For Collins' quantifier elimination method, the factor corresponding to x_k is $2^{O(n_k)}$ (note that $2^{O(n)} = 2^{O(n_0)} \cdots 2^{O(n_\omega)}$). For the method introduced by the author, the factor is $O(n_k)$. The factor corresponding to Grigor'ev's decision method is approximately $O(n)^4$ independently of the number of variables in x_k. In that method a vector with few variables can potentially create as large of a factor as one with many variables. Similarly, the factor corresponding to the quantifier elimination method of Heintz, Roy and Solernó is $O(n)^{O(1)}$ independently of x_k.

Many interesting formulae have blocks of variables of various sizes. For example, sentences asserting continuity generally have $(\forall \epsilon)(\exists \delta)$, along with large blocks of variables. The exponent $E = \prod O(n_k)$ is especially relevant for such sentences, as the contribution to E from the smaller blocks is modest.

For the record, the quantifier elimination method in (Renegar 1992a, 1992b, 1992c) produces a quantifier-free formula of the form

$$\bigvee_{i=1}^{I} \bigwedge_{j=1}^{J_i} (h_{ij}(z) \Delta_{ij} 0)$$

where $I \le (md)^E$, $J_i \le (md)^{E/n_0}$, $E = \prod_{k=0}^{\omega} O(n_k)$, the degree of h_{ij} is at most $(md)^{E/n_0}$ and the Δ_{ij} are standard relations $(\ge, >, =, \ne, <, \le)$. If the coefficients of the original formula are integers of bit length at most L, the coefficients of the polynomials h_{ij} will be integers of bit length at most $(L + n_0)(md)^{E/n_0}$.

Results of Weispfenning (1988), and Davenport and Heintz (1988), show
that the double exponential dependence on ω of the above bound on the degrees
of the polynomials h_{ij} cannot be improved in the worst case.

Fischer and Rabin (1974a) proved an exponential worst-case lower bound for
decision methods. However, the lower bound is exponential only in the number
of quantifier alternations and is only singly exponential in that. A tremendous
gap remains between the known upper and lower bounds for decision methods.

In closing this section we mention that work of Canny (1989, 1988, 1990) has
been especially influential in this area in recent years, both for the techniques he
has developed and employed and for the connections he has established between
the area and robotics. Work of Grigor'ev and Vorobjov (1988) has also been
very influential. Recent work of Pedersen (1991a) is especially relevant.

3   Discussion of Ideas Behind the Algorithms

3.1

We now discuss some of the main ideas behind the algorithms with the best
complexity bounds. Our discussion is rather loose. At times it is not quite
correct, i.e., *some of the assertions that follow are simply not quite true*. However, the discussion is *close in spirit* to a rigorous development. Only by hiding
technical details can we present a comprehensible discussion that can be read in
a reasonable amount of time. The reader can find all of the details in (Renegar
1992a, 1992b, 1992c).

3.2

We take as our starting point a fundamental result due to Ben-Or, Kozen and
Reif regarding univariate polynomials.

Let $g_1, \ldots, g_m : \mathbb{R} \to \mathbb{R}$ be real univariate polynomials. To each $x \in \mathbb{R}$
there is associated a sign vector $\sigma(x) \in \mathbb{R}^m$ whose i-th coordinate is defined as
follows:
$$\sigma_i(x) := \begin{cases} 1 & \text{if } g_i(x) > 0, \\ 0 & \text{if } g_i(x) = 0 \\ -1 & \text{if } g_i(x) < 0. \end{cases}$$

The sign vectors of $g_1, \ldots, g_m$ are the vectors $\sigma(x)$ obtained as x varies over
$\mathbb{R}$. If each of the univariate polynomials g_i is of degree at most d, then the set
of sign vectors has at most $2md + 1$ elements because there are at most md
distinct roots and $md + 1$ open intervals defined by them.

The entire set of sign vectors of $g_1, \ldots, g_m$ can be constructed efficiently
by algorithms based on Sturm sequence computations. In fact, they can be
constructed quickly in parallel (i.e., in polylogarithmic time), as was shown by
Ben-Or et al. (1986). We take this fundamental result as a "given" and do not
elaborate except to mention that the ideas developed in (Ben-Or et al. 1986)
can be found throughout the real decision method literature that followed its

appearance. We refer to the algorithm of Ben-Or, Kozen and Reif as the BKR algorithm.

A simple corollary is the fact that there exists an efficient decision method for sentences with exactly one variable (and hence one quantifier), i.e., an efficient decision method for univariate sentences in the *existential theory of the reals*. This is because it is trivial to determine if a sentence with one quantifier is true if one knows the sign vectors of the atomic predicate polynomials.

3.3

Our first step in designing general decision and quantifier elimination methods is to provide a method for efficiently computing all sign vectors of multivariate polynomials. The approach is to efficiently reduce the set of multivariate polynomials whose sign vectors are to be determined to a set of univariate polynomials with the same sign vectors; then the BKR algorithm can be applied.

There are various ways to efficiently reduce multivariate polynomials to appropriate univariate ones. One approach is based on the effective Nullstellensatz (Brownawell 1987). Such an approach was used by Heintz et al. (1990). Another approach is based on so-called *u-resultants*. This is the approach that we discuss; it has led to the best complexity bounds.

For the moment we set aside the problem of efficiently computing the sign vectors for a set of multivariate polynomials and present a brief, but not quite accurate, discussion of *u*-resultants. Renegar (1992a, 1992b, 1992c) presents a completely accurate, but lengthy, discussion.

3.4

The *u*-resultant is an algebraic construction that allows one to reduce the problem of finding the zeros of a system of multivariate polynomials to the problem of finding the zeros of a univariate polynomial.

Let $f : \mathbb{R}^n \to \mathbb{R}^n$ be a system of n polynomials in n variables and let $F : \mathbb{R}^{n+1} \to \mathbb{R}^n$ denote the homogenization of f, i.e., the terms of the i-th coordinate polynomial F_i of F are obtained by multiplying the terms of f_i by the appropriate power of an additional variable x_{n+1} so as to make all terms have the same degree as that of the highest degree term in f_i. Then $F^{-1}(0)$, the zero set of F, consists of lines through the origin.

Assume that $F^{-1}(0)$ consists of only finitely many such *zero lines*. If one knows all of these zero lines, then it is a trivial matter to compute all of the zeros of f. Thus, the problem of finding the zeros of f reduces to the problem of finding the zero lines of F.

From the coefficients of F, one can construct the *u*-resultant $R : \mathbb{R}^{n+1} \to \mathbb{R}$ of F. This is a single homogeneous polynomial with the same domain as F, i.e., $\mathbb{R}^{n+1}$. If F has infinitely many zero lines, then the *u*-resultant is identically zero. However, if F has only finitely many zero lines then the zero set of the *u*-resultant, $R^{-1}(0)$, consists of those hyperplanes in $\mathbb{R}^{n+1}$ that contain the origin and are orthogonal to the zero lines of F; so each zero line is associated with a hyperplane in $R^{-1}(0)$. It is this property that allows one to reduce the problem of finding the zero lines of F to the problem of finding the zeros of a univariate polynomial, as we now show.

Consider an arbitrary point x (other than the origin) on one of the hyperplanes of $R^{-1}(0)$. From calculus, the gradient of R at x, $\nabla R(x)$, is perpendicular to the hyperplane and, hence, lies in a zero line of F. Thus, if we could merely get a point on each of the hyperplanes of $R^{-1}(0)$, and then compute the gradient of R at those points, we would have all of the zero lines of F. (We are glossing over subtleties like hyperplanes occurring with multiplicities, in which case the gradient will be the zero vector.)

The fact that the hyperplanes are n-dimensional makes finding points on them relatively easier than finding one-dimensional zero lines of F directly.

Consider a line $\{t\alpha + \beta; t \in \mathbb{R}\}$ that probes $\mathbb{R}^{n+1}$ in search of hyperplanes; here, α and β are fixed vectors in $\mathbb{R}^{n+1}$. Now consider the restriction of the u-resultant to this line, that is, consider the univariate polynomial $t \to R(t\alpha + \beta)$. Then t' is a zero of this univariate polynomial iff $t'\alpha + \beta$ is a point on one of the hyperplanes. Assuming, for simplicity, that the probe line is in general position relative to the hyperplanes, so that it intersects each of the hyperplanes, there is thus a corresponding zero of the univariate polynomial for each of the hyperplanes. (In a complete development one needs a method of choosing α and β to guarantee that the probe line is indeed in general position.)

We have reduced the problem of finding all of the zero lines of F to the problem of finding the zeros of a univariate polynomial, assuming F has only finitely many zero lines. In short, letting

$$
\begin{aligned}
p(t) &:= R(t\alpha + \beta) \quad p: \mathbb{R} \to \mathbb{R}, \\
q(t) &:= \nabla R(t\alpha + \beta) \quad q: \mathbb{R} \to \mathbb{R}^{n+1},
\end{aligned}
$$

where α and β are fixed vectors in $\mathbb{R}^{n+1}$, we have that $p(t') = 0$ if and only if $q(t')$ is a zero line of F.

Of course one can carry out the construction of the polynomials p and q efficiently if one can construct the u-resultant efficiently. Constructing the u-resultant essentially amounts to computing the determinants of matrices. As it is well known how to compute determinants efficiently, the polynomials p and q can be computed efficiently. Again, we refer to (Renegar 1992a, 1992b, 1992c) for details.

The u-resultant has been around for a long time, e.g., one can find it discussed in earlier editions of van der Waerden (1949) in the chapter on elimination theory. (Unfortunately, that chapter was eliminated in later editions.)

3.5

Now we simplify again, but not in a serious way. For our interests, the last section can be summarized as follows: $p(t') = 0$ if and only if $q(t')$ is a zero line of F, where F is the homogenization of $f : \mathbb{R}^n \to \mathbb{R}^n$. Rather than this, we will pretend the last section can be summarized in another way: $p(t') = 0$ iff $q(t')$ is a zero of f. In other words, we pretend the polynomials p and q give us the zeros of f directly rather than give us something that has to be dehomogenized. Our intent here is simply to make the following discussion easier to follow. We want to present the discussion in terms of the geometry of the atomic predicate polynomials occurring in a sentence or formula rather than in terms of the geometry of their homogenizations, but by doing so we are not quite honest (although we are close to being honest).

3.6

Now we show the relevance of u-resultants for the problem of computing the sign vectors of a set of multivariate polynomials. We begin by considering a more structured problem, namely, given a polynomial system $f : \mathbb{R}^n \to \mathbb{R}$ with only finitely many zeros, and given polynomials $g_1, \ldots, g_m : \mathbb{R}^n \to \mathbb{R}$, compute the sign vectors of $g_1, \ldots, g_m$ which occur at the zeros of f.

The more structured problem is easily solved. Letting $p(t)$ and $q(t)$ be as in the last subsection, we can apply the BKR algorithm to compute all sign vectors of the univariate polynomials

$$p(t), g_1(q(t)), \ldots, g_m(q(t)).$$

We can then search through the sign vectors and find those whose first coordinate is zero. Clearly, the remaining coordinates of such a sign vector yield a sign vector for $g_1, \ldots, g_m$ at a zero of f; moreover, all sign vectors of $g_1, \ldots, g_m$ at zeros of f are obtained in this way.

This motivates an approach to computing all sign vectors of $g_1, \ldots, g_m$. First construct systems $f : \mathbb{R}^n \to \mathbb{R}^n$ at whose zeros all of the sign vectors of $g_1, \ldots, g_m$ occur, i.e., for each sign vector at least one of the systems has a zero at which that is indeed the sign vector. Then proceed as above.

3.7

We are now motivated to consider the following problem: Given polynomials $g_1, \ldots, g_m : \mathbb{R}^n \to \mathbb{R}$ construct systems $f : \mathbb{R}^n \to \mathbb{R}^n$ at whose zeros all of the sign vectors occur. We require each of the systems to have only finitely many zeros; otherwise, the polynomials $p(t)$ and $q(t)$ will be identically zero as the u-resultant will be identically zero; if $p(t)$ and $q(t)$ are identically zero, the preceding approach flounders.

The polynomials g_i partition $\mathbb{R}^n$ into maximal connected components where two points are in the same connected component only if they have the same sign vector. This is shown schematically in Figure 1. The zero sets of the polynomials determine the boundaries of the connected components; in the figure the components are 0-, 1- and 2-dimensional. We refer to the partition as the *connected sign partition* for $g_1, \ldots, g_m$.

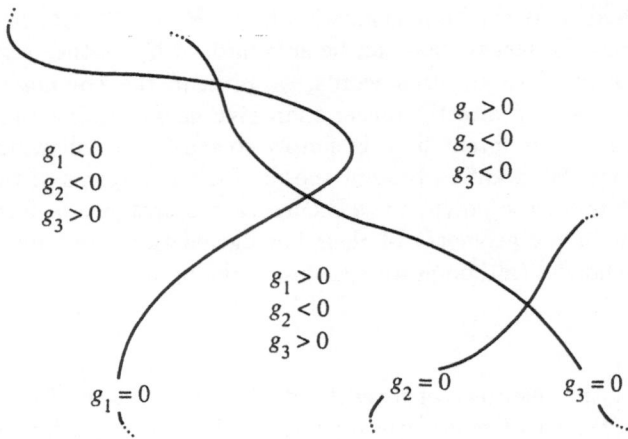

$g_1 < 0$
$g_2 < 0$
$g_3 > 0$

$g_1 > 0$
$g_2 < 0$
$g_3 < 0$

$g_1 > 0$
$g_2 < 0$
$g_3 > 0$

$g_1 = 0$ $g_2 = 0$ $g_3 = 0$

Figure 1

To construct systems $f : \mathbb{R}^n \to \mathbb{R}^n$ at whose zeros all of the sign vectors occur, it certainly suffices to construct systems $f : \mathbb{R}^n \to \mathbb{R}^n$ which have zeros in all of the connected components.

We again change focus slightly. Consider the following problem: Given polynomials $g_1, \ldots, g_m : \mathbb{R}^n \to \mathbb{R}$ construct systems $f : \mathbb{R}^n \to \mathbb{R}^n$ with zeros in the *closures* of all of the connected components. As is indicated in Section 3.8, for our purposes it actually suffices to solve this apparently easier problem. More specifically, given arbitrary polynomials $g_1, \ldots, g_m : \mathbb{R}^n \to \mathbb{R}^n$, it suffices to be able to compute non-vanishing pairs (p, q), $p : \mathbb{R} \to \mathbb{R}$, $q : \mathbb{R} \to \mathbb{R}^n$, with the property that for each component of the connected sign partition for $g_1, \ldots, g_m$, there exists t' for which one of the pairs (p, q) satisfies $p(t') = 0$ and $q(t')$ is in the closure of the component.

(The following discussion is somewhat technical and represents the hardest parts of (Renegar 1992a, 1992b, 1992c); the reader may wish to skip ahead to Section 3.8. For the reader who does skip ahead we remark that in the following discussion we construct pairs (p, q) with the property stated in the last sentence of the previous paragraph. The construction naturally leads to the following indexing of the pairs: $(\bar{p}_A, \bar{q}_A)$ where $A \subseteq \{1, \ldots, m\}$ and $\#A \leq n$, i.e., the number of elements in A does not exceed n.)

A naive solution to the problem of constructing systems with zeros in all of the closures can be obtained via calculus. Noting that the closure of each connected component has a point closest to the origin (i.e., a minimizer of $\sum_j x_j^2$ restricted to the closure) standard arguments regarding necessary conditions for optimality lead to the following: For each connected component there exists a subset $A \subseteq \{1, \ldots, m\}$ such that the system $f_A : \mathbb{R}^n \to \mathbb{R}^n$ defined by

$$f_A(x) := \nabla[\det(M_A^T(x) M_A(x)) + \sum_{i \in A} (g_i(x))^2] \tag{7}$$

has a zero in the closure of the component; ∇ denotes *the gradient of* and $M_A(x)$ is the matrix whose rows are the polynomials $\nabla g_i(x)$, for $i \in A$, and the identity polynomial $x \mapsto x$. (We do not digress to discuss necessary conditions for optimality.)

There are undoubtedly numerous elementary ways to construct systems which have zeros in the closures of each of the connected components, each way having shortcomings for our purposes. The rather strange looking systems (7) we focus on have some especially useful properties. We motivate discussion of these useful properties by discussing the systems' shortcomings.

One glaring shortcoming of the systems f_A is simply that there are too many of them, i.e., 2^m. If for each system f_A we formed the corresponding polynomials $p_A(t)$ and $q_A(t)$ and applied the BKR algorithm to the univariate polynomials $g_1(q_A(t)), \ldots, g_m(q_A(t))$, we would end up with an algorithm for the existential theory of the reals requiring at least 2^m operations. However, the number of atomic predicate polynomials does not appear exponentially in the best complexity bounds for decision methods, so use of all of the systems f_A is certainly not the way to proceed.

Another shortcoming of the systems f_A is that for some polynomials $g_1, \ldots,$ g_m they may have infinitely many zeros. Consequently, the corresponding u-resultants, and hence the corresponding polynomials p_A and q_A, may be identically zero, in which case we get no information from the sign vectors of the univariate polynomials $g_1(q_A(t)), \ldots, g_m(q_A(t))$, i.e., our approach flounders.

A way to circumvent these shortcomings is to slightly alter our approach to computing all of the sign vectors of $g_1, \ldots, g_m$, the alteration depending heavily on special properties of u-resultants and the systems f_A. Briefly, as these ideas will be developed more fully in the following paragraphs, we carefully perturb the polynomials $g_1, \ldots, g_m$, construct the exactly analogous systems f_A for the perturbed polynomials, construct the pairs (p_A, q_A) from these, and from these construct pairs $(\bar{p}_A, \bar{q}_A)$ which are something like limits of the pairs (p_A, q_A) as the perturbation goes to zero; the true limits of the pairs (p_A, q_A) may be identically zero, but the pairs $(\bar{p}_A, \bar{q}_A)$ will not be. The *limit* pairs $(\bar{p}_A, \bar{q}_A)$ have the property that for each connected component of the sign partition of $g_1, \ldots, g_m$, there exists $A \subseteq \{1, \ldots, m\}$, $\#A \leq n$, such that for some $t', \bar{p}_A(t') = 0$ and $\bar{q}_A(t')$ is in the closure of the connected component. Note that this circumvents both of the previously mentioned shortcomings! Even though the systems f_A constructed directly from $g_1, \ldots, g_m$ may have infinitely many zeros so that we cannot construct useful pairs (p_A, q_A) from their u-resultants, we are still able to construct useful pairs indirectly! Moreover, rather than having 2^m pairs to contend with, we have $\binom{m}{n} = m^{O(n)}$ pairs to contend with, a major improvement.

Now we elaborate a bit. The perturbed polynomials we use are defined as follows:

$$g_i(x, \delta) := (1 - \delta) g_i(x) + \delta \left(1 + \sum_j i^j x_j^d \right)$$

where d is the maximal degree occurring among the unperturbed polynomials; the additional variable δ is a perturbation parameter; when $\delta = 0$ we have $x \mapsto g_i(x, \delta) = g_i(x)$.

Viewing δ as fixed consider the polynomials

$$x \mapsto g_1(x, \delta), \ldots, x \mapsto g_m(x, \delta).$$

Just as we defined the systems $f_A(x)$ from the polynomials $g_1, \ldots, g_m$, we can define the systems $x \mapsto f_A(x, \delta)$; simply substitute $g_i(x, \delta)$ for $g_i(x)$ in (7), the gradients referring to derivatives only in the x-variables.

Continuing to view δ as fixed, we construct the pairs of polynomials $t \mapsto p_A(t, \delta)$, $t \mapsto q_A(t, \delta)$ from the u-resultants of the systems $x \mapsto f_A(x, \delta)$. The careful choice of the perturbed polynomials and the definition of the systems $x \mapsto f_A(x, \delta)$ allow one to prove (something similar to) the following: for each sufficiently small $\delta > 0$ and for each component of the connected sign partition of $x \mapsto g_1(x, \delta), \ldots, x \mapsto g_m(x, \delta)$ there exists $A \subseteq \{1, \ldots, m\}$, $\#A \leq n$, and there exists $t' \in \mathbb{R}$ such that $p_A(t') = 0$ and $q_A(t')$ is in the closure of the connected component.

We would like the same conclusion to be true for $\delta = 0$. However, it may not be true for $\delta = 0$ because the systems $x \mapsto f_A(x, 0)$ may have infinitely many zeros causing the corresponding pairs $t \mapsto p_A(t, 0)$, $t \mapsto q_A(t, 0)$ to be identically zero. This problem can be handled as follows.

Tracing through definitions (all of which can be found in (Renegar 1992a, 1992b, 1992c)) one finds the pairs $p_A(t, \delta)$, $q_A(t, \delta)$ to be polynomials in δ as well as in t. Expanding in powers of δ,

$$p_A(t, \delta) = \sum_i \delta^i p_A^{[i]}(t), \tag{8}$$

$$q_A(t, \delta) = \sum_i \delta^i q_A^{[i]}(t), \tag{9}$$

let $\bar{p}_A(t) := p_A^{[i']}(t)$ where the polynomial $p_A^{[i]}(t)$ is identically zero for all $i < i'$, but not for i'; similarly, let $\bar{q}_A(t) := q_A^{[i'']}(t)$ where the polynomial system $q_A^{[i]}(t)$ is identically zero for all $i < i''$, but not for i''.

Relying on the definition of the perturbed polynomials, now considering the perturbation parameter $\delta > 0$ to tend to zero, one can prove that (something very similar to) the following is true: for each component of the sign partition of the original polynomials $g_1, \ldots, g_m$, there exists $A \subseteq \{1, \ldots, m\}$, $\#A \leq n$, and there exists t' such that $\bar{p}_A(t') = 0$ and $\bar{q}_A(t')$ is in the closure of the connected component.

The idea of combining u-resultants and perturbations can be found in (Canny 1990), which focuses on the problem of solving degenerate systems of polynomial equations $f(x) = 0$ where $f : \mathbb{C}^n \to \mathbb{C}^n$.

3.8

Recall that our real goal is to construct all of the sign vectors of $g_1, \ldots, g_m$. If in the conclusion of the next-to-last paragraph the words "the closure of" did not appear we would be in a position to accomplish the goal: simply apply the BKR algorithm to the univariate polynomials $t \mapsto g_1(\bar{q}_A(t)), \ldots, t \mapsto g_m(\bar{q}_A(t))$ for all $A \subseteq \{1, \ldots, m\}$ satisfying $\#A \leq n$.

However, the words "the closure of" certainly do appear, but this problem is easily remedied. From the polynomials $g_1, \ldots, g_m : \mathbb{R}^n \to \mathbb{R}$ one can easily construct polynomials $\tilde{g}_1, \ldots, \tilde{g}_m : \mathbb{R}^{n+1} \to \mathbb{R}$ with the property that for each connected component C_1 in the sign partition of $g_1, \ldots, g_m$, there exists a connected component C_2 in the sign partition of $\tilde{g}_1, \ldots, \tilde{g}_m$ such that C_1 is the projection of the closure of C_2 onto $\mathbb{R}^n$. (We do not provide details here but, as an example, note that the positive real numbers are the projection of the closed semi-algebraic set $\{(x,y); xy \geq 1, x \geq 0, y \geq 0\}$.) Consequently, if we construct the analogous pairs $\bar{p}_A : \mathbb{R} \to \mathbb{R}, \bar{q}_A : \mathbb{R} \to \mathbb{R}^{n+1}$ for $\tilde{g}_1, \ldots, \tilde{g}_m$, and discard the last coordinate polynomial of $\bar{q}_A$, we obtain appropriate pairs for $g_1, \ldots, g_m$. Henceforth, we use the notation $(\bar{p}_A, \bar{q}_A)$ to denote pairs for $g_1, \ldots, g_m$ obtained in this way, so we no longer need to refer to the closure of the components.

3.9

The pairs $(\bar{p}_A, \bar{q}_A)$ constructed for $g_1, \ldots, g_m$ have another property to be highlighted; the coefficients of the pairs are polynomials in the coefficients of $g_1, \ldots, g_m$. If one fixes m, n, and d and considers the vector space of tuples $g_1, \ldots, g_m$ of m polynomials $g_i : \mathbb{R}^n \to \mathbb{R}$ of degree at most d (identifying polynomials with their coefficients), one can replace the univariate polynomial pairs $(\bar{p}_A(t), \bar{q}_A(t))$ for fixed $g_1, \ldots, g_m$, with pairs $(\bar{p}_A(g_1, \ldots, g_m, t), q_A(g_1, \ldots, g_m, t))$ for variable $g_1, \ldots, g_m$; to obtain the corresponding univariate polynomials for particular $g_1, \ldots, g_m$ just substitute the coefficients of $g_1, \ldots, g_m$. In particular, *construction of the pairs can be accomplished by straight-line programs*, with input being the coefficients of $g_1, \ldots, g_m$. This is very important in obtaining the best complexity bounds.

(To be more accurate, to obtain the pairs as straight-line programs one needs to include all pairs $(p_A^{[i]}(t), q_A^{[i]}(t))$ where $p_A^{[i]}$ and $q_A^{[i]}$ are as in (8) and (9) rather than simply $\bar{p}_A = p_A^{[i']}$ and $\bar{q}_A = \bar{q}_A^{[i'']}$ although for simplicity we ignore this in what follows.)

Something similar to the above pairs are constructed in (Heintz et al. 1990) by using the effective Nullstellensatz rather than u-resultants. However, the construction there is not by straight-line programs.

3.10

To simplify what follows we consider the following to be established: we can construct a single pair (rather than a reasonably small set of pairs) of polynomials $p(g_1, \ldots, g_m, t)$, $q(g_1, \ldots, g_m, t)$ with the property that for any fixed $g_1, \ldots, g_m : \mathbb{R}^n \to \mathbb{R}$ of degree at most d and any component of the connected sign partition of $g_1, \ldots, g_m$ there exists t' such that $p(g_1, \ldots, g_m, t') = 0$ and $q(g_1, \ldots, g_m, t')$ is in the component.

Whenever we write $p(t), q(t)$, we are viewing $g_1, \ldots, g_m$ as fixed as will be clear from context.

3.11

We now have an efficient decision method for the existential theory of the reals, i.e., a method for deciding sentences with only one quantifier. Simply apply the BKR algorithm to the univariate polynomials

$$g_1(q(t)), \ldots, g_m(q(t))$$

where $g_1, \ldots, g_m$ are the atomic predicate polynomials in the sentence, thus constructing all sign vectors for $g_1, \ldots, g_m$. From these, the truth or falsity of the sentence is easily determined.

3.12

We are now in position to describe a quantifier elimination method which works inductively.

First, assume we wish to obtain a quantifier-free formula equivalent to a formula $(\exists x) P(z, x)$ with a single quantifier; here, z and x are vectors of variables, z being the free variables. The variables z can be viewed as the coefficients of the atomic predicate polynomials; different values for z yield different sentences. Viewing z as fixed, thereby obtaining a sentence with a single quantifier, we could apply the decision method of Section 3.11 to determine if the sentence is true.

From this viewpoint the decision method for the existential theory of the reals is a method for deciding if sentences corresponding to different values of z are true; input to the algorithm is simply specific values for z. *Unrolling* the algorithm, which requires a close examination of the BKR algorithm, one obtains an algebraic decision tree with input z as depicted in Figure 2.

Each node of the tree is associated with a polynomial in z. Upon input z, the polynomial at the initial node is evaluated, and the value is compared to zero. Depending on the comparison, one branches to the left or to the right, thereby determining the next node and polynomial evaluation to be made, and so on until a leaf is reached.

Each leaf is assigned the value *True* or *False*. All inputs z arriving at a leaf marked True, correspond to sentences which are true. Similarly, all inputs z arriving at a leaf marked False, correspond to sentences that are false.

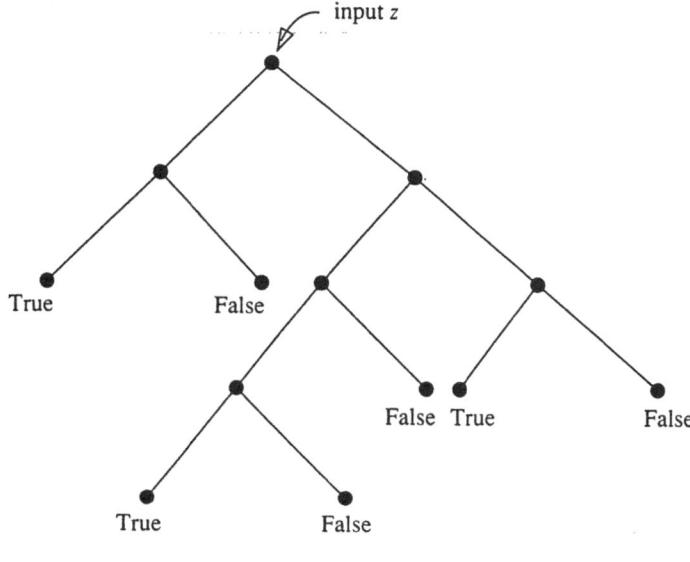

Figure 2

Here is a quantifier elimination method for the sentence: First, unroll the decision method to obtain an algebraic decision tree with input z. Next, determine all paths from the root that lead to leaves marked True. For each of these paths p, the decision method follows that path if and only if the corresponding sequence of polynomial inequalities and equalities are satisfied, i.e., z must satisfy the appropriate conjunction of polynomial equalities and inequalities

$$\bigwedge_{i \in I_p} (h_i(z) \Delta_i 0),$$

where I_p is an index set for path p and the Δ_i are standard relations ($\geq, >, =, \neq, <, \leq$). Taking the disjunction over all paths p leading to leaves marked True, one obtains a quantifier-free formula

$$\bigvee_p \bigwedge_{i \in I_p} (h_i(z) \Delta_i 0),$$

which is equivalent to $(\exists x) P(z, x)$.

3.13

The problem with this quantifier elimination method for formulae with a single quantifier is that there are too many paths in the tree; naively unrolling the decision method with input z leads to a bad complexity bound. However, this can be avoided with close examination of the BKR algorithm.

One can prove that most paths in the decision tree are not followed for any input z, although which paths are not followed is a function of the particular formula.

Rather than unroll the decision method naively, we can use the decision method to unroll itself. First, assuming the polynomial associated with the root node is $h_0(z)$ and one branches to the left if and only if $h_0(0) < 0$, we determine if the sentence $(\exists z)(h_0(z) < 0)$ is true. If it is false we know there is no need to investigate paths beginning with the edge to the left of the initial node and so we can *prune* it (and all paths containing it). Continuing in the obvious manner, pruning as we unroll the tree, we do indeed obtain a fairly efficient quantifier elimination method for formulae with a single quantifier.

3.14

To obtain a general quantifier elimination method we can simply proceed inductively on the blocks of variables in a formula

$$(Q_1 x_1) \cdots (Q_\omega x_\omega) P(z, x_1, \ldots, x_\omega).$$

We first eliminate the block x_ω from the formula

$$(Q_\omega x_\omega) P(z, x_1, \ldots, x_\omega)$$

to obtain an equivalent quantifier-free formula $P'(z, x_1, \ldots, x_{\omega-1})$. We then eliminate the block $x_{\omega-1}$ from the formula

$$(Q_{\omega-1} x_{\omega-1}) P'(z, x_1, \ldots, x_{\omega-1})$$

to obtain an equivalent quantifier-free formula $P''(z_1, x_1, \ldots, x_{\omega-2})$, and so on, until we have finally eliminated all of the quantified variables in the original formula.

This quantifier elimination method is fairly efficient. In terms of the notation introduced in Section 2 it readily yields a complexity bound with $E = O(n)^{O(\omega)}$ where n is the total number of variables in the formula. This is the bound obtained by Heintz et al. (1990) by a somewhat different approach relying on the effective Nullstellensatz rather than u-resultants.

In the following sections we describe further ideas that lead to a method with the best complexity bound, i.e., $E = \Pi O(n_k)$. The speed-up is strongly dependent on the fact that the pairs (p, q) can be computed by straight-line programs as we discussed earlier.

3.15

Our decision method for the existential theory of the reals was essentially obtained by replacing the block of variables by a single variable, that is, the sentence $(\exists x)P(x)$ with atomic predicate polynomials $g_1(x), \ldots, g_m(x)$ was essentially reduced to the equivalent univariate sentence $(\exists t)P(t)$ obtained by substituting $g_i(q(t))$ for $g_i(x)$.

It is natural to attempt to extend this idea to sentences involving many quantifiers, reducing each block of variables to a single variable. We discuss a way to do this for sentences involving two quantifiers. The generalization to arbitrary sentences is fairly straightforward; complete details can be found in (Renegar 1992a, 1992b, 1992c).

Consider a sentence $(\exists y)(\forall x)P(y, z)$ with atomic predicate polynomials $g_1(y, x), \ldots, g_m(y, x)$. To each fixed y there corresponds the set of sign vectors of the polynomials in x,

$$x \mapsto g_1(y, x), \ldots, x \mapsto g_m(y, x).$$

Varying y we obtain a family of sets of sign vectors.

It is easy to see that if we knew the entire family then we could decide if the sentence is true. With this as motivation, we proceed.

3.16

Consider the special case of a sentence with two quantifiers $(\exists y)(\forall x)P(y, x)$ for which x consists of a single variable; to distinguish this special case we write $(\exists y)(\forall t)P(y, t)$. Momentarily, we reduce a general sentence with two quantifiers to the special case.

For the special case, to each y there corresponds the sign vectors of the univariate polynomials $t \mapsto g_1(y, t), \ldots, t \mapsto g_m(y, t)$. We aim to construct the family of all such sets of sign vectors obtained as y varies over its domain.

The polynomials $g_1, \ldots, g_m$ partition (y, t)-space as is shown schematically in Figure 3, i.e., the connected sign partition of $g_1, \ldots, g_m$. When we focused on sentences with a single quantifier, we aimed at obtaining a point in each of the connected components of the partition. Now our goal is different.

For fixed y, the univariate polynomials $t \mapsto g_1(y, t), \ldots, t \mapsto g_m(y, t)$ are simply the restrictions of the polynomials $g_1, \ldots, g_m$ to the line through $(y, 0)$ that is parallel to the t-axis. It is easily seen that the sign vectors of the univariate polynomials are completely determined by which components of the connected sign partition of $g_1, \ldots, g_m$ the line passes through. Now viewing y as varying continuously, the set of sign vectors of the univariate polynomials $t \mapsto g_1(y, t), \ldots, t \mapsto g_m(y, t)$ can change only when y passes through one of the points indicated in Figure 3.

We thus see that y-space is partitioned into maximal connected components with the property that if y' and y'' are in the same component, then the set of sign vectors for the univariate polynomials $t \mapsto g_1(y', t), \ldots, t \mapsto g_m(y', t)$ is the same as for the univariate polynomials $t \mapsto g_1(y'', t), \ldots, t \mapsto g_m(y'', t)$. Hence, if for each component of this partition of y-space we could construct a point y' in it, then we could apply the BKR algorithm to the resulting sets

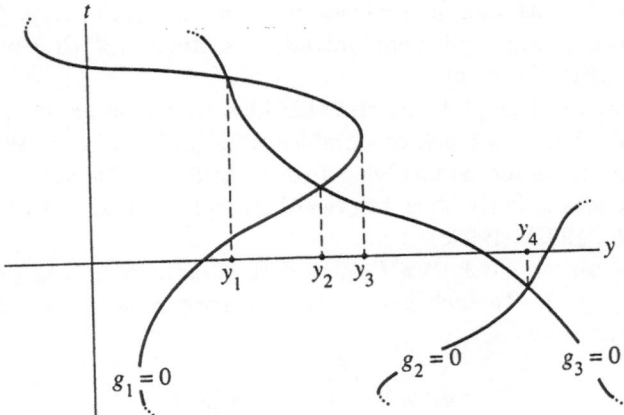

Figure 3

of univariate polynomials $t \mapsto g_1(y', t), \ldots, t \mapsto g_m(y', t)$ to compute the entire family of sets of sign vectors that we desire, and from those determine if the sentence is true.

3.17

We now construct the partition of y-space, that is, we present a method for computing polynomials $h_1(y), \ldots, h_\ell(y)$ with the property that if y' and y'' are in the same component of the connected sign partition of $h_1, \ldots, h_\ell$, then the set of sign vectors for the univariate polynomials $t \mapsto g_1(y, t), \ldots, t \mapsto g_m(y, t)$ is the same for $y = y'$ and $y = y''$. The method figures prominently in Collins' work (Collins 1975) and is very well-known. (We oversimplify the method to avoid technicalities, but our oversimplification is close in spirit to the true method.)

The breakpoints in Figure 3 fall into two categories. The first category consists of points which are the projection onto y-space of intersections of the form $\{(y, t); g_i(y, t) = 0\} \cap \{(y, t); g_j(y, t) = 0\}$ for some $i \neq j$. The points y_1, y_2 and y_4 in the figure all fall into this category.

The second category consists of points which are the projection onto y-space of intersections of the form $\{(y, t); g_i(y, t) = 0\} \cap \{(y, t); \frac{d}{dt}g_i(y, t) = 0\}$. The point y_3 in the figure falls into this category.

A point y is in the first category only if the Sylvester resultant of the univariate polynomials $t \mapsto g_i(y, t)$ and $t \mapsto g_j(y, t)$ is zero at y for some $i \neq j$; the Sylvester resultant is a polynomial in the coefficients of the univariate polynomials, which here means that it is a polynomial in y.

Similarly, a point y is in the second category only if the Sylvester resultant of the univariate polynomials $t \mapsto g_i(y, t)$ and $t \mapsto \frac{d}{dt}g_i(y, t)$ is zero at y for some i.

Collecting the Sylvester resultant polynomials, we obtain the desired polynomials $h_1(y), \ldots, h_\ell(y)$ which appropriately partition y-space. (Again, to be honest, the construction of appropriate polynomials h_i is a bit more complicated.)

3.18

Polynomials analogous to the h_i can be readily obtained for general sentences $(\exists y)(\forall x)P(y, x)$ involving two quantifiers, where x is not required to consist of a single variable.

First, viewing y as the *coefficients* of the atomic predicate polynomials $g_1, \ldots, g_m$, from the polynomials $x \mapsto g_1(y, x), \ldots, x \mapsto g_m(y, x)$ we construct the pair $p(y, t), q(y, t)$; for each component of the connected sign partition of $x \mapsto g_1(y, x), \ldots, x \mapsto g_m(y, x)$ there exists t' such that $p(y, t') = 0$ and $q(y, t')$ is in the component; the pair $p(y, t), q(y, t)$ are polynomials in y as well as in t.

Then we apply the preceding Sylvester matrix determinant computations to the univariate polynomials

$$t \mapsto g_1(y, q(y, t)), \ldots, t \mapsto g_m(y, q(y, t)),$$

thereby obtaining polynomials $h_1(y), \ldots, h_\ell(y)$ with the following property: If y' and y'' are in the same component of the connected sign partition of $h_1, \ldots, h_\ell$, then the set of sign vectors for the polynomials $x \mapsto g_1(y, x), \ldots, x \mapsto g_m(y, x)$ is the same for $y = y'$ and $y = y''$.

This is where it is crucial in designing the most efficient decision and quantifier-elimination methods that the pair $p(y, t), q(y, t)$ be constructed by straight-line program, that is, the same pair works for all y (actually, recalling we have suppressed the subscripts A, the same reasonably "small" set of pairs suffices for all y). The number ℓ of polynomials $h_1, \ldots, h_\ell$ does not depend on the number of variables in y. Methods whose construction of similar pairs requires branching on the coefficients y, lead to polynomials $h_1, \ldots, h_\ell$ where ℓ depends exponentially on the number of variables in y. If our ℓ did depend exponentially on the number of variables in y, the approach we are now discussing would again lead to a complexity bound with $E = O(n)^{O(\omega)}$ rather than with $E = \prod_k O(n_k)$.

3.19

For an arbitrary formula $(\exists y)(\forall y)P(y, x)$ we have now described a method for constructing $h_1, \ldots, h_\ell$, which partition y-space in a useful way; if for each connected component of this partition we could construct a point y' in it, then we could apply the BKR algorithm to the resulting sets of univariate polynomials $t \mapsto g_1(y', q(y', t)), \ldots, t \mapsto g_m(y', q(y', t))$ to compute the entire family of sets of sign vectors that we desire, i.e., the family of sets of sign vectors for $x \mapsto g_1(y, x), \ldots, x \mapsto g_m(y, x)$ obtained as y varies.

Motivated by this, from the polynomials $h_1, \ldots, h_\ell$ we construct a corresponding pair $\bar{p}(s), \bar{q}(s)$; for each component of the connected sign partition of $h_1, \ldots, h_\ell$, there exists s' such that $\bar{p}(s') = 0$ and $\bar{q}(s')$ is in the component. Moreover, we already know that for each y and each component of the connected sign partition of $x \mapsto g_1(y, x), \ldots, x \mapsto g_m(y, x)$ there exists t' such that $p(y, t') = 0$ and $q(y, t')$ is in the component.

Composing these facts, we have that for each set of sign vectors for polynomials of the form

$$x \mapsto g_1(y, x), \ldots, x \mapsto g_m(y, x),$$

there exists s' such that $\bar{p}(s') = 0$ and the set is identical to the set of sign vectors for the polynomials

$$x \mapsto g_1(\bar{q}(s'), x), \ldots, x \mapsto g_m(\bar{q}(s'), x).$$

Moreover, for each sign vector in this set there exists t' such that $p(\bar{q}(s'), t') = 0$; and the sign vector is identical to that for the vector

$$g_1(\bar{q}(s'), q(\bar{q}(s'), t')), \ldots, g_m(\bar{q}(s'), q(\bar{q}(s'), t')),$$

i.e., has the same positive, zero and negative coordinates.

3.20

In essence, we have now reduced the sentence $(\exists y)(\forall x) P(y, x)$ to the bivariate sentence $(\exists s)(\forall t) P(s, t)$ obtained by replacing the atomic predicate polynomials $g_i(y, x)$ with the polynomials $g_i(\bar{q}(s), q(\bar{q}(s), t))$. Directly generalizing this approach one can obtain a procedure for replacing each block of variables in an arbitrary sentence by a single variable. Then simply applying the quantifier elimination method of Section 3.14 to the resulting sentence in many fewer variables, one obtains a decision method with a complexity bound for which $E = \prod O(n_k)^{O(1)}$.

However, this approach is not good for quantifier elimination because the block of free variables cannot be replaced by a single variable. Consequently, only replacing each block of quantified variables by a single variable and then applying the quantifier elimination method of Section 3.14, one obtains a complexity bound with $E = (n_0)^{O(\omega)} \prod_{k=1}^{\omega} O(n_k)^{O(1)}$, where n_0 is the number of free variables. Moreover, even for the decision problem alone, I see no way of refining the above approach to obtain a bound with $E = \prod O(n_k)$, rather than with $E = \prod O(n_k)^{O(1)}$.

3.21

The key to obtaining the most efficient decision and quantifier elimination methods is to also rely on the first polynomial in the pair $(\bar{p}, \bar{q})$; note that it did not appear in the approach of the previous subsection.

The reliance is via Thom's lemma (cf. Coste and Roy 1988) which asserts that if $p(t)$ is a univariate polynomial of degree d, then no two of the sign vectors of the set of derivative polynomials $p, p', \ldots, p^{(d-1)}$ for which the first coordinate is zero are identical, i.e., if $p(t') = 0 = p(t'')$ then the sign of $p^{(i)}(t')$ differs from that of $p^{(i)}(t'')$ for some i. The point of this is that the sign vector of the polynomial and its derivatives at a root provides a signature for the root that allows it to be distinguished from the other roots.

Again consider an arbitrary sentence $(\exists y)(\forall x)P(y,x)$ with two quantifiers, and let $(p,q), (\bar{p},\bar{q})$ be as before. Consider the set of sign vectors of the system of bivariate polynomials,

$$
\begin{aligned}
&\frac{d^i}{ds^i}\bar{p}(s) && i = 0, \ldots, D, \\
&g_i(\bar{q}(s), q(\bar{q}(s), t)) && i = 1, \ldots, m,
\end{aligned}
\tag{10}
$$

where D is at least as large as the degree of $\bar{p}$. As we already know from earlier sections, the entire set of sign vectors for these polynomials can be computed efficiently.

We claim that from this set of sign vectors one can easily construct the family of sets of sign vectors for the polynomials $x \mapsto g_1(y,x), \ldots, x \mapsto g_m(y,x)$ obtained as y varies. To see this first recall two observations recorded in Section 3.19: (i) for each component of the connected sign partition of $h_1, \ldots, h_\ell$ there exists s' such that $\bar{p}(s') = 0$ and $\bar{q}(s')$ is in the component; (ii) for each y and each component of the connected sign partition of $x \mapsto g_1(y,x), \ldots, x \mapsto g_m(y,x)$ there exists t' such that $p(y,t') = 0$ and $q(y,t')$ is in the component.

Proceed as follows. First partition the set of sign vectors for the set of polynomials (10) according to the coordinates corresponding to the polynomials $\frac{d^i}{ds^i}\bar{p}(s)$, $i = 0, \ldots, D$, grouping those vectors together for which the coordinates are identical and discarding all vectors for which the first of the coordinates is not zero. By Thom's lemma, for each group there exists a root s' of $\bar{p}$ such that the sign vectors in the group are precisely those occurring at points of the form (s', t). Hence, by (i) and (ii) above, for each y there exists one of the groups such that the sign vectors of $x \mapsto g_1(y,x), \ldots, x \mapsto g_m(y,x)$ are identical to the sign vectors obtained by truncating the sign vectors in the group, where the truncation leaves only the coordinates corresponding to the polynomials

$$
g_1(\bar{q}(s), q(\bar{q}(s), t)), \ldots, g_m(\bar{q}(s), q(\bar{q}(s,t))).
$$

Conversely, for each of the groups, the truncated sign vectors are precisely the sign vectors of

$$
x \mapsto g_1(y,x), \ldots, x \mapsto g_m(y,x)
$$

for some y.

We thereby obtain another approach to determining if a sentence with two quantifiers is true. This approach directly reduces the general decision problem to the problem of determining signs of multivariate polynomials, and allows us to avoid unrolling any algorithms.

3.22

The above procedure generalizes in a straightforward manner to arbitrary sentences

$$(Q_1 x_1) \ldots (Q_\omega x_\omega) P(x_1, \ldots, x_\omega).$$

First viewing $(x_1, \ldots, x_{\omega-1})$ as coefficients of the atomic predicate polynomials, one constructs a pair $(p^{[\omega-1]}, q^{[\omega-1]})$ and polynomials $h_i^{[\omega-1]}$, which partition $(x_1, \ldots, x_{\omega-1})$-space. Then, viewing $(x_1, \ldots, x_{\omega-2})$ as coefficients of the polynomials $h_i^{[\omega-1]}$, one constructs a pair $(p^{[\omega-2]}, q^{[\omega-2]})$ and polynomials $h_i^{[\omega-2]}$, which partition $(x_1, \ldots, x_{\omega-2})$-space, and so on.

Next, one computes the sign vectors of the polynomials

$$\frac{d_i}{dt_1^i} p^{[1]}(t_1) \quad i = 0, \ldots, D$$

$$\frac{d_i}{dt_2^i} p^{[2]}(q^{[1]}(t_1), t_2) \quad i = 0, \ldots, D$$

$$\vdots$$

$$\frac{d_i}{dt_{\omega-1}^i} p^{[\omega-1]}(q^{[1]}(t_1), q^{[2]}(q^{[1]}(t_1), t_2), \ldots,$$
$$q^{[\omega-2]}(q^{[1]}(t_1), q^{[2]}(q^{[1]}(t_1), t_2), \ldots, t_{\omega-2}), t_{\omega-1}) \quad i = 0, \ldots, D$$
$$g_i(q^{[1]}(t_1), q^{[2]}(q^{[1]}(t_1), t_2), \ldots,$$
$$q^{[\omega-1]}(q^{[1]}(t_1), q^{[2]}(q^{[1]}(t_1), t_2), \ldots, t_{\omega-1}), t_\omega)) \quad i = 1, \ldots, m.$$

These are the polynomials obtained by first substituting $q^{[\omega]}(x_1, \ldots, x_{\omega-1}, t_\omega)$ for x_ω in the polynomials $(d^i/dt_j)p(x_1, \ldots, x_{j-1}, t_j)$ and $g_i(x_1, \ldots, x_\omega)$. Then, one substitutes $q^{[\omega-1]}(x_1, \ldots, x_{\omega-2}, t_{\omega-2})$ for $x_{\omega-2}$ in the resulting polynomials, and so on. From these sign vectors one is able to determine if the original sentence is true. (See Renegar 1992a, 1992b, 1992c, for the details.)

3.23

The principal advantage to the above decision method in contrast to the one obtained in Section 3.20 by replacing each block of variables with a single variable and then applying the quantifier elimination method of Section 3.14 is that the above method does not require any algorithm to be unrolled into an algebraic decision tree; the quantifier elimination method of Section 3.14 was obtained by repeated unrolling, once each time a block of quantified variables was eliminated.

To obtain a quantifier elimination method for an arbitrary formula

$$(Q_1 x_1) \cdots (Q_\omega x_\omega) P(z, x_1, \ldots, x_\omega)$$

with free variables z, we can unroll our new decision method into an algebraic decision tree in z, *pruning* branches (as in Section 3.13) as we unroll. Hence, to obtain a quantifier elimination method, we need only unroll once as opposed to once for each block of quantified variables. In this manner one is led to a quantifier elimination method with complexity bound for which $E = \prod_{k=0}^{\omega} O(n_k)^{O(1)}$.

With considerable attention to detail, the method can be refined to provide a complexity bound for which $E = \prod O(n_k)$, the best available; once again, we remark that the refinements are in (Renegar 1992a, 1992b, 1992c).

Acknowledgments

This research was supported by NSF Grant No. DMS-8800835.

An Improved Projection Operation for Cylindrical Algebraic Decomposition

Scott McCallum

1 Introduction

A key component of the cylindrical algebraic decomposition (cad) algorithm is the projection (or elimination) operation: the *projection* of a set A of r-variate integral polynomials, where $r \geq 2$ is defined to be a certain set $\mathrm{PROJ}(A)$ of $(r-1)$-variate integral polynomials. The property of the map PROJ of particular relevance to the cad algorithm is that, for any finite set A of r-variate integral polynomials, where $r \geq 2$, if S is any connected subset of $(r-1)$-dimensional real space $\mathbb{R}^{(r-1)}$ in which every element of $\mathrm{PROJ}(A)$ is invariant in sign then the portion of the zero set of the product of those elements of A which do not vanish identically on S that lies in the cylinder $S \times \mathbb{R}$ over S consists of a number (possibly 0) of disjoint "layers" (or sections) over S in each of which every element of A is sign-invariant: that is, A is "delineable" on S. It follows from this property that, for any finite set A of r-variate integral polynomials, $r \geq 2$, any decomposition of $\mathbb{R}^{(r-1)}$ into connected regions such that every polynomial in $\mathrm{PROJ}(A)$ is invariant in sign throughout every region can be extended to a decomposition of $\mathbb{R}^r$ (consisting of the union of all of the above-mentioned layers and the regions in between successive layers, for each region of $\mathbb{R}^{(r-1)}$) such that every polynomial in A is invariant in sign throughout every region of $\mathbb{R}^r$.

This paper is concerned with a refinement to the projection operation in the cad algorithm. Collins (1975) observed that a smaller projection suffices for a set A of bivariate integral polynomials. Provided that the elements of A have positive degree in the main variable and are primitive, squarefree and pairwise relatively prime, it suffices to define $\mathrm{PROJ}(A)$ to be the set of all nonzero univariate leading coefficients of the elements of A, the univariate discriminants of the elements of A, and the univariate resultants of pairs of distinct elements of A. McCallum (1988) proved that a similar (though not exactly the same) kind of simplification can be made to the projection of a set of trivariate polynomials. The present paper partially extends the result of (McCallum 1988) to an arbitrary number of variables. The present paper, foreshadowed in (McCallum 1988), is based on the author's PhD thesis (McCallum 1984).

A discussion, in the context of the piano mover's problem, about improvements to the projection operation in the cad algorithm can be found in (Benedetti and Risler 1990). Another improvement to the original projection operation can be found in (Hong 1990a).

Section 2 of this paper provides certain important definitions and some background mathematical material that may be helpful to the reader. Section 3 contains the statements of the main theorems of the paper. Sections 4 and 5 contain the proofs of the theorems stated in Section 3. In Section 6 we present cad algorithms which use the improved projection operation. Section 7 reports some examples.

2   Background Material

Throughout this section let K denote either the field $\mathbb{R}$ of all real numbers or the field $\mathbb{C}$ of all complex numbers. A function $f : U \to K$ from an open subset U of K^n into K is said to be *analytic* (in U) if it has a multiple power series representation about each point of U. If $c \in K^n$, then a *neighborhood* of c is an open subset W of K^n containing c. The *polydisc* in $\mathbb{C}^n$ about the point $c = (c_1, \ldots, c_n)$ of *polyradius* $r = (r_1, \ldots, r_n)$, where each $r_i > 0$, is the set of points $z = (z_1, \ldots, z_n)$ in $\mathbb{C}^n$ satisfying $|z_i - c_i| < r_i$, for each i. If Δ is the polydisc in $\mathbb{C}^n$ about c of polyradius r and $1 \le s \le n$, then we denote by $\Delta^{(s)}$ the polydisc in $\mathbb{C}^s$ defined by $|z_i - c_i| < r_i$, for $1 \le i \le s$. Let Δ be a polydisc about 0 in $\mathbb{C}^{n-1}$, where $n \ge 2$, and let R be the ring of all analytic functions $f(z_1, \ldots, z_{n-1})$ in Δ. An element of the polynomial ring $R[z_n]$ is called a *pseudo-polynomial* in Δ. Let z denote the $(n-1)$-tuple $(z_1, \ldots, z_{n-1})$. A monic pseudo-polynomial in Δ

$$h(z, z_n) = z_n^m + a_1(z) z_n^{m-1} + \cdots + a_m(z)$$

of positive degree m, such that $a_i(0) = 0$ for each i, $1 \le i \le m$, is called a *Weierstrass polynomial* in Δ.

Let $f(z_1, \ldots, z_n)$ be an analytic function defined in some open set U of K^n. Let $p = (p_1, \ldots, p_n)$ be a point of U. If f and all its partial derivatives (pure and mixed) of all orders vanish at p, then we say f has infinite *order* at p, and write $\mathrm{ord}_p f = \infty$; we also say that f has infinite order in each of its variables at p. If, on the other hand, either f or some partial derivative of f of some order does not vanish at p then we say that f has *order* k at p, and write $\mathrm{ord}_p f = k$, provided that k is the least non-negative integer such that some partial derivative of total order k does not vanish at p; we also say that f has order k_i at p in z_i provided that k_i is the least non-negative integer such that some partial derivative of f of order k_i in z_i does not vanish at p. Thus, for example, $\mathrm{ord}_p f = 0$ if $f(p) \neq 0$. Also, the order of f at p is greater than or equal to the order of f at p in each of its variables z_i. An analytic function defined in an open set U of K^n is said to be *order-invariant* in a subset S of U provided that the order of f is the same at every point of S. We remark that if $K = \mathbb{R}$ and the analytic function $f : U \to K$ is order-invariant in the connected subset S of U then f is sign-invariant in S. Consider the following example: let $f(x,y) = xy$, let C_f be the union of the x-axis and the y-axis and let $S = C_f - \{(0,0)\}$. Then f is order-invariant in S but not in C_f.

The reader is referred to any of the texts by Gunning and Rossi (1965), Bochner and Martin (1948) or Kaplan (1966) for a detailed discussion of the basic properties of analytic functions.

An *analytic submanifold* of $\mathbb{R}^n$ of dimension s is a nonempty set S in $\mathbb{R}^n$ which "looks locally like real s-space $\mathbb{R}^s$"; that is, for every point p of s, there is an analytic coordinate system about p with respect to which S is locally the intersection of $n - s$ coordinate hyperplanes. The only kind of submanifold we shall consider in this paper is the analytic kind. Thus, we shall henceforth omit the term "analytic" when referring to submanifolds: all submanifolds will be understood to be analytic. For example, the unit sphere S^{n-1} about the origin in $\mathbb{R}^n$ is a submanifold of $\mathbb{R}^n$ of dimension $n - 1$.

Nonempty open subsets of $\mathbb{R}^n$ are submanifolds – in fact the nonempty open subsets are the n-dimensional submanifolds. But lower-dimensional submanifolds are not open. One can, however, define in a natural way the notion of an analytic function from an s-dimensional submanifold S into $\mathbb{R}$, where $0 \leq s \leq n$. A function $f : S \to \mathbb{R}$ is said to be analytic if for every point p of S, there is a coordinate system about p with respect to which S looks locally like $\mathbb{R}^s$ and f looks locally like an analytic function from $\mathbb{R}^s$ into $\mathbb{R}$. One can show that the definition of analyticity is independent of any particular coordinate system for the submanifold. The reader can consult (Whitney 1972, appendix II) or (McCallum 1984, chapter 2) for precise definitions and basic properties of submanifolds and analytic functions defined in submanifolds. McCallum (1988) contains a brief summary of some basic properties of submanifolds and applies the concepts in developing an improved projection operation for cad in the three variable case.

3   Statements of Theorems about Improved Projection Map

Let R be any commutative ring with identity element. Let $R[x_1, \ldots, x_r]$ denote the ring of all polynomials in $x_1, \ldots, x_r$ with coefficients in R. We regard the elements of this ring as polynomials in x_r over $R[x_1, \ldots, x_{r-1}]$. Thus, for example, the degree $\deg f$ of a polynomial f in this ring means the degree of f in x_r. Let $\mathbb{Z}$ denote the domain of all rational integers. A set A of polynomials in $\mathbb{Z}[x_1, \ldots, x_r]$ is said to be a *squarefree basis* if the elements of A have positive degree, and are primitive, squarefree and pairwise relatively prime. Let A be a squarefree basis in $\mathbb{Z}[x_1,\ldots,x_r]$, where $r \geq 2$. We define (cf. McCallum 1988) the improved projection $P(A)$ of A to be the union of the set of all nonzero coefficients of the elements of A, the set of all discriminants of elements f of A and the set of all resultants of pairs f, g of distinct elements of A. (In Section 6 we shall slightly modify the definition of P.)

We define a slight variant of a basic concept from (Collins 1975) and (McCallum 1988). Let x denote the $(r-1)$-tuple $(x_1,\ldots,x_{r-1})$. An r-variate polynomial $f(x, x_r)$ over the reals is said to be *analytic delineable* on a submanifold S (usually connected) of $\mathbb{R}^{r-1}$ if

1. the portion of the real variety of f that lies in the cylinder $S \times \mathbb{R}$ over S consists of the union of the graphs of some $k \geq 0$ analytic functions $\theta_1 < \cdots < \theta_k$ from S into $\mathbb{R}$; and
2. there exist positive integers $m_1,\ldots,m_k$ such that for every $a \in S$, the multiplicity of the root $\theta_i(a)$ of $f(a, x_r)$ (considered as a polynomial in x_r alone) is m_i.

A remark: if f has no zeros in $S \times \mathbb{R}$, then f is analytic delineable on S as we may take $k = 0$ in the definition. An example: if $f(x_1, x_2) = x_2^3 + x_1^2 x_2$ and S is $\mathbb{R}^1$, then condition 1 of the definition is satisfied (with $k = 1$ and $\theta_1(x_1) = 0$ for all x_1), but condition 2 is not satisfied. Suppose that the r-variate polynomial f is analytic delineable on the connected s-dimensional submanifold S in $\mathbb{R}^{r-1}$. Then the θ_i in condition 1 of the definition of analytic delineability are sometimes called the *real root functions* of f on S, the graphs of the θ_i are called the *f-sections* over S, and the regions between successive f-sections are called *f-sectors* over S. Each f-section over S is a connected s-dimensional submanifold of $\mathbb{R}^r$ (McCallum 1984, theorem 2.2.3), and each f-sector over S is a connected $(s+1)$-dimensional submanifold of $\mathbb{R}^r$ (McCallum 1984, theorem 2.2.4).

We recall some basic concepts from (Collins 1975) and (McCallum 1988). An r-variate polynomial $f(x, x_r)$ over $\mathbb{R}$ is said to *vanish identically* on a subset S of $\mathbb{R}^{r-1}$ if $f(p, x_r) = 0$ for every point p of S. An r-variate polynomial $f(x, x_r)$ over $\mathbb{R}$ is said to be *degree-invariant* on a subset S of $\mathbb{R}^{r-1}$ if the degree of $f(p, x_r)$ (as a polynomial in x_r) is the same for every point p of S.

We are now ready to state the main theorem of this paper. We suggest that the reader reviews the discussion in Section 2 about the order of an analytic function at a point before reading the statement of the theorem.

Theorem 1 *Let A be a finite squarefree basis of integral polynomials in r variables, where $r \geq 2$. Let S be a connected submanifold of $\mathbb{R}^{r-1}$. Suppose that each element of $P(A)$ is order-invariant in S. Then each element of A either vanishes identically on S or is analytic delineable on S, the sections over S of the elements of A which do not vanish identically on S are pairwise disjoint, and each element of A which does not vanish identically on S is order-invariant in every such section.*

Theorem 1 is a generalization to arbitrary r of theorem 3.1 of (McCallum 1988). Theorem 1 can be quite readily derived from Theorem 2 (to be stated shortly). A proof that Theorem 1 follows from Theorem 2 can be obtained by slightly modifying the proof of theorem 3.1 presented on p. 148 of (McCallum 1988) (in modifying the proof of theorem 3.1 of (McCallum 1988) one would set f to be the product of those elements f_i of A which do not vanish identically on S).

Theorem 2 *Let $r \geq 2$. Let $f(x, x_r)$ be a polynomial in $\mathbb{R}[x, x_r]$ of positive degree. Let $D(x)$ be the discriminant of $f(x, x_r)$ and suppose that $D(x)$ is a nonzero polynomial. Let S be a connected submanifold of $\mathbb{R}^{r-1}$ on which f is degree-invariant and does not vanish identically, and in which D is order-invariant. Then f is analytic delineable on S and is order-invariant in each f-section over S.*

Theorem 2 can, in turn, be derived from Theorem 3 (to be stated shortly). A proof that Theorem 2 follows from Theorem 3 can be obtained by generalizing in a straightforward way the proof of theorem 3.2 presented in section 4 of (McCallum 1988). We remark that Theorem 2 appears as theorem 3.2.1 of (McCallum 1984) and as theorem D.5.1 of (Benedetti and Risler 1990). Before reading the statement of the next theorem, the reader is advised to review the definitions of polydisc and Weierstrass polynomial given in Section 2. In the statements of the next two theorems, z denotes the $(r-1)$-tuple $(z_1, \ldots, z_{r-1})$.

Theorem 3 (Zariski) *Let $r \geq 3$. Let $h(z, z_r)$ be a Weierstrass polynomial of degree $m \geq 1$ in the polydisc Δ_1 about 0 in $\mathbb{C}^{r-1}$, and assume that for every fixed z in Δ_1, every root of $h(z, z_r)$ (considered as a polynomial in z_r alone) is contained in the disc $|z_r| < \epsilon$. Let $F(z)$ be the discriminant of $h(z, z_r)$, and assume that F does not vanish identically. Let $1 \leq s \leq r - 2$, let*

$$T^* = \{z = (z_1, \ldots, z_{r-1}) \in \mathbb{C}^{r-1} | z_{s+1} = 0 \wedge \cdots \wedge z_{r-1} = 0\}$$

and assume that F is order-invariant in $T^* \cap \Delta_1$. Then there exists a polydisc $\Delta_2 \subseteq \Delta_1$ about 0 and an analytic function $\psi(z_1, \ldots, z_s)$ from $\Delta_2^{(s)}$ into the disc $|z_r| < \epsilon$ such that for every fixed $z = (z_1, \ldots, z_s, 0, \ldots, 0)$ in $T^* \cap \Delta_2$, $h(z, z_r)$ (as a polynomial in z_r) has exactly one root (necessarily of multiplicity m), namely $\psi(z_1, \ldots, z_s)$, in the disc $|z_r| < \epsilon$, and such that h is order-invariant in the set

$$G^* = \{(z, z_r) \in \mathbb{C}^r | z = (z_1, \ldots, z_s, 0, \ldots, 0) \in T^* \cap \Delta_2 \wedge z_r = \psi(z_1, \ldots, z_s)\}.$$

Theorem 3, stated and proved using notation and terminology rather different from ours, can be found in (Zariski 1975). An alternative exposition of elements of the proof of Theorem 3 can be found in appendix D.5 of (Benedetti and Risler 1990). We present a complete proof of Theorem 3, along the lines sketched by (Zariski 1975), in Section 4. At a crucial point in the proof, Theorem 4 (stated below) will be used! In the statement of Theorem 4 the symbol $\tilde{z}$ denotes the $(r-2)$-tuple $(z_1, \ldots, z_{r-2})$; thus $z = (\tilde{z}, z_{r-1})$.

Theorem 4 (Zariski) *Let $r \geq 3$. Let $h(z, z_r)$ be a Weierstrass polynomial of degree $m \geq 1$ in the polydisc Δ_1 about 0 in $\mathbb{C}^{r-1}$, and assume that for every fixed z in Δ_1, all the roots of $h(z, z_r)$ (as a polynomial in z_r) are contained in the disc $|z_r| < \epsilon$. Let $F(z)$ be the discriminant of $h(z, z_r)$ and suppose that there exists a function $N(z)$, analytic and non-vanishing in Δ_1, and an integer $k \geq 0$, such that*

$$F(z) = z_{r-1}^k N(z)$$

for all $z = (\tilde{z}, z_{r-1})$ in Δ_1. Let H^* be the hyperplane $\{z = (\tilde{z}, z_{r-1}) \in \mathbb{C}^{r-1} | z_{r-1} = 0\}$ in $\mathbb{C}^{r-1}$. Then there exists an analytic function $\psi(\tilde{z})$ from $\Delta_1^{(r-2)}$ into the disc $|z_r| < \epsilon$ such that for every fixed $z = (\tilde{z}, 0)$ in $H^* \cap \Delta_1$, $h(z, z_r)$ (as a polynomial in z_r) has exactly one root (necessarily of multiplicity m), namely $\psi(\tilde{z})$, in the disc $|z_r| < \epsilon$, and such that h is order-invariant in the set

$$G^* = \{(\tilde{z}, z_{r-1}, z_r) \in \mathbb{C}^r | (\tilde{z}, z_{r-1}) \in H^* \cap \Delta_1 \wedge z_r = \psi(\tilde{z})\}.$$

Theorem 4 is implied by theorem 4.5 in (Zariski 1965). An alternative exposition of the proof of Theorem 4 can be found in appendix D.4 of (Benedetti and Risler 1990). A proof of Theorem 4, rather different from arguments given by (Zariski 1965) and from the proof presented in (Benedetti and Risler 1990), is presented in Section 5.

4   Proof of Theorem 3 (and Lemmas)

We define some notation used throughout this section first of all. Let $0 \leq s \leq t \leq n$. We associate with s, t and n a *quadratic transformation* $Q_{s,t,n} : \mathbb{C}^n \rightarrow \mathbb{C}^n$ defined by

$$Q_{s,t,n}(z_1, \ldots, z_n) = (z_1, \ldots, z_s, z_{s+1} z_t, z_{s+2} z_t, \ldots, z_{t-1} z_t, z_t, \ldots, z_n).$$

(Such a transformation is sometimes called a *monoidal transformation* in the literature. See (Abhyankar 1990) for general information about quadratic and monoidal transformations.) Observe that if $s + 1 > t - 1$ then $Q_{s,t,n}$ is the identity map. We associate with s and n an s-dimensional linear subspace $T_{s,n}^*$ of $\mathbb{C}^n$ as follows:

$$T_{s,n}^* = \{(z_1, \ldots, z_n) \in \mathbb{C}^n | z_{s+1} = 0 \wedge \cdots \wedge z_n = 0\}.$$

We associate with $n > 0$ a hyperplane H_n^* in $\mathbb{C}^n$ by setting

$$H_n^* = T_{n-1,n}^*.$$

We remark that $T_{0,n}^*$ is the origin in $\mathbb{C}^n$ and $T_{n,n}^* = \mathbb{C}^n$. In Lemma 1 z denotes the n-tuple $(z_1, \ldots, z_n)$.

Lemma 1 *Let $n \geq 1$ and let $F(z) = F(z_1, \ldots, z_n)$ be analytic and not identically vanishing in the polydisc Δ about 0 in $\mathbb{C}^n$. Let $0 \leq s \leq n$, let Q be the* *quadratic transformation* $Q_{s,n,n} : \mathbb{C}^n \to \mathbb{C}^n$ *and let* $T^* = T^*_{s,n}$. *Suppose that* F *is order-invariant in* $T^* \cap \Delta$ *and that* $\partial^k F / \partial z_n^k(0) \neq 0$, *where* $k = \mathrm{ord}_0 F$. *Then there exists a polydisc* Δ' *about* 0 *in* $\mathbb{C}^n$ *and a function* $N(z) = N(z_1, \ldots, \dot{z}_n)$, *analytic and non-vanishing in* Δ', *such that* $Q(\Delta') \subseteq \Delta$ *and*

$$F(Q(z_1, \ldots, z_n)) = z_n^k N(z_1, \ldots, z_n)$$

for all $(z_1, \ldots, z_n)$ *in* Δ'.

Proof. In this proof we denote any s-tuple of variables of the form $(v_1, \ldots, v_s)$ by $\tilde{v}$: thus $v = (\tilde{v}, v_{s+1}, \ldots, v_n)$.

Let the power series expansion for $F(z)$ about 0 be

$$\sum_{i_1, \ldots, i_n \geq 0} f_{i_1, \ldots, i_n} z_1^{i_1} \cdots z_n^{i_n}. \tag{1}$$

Series (1) is absolutely convergent in Δ (by theorem 3 in chapter II of (Bochner and Martin 1948)).

Series (1) can be arranged as an iterated series thus:

$$\sum_{i_{s+1}, \ldots, i_n \geq 0} \left(\sum_{\tilde{i} \geq 0} f_{\tilde{i}, i_{s+1}, \ldots, i_n} z_1^{i_1} \cdots z_s^{i_s} \right) z_{s+1}^{i_{s+1}} \cdots z_n^{i_n}. \tag{2}$$

A brief discussion of iterated series appears on p. 173 of (Kaplan 1966). For each fixed $i_{s+1}, \ldots, i_n$, the inner series in Series (2) is absolutely convergent in $\Delta^{(s)}$. For $i_{s+1}, \ldots, i_n \geq 0$ and $\tilde{z} \in \Delta^{(s)}$, let $p_{i_{s+1}, \ldots, i_n}(\tilde{z})$ denote the sum of this inner series: then, by the s-variable analogue of theorem 56 of (Kaplan 1966), the function $p_{i_{s+1}, \ldots, i_n}$ is analytic in $\Delta^{(s)}$. Let $\delta = (\tilde{\delta}, \delta_{s+1}, \ldots, \delta_n)$ be the polyradius of Δ. Then, for each $\tilde{z}$ in $\Delta^{(s)}$, the outer series

$$\sum_{i_{s+1}, \ldots, i_n \geq 0} p_{i_{s+1}, \ldots, i_n}(\tilde{z}) z_{s+1}^{i_{s+1}} \cdots z_n^{i_n} \tag{3}$$

is absolutely convergent in the polydisc $|z_{s+1}| < \delta_{s+1}, \ldots, |z_n| < \delta_n$, with sum $F(\tilde{z}, z_{s+1}, \ldots, z_n) = F(z)$.

For $\tilde{z} \in \Delta^{(s)}$ and $l \geq 0$, define the homogeneous polynomial $P_{\tilde{z}, l}(z_{s+1}, \ldots, z_n)$ of degree l (if nonzero) by:

$$P_{\tilde{z}, l}(z_{s+1}, \ldots, z_n) = \sum_{i_{s+1} + \cdots + i_n = l} p_{i_{s+1}, \ldots, i_n}(\tilde{z}) z_{s+1}^{i_{s+1}} \cdots z_n^{i_n}.$$

Then, for every $\tilde{z} \in \Delta^{(s)}$ and every l, $0 \leq l < k$, $P_{\tilde{z}, l}(z_{s+1}, \ldots, z_n) = 0$, by order-invariance of F in $T^* \cap \Delta$. Hence, for each fixed $\tilde{z} \in \Delta^{(s)}$, we have the following power series representation for $F(z) = F(\tilde{z}, z_{s+1}, \ldots, z_n)$:

$$F(z) = P_{\tilde{z}, k}(z_{s+1}, \ldots, z_n) + P_{\tilde{z}, k+1}(z_{s+1}, \ldots, z_n) + \cdots, \tag{4}$$

obtained by suitably grouping terms in Series (3), which is valid in the polydisc $|z_{s+1}| < \delta_{s+1}, \ldots, |z_n| < \delta_n$.

Let δ' satisfy $Q(\delta') = \delta$ (δ' exists because $\delta_n \neq 0$). Let Δ_1' be the polydisc of polyradius δ' about 0 in $\mathbb{C}^n$. We have $Q(\Delta_1') \subseteq \Delta$. Let $z = (z_1, \ldots, z_n)$ be a point of Δ_1'. Then we have

$$F(Q(z)) = z_n^k (P_{\bar{z},k}(z_{s+1}, \ldots, z_{n-1}, 1) + z_n P_{\bar{z},k+1}(z_{s+1}, \ldots, z_{n-1}, 1) + \cdots),$$

where the series appearing in parentheses on the right-hand side of the above equation is absolutely convergent: define $N(z)$ to be the sum of this series. By the n-variable analogue of theorem 56 of (Kaplan 1966), $N(z)$ is analytic in Δ_1'. Now $N(0) = P_{\bar{0},k}(0, \ldots, 0, 1) = f_{0,\ldots,0,k} \neq 0$, by hypothesis. Therefore, by continuity of N, there exists a polydisc $\Delta' \subseteq \Delta_1'$ about 0 such that $N \neq 0$ throughout Δ'. ∎

Remark 1 *The hypothesis of Lemma 1 that $\partial^k F / \partial z_n^k(0) \neq 0$ is essential to the conclusions of the lemma. This can be seen by considering the function $F(x, y, z) = z^2 + xz - y$, which is order-invariant in the x-axis in $\mathbb{C}^3$; $\mathrm{ord}_0 F = 1$, but $\partial F / \partial z(0) = 0$. We have*

$$F(Q(x, y, z)) = F(x, yz, z) = z(z + x - y),$$

but the function $z + x - y$ vanishes at the origin.

In Lemma 2 we denote any $(n-1)$-tuple of variables of the form $(v_1, \ldots, v_{n-1})$ by v.

Lemma 2 *Let $n \geq 3$, let $1 \leq s \leq n - 2$, let Q be the quadratic transformation $Q_{s,n-1,n} : \mathbb{C}^n \to \mathbb{C}^n$ and let T^* be the s-dimensional linear subspace $T_{s,n-1}^*$ of $\mathbb{C}^{n-1}$. Let (w, w_n) be a point of $T^* \times \mathbb{C}$, let U be a neighborhood of (w, w_n) in $\mathbb{C}^n$ such that $Q(U) \subseteq U$, and let $h(z, z_n) = h(z_1, \ldots, z_n)$ be analytic in U. Then there exists a point (w', w_n') in U such that $Q(w', w_n') = (w, w_n)$ and*

$$\mathrm{ord}_{(w,w_n)} h = \mathrm{ord}_{(w',w_n')} hQ,$$

where hQ denotes the composite of h and Q.

Proof. Observe that for every point (z', z_n') in U satisfying $Q(z', z_n') = (w, w_n)$ we have

$$\mathrm{ord}_{(w,w_n)} h \leq \mathrm{ord}_{(z',z_n')} hQ$$

by theorem 2.1 of (McCallum 1988). Hence it suffices to show that there exists some point $(w', w_n') \in U$ satisfying $Q(w', w_n') = (w, w_n)$ for which

$$\mathrm{ord}_{(w,w_n)} h \geq \mathrm{ord}_{(w',w_n')} hQ.$$

In order to reduce somewhat the complexity of the notation we shall present a proof of the required property for the special case $n = 4$ and $s = 1$. A general proof can be obtained by extending the proof given in a straightforward way. For (z_1, z_2, z_3, z_4) we shall use (x, y, z, w) and for $(w_1, w_2, w_3, w_4) = (w_1, 0, 0, w_4)$ we shall use $(a, 0, 0, d)$.

Let the power series expansion for h about $(a, 0, 0, d)$ be

$$\sum_{i,j,k,l \geq 0} h_{i,j,k,l}(x - a)^i y^j z^k (w - d)^l. \tag{5}$$

Let t be the order of h at $(a, 0, 0, d)$. If $t = \infty$ then we take $(a', b', c', d') = (a, 0, 0, d)$ and we are done. Suppose, on the other hand, that $t < \infty$. Then there is some 4-tuple (i_0, j_0, k_0, l_0) of non-negative integers whose sum is t and for which $h_{i_0,j_0,k_0,l_0} \neq 0$. Now the power series expansion of hQ about $(a, 0, 0, d)$ can be obtained by formal substitution of the power series expansions about $(a, 0, 0, d)$ of the component functions of Q into Series (5), followed by simplification (Bochner and Martin 1948, p. 33). Hence, where $p = j_0 + k_0$, the non-zero term

$$h_{i_0,j_0,k_0,l_0}(x - a)^{i_0} y^{j_0} z^{k_0} (w - d)^{l_0}$$

in Series (5) gives rise to the non-zero term

$$h_{i_0,j_0,k_0,l_0}(x - a)^{i_0} y^{j_0} z^p (w - d)^{l_0}$$

in the power series expansion of hQ about $(a, 0, 0, d)$. Let the power series expansion for hQ about $(a, 0, 0, d)$ be

$$\sum_{m,n,r,s \geq 0} h'_{m,n,r,s}(x - a)^m y^n z^r (w - d)^s. \tag{6}$$

Let $\Delta' \subseteq U$ be a polydisc about $(a, 0, 0, d)$ in which Series (6) is absolutely convergent and let $\delta' = (\delta'_1, \ldots, \delta'_4)$ be the polyradius of Δ'. Series (6) can be arranged as an iterated series thus:

$$\sum_{m,r,s \geq 0} \left(\sum_{n \geq 0} h'_{m,n,r,s} y^n \right)(x - a)^m z^r (w - d)^s. \tag{7}$$

A brief discussion of iterated series appears on p.173 of (Kaplan 1966). For each fixed $m, r, s \geq 0$, the inner series in Series (7) is absolutely convergent in the disc $|y| < \delta'_2$. For $m, r, s \geq 0$ and $|y| < \delta'_2$ let $p'_{m,r,s}(y)$ denote the sum of the inner series in Series (7). By a well-known result on convergence each function $p'_{m,r,s}$ is analytic in the disc $|y| < \delta'_2$. Now $h'_{i_0,j_0,p,l_0} = h_{i_0,j_0,k_0,l_0} \neq 0$. Hence p'_{i_0,p,l_0} is not identically vanishing. Hence there exists $b' \in \mathbb{C}$ such that $|b'| < \delta'_2$ and

$$p'_{i_0,p,l_0}(b') \neq 0. \tag{8}$$

Let p' denote the left-hand side of Inequation (8). Let $a' = a$ and let $d' = d$. Then $(a', b', 0, d') \in \Delta'$ and $Q(a', b', 0, d') = (a, 0, 0, d)$. The power series expansion of hQ about $(a', b', 0, d')$ contains the term

$$p'(x - a')^{i_0}(y - b')^0 z^p (w - d')^{l_0}.$$

Hence, since $p' \neq 0$ and the total degree of this term is $i_0 + p + l_0 = t$, we have

$$t \geq \mathrm{ord}_{(a',b',0,d')} hQ.$$

∎

4.1 Proof of Theorem 3

In this proof we shall denote any s-tuple of variables of the form $(v_1, \ldots, v_s)$ by $\tilde{v}$, and any $(r-1)$-tuple of variables of the form $(v_1, \ldots, v_{r-1})$ by v: thus $v = (\tilde{v}, v_{s+1}, \ldots, v_{r-1})$. Let $\mathrm{ord}_0 F = k$. We may assume without loss of generality that $\partial^k F / \partial z_{r-1}^k(0) \neq 0$. (If this is not the case then we can find suitable $\lambda_1, \ldots, \lambda_{r-2} \in \mathbb{C}$ such that, where L is the invertible linear transformation

$$L(z_1, \ldots, z_{r-1}) = (z_1 + \lambda_1 z_{r-1}, \ldots, z_{r-2} + \lambda_{r-2} z_{r-1}, z_{r-1})$$

of $\mathbb{C}^{r-1}$ fixing T^* – indeed the hyperplane $z_{r-1} = 0$ – pointwise, the transform $\bar{F} = F_o L^{-1}$ of F by L satisfies $\partial^k \bar{F} / \partial z_{r-1}^k(0) \neq 0$.)

Let Q be the quadratic transformation $Q_{s,r-1,r-1} : \mathbb{C}^{r-1} \to \mathbb{C}^{r-1}$. By Lemma 1 (applied with $n = r - 1$ and $\Delta = \Delta_1$), there exists a polydisc Δ_1' about 0 in $\mathbb{C}^{r-1}$ and a function $N'(z) = N'(z_1, \ldots, z_{r-1})$, analytic and nonvanishing in Δ_1', such that $Q(\Delta_1') \subseteq \Delta_1$ and

$$F(Q(z)) = z_{r-1}^k N'(z)$$

for all $z = (z_1, \ldots, z_{r-1})$ in Δ_1'. For each $z \in \Delta_1'$, define

$$h'(z, z_r) = h(Q(z), z_r).$$

Then $h'(z, z_r)$ is a Weierstrass polynomial of degree m in Δ_1' such that for every fixed z in Δ_1', every root of $h'(z, z_r)$ (as a polynomial in z_r alone) is contained in the disc $|z_r| < \epsilon$. Let $F'(z)$ be the discriminant of $h'(z, z_r)$. Then

$$F'(z) = F(Q(z))$$

for every $z \in \Delta_1'$. Let H^* be the hyperplane H_{r-1}^* in $\mathbb{C}^{r-1}$. (That is, H^* is defined by $z_{r-1} = 0$.) By Theorem 4 (applied with $h = h'$, $\Delta_1 = \Delta_1'$, $F = F'$ and $N = N'$), there exists an analytic function $\psi'(z_1, \ldots, z_{r-2})$ from $\Delta_1'^{(r-2)}$ into the disc $|z_r| < \epsilon$ such that for every fixed $z = (z_1, \ldots, z_{r-2}, 0) \in H^* \cap \Delta_1'$, $h'(z, z_r)$ (as a polynomial in z_r) has exactly one root (necessarily of multiplicity m), namely $\psi'(z_1, \ldots, z_{r-2})$, in the disc $|z_r| < \epsilon$, and such that h' is order-invariant in the set

$$K^* = \{(z, z_r) \in \mathbb{C}^r | z \in H^* \cap \Delta_1' \wedge z_r = \psi'(z_1, \ldots, z_{r-2})\}.$$

Let $\nu' = (\nu_1', \ldots, \nu_{r-1}')$ be the polyradius of Δ_1', let $\nu = (\nu_1', \ldots, \nu_{r-2}', \min(\nu_{r-1}', 1))$ and let Δ_2 be the polydisc of polyradius ν about 0 in $\mathbb{C}^{r-1}$. Define $\psi : \Delta_2^{(s)} \to \mathbb{C}$ by

$$\psi(\tilde{z}) = \psi'(\tilde{z}, 0, \ldots, 0).$$

We shall show that ψ satisfies the properties asserted of it in the conclusion of Theorem 3. It is clearly the case that, for all $z = (\tilde{z}, 0, \ldots, 0) \in T^* \cap \Delta_2$, $h(z, z_r)$ (as a polynomial in z_r) has exactly one root (necessarily of multiplicity m), namely $\psi(\tilde{z})$, in the disc $|z_r| < \epsilon$. It remains to show that h is order-invariant in the set G^* defined in the statement of the theorem.

Let (w, w_r) be an arbitrary point of G^*. I shall show that there exists a point (w', w'_r) of K^* such that

$$\text{ord}_{(w, w_r)} h = \text{ord}_{(w', w'_r)} h'.$$

The order-invariance of h in G^* will then follow by the order-invariance of h' in K^*, given by Theorem 4. Let $\hat{Q} = Q_{s,r-1,r} : \mathbb{C}^r \to \mathbb{C}^r$ and let $U = \Delta_2 \times \mathbb{C}$. Then $\hat{Q}(U) \subseteq U$. Moreover $(w, w_r) \in (T^* \times \mathbb{C}) \cap U$ (since $w \in T^* \cap \Delta_2$, by definition of G^*). Hence, by Lemma 2 (applied with $n = r$, $Q = \hat{Q}$), there exists a point $(w', w'_r) \in U$ such that $\hat{Q}(w', w'_r) = (w, w_r)$ and

$$\text{ord}_{(w, w_r)} h = \text{ord}_{(w', w'_r)} h'.$$

I claim that $(w', w'_r) \in K^*$. The reason is that $w' \in H^* \cap \Delta'_2$ and

$$h'(w, z_r) = h(Q(w), z_r) = h(Q(w'), z_r) = h'(w', z_r).$$

The order-invariance of h in G^* now follows by the order-invariance of h' in K^*, given by Theorem 4.

5   Proof of Theorem 4 (and Lemmas)

Recall that in Theorem 4 the symbol $\tilde{z}$ denotes the $(r-2)$-tuple $(z_1, \ldots, z_{r-2})$ and the symbol z denotes the $(r-1)$-tuple $(z_1, \ldots, z_{r-1})$: thus $z = (\tilde{z}, z_{r-1})$. Theorem 4 has a two-part conclusion:

1. that the root $z_r = 0$ of $h(0, z_r)$ does not "split" into many roots as $z = 0$ is perturbed a little within H^* (in the sense of section 4 of (McCallum 1988)); and

2. that, assuming part 1 of the conclusion, h is order-invariant in G^* (which can be thought of as the graph of the root function ψ for h in H^*, near 0.)

The first part of the conclusion of Theorem 4 follows by a straightforward generalization to r variables of the proof of Theorem 3.3 of (McCallum 1988) (the conclusion of which is essentially part 1 of Theorem 4 in the case $r = 3$). We shall make some remarks concerning the development of such a generalization.

(a) In a proof of Theorem 4, part 1, H^* would play the role of the complex x-axis T^* in $\mathbb{C}^2$, defined in the statement of Theorem 3.3 of (McCallum 1988, p. 147).

(b) The first step in the proof of Theorem 3.3 of (McCallum 1988) is an application of lemma A.6 of (McCallum 1988). In a proof of Theorem 4, part 1, however, it is not necessary to apply a generalization of lemma A.6 of (McCallum 1988). Instead one sees immediately from the hypotheses of Theorem 4 that the zero set of F in Δ_1 is either empty (in case $k = 0$) or equal to $H^* \cap \Delta_1$ (in case $k > 0$).

(c) A generalization of the proof of theorem 3.3 of (McCallum 1988) would show that for every fixed $(\tilde{z}, 0) \in H^* \cap \Delta_1$, $h(\tilde{z}, 0, z_r)$ (as a polynomial in z_r) has exactly one root (necessarily of multiplicity m), say $\psi(\tilde{z})$, in the disc $|z_r| < \epsilon$. We see that ψ is analytic in $\Delta_1^{(r-2)}$ as follows. For $\tilde{z} \in \Delta_1^{(r-2)}$, we have

$$h(\tilde{z}, 0, z_r) = (z_r - \psi(\tilde{z}))^m.$$

Hence, where $a_1(\tilde{z}, z_{r-1})$ is the coefficient of z_r^{m-1} in h, we have $a_1(\tilde{z}, 0) = -m\psi(\tilde{z})$. The analyticity of ψ now follows from that of a_1.

For the next two lemmas we use the notation $\tilde{v}$ to denote any $(r - 2)$-tuple $(v_1, \ldots, v_{r-2})$: thus $v = (\tilde{v}, v_{r-1})$.

Lemma 3 *Let $r \geq 2$. Let $h(z, z_r)$ be a Weierstrass polynomial of degree $m \geq 1$ in the polydisc Δ_1 about 0 in $\mathbb{C}^{r-1}$, and assume that for every fixed z in Δ_1, every root of $h(z, z_r)$ (as a polynomial in z_r) is contained in the disc $|z_r| < \epsilon$. Assume that h is irreducible. Let $F(z)$ be the discriminant of $h(z, z_r)$ and suppose that there exists a function $N(z)$, analytic and non-vanishing in Δ_1, and an integer $k \geq 0$, such that*

$$F(z) = z_{r-1}^k N(z)$$

for all $z \in \Delta_1$. Let $\delta = (\delta_1, \ldots, \delta_{r-1})$ be the polyradius of Δ_1, let $\rho = (\delta_1, \ldots, \delta_{r-2}, \delta_{r-1}^{1/m})$, and let Δ_1' be the polydisc about 0 of polyradius ρ in $\mathbb{C}^{r-1}$. Then there is an analytic function $\phi : \Delta_1' \rightarrow \mathbb{C}$ such that for every $z = (\tilde{z}, z_{r-1}) \in \Delta_1$ and every z_r in the disc $|z_r| < \epsilon$, $h(z, z_r) = 0$ if and only if there exists u, $|u| < \delta_{r-1}^{1/m}$, such that

$$
\begin{aligned}
z_{r-1} &= u^m, \\
z_r &= \phi(\tilde{z}, u).
\end{aligned}
\tag{9}
$$

Proof. The case $r = 2$ of this lemma is theorem 10A in chapter 1 of (Whitney 1972). The proof given by (Whitney 1972), which uses the method of analytic continuation, readily generalizes to yield a proof of the case $r > 2$ as well. (section 5 of (McCallum 1988) contains a closely related application of the method of analytic continuation.) ∎

Remark 2 *With h and ϕ as in the above lemma, we shall say that pair of equations (9) defines a* parameterization *for the zeros of h (over Δ_1).*

Let K be any field. We shall denote by $K(x)^*$ the field of all fractional power series in the indeterminate x with coefficients in K (see chapter IV of (Walker 1978)). The field $K(x)^*$ is sometimes called the field of all *Puiseux series* in x over K. Every non-zero element θ of $K(x)^*$ can be expressed, essentially uniquely, in the form

$$\theta = a_h x^{h/m} + a_{h+1} x^{(h+1)/m} + \cdots \tag{10}$$

where $m \geq 1$, h is an integer (possibly negative), the a_i are elements of K and $a_h \neq 0$. Where θ is given by Eq. (10) and $a_h \neq 0$ we define the *order* of θ, denoted by $O(\theta)$, to be the rational number h/m. We define the order of zero to be ∞.

There is a close connection between parameterizations and fractional power series. This connection will be exploited in the next lemma.

Lemma 4 *Let $r \geq 3$. Let $h(z, z_r)$ be a Weierstrass polynomial of degree $m \geq 1$ in the polydisc Δ_1 about 0 in $\mathbb{C}^{r-1}$, and let $F(z)$ be the discriminant of $h(z, z_r)$. Assume that h and F satisfy the hypotheses of Lemma 3. Let*

$$
\begin{aligned}
z_{r-1} &= u^m, \\
z_r &= \phi(\tilde{z}, u)
\end{aligned}
$$

define a parameterization for the zeros of h over Δ_1, as in the statement of Lemma 3. Suppose that the order m_1 of ϕ at 0 in u is non-zero. Then

$$\operatorname{ord}_{(\tilde{w},0,0)} h = \min(m, m_1)$$

for every $\tilde{w}$ in $\Delta_1^{(r-2)}$.

Proof. Suppose first that $m_1 = \infty$. Then $\phi(\tilde{z}, u) = 0$ for every $\tilde{z}$ in $\Delta_1^{(r-2)}$ and every u such that $|u| < \delta_{r-1}^{1/m}$ (where $\delta = (\delta_1, \ldots, \delta_{r-1})$ is the polyradius of Δ_1). Then, for all $z \in \Delta_1$, $h(z, z_r) = 0$ implies $z_r = 0$. Hence $h(z, z_r) = z_r^m$ (and in fact $m = 1$ by hypothesis). Thus $\operatorname{ord}_{(\tilde{z},0,0)} h = m$ for all $\tilde{z} \in \Delta_1^{(r-2)}$. Henceforth suppose that $m_1 < \infty$.

In order to reduce somewhat the complexity of the notation, we shall prove the desired relation in the special case $r = 3$. A proof of the desired result for the case $r > 3$ can be obtained by generalizing the proof given for $r = 3$ in a straightforward way. We shall use (x, y, z) for (z_1, z_2, z_3): thus x denotes $\tilde{z}$. Let the power series expansion for ϕ about 0 be arranged as follows:

$$c_1(x) u^{m_1} + c_2(x) u^{m_2} + \cdots$$

where $0 < m_1 < m_2 < \cdots$ and the $c_i(x)$ are non-zero power series in x (there is a non-empty finite or infinite sequence of the $c_i(x)$, and each $c_i(x)$ is absolutely convergent in the disc $|x| < \delta_1$). Let K be the quotient field of the power series domain $\mathbb{C}\,[[x]]$. Let ζ be a primitive m-th root of unity. For $1 \le i \le m$, let θ_i denote the element

$$c_1(x)\zeta^{(i-1)m_1}y^{m_1/m} + c_2(x)\zeta^{(i-1)m_2}y^{m_2/m} + \cdots$$

of $K(y)^*$. Regard $h(x, y, z)$ as an element of $K(y)^*[z]$ by replacing each coefficient of h by its power series expansion about the origin (arranged as a power series in y with coefficients in K). Now the power series in x and u (that is, in $K[[u]]$) obtained by expanding

$$h(x, u^m, c_1(x)u^{m_1} + c_2(x)u^{m_2} + \cdots) \tag{11}$$

is the zero power series. Hence, for $1 \le i \le m$, we have $h(x, y, \theta_i) = 0$, by setting $u = \zeta^{(i-1)}y^{1/m}$ in Expression (11). It is not difficult to see that the θ_i, $1 \le i \le m$, are distinct elements of $K(y)^*$. Hence the θ_i, $1 \le i \le m$, constitute the complete set of Puiseux roots of $h(x, y, z)$. Thus

$$h(x, y, z) = \prod_{i=1}^{m}(z - \theta_i).$$

Therefore, where

$$h(x, y, z) = z^m + a_1(x, y)z^{m-1} + \cdots + a_m(x, y),$$

we have

$$
\begin{aligned}
a_1(x, y) &= -(\theta_1 + \cdots + \theta_m), \\
a_2(x, y) &= + \sum_{1 \le i < j \le m} \theta_i\theta_j, \\
&\;\;\vdots \\
a_m(x, y) &= (-1)^m \theta_1 \cdots \theta_m.
\end{aligned}
$$

We have $O(\theta_i) = m_1/m$, for $1 \le i \le m$. Hence $O(a_j) \ge jm_1/m$, for $1 \le j \le m$. This implies that, for $1 \le j \le m$, the order in y of a_j at the origin is greater than or equal to jm_1/m. It follows that, for each point a of the disc $|x| < \delta_1$, the order in y of a_j at $(a, 0)$ is at least jm_1/m. Hence

$$\mathrm{ord}_{(a,0)}a_j \ge jm_1/m. \tag{12}$$

We now consider separately the two cases $m \le m_1$ and $m > m_1$. Suppose first that $m \le m_1$. By Inequality (12), $\mathrm{ord}_{(a,0)}a_j \ge j$. Hence

$$\mathrm{ord}_{(a,0,0)}(a_j(x, y)z^{m-j}) \ge j + m - j = m.$$

Therefore

$$\mathrm{ord}_{(a,0,0)}h = m = \min(m, m_1).$$

Suppose on the other hand that $m > m_1$. Now $O(a_m) = m_1$: indeed a_m contains the term $(-1)^{(m-1)m_1} c_1(x)^m y^{m_1}$. Therefore, by Lemma 5,

$$\operatorname{ord}_{(a,0)}(a_m(x,y)) = m_1.$$

But, for $1 \le j < m$,

$$\begin{aligned}
\operatorname{ord}_{(a,0,0)}(a_j(x,y)z^{m-j}) &\ge jm_1/m + m - j \\
&= m - j(1 - m_1/m) \\
&> m - m(1 - m_1/m) \\
&= m_1.
\end{aligned}$$

Hence

$$\operatorname{ord}_{(a,0,0)} h = m_1 = \min(m, m_1).$$

∎

We are now in a position to present the *Proof of Theorem 4 (Part 2)*. We assume that the first part of the conclusion of Theorem 4 has been established (see remarks about this earlier in this section). It remains to establish part 2 of the conclusion of Theorem 4, namely, that h is order-invariant in the set G^* (which can be thought of as the graph of the root function ψ for h in H^*, near 0).

We first observe that there is no loss of generality in assuming that $h(\tilde{z}, 0, z_r) = z_r^m$ for all $\tilde{z} \in \Delta_1^{(r-2)}$, and equivalently, that $\psi(\tilde{z}) = 0$ for all $\tilde{z} \in \Delta_1^{(r-2)}$. For if this is not true of h then we can consider the Weierstrass polynomial

$$h'(\tilde{z}, z_{r-1}, z_r) = h(\tilde{z}, z_{r-1}, z_r + \psi(\tilde{z}))$$

in Δ_1, for which $h'(\tilde{z}, 0, z_r) = z_r^m$, and replace h by h'.

Let us write h as a product of irreducible Weierstrass polynomials $h_1 \ldots h_n$ (where each h_i has positive degree). Let $1 \le i \le k$ and let $G(z)$ be the discriminant of $h_i(z, z_r)$. Recall that by hypothesis the discriminant $F(z)$ of $h(z, z_r)$ has the form

$$F(z) = z_{r-1}^k N(z)$$

where $N(z)$ is analytic and non-vanishing in Δ_1. Now $F(z) = G(z)H(z)$, for some analytic $H(z)$, by lemma A.1 of (McCallum 1988). Therefore, by the Weierstrass preparation theorem (theorem 62 of chapter 9 of (Kaplan 1966)), there exists a function $U(z)$, analytic and non-vanishing in Δ_1, and an integer $l \ge 0$ such that

$$G(z) = z_{r-1}^l U(z)$$

for all $z \in \Delta_1$ (actually only certain elements of the proof of the Weierstrass preparation theorem are required for this application). Let p be the degree of h_i. By Lemma 3 there exists a parameterization

$$\begin{aligned}
z_{r-1} &= u^p, \\
z_r &= \phi(\tilde{z}, u)
\end{aligned}$$

for the zeros of h_i over Δ_1. Let p_1 be the order of ϕ in u at the origin. It follows from the property $h(\tilde{z}, 0, z_r) = z_r^m$ (all $\tilde{z}$) that p_1 is non-zero. Hence by Lemma 4

$$\text{ord}_{(\tilde{w},0,0)} h_i = \min(p, p_1)$$

for all $\tilde{w} \in \Delta_1^{(r-2)}$. Hence h_i is order-invariant in G^*. Therefore, since $h = h_1 \ldots h_n$, h is order-invariant in G^*, by lemma A.3 of (McCallum 1988).

6   CAD Construction Using Improved Projection

Let A be a finite set of r-variate integral polynomials. An *A-invariant cylindrical algebraic decomposition (cad)* of $\mathbb{R}^r$ partitions $\mathbb{R}^r$ into a finite collection of cylindrically arranged semialgebraic cells in each of which every polynomial in A is sign-invariant. A precise definition of cad is given in section 2 of (Arnon et al. 1984a). The original cad algorithm presented on pp. 151–154 of (Collins 1975), and summarized on p. 875 of (Arnon et al. 1984a), accepts as input a finite set A of r-variate integral polynomials, where $r \geq 1$, and yields as output a description of an A-invariant cad D of $\mathbb{R}^r$. The description of D takes the form of a list of cell indices and sample points for the cells of D (every cell is assigned an index which indicates its position within the cylindrical structure of D, as explained in section 4 of (Arnon et al. 1984a)).

In this section we present cad algorithms which use the improved projection operation. We slightly modify the definition of the improved projection map P. We shall use terminology from (Collins 1975), pp. 145 and 146. Let $r \geq 2$ and let A be a set of r-variate integral polynomials. Suppose first that the elements of A are ample, primitive, irreducible polynomials of positive degree. Then $P(A)$ is defined exactly as in Section 3. Suppose now that A is an arbitrary set of r-variate integral polynomials. Then $P(A)$ is redefined to be the union of the set cont(A) of nonzero nonunit contents of the elements of A and the set $P(B)$, where B is the finest squarefree basis for (that is, the set of all ample irreducible factors of) the set prim(A) of all primitive parts of the elements of A which have positive degree. We further define $P^0(A)$ to be A itself, and $P^k(A)$ to be $P(P^{k-1}(A))$, for $1 \leq k \leq r - 1$.

Remark 3 *Squarefree basis computation is optional in the original cad algorithm (Collins 1975). However squarefree basis computation is essential in the algorithms presented in this section because of one of the hypotheses of Theorem 1, namely, that "A be a finite squarefree basis". We have redefined $P(A)$ using the concept of the finest squarefree basis for a set of primitive polynomials of positive degree, because experience has shown that this likely leads to the most efficient cad algorithm on average.*

6.1   CAD Computation for Well-Oriented Sets of Polynomials

A set A of r-variate integral polynomials, where $r \geq 1$, is said to be *well-oriented* if whenever $r > 1$ the following two conditions hold:

WO1. for every $f \in \operatorname{prim}(A)$, $f(a, x_r) = 0$ for at most a finite number of points $a \in \mathbb{R}^{r-1}$; and

WO2. $P(A)$ is well-oriented.

That is, A is well-oriented if and only if for every k, $0 \leq k < r-1$, and for every $f \in \operatorname{prim}(P^k(A))$, $f(a, x_{r-k}) = 0$ for at most a finite number of points $a = (a_1, \ldots, a_{r-k-1})$ of $\mathbb{R}^{r-k-1}$.

It is not difficult to see that if $1 \leq r \leq 3$ then *every* set of r-variate integral polynomials is well-oriented (lemma A.2 of (McCallum 1988)). Furthermore, if $r > 3$, there is reason to suspect that a finite set of random r-variate integral polynomials $f(x_1, \ldots, x_r)$ such that, for each i, $4 \leq i \leq r$, the degree of f in x_i is at least $i - 2$ is likely to be well-oriented (section 5.1, (McCallum 1984)).

The question arises as to how one determines whether a given set of polynomials is well-oriented or not. When the number of variables does not exceed three, the set is always well-oriented. In the case of four variables one can examine the primitive part of each input polynomial to determine whether its coefficients vanish simultaneously at at most finitely many points in three-space. In general, one must examine not only the input set A, but also $P(A)$, $P^2(A)$, and so on. It is sometimes easy to see whether the coefficients of a polynomial $f(x_1, \ldots, x_r)$ have at most finitely many common zeros (for example, whenever a coefficient is a nonzero constant). Computation of successive resultants of some or all of the pairs of the coefficients (eliminating $x_{r-1}, \ldots, x_2$ in that order) may reveal that the common zeros of the coefficients have only finitely many distinct x_1-coordinates. If it can be determined in a similar way that, for each i, $1 \leq i \leq r-1$, the common zeros have only finitely many distinct x_i-coordinates, then there are only finitely many common zeros. An algorithm which can often determine whether a given set of $(r-1)$-variate integral polynomials has at most finitely many common zeros can be based upon resultant computation as sketched above: we shall denote such an algorithm by IPFZT. (In fact the SAC-2 computer algebra system has an algorithm of this kind and with this name.) There are cases in which IPFZT cannot distinguish the well-oriented from the non well-oriented case. In any event, the algorithm CADW which we shall shortly present provides an infallible test for well-orientedness.

Remark 4 *It is not difficult to show that, for any finite set A of r-variate integral polynomials, one can construct an invertible linear transformation T (in which every entry of the corresponding matrix is an integer) of $\mathbb{R}^r$ such that the transform of A by T is well-oriented. Details can be found in (McCallum 1984).*

Definition 1 *Let $r > 1$, let $f \in \mathbb{Z}[x_1, \ldots, x_r]$ and let $\alpha \in \mathbb{R}^{r-1}$. An element $p \in \mathbb{Z}[x_1, \ldots, x_r]$ is called a* delineating *polynomial for f on α if the following three conditions hold:*

DP1. *p is a partial derivative of f (pure or mixed) of some total order $k \geq 0$;*
DP2. *$p(\alpha, x_r) \neq 0$; and*
DP3. *for every l, $0 \leq l < k$, and for every partial derivative q of f (pure or mixed) of total order l, $q(\alpha, x_r) = 0$.*

Remark 5 *Let f be a nonzero r-variate integral polynomial and let $\alpha \in \mathbb{R}^{r-1}$. Then f has a delineating polynomial on α.*

Definition 2 *Let A be a set of r-variate integral polynomials and let $\alpha \in \mathbb{R}^{r-1}$. A set $\hat{A}$ of r-variate integral polynomials is called a* delineating *set for A on α if, for every $f \in A$, $\hat{A}$ contains a delineating polynomial for f on α , and, for every $p \in \hat{A}$, p is a delineating polynomial on α for some $f \in A$.*

$$\text{CADW}(r, A; w, I, S);$$

Inputs: $r \geq 1$. A is a finite set of r-variate integral polynomials.
Outputs: If A is well-oriented then $w = true$ and I, S are lists of the indices and sample points respectively of the cells comprising an A-invariant cad D of R^r such that every cell in D is a submanifold of $\mathbb{R}^r$ and each polynomial in A is order-invariant in each cell of D; otherwise $w = false$.

(1) *[$r = 1$.]* If $r > 1$ then go to 2. Set $B \leftarrow$ the finest squarefree basis for prim(A). Set $w \leftarrow true$. Set $I \leftarrow$ the empty list. Set $S \leftarrow$ the empty list. Isolate the real roots of the product of the polynomials in B. Construct the indices of the cells of D (as described by (Arnon et al. 1984a)), adding them to I. Construct sample points for the cells of D (as described by (Arnon et al. 1984a)), adding them to S. Exit.

(2) *[$r > 1$.]* Set $B \leftarrow$ the finest squarefree basis for prim(A). Set $P \leftarrow P(A) = \text{cont}(A) \cup P(B)$. Call CADW recursively with inputs $r - 1$ and P to obtain outputs w', I' and S'. If $w' = false$ then set $w \leftarrow false$ and exit. Otherwise I' and S' specify a P-invariant cad D' of $\mathbb{R}^{r-1}$ such that every cell of D' is a submanifold of $\mathbb{R}^{r-1}$ and each polynomial in P is order-invariant in each cell of D'. Set $w \leftarrow true$. Set $I \leftarrow$ the empty list. Set $S \leftarrow$ the empty list. For each cell c of D', let $i \in I'$ be the index of c, let $\alpha \in S'$ be the sample point for c and carry out the following sequence of steps: first, if the dimension of c is positive and there exists $h \in B$ such that $h(\alpha, x_r) = 0$ then set $w \leftarrow false$ and exit; second, if the dimension of c is zero then set $\hat{B} \leftarrow$ a delineating set for B on c else set $\hat{B} \leftarrow B$; third, set $f^*(x_r) \leftarrow \prod_{\hat{h} \in \hat{B}} \hat{h}(\alpha, x_r)$ ($f^*(x_r)$ is constructed using exact arithmetic in $\mathbb{Q}(\alpha)$, (Loos 1982a)); fourth, isolate the real roots of $f^*(x_r)$ (Loos 1982a, section 2); fifth, use i, α and the isolating intervals for the roots of $f^*(x_r)$ to construct cell indices and sample points (as described by (Arnon, Collins,

and McCallum 1984a)) for the sections and sectors over c of the polynomials in $\hat{B}$; sixth, add the new cell indices to I and the new sample points to S. Exit.

It is straightforward to prove the validity of CADW using Theorem 1:

Theorem 5 *Algorithm* CADW *is correct.*

Proof. The proof is by induction on r. The algorithm clearly produces correct output, given valid input, when $r = 1$. Let $r > 1$ and assume that the algorithm always produces correct output when given any finite set of $(r-1)$-variate integral polynomials. Let A be a finite set of r-variate integral polynomials, and suppose that (r, A) is presented as input to CADW. Let B be the finest squarefree basis for prim(A). Let $P = P(A) = \text{cont}(A) \cup P(B)$. Now CADW activates itself recursively with input $(r-1, P)$. Denote the output of this recursive call to CADW by (w', I', S').

Suppose first that A is well-oriented. Then, by *WO2*, P is well-oriented. Hence, by induction hypothesis, $w' = true$ and I', S' are lists of cell indices and sample points respectively for a P-invariant cad D' of $\mathbb{R}^{r-1}$ such that every cell in D' is a submanifold of $\mathbb{R}^{r-1}$ and each element of P is order-invariant in each such cell. Let c be a cell of D' and let $\alpha \in S'$ be the sample point for c. If the dimension of c is zero then let $\hat{B}$ be a delineating set for B on c and otherwise let $\hat{B} = B$.

Suppose that the dimension of c is positive. Then c has an infinite number of points. Hence, by *WO1*, no element of prim(A) vanishes identically on c. Therefore no element of $\hat{B} = B$ vanishes identically on c. Hence the final value of w will be true. Moreover, by Theorem 1, each element of $\hat{B}$ is analytic delineable on c, the sections over c of the elements of $\hat{B}$ are pairwise disjoint and each element of $\hat{B}$ is order-invariant in every such section. Thus the sections and sectors over c of the elements of $\hat{B}$ comprise a stack Σ over c in the sense of section 2 of (Arnon et al. 1984a). Moreover every cell in this stack is a submanifold of $\mathbb{R}^r$ (by remarks made in Section 3) in which every element of B is order-invariant.

Suppose on the other hand that the dimension of c is zero: $c = \{\alpha\}$. The roots of the polynomials $\hat{h}(\alpha, x_r)$, $\hat{h} \in \hat{B}$, determine a stack Σ over c (each $\hat{h}(\alpha, x_r) \neq 0$ by *DP2*). The cells of Σ are naturally submanifolds of $\mathbb{R}^r$. Let σ be a sector of Σ and let $h \in B$. Then $\hat{B}$ contains a delineating polynomial say $\hat{h}$ for h on α. Since σ is a sector of Σ, $\hat{h}(\alpha, \beta) \neq 0$, for all $(\alpha, \beta) \in \sigma$. By *DP1*, $\hat{h}$ is a partial derivative of h (pure or mixed) of some total order say $k \geq 0$. By *DP3*, for every partial derivative q of h of total order l such that $0 \leq l < k$, we have $q(\alpha, x_r) = 0$. It follows that $\text{ord}_{(\alpha, \beta)} h = k$ for all $(\alpha, \beta) \in \sigma$: so h is order-invariant in σ. Hence every element of B is order-invariant in each cell of Σ.

To summarize: we've shown that, whatever the dimension of c, the roots of the polynomials in $\hat{B}$ determine a stack Σ over c comprising submanifolds such that each element of B is order-invariant in every cell of Σ. Now let $f \in A$.

Then $f = \pm gp$, where g is the content of f and p is the primitive part of f. Then g is order-invariant in the whole cylinder over c, and a fortiori in each cell of Σ, since g is order-invariant in c. Now $p = h_1^{e_1} \ldots h_k^{e_k}$, for some $k \geq 0$, some $h_i \in B$ and some $e_i > 0$. Therefore, by lemma A.3 of (McCallum 1988), f is order-invariant in each cell of Σ.

It remains to deal with the case in which A is not well-oriented: suppose that this is instead the case. Suppose that P is not well-oriented. Then, by induction hypothesis, $w' = \textit{false}$. Hence CADW sets w equal to $\textit{false}$. Suppose, on the other hand, that P is well-oriented. Then, by induction hypothesis, $w' = \textit{true}$ and I', S' are lists of cell indices and sample points respectively for a P-invariant cad D' of $\mathbb{R}^{r-1}$, comprising submanifolds order-invariant for P. But, since A is not well-oriented and P is well-oriented, there exists $f \in \operatorname{prim}(A)$ such that $f(\alpha, x_r) = 0$ for an infinite number of points $\alpha \in \mathbb{R}^{r-1}$. Hence there exists $h \in B$ such that $h(\alpha, x_r) = 0$ for an infinite number of points $\alpha \in \mathbb{R}^{r-1}$. Hence there exists a cell $c \in D'$ of positive dimension such that $h(\alpha, x_r) = 0$ for all $\alpha \in c$. Hence CADW will set w equal to $\textit{false}$. ∎

Remark 6 *Let A be a set of four-variate integral polynomials. Suppose that one wishes to construct an A-invariant cad of $\mathbb{R}^4$, but that one does not care whether the cad is order-invariant for A. Then, despite the fact that A might fail to be well-oriented, one can use a suitable modification of CADW to construct an A-invariant cad of $\mathbb{R}^4$.*

Definition 3 *Let f be an r-variate integral polynomial. Let c be a cell in $\mathbb{R}^{r-1}$. Then c is said to be a* nullifying cell *for f if the primitive part of f vanishes identically on c.*

We shall briefly describe a variant of CADW which, given any finite set A of r-variate integral polynomials, whether well-oriented or not, constructs a description of an A-invariant cad of $\mathbb{R}^r$ comprising submanifolds, each order-invariant for A. The strategy of the variant of CADW is as follows. Let $r > 1$ and let A be a set of r-variate integral polynomials. Set B equal to the finest squarefree basis for $\operatorname{prim}(A)$. Set $\bar{B}$ equal to B. Each polynomial $h \in B$ must be examined in turn to try to determine whether its coefficients have at most finitely many common zeros. An algorithm such as IPFZT can be used for this purpose. If it is not determined that the coefficients of h have at most finitely many common zeros (in which case h could have a nullifying cell of positive dimension), then the set $\bar{B}$ must be enlarged by adding to it all of the non-constant partial derivatives of h (pure and mixed) of all positive orders. Next we form the improved projection $P(\bar{B})$ of $\bar{B}$, set P equal to $\operatorname{cont}(A) \cup P(\bar{B})$, and recursively construct a cad D' of $\mathbb{R}^{r-1}$ comprising submanifolds, each order-invariant for P. Let c be a cell of D'. If the dimension of c is zero then let $\hat{B}$ be a delineating set for B on c and otherwise let $\hat{B} = \bar{B}$. Suppose that the dimension of c is positive. By Theorem 1, each element of $\bar{B}$ either vanishes identically on c or is analytic delineable on c, the sections of those elements of

$\bar{B}$ which don't vanish identically on c are pairwise disjoint and each element of $\bar{B}$ which does not vanish identically on c is order-invariant in each such section. It follows that the roots of those elements of $\bar{B}(=\hat{B})$ which don't vanish identically on c determine a stack over c comprising submanifolds in each of which every element of B is order-invariant. Suppose that the dimension of c is zero. Then the roots of the elements of $\hat{B}$ determine a stack over c comprising submanifolds in each of which every element of B is order-invariant. It suffices, then, whatever the dimension of c, to construct cell indices and sample points for the sections and sectors of those elements of $\hat{B}$ which don't vanish identically on c in the usual way. The problem with this kind of approach is that it tends to defeat the purpose of reducing the size of the projection set!

6.2   Quantifier Elimination Using Improved Projection

An important application of cad computation is quantifier elimination for elementary real algebra (Collins 1975). Collins and Hong (1991) introduced the method of *partial cad construction* for quantifier elimination. This method is based on the simple observation that we can often solve a quantifier elimination problem by means of a partially built cad.

We shall describe a variant of the algorithm QEPCAD from (Collins and Hong 1991):

$$\text{QEPCADW}(F^*; w, F')$$

Inputs: F^* is a quantified formula $(Q_{f+1}x_{f+1})\ldots(Q_r x_r)F(x_1,\ldots,x_r)$, where $0 \le f < r$ and F is a quantifier-free formula.
Outputs: If the set of polynomials occurring in F^* is well-oriented then $w = true$ and $F' = F'(x_1,\ldots,x_f)$ is a quantifier-free formula equivalent to F^*; otherwise $w = false$.

We shall indicate the modifications to QEPCAD presented on p. 309 of (Collins and Hong 1991) that are required to obtain QEPCADW. Step 1 of QEPCAD is the projection phase of the algorithm. In this phase QEPCADW should use the improved projection operation P (described at the beginning of Section 6). Step 5 of QEPCAD a key part of the stack construction phase, prescribes the construction of the stack over a chosen cell c. Algorithm CCHILD actually accomplishes this construction. Algorithm CCHILD must be modified so that, in case c has positive dimension, it will detect whether some basis polynomial vanishes identically on c and, if so, set w equal to *false*. It is further necessary to incorporate the use of delineating sets into CCHILD, along the lines in which delineating sets are used in CADW (Section 6.1).

7   Examples

The most recent implementation of cad-based quantifier elimination is due to Hong (1990b). The program was written using the ALDES programming language and the SAC-2 computer algebra system. Collins and Encarnación have since incorporated several new features into the program. It is possible to select the options for running the program in such a way that the program is essentially equivalent to an implementation of a version of the full cad algorithm CAD (Arnon et al. 1984a) which uses an improved projection operation, known to be always valid, due to Hong (1990a). Also, it is possible to select the options for running the program so that the program is essentially equivalent to an implementation of the full cad algorithm CADW described in Section 6.1. (Actually, this is not yet completely true, although it is close to the truth; it is anticipated that the statement will be completely true in the very near future.) For the rest of this section, for simplicity, we shall denote these modes of operation of Hong's program by CAD and CADW, respectively.

Furthermore, two sets of options for running Hong's program can be selected so that, for the first set of options, the program is essentially equivalent to a version of algorithm QEPCAD (Collins and Hong 1991) which uses Hong's improved projection operation (Hong 1990a), and for the second set of options, the program is essentially equivalent to algorithm QEPCADW sketched in Section 6.2. (Again, this is not yet completely true, although it is close to the truth; it is anticipated that the statement will be completely true in the very near future.) For the rest of this section, for simplicity, we shall denote these two modes of operation of Hong's program by QEPCAD and QEPCADW, respectively.

As noted by Collins (1975), McCallum (1988) and Hong (1990b) there are often elementary improvements that can be made to the projection sets. For example, whenever certain subsets of a projection set can be shown to have at most a finite number of common real zeros, then certain polynomials can be omitted from the projection set. Such elementary improvements have been incorporated into the program modes discussed in this section.

We applied the program modes CAD, CADW, QEPCAD and QEPCADW to three different problems. The results of the experiments are discussed in the three subsections which follow. Computing times were measured on a DEC-station 5000/200 with an R3000 risc-processor. Four megabytes of memory were made available for each experiment.

7.1   Catastrophe Surface and Sphere

Two well-known surfaces are the unit sphere centered at the origin

$$f(x, y, z) = z^2 + y^2 + x^2 - 1 = 0$$

and the catastrophe surface

$$g(x, y, z) = z^3 + xz + y = 0.$$

Let A be the set $\{f, g\}$. Several authors (McCallum 1988; Arnon 1988b; Collins 1997) have reported on the application of one or more variants of the cad algorithm to the set A.

We first report on the application of the full cad algorithms CAD and CADW to the set A. A sample comparison is provided in Table 1. In Table 1 the characteristic "total no. of projection polys" refers to the total number of projection polynomials, univariate and bivariate, computed by the algorithm; the characteristic "total no. of projection factors" refers to the total number of irreducible projection factors (that is, basis polynomials), univariate, bivariate and trivariate, computed by the algorithm. A diagram illustrating the $P(A)$-invariant cad of the plane constructed by CADW is given on p. 156 of (McCallum 1988).

Table 1: Comparison of CAD and CADW for the catastrophe surface and sphere

Characteristic	CAD	CADW
total no. of projection polys	15	9
total no. of projection factors	16	14
no. of cells in $\mathbb{R}^3$	1393	971
total no. of cells	1661	1159
total time (seconds)	27.4	16.2

We now consider the following quantifier elimination problem:

$$(\exists x)(\exists y)(\exists z)[x^2 + y^2 + z^2 - 1 = 0 \wedge z^3 + xz + y = 0].$$

We applied the program modes QEPCAD and QEPCADW to this problem. A sample comparison between the algorithms is presented in Table 2.

Table 2: Comparison of QEPCAD and QEPCADW for the decision problem

Characteristic	QEPCAD	QEPCADW
no. of cells in $\mathbb{R}^3$ computed	30	30
total no. of cells computed	79	73
time for projection phase (ms)	1183	1050
time for stack constr. phase (ms)	167	167
total time (ms)	1400	1250

Comparing Table 1 with Table 2 we see that there is much time to be saved by using a partial rather than a full cad for this quantifier elimination problem.

7.2 Quartic Polynomial

We consider the following quantifier elimination problem:

$$(\forall x)[x^4 + px^2 + qx + r \geq 0].$$

That is, we seek necessary and sufficient algebraic conditions on p, q, r such that the quartic polynomial $x^4 + px^2 + qx + r$ is positive semidefinite. This problem has been studied by Arnon and Mignotte (1988).

We applied program modes QEPCAD and QEPCADW to this problem. We also applied program modes CAD and CADW to the set A consisting of the quartic polynomial alone. A sample comparison is provided in Table 3. Times are reported in milliseconds.

Table 3: Comparison of algorithms for quartic problem

Characteristic	CAD	CADW	QEPCAD	QEPCADW
total no. proj. polys	17	4	17	4
total no. proj. factors	8	5	8	5
no. cells in $\mathbb{R}^4$	489	213	489	213
total no. cells	636	284	636	284
time proj. phase	1383	83	1350	100
time stack constr. phase	6799	1467	6534	1416
time formula phase	n/a	n/a	233	17
total time	8816	1567	8800	1550

Program mode QEPCADW was unable to construct an equivalent quantifier-free formula, though it did provide separate necessary and sufficient conditions for positive semidefiniteness. Solution formula construction for this problem is discussed by Collins (1997). The solution formula constructed by QEPCAD was:

$$[[\delta \geq 0 \wedge L < 0] \vee$$

$$[9q^2 + 2p^3 \leq 0 \wedge \delta \geq 0 \wedge L \leq 0] \vee$$

$$[27q^2 + 8p^3 > 0 \wedge \delta \geq 0]]$$

where

$$\delta = 256r^3 - 128p^2r^2 + 144pq^2r +$$

$$16p^4r - 27q^4 - 4p^3q^2$$

and $L = 8pr - 9q^2 - 2p^3$.

7.3 Solotareff's Problem

We consider a special case of the Solotareff approximation problem (Achieser 1956). This problem is discussed by Collins (1997). The input formula is

$$(\exists b)(\exists u)(\exists v)[[r^2 - 24r + 16 < 0 \vee$$

$$[r > 1 \wedge r^2 - 24r + 16 \geq 0]]$$

$$\wedge -1 < u \wedge u < v \wedge v < 1 \wedge$$

$$u^4 + ru^3 - au^2 - bu - (1-a) = r - b \wedge$$

$$v^4 + rv^3 - av^2 - bv - (1-a) = -r + b \wedge$$

$$4u^3 + 3ru^2 - 2au - b = 0 \wedge$$

$$4v^3 + 3rv^2 - 2av - b = 0 \wedge r - b > 0].$$

We applied program modes QEPCAD and QEPCADW to this problem. The computations were carried out as far as the construction of the cad of $\mathbb{R}^1$. Table 4 provides a sample comparison between the two algorithms for this problem.

Table 4: Comparison of QEPCAD and QEPCADW for Solotareff's problem

Characteristic	QEPCAD	QEPCADW
total no. of projection polys	529	204
total no. of projection factors	361	140
no. of cells in $\mathbb{R}^1$	923	215
time for projection phase	33066	6183
time for cad of $\mathbb{R}^1$	7234	867

We also applied both QEPCAD and QEPCADW to the formula obtained by setting r equal to 2 in the above input formula. The programs were able to solve this special case of the Solotareff problem. A sample comparison is provided in Table 5. The programs constructed the same solution formula, namely

$$81a^3 - 180a^2 + 448a - 432 = 0.$$

Table 5: Comparison of QEPCAD and QEPCADW for Solotareff's problem with $r = 2$

Characteristic	QEPCAD	QEPCADW
total no. of projection polys	58	39
total no. of projection factors	40	32
no. of cells in $\mathbb{R}^4$	439	317
total no. of cells	6154	4032
time for projection phase	1300	367
time for stack constr. phase	38815	23534
time solution formula phase	50	34
total time	45967	27217

Consideration of examples such as these suggests that, while the improved projection yields only a rather modest improvement in the observed computing times for small examples, the benefit to be gained from using the improved projection increases significantly with increasing problem size.

8 Appendix

We present a result used in establishing Lemma 4.

Lemma 5 *Let $h(x, y, z)$ be a Weierstrass polynomial of degree $m \geq 1$ in the polydisc Δ about 0 in $\mathbb{C}^2$. Let (δ_1, δ_2) be the polyradius of Δ. Let $F(x, y)$ be the discriminant of $h(x, y, z)$. Assume that h and F satisfy the hypotheses of Lemma 3. Let $c_1(x), c_2(x), \ldots$ be a non-empty finite or infinite sequence of non-zero elements of the power series domain $\mathbb{C}[[x]]$, and suppose that each $c_i(x)$ is absolutely convergent in the disc $|x| < \delta_1$. Let ζ be a primitive m-th root of unity. Let K be the quotient field of $\mathbb{C}[[x]]$. Suppose that the elements*

$$\theta_i(x) = c_1(x)\zeta^{(i-1)m_1}y^{m_1/m} + c_2(x)\zeta^{(i-1)m_2}y^{m_2/m} + \cdots$$

of $K(y)^*$, $1 \leq i \leq m$, where $0 < m_1 < m_2 < \cdots$, are distinct roots of $h(x, y, z)$, considered as a polynomial in z over $K(y)^*$. Suppose that $m > m_1$. Then $c_1(a) \neq 0$ for every a in the disc $|x| < \delta_1$.

Proof. Let $|a| < \delta_1$. Let $g_a(y, z)$ be the Weierstrass polynomial $h(a, y, z)$ in the disc $|y| < \delta_2$. Regard g_a as a polynomial in z over the field $\mathbb{C}(y)^*$ by replacing each coefficient of g_a by its power series expansion about 0. Then, by hypothesis, the elements $\theta_i(a)$ of $\mathbb{C}(y)^*$, $1 \leq i \leq m$, constitute the complete set of Puiseux roots of $g_a(y, z)$. For $1 \leq i < j \leq m$, let $\eta_{i,j}(a) = \theta_i(a) - \theta_j(a)$: then

$$\eta_{i,j}(a) = (\zeta^{(i-1)m_1} - \zeta^{(j-1)m_1})c_1(a)y^{m_1/m} + (\zeta^{(i-1)m_2} - \zeta^{(j-1)m_2})c_2(a)y^{m_2/m} + \cdots$$

Let $k_{i,j}$ be the smallest positive integer k such that

$$\zeta^{(i-1)m_k} - \zeta^{(j-1)m_k} \neq 0.$$

Then

$$O(\eta_{i,j}(a)) \geq m_{k_{i,j}}/m. \tag{13}$$

We shall show that in fact we have equality here. Let $D_a(y) = F(a, y)$. By hypothesis

$$F(a, y) = y^k N(a, y)$$

for all y with $|y| < \delta_2$. Hence

$$\mathrm{ord}_0 D_a = k. \tag{14}$$

Now, regarding D_a as an element of $\mathbb{C}(y)^*$, we have

$$D_a = \prod_{1 \leq i < j \leq m} (\theta_i(a) - \theta_j(a))^2 = \prod_{1 \leq i < j \leq m} \eta_{i,j}(a)^2.$$

Hence, by Eq. (14), we have

$$k = 2 \sum_{1 \leq i < j \leq m} O(\eta_{i,j}(a)). \tag{15}$$

Let $c(a)$ be the product of the $c_{k_{i,j}}(a)$, taken over all i, j, with $1 \leq i < j \leq m$. Then the analytic function c so defined in the disc $|x| < \delta_1$ is not identically zero, by the identity theorem (theorem I-6, (Gunning and Rossi 1965)). Therefore there exists a point a' in the disc $|x| < \delta_1$ such that $c(a') \neq 0$, hence such that

$$O(\eta_{i,j}(a')) = m_{k_{i,j}}/m \tag{16}$$

for all i, j with $1 \leq i < j \leq m$. Now Eq. (15) holds with a replaced by a'. Hence, by inequality (13) and Eqs. (15) and (16),

$$O(\eta_{i,j}(a)) = m_{k_{i,j}}/m \tag{17}$$

for all i, j, with $1 \leq i < j \leq m$.

The proof of the lemma may now be completed as follows. We have $\zeta^{m_1} \neq 1$, since $0 < m_1 < m$ and ζ is a primitive m-th root of unity. Thus $\zeta^{0.m_1} - \zeta^{1.m_1} \neq 0$, and hence $k_{1,2} = 1$. Therefore, by Eq. (17),

$$O(\eta_{1,2}(a)) = m_1/m.$$

Hence $c_1(a) \neq 0$. ∎

Algorithms for Polynomial Real Root Isolation

J. R. Johnson

1 Introduction

This paper summarizes results obtained in the author's Ph.D. thesis (Johnson 1991). Improved maximum computing time bounds are obtained for isolating the real roots of an integral polynomial. In addition to the theoretical results a systematic study was initiated comparing algorithms based on Sturm sequences, the derivative sequence, and Descartes' rule of signs. The algorithm with the best theoretical computing time bound is the coefficient sign variation method, an algorithm based on Descartes' rule of signs. Moreover, the coefficient sign variation method typically outperforms the other algorithms in practice; however, we exhibit classes of input polynomials for which each algorithm is superior.

Section 2 reviews the mathematical foundation of the root isolation algorithms we investigate. Section 3 outlines three different types of root isolation algorithms: one which uses Sturm sequences, one which uses the derivative sequence and Rolle's theorem, and one based on a sequence of polynomial transformations and Descartes' rule of signs. These algorithms have already appeared in the literature; however, several variations were investigated and some improvements were incorporated. Section 4 presents the best known computing time bounds for these algorithms, and Section 5 reports on an empirical investigation of the performance of the algorithms for different classes of input polynomials.

2 Preliminary Mathematics

2.1 Root Counting Theorems

In this section we present some theorems for determining the number of real roots in an interval. Each of these theorems serves as the basis of a different real root isolation algorithm. Proofs of well-known theorems are not given, and can be found in (Johnson 1991) or standard texts on the Theory of Equations such as (Uspensky 1948) or (Burnside and Panton 1899).

The first theorem, due to Sturm, gives the exact number of real roots in an interval and it is the basis of the first algorithm. Unfortunately the application of this theorem is costly, so alternative, less precise, theorems are sought.

The second theorem, due to Budan and Fourier, gives a upper bound for the number of real roots in an interval, and its application is less costly than Sturm's theorem. Although it may not give the exact answer, this deficiency can be remedied in two ways. The first uses an immediate corollary of the theorem of Budan and Fourier called Descartes' rule of signs. The key to its successful application is the use of polynomial transformations which move the roots apart. When the roots have been sufficiently separated Descartes' rule will yield the exact result. The second way of removing the shortcomings of the theorem of Budan and Fourier is to ensure that there are no roots of any of the derivatives in the interval in question. If this is the case, the application of the theorem of Budan and Fourier reduces to the intermediate value theorem and Rolle's theorem, and leads to an inductive algorithm which obtains isolating intervals for a polynomial from isolating intervals of its derivative.

Sturm's theorem relies on a sequence of polynomial called a Sturm sequence.

Definition 1 (Sturm Sequence) *A sequence of polynomials $A_1(x) = A(x)$, $A_2(x), \ldots, A_r(x)$ is a Sturm sequence for the interval $[a, b]$ if the following properties hold.*

1. *If $\alpha \in [a, b]$ and $A(\alpha) = 0$, then there exists an $\epsilon > 0$ such that $A_1(x)A_2(x) < 0$ for $x \in (\alpha - \epsilon, \alpha)$ and $A_1(x)A_2(x) > 0$ for $x \in (\alpha, \alpha + \epsilon)$.*
2. *If $\alpha \in [a, b]$ and $A_k(\alpha) = 0$ for some $0 < k < r$, then $A_{k-1}(\alpha)A_{k+1}(\alpha) < 0$.*
3. *For all $x \in [a, b]$, $A_r(x) \neq 0$.*

Theorem 1 (Sturm's Theorem) *Let $A(x)$ be a polynomial with real coefficients. Let $A(x) = A_1(x), A_2(x), \ldots, A_r(x)$ be a Sturm sequence for the interval $[a, b]$. Let V_x be the number of sign variations in the Sturm sequence evaluated at x. Then $V_a - V_b$ is equal to the number of distinct roots of $A(x)$ in the interval $(a, b]$.*

A negative polynomial remainder sequence (PRS) can be used to construct a Sturm sequence.

Definition 2 (Negative PRS) *A negative PRS is a PRS $A_1, A_2, \ldots, A_r$, $A_{r+1} = 0$ defined by $e_i A_i(x) = Q_i(x)A_{i+1}(x) + f_i A_{i+2}(x)$, with $e_i f_i < 0$.*

Theorem 2 *If $A(x)$ is a squarefree polynomial with real coefficients, then a negative PRS for $A(x)$ and $A'(x)$ is a Sturm sequence for any interval.*

The following lemma, from (Johnson 1991) shows how to convert any PRS into a negative PRS. This lemma allows the computation of the PRS to be done independently of the sign corrections necessary to apply Sturm's theorem. In particular, it allows the use of modular algorithms based on subresultants.

Lemma 1 *Let $\{A_1(x), A_2(x), \ldots, A_r(x), A_{r+1}(x) = 0\}$ be a PRS defined by $e_i A_i(x) = Q_i(x) A_{i+1}(x) + f_i A_{i+2}(x)$. If $\bar{A}_i(x) = s_i A_i(x)$, where $s_1 = 1$, $s_2 = 1$, and $s_{i+2} = -\text{sign}(e_i)\text{sign}(f_i)\text{sign}(s_i)$ for $i = 1, \ldots, r-1$, then $\{\bar{A}_1(x), \bar{A}_2(x), \ldots, \bar{A}_r(x), \bar{A}_{r+1}(x) = 0\}$ is a negative PRS for $A_1(x)$ and $A_2(x)$.*

A theorem due to Budan and Fourier gives a similar but weaker test than Sturm's theorem. The theorem of Budan and Fourier gives a upper bound on the number of roots in an interval. This theorem can be stated in two equivalent ways. The first is due to Fourier and the second is due to Budan. The following theorem is a slightly stronger version of Fourier's theorem and is better suited to our algorithmic purposes. The stronger version holds for left-open and right-closed intervals and in some cases does not require the entire derivative sequence.

Theorem 3 (Fourier's Theorem) *Let $A(x)$ be a polynomial with real coefficients. Assume that for all $x \in [a, b]$ the k-th derivative, $A^{(k)}(x) \neq 0$. Let V_x be the number of sign variations in the partial derivative sequence $A(x), A^{(1)}(x), \ldots, A^{(k)}(x)$. Then $V_a - V_b = r + 2h$ where $h \geq 0$ and r is the number of real roots, multiplicities counted, in the interval $(a, b]$.*

Let $\text{var}(A(x))$ denote the number of sign variations in the non-zero coefficients of $A(x)$.

Corollary 1 (Budan) *Let $A(x)$ be a polynomial with real coefficients. Let $a < b$ be real numbers. Then $\text{var}(A(x + a)) - \text{var}(A(x + b)) = r + 2h$, where $h \geq 0$ and r is equal to the number of roots of $A(x)$ in $(a, b]$.*

An important corollary of the theorem of Budan and Fourier is Descartes' rule of signs, which relates the number of positive roots to the number of coefficient sign variations.

Theorem 4 (Descartes' Rule of Signs) *Let $A(x)$ be a polynomial with real coefficients. Then $\text{var}(A(x)) = r + 2h$, where $h \geq 0$ and r is the number of positive roots of $A(x)$.*

We highlight two special cases which give exact information.

Corollary 2 *If $\text{var}(A(x)) = 0$ then $A(x)$ has no positive real roots.*

Corollary 3 *If $\text{var}(A(x)) = 1$ then $A(x)$ has exactly one positive real root.*

The following theorem gives a partial converse of Corollary 2.

Theorem 5 *Let $A(x)$ be a polynomial with real coefficients. If $A(x)$ does not have any roots with positive real parts then $\text{var}(A(x)) = 0$.*

The converse of Corollary 3 is not true. For example, $A(x) = (x - 1)(x - i)(x + i) = x^3 - x^2 + x - 1$ has one positive real root and three variations. However, the converse is true if the remaining roots are far enough away. The following theorem is due to Vincent. Corollary 4 is due to Collins and Akritas (1976).

Theorem 6 (Vincent) *Let $A(x)$ be a squarefree polynomial with real coefficients of degree m. If $A(x)$ has exactly one positive real root and all others are inside a circle of radius*

$$e_m = (1 + \frac{1}{m})^{1/(m-1)} - 1$$

centered at -1 then $\mathrm{var}(A(x)) = 1$.

Proof. See (Collins and Loos 1982b) or (Uspensky 1948). ∎

Corollary 4 *Let $A(x)$ be a squarefree polynomial with real coefficients of degree m. If $A(x)$ has a single real root in the interval $(0, 1)$ and the remaining roots are outside a circle about the origin of radius m^2, then* $\mathrm{var}((x + 1)^m A(1/(x + 1))) = 1$.

Proof. See (Collins and Loos 1982b). ∎

The following theorem, due to Collins and Johnson, strengthens Corollary 4 so that the remaining roots only need to be outside a circle of constant radius.

Theorem 7 *Let $A(x)$ be a squarefree polynomial with real coefficients of degree m. If $A(x)$ has a single root in the interval $(0, 1)$ and all of the remaining real and complex roots are outside the circles of radius 1 centered at $(0, 0)$ and $(1, 0)$, then* $\mathrm{var}((x + 1)^m A(1/(x + 1))) = 1$.

Proof. See (Collins and Johnson 1989b). ∎

Another way of getting exact information from the theorem of Budan and Fourier is to apply the theorem to intervals containing no roots of the derivative. In particular, if for all $x \in [a, b]$ $A'(x) \neq 0$ then $A(x)$ has a root in (a, b) if and only if $A(a)$ and $A(b)$ have opposite signs. This leads to the following theorem.

Theorem 8 (Root Counting Based on Rolle's Theorem) *Let $A(x)$ be a squarefree polynomial with real coefficients. Let $\alpha_1, \ldots, \alpha_r$ be the real roots of $A'(x)$. Then the number of sign variations in $A(-\infty), A(\alpha_1), \ldots, A(\alpha_r), A(\infty)$ is equal to the number of real roots of $A(x)$.*

This theorem can be applied inductively to the derivative sequence of a polynomial. The root isolation algorithms in Section 3 use this inductive process and Theorem 8 to recursively obtain isolating intervals for a polynomial from isolating intervals of its derivative.

2.2 Polynomial Root Bounds

The root isolation algorithms occurring in this paper begin by computing a root bound. There are many theorems providing a bound on the roots of a polynomial. The bound we use is due to Knuth (1969, section 4.6.2, exercise 20). Knuth's bound is modified so that a bound is obtained for the positive roots of a polynomial.

Theorem 9 (Positive Root Bound) *Let $A(x) = \sum_{i=0}^{m} a_i x^i$ be a polynomial with real coefficients and $a_m > 0$. If*

$$B = \max_{a_{m-k} < 0, \ 1 \leq k \leq m} (a_{m-k}/a_m)^{1/k},$$

then $2B$ is a root bound for $A(x)$.

Proof. Assume $A(\alpha) = 0$ and $\alpha > B$. Then

$$\alpha^m \leq (1/a_m) \sum_{a_i < 0, \ 0 \leq i < m} a_i \alpha^i, \tag{1}$$

which implies

$$1 \leq (1/a_m) \sum_{a_i < 0, \ 0 \leq i < m} a_i \alpha^{i-m}. \tag{2}$$

Setting $k = m - i$ we obtain

$$1 \leq (1/a_m) \sum_{a_{m-k} < 0, \ 1 \leq k \leq m} \frac{a_{m-k}}{\alpha^k} < \sum_{k=1}^{\infty} \left(\frac{B}{\alpha}\right)^k, \tag{3}$$

which implies that

$$1 < \frac{B}{\alpha - B},$$

which proves the theorem. ∎

The algorithm IUPPRB (Integral Univariate Polynomial Positive Root Bound) uses this theorem to compute a bound on the positive roots of a polynomial. A bound on the negative roots of $A(x)$ can be obtained by applying IUPPRB to $A(-x)$. For many polynomials better bounds can be obtained by combining positive and negative root bounds instead of directly applying Knuth's result to obtain a bound on all of the real roots.

2.3 Minimum Root Separation

The following theorem gives a bound on the product of the distances between certain roots of a polynomial. This is a generalization of a theorem due to Davenport (1985), which in turn is a generalization of a theorem due to Mahler

(1964). The improved root separation theorems are crucial to obtaining improved computing time bounds for real root isolation algorithms.

Let $A(x) = \sum_{i=0}^{m} a_i x^i = a_m \prod_{i=1}^{m}(x - \alpha_i)$. The measure of $A(x)$, $M(A(x))$, is equal to $|a_m| \prod_{i=1}^{m} \max(1, |\alpha_i|)$. Landau's inequality states that $M(A(x)) \leq |A(x)|_2$, where $|A(x)|_2$ is the Euclidean norm of the coefficients of $A(x)$. The discriminant of $A(x)$, $\operatorname{disc}(A(x))$ is defined to be $a_m^{2m-2} \prod_{i<j}(\alpha_i - \alpha_j)$.

Theorem 10 (Davenport) *Let $A(x) = \sum_{i=0}^{m} a_i x^i = a_m \prod_{i=1}^{m}(x - \alpha_i)$ be a polynomial with complex coefficients. Let $\beta_1, \ldots, \beta_k$ be a subset of the roots of $A(x)$ with $|\beta_i| \leq |\alpha_i|$ and $\beta_i \notin \{\alpha_1, \ldots, \alpha_i\}$. Let $D = |\operatorname{disc}(A(x))|$ and $M = M(A(x))$ the measure of $A(x)$. Then*

$$\prod_{i=1}^{k} |\alpha_i - \beta_i| \geq 3^{k/2} D^{1/2} M^{-m+1} m^{-k-m/2}.$$

Proof. We can assume $A(x)$ is squarefree since if $A(x)$ is not squarefree its discriminant is zero and the theorem is trivially true. Therefore $\alpha_1, \ldots, \alpha_m$ are distinct. Let V be the Vandermonde matrix $V(\alpha_1, \ldots, \alpha_m)$. Using the well known formula for the determinant of a Vandermonde matrix (see Uspensky 1948), $D^{1/2} = |a_m|^{m-1} |\det(V)|$. For $i = 1, 2, \ldots, k$, subtract row $j(i)$ from row i, where $\beta_i = \alpha_{j(i)}$, obtaining the matrix

$$V' = \begin{pmatrix} \alpha_1^{m-1} - \beta_1^{m-1} & \cdots & \alpha_1 - \beta_1 & 0 \\ \vdots & \ddots & \vdots & \vdots \\ \alpha_k^{m-1} - \beta_k^{m-1} & \cdots & \alpha_k - \beta_k & 0 \\ \alpha_{k+1}^{m-1} & \cdots & & \alpha_k & 1 \\ \vdots & \ddots & \vdots & \vdots \\ \alpha_m^{m-1} & \cdots & & \alpha_m & 1 \end{pmatrix}.$$

Since $\alpha_i^{h+1} - \beta_i^{h+1} = (\alpha_i - \beta_i) \sum_{j=0}^{h} \alpha_i^{h-j} \beta_i^j$,

$$D^{1/2}/|a_m|^{m-1} = \left\{ \prod_{i=1}^{k} |\alpha_i - \beta_i| \right\} |\det(V'')|, \tag{4}$$

where

$$V'' = \begin{pmatrix} \sum_{j=0}^{m-2} \alpha_1^{m-2-j} \beta_1^j & \cdots & 1 & 0 \\ \vdots & \ddots & \vdots & \vdots \\ \sum_{j=0}^{m-2} \alpha_k^{m-2-j} \beta_k^j & \cdots & 1 & 0 \\ \alpha_{k+1}^{m-1} & \cdots & \alpha_k & 1 \\ \vdots & \ddots & \vdots & \vdots \\ \alpha_m^{m-1} & \cdots & \alpha_m & 1 \end{pmatrix}.$$

The desired inequality is obtained by applying Hadamard's inequality (see Minc and Marcus 1964) to $|\det(V'')|$. To apply Hadamard's inequality we need a bound on the Euclidean norms of the rows of V''. Since $|\beta_i| \leq |\alpha_i|$

$$\left| \sum_{j=0}^{h} \alpha_i^{h-j} \beta_i^j \right| \leq \sum_{j=0}^{h} |\alpha_i^{h-j} \beta_i^j|$$

$$\leq \sum_{j=0}^{h} |\alpha_i|^h = (h+1)|\alpha_i|^h,$$

and for $1 \leq i \leq k$,

$$|V_i''|_2 \leq \max(1, |\alpha_i|)^{m-2} \left(\sum_{h=1}^{m-1} h^2 \right)^{1/2} \leq (m^3/3)^{1/2} \max(1, |\alpha_i|)^{m-1},$$

where V_i'' is the i-th row of V''. For $i > k$, $|V_i''|_2 \leq m^{1/2} \max(1, |\alpha_i|)^{m-1}$. Hadamard's inequality then implies

$$|\det(V'')| \leq \prod_{i=1}^{k} \left(m^{3/2}/\sqrt{3} \right) \max(1, |\alpha_i|)^{m-1} \prod_{i=k+1}^{m} m^{1/2} \max(1, |\alpha_i|)^{m-1}$$

$$= 3^{-k/2} m^{m/2+k} (M/|a_m|)^{m-1}.$$

Inserting this into Eq. (4) completes the proof. ∎

Corollary 5 (Integral Polynomials) *Assume the same hypotheses as Theorem 10. Furthermore assume that $A(x)$ has integral coefficients, and let $d = |A(x)|_2$. Then*

$$\prod_{i=1}^{k} |\alpha_i - \beta_i| \geq 3^{k/2} d^{-m+1} m^{-k-m/2}.$$

Proof. If $A(x)$ is a squarefree integral polynomial, its discriminant is a non-zero integer, so $D^{1/2} \geq 1$. The proof is completed since $M \leq |A(x)|_2$ by Landau's inequality. ∎

Corollary 6 (Mahler's Root Separation Theorem) *Let $A(x)$ be a square-free polynomial with complex coefficients. Then*

$$\text{sep}(A(x)) \geq \sqrt{3} m^{-1-m/2} D^{1/2} d^{-m+1}.$$

If $A(x)$ is an integral polynomial, then

$$\text{sep}(A(x)) \geq \sqrt{3} m^{-1-m/2} d^{-m+1}.$$

Proof. Apply Theorem 10 and Corollary 5 with $k = 1$ and α_1 and β_1, with $|\alpha_1| \leq |\beta_1|$, the two closest roots ∎

Corollary 7 (Product of Real Roots) *Let* $\gamma_1, \ldots, \gamma_{k+1}$ *be any of the real roots of* $A(x)$ *with* $\gamma_1 < \gamma_2 < \cdots < \gamma_{k+1}$. *Then*

$$\prod_{i=1}^{k} |\gamma_i - \gamma_{i+1}| \geq 3^{k/2} D^{1/2} d^{-m+1} m^{-k-m/2}.$$

Proof. Assume $\gamma_1 < \cdots < \gamma_j < 0 < \gamma_{j+1} < \cdots < \gamma_k$. We can further assume that $|\gamma_{j+1}| \geq |\gamma_j|$ since we can multiply all of the roots by -1 if this assumption is not true. Let $\alpha_i = \gamma_i$ and $\beta_i = \gamma_{i+1}$ for $1 \leq i < j$. Let $\alpha_i = \gamma_{k+j-i}$ and $\beta_i = \gamma_{k+j-i-1}$ for $j \leq i < k$ and apply Theorem 10. The theorem is proved since $\prod_{i=1}^{k} |\alpha_i - \beta_i| = \prod_{i=1}^{k} |\gamma_i - \gamma_{i+1}|$. ∎

Corollary 8 (Product of Complex Conjugates) *Let* $\gamma_1, \gamma_2, \ldots, \gamma_{2k-1}, \gamma_{2k}$ *be* k *pairs of complex conjugate roots, with* $\gamma_{2i} = \overline{\gamma_{2i-1}}$, *the complex conjugate of* γ_{2i-1}. *Then*

$$\prod_{i=1}^{k} |\gamma_{2i-1} - \gamma_{2i}| \geq 3^{k/2} D^{1/2} d^{-m+1} m^{-k-m/2}.$$

Proof. Let $\alpha_i = \gamma_{2i-1}$ and $\beta_i = \gamma_{2i}$ for $1 \leq i \leq k$ and apply Theorem 10. ∎

3 Algorithms

3.1 Sturm Sequence Algorithm

Sturm's theorem suggests an algorithm for isolating the real roots of a polynomial. First a Sturm sequence is computed (this can easily be done since any negative PRS for a polynomial and its derivative is a Sturm sequence). After a Sturm sequence is computed Sturm's theorem is applied to an initial interval which contains all of the real roots. If no roots are detected the algorithm is done. If a single root is detected, an isolating interval is obtained. If more than one root is in the interval, the interval is bisected and the algorithm is recursively applied to the left and right subintervals. Algorithms based on this approach were first implemented and analyzed by Heindel (1970) in his Ph.D. thesis.

Heindel used the primitive PRS, modified to compute a negative PRS, as a Sturm sequence. The algorithm IPPNPRS (Integral polynomial primitive negative Polynomial Remainder Sequence) computes the negative primitive PRS of two polynomials. Alternatively, the subresultant PRS (Brown 1978b), in conjunction with Lemma 1, can be used to compute a Sturm sequence. Moreover, the chain of subresultants can be computed with a modular algorithm; from which those belonging to the subresultant PRS can be extracted.

Algorithms IPRIST and IPIISS follow the outline above; however, IPRIST isolates the positive and negative roots separately, making use of a positive and negative root bound.

$$L \leftarrow \text{IPRIST}(A(x))$$

Algorithm IPRIST (Integral polynomial real root isolation using Sturm sequences).
Input: $A(x)$ is a squarefree integral polynomial.
Output: $L = (I_1, \ldots, I_r)$ is a list of standard isolating intervals for all of the real roots of $A(x)$.

(1) [Compute Sturm sequence.] $\bar{A}(x) \leftarrow \text{pp}(A(x))$; $\bar{A}'(x) \leftarrow \text{pp}(A'(x))$; $S \leftarrow \text{IPPNPRS}(\bar{A}, \bar{A}')$.
(2) [Isolate positive roots.] $b_2 \leftarrow \text{IUPPRB}(\bar{A})$; $v_2 \leftarrow \text{var}(\text{IPLEV}(S, b_2))$; $v_0 \leftarrow \text{var}(\text{IPLEV}(S, 0))$; $L_2 \leftarrow \text{IPIISS}(S, 0, b_2, v_0, v_2)$.
(3) [Isolate non-positive roots.] $b_1 \leftarrow \text{IUPPRB}(\bar{A}(-x))$; $v_1 \leftarrow \text{var}(\text{IPLEV}(S, -b_1))$; $L_2 \leftarrow \text{IPIISS}(S, -b_1, 0, v_1, v_0)$.
(4) [Combine intervals.] $L \leftarrow \text{concat}(L_1, L_2)$

$$L \leftarrow \text{IPIISS}(S, a, b, v_1, v_2)$$

Algorithm IPIISS (Integral polynomial isolating interval search using Sturm sequence). *Inputs*: $S = (A_1(x), A_2(x), \ldots, A_r(x))$ is a list of integral polynomials. S is a Sturm sequence for $A_1(x)$. $a < b$ are binary rational numbers. $v_1 = \text{var}(S, a)$ and $v_2 = \text{var}(S, b)$.
Output: $L = (I_1, \ldots, I_r)$ is a list of standard isolating intervals for the real roots of $A_1(x)$ in the interval $(a, b]$.

(1) [Base Case.] if $v_1 - v_2 = 0$ then { $L \leftarrow ()$; return } else if $v_1 - v_2 = 1$ then { $L \leftarrow ((a, b))$; return }.
(2) [Evaluate Sturm Sequence at midpoint.] $c \leftarrow (a + b)/2$; ; $v \leftarrow \text{var}(\text{IPLEV}(S, c))$; .
(3) [Check left subinterval $(a, c]$.] $L_1 \leftarrow \text{IPIISS}(S, a, c, v_1, v)$.
(4) [Check right subinterval $(c, b]$.] $L_2 \leftarrow \text{IPIISS}(S, c, b, v, v_2)$.
(5) [Combine.] $L \leftarrow \text{concat}(L_1, L_2)$

Algorithms IPRIST and IPIISS use the subalgorithm IPLEV to evaluate a Sturm sequence at a binary rational number (a rational number whose denominator is a power of two). IPLEV uses a modification of Horner's method to evaluate $\text{sign}(A_i(x))$ and returns the list $(\text{sign}(A_1(a)), \ldots, \text{sign}(A_r(a)))$. Let $a = e/f$ and $m = \deg(A(x))$. Since $A_i(x)$ is an integral polynomial, rational arithmetic can be avoided in the computation of $\text{sign}(A_i(a))$, by computing $f^m A(e/f)$. The computation of $f^m A(e/f)$ can be performed using a scheme similar to Horner's method.

$$L \leftarrow \text{IPLEV}(S, a)$$

Algorithm IPLEV (Integral polynomial list evaluation of signs).

Inputs: $S = (A_1(x), A_2(x), \ldots, A_t(x))$ is a list of integral polynomials and a a binary rational number.

Output: $L = (s_1, \ldots, s_t)$ where $s_i = \text{sign}(A_i(a))$.

Let $\{A_1(x), \ldots, A_r(x), A_{r+1}(x) = 0\}$ be a PRS of integral polynomials, with degree sequence $\{m_1, \ldots, m_r\}$, defined by $e_i A_i(x) = Q_i(x) A_{i+1}(x) = f_i A_{i+2}(x)$. Schwartz and Sharir (1983b) suggested using the following recurrence relation to evaluate the PRS at a point $a = e/f$. Let $c_i = A_i(a)$. Then

$$c_i = \frac{Q_i(a)c_{i+1} + f_i c_{i+2}}{e_i} \tag{5}$$

$$c_{r-1} = \frac{Q_{r-1}(a)c_r}{e_{r-1}}$$

$$c_r = A_r(a).$$

Using this relation, all $A_i(a)$ can be computed from $A_r(a)$, the quotient sequence $Q = \{Q_{r-1}(x), \ldots, Q_1(x)\}$, and the similarity coefficients $E = \{e_{r-1}, \ldots, e_1\}$ and $F = \{f_{r-2}, \ldots, f_1\}$.

The algorithm IPPSQSEQ (Integral polynomial primitive Sturm quotient sequence) computes $A_r(x)$, Q, E, and F using the primitive Sturm sequence. In this case $Q_i(x) = \text{pquot}(A_i(x), A_{i+1}(x))$, $\deg(Q_i(x)) = \delta_i = m_i - m_{i+1}$, $e_i = \text{ldcf}(A_{i+1}(x))^{\delta_i}$, and $f_i = \pm\text{cont}(\text{prem}(A_i(x), A_{i+1}(x)))$, where prem is the pseudo-remainder, where pquot is the pseudo-quotient, and cont is the content.

The recurrence relation in Eq. (5) involves rational arithmetic with costly integers gcd operations. This can be remedied by computing $\bar{c}_i = f^{m_i} A_i(e/f)$ instead of c_i.

$$\bar{c}_i = \frac{f^{\delta_i} Q_i(e/f)\bar{c}_{i+1} + f^{\delta_i + \delta_{i+1}} f_i \bar{c}_{i+2}}{e_i}, \tag{6}$$

which only involves integer arithmetic. This suffices for computing the sign of $A_i(a)$ since $\text{sign}(c_i) = \text{sign}(\bar{c}_i)$. Algorithm IPLREV (Integral polynomial list recursive evaluation of signs) implements this recurrence relation.

This method seems preferable to directly evaluating the PRS since the sum of the degrees of the quotient sequence is m whereas the sum the degrees of the remainder sequence is m^2. Moreover, if the PRS is normal then each quotient polynomial $Q_i(x)$ is linear. However, this is not the case if the size of a is small compared to $mL(d)$.

Table 1 compares the theoretical computing times of IPLREV and IPLEV (proofs can be found in Johnson 1991).

Table 1 suggests that IPLEV performs better in practice where $L(a)$ is likely to be codominant with 1 or $L(d)$, but IPLREV is better in the worst case. Algorithm IPRISTR (Integral polynomial root isolation using Sturm sequences and recursive evaluation) is the same as IPRIST except that IPLREV is used instead of IPLEV.

Table 1: Comparison of IPLEV and IPLREV

L(a)	IPLEV	IPLREV
$L(a) \sim 1$	$m^3 L(d)$	$m^3 L(d)^2$
$L(a) \sim L(d)$	$m^3 L(d)^2$	$m^3 L(d)^2$
$L(a) \sim mL(d)$	$m^5 L(d)^2$	$m^4 L(d)^2$

3.2   Derivative Sequence Algorithm

A major problem with the Sturm sequence algorithm is that the sizes of the
coefficients of the polynomials in the Sturm sequence are typically very large,
and hence it can be costly to compute and evaluate Sturm sequences. Therefore,
alternative methods are sought which do not involve polynomials with such
large sized coefficients. One possibility is to use the derivative sequence instead
of a Sturm sequence and Fourier's theorem instead of Sturm's theorem. This is
promising since the sizes of the coefficients of the primitive derivative sequence
are typically much smaller than those of the primitive Sturm sequence. Thus
we can try to replace the use of Sturm sequences in the algorithm IPRIST by
the derivative sequence. Even though Fourier's theorem only gives an upper
bound on the number of roots in an interval, it gives the exact answer if zero or
one variations are obtained. Therefore, this algorithm proceeds exactly as the
Sturm sequence algorithm does. If $V_a - V_b = 0$, the interval $(a, b]$ is discarded,
if $V_a - V_b = 1$, $(a, b]$ is an isolating interval, and if $V_a - V_b > 1$, the interval
$(a, b]$ is bisected.

Unfortunately, this does not lead to an algorithm since it may never terminate. For example, consider $A(x) = x^2 + x + 1$ and the interval $(-1, 0]$.
The derivative sequence is $A(x)$, $A^{(1)}(x) = 2x + 1$, $A^{(2)}(x) = 2$. Let $V_a =$
$\text{var}(\text{sign}(A(a)), \text{sign}(A^{(1)}(a)), \text{sign}(A^{(2)}(a)))$. Then $V_{-1} - V_0 = 2$ and I must
be bisected. The left subinterval, $(-1, -1/2]$ must also be bisected since
$V_{-1} - V_{-1/2} = 2$. Moreover, $V_a - V_{-1/2} = 2$ for all $a < -1/2$, so the subinterval with $-1/2$ as an endpoint will always be bisected and the algorithm will
never determine that there are no roots in the interval $(-1, 0]$. In fact, any
time $V_a - V_b$ is greater than the number of roots in $(a, b]$ the algorithm will not
terminate since $V_a - V_b = (V_a - V_c) + (V_c - V_b)$ for all $c \in (a, b)$ and at least
one of the subintervals will contain more variations than roots.

Despite the failure of this approach, the derivative sequence can still be used
to derive a valid algorithm. One possible approach was presented by Collins
and Loos (1976). Their algorithm inductively obtains isolating intervals for a
polynomial from isolating intervals for its derivative. The algorithm is based
on Rolle's theorem which implies that any interval containing at most one root
of the derivative can contain at most two roots of the polynomial. Using this
basic idea, the algorithm does a case by case analysis, based on the signs of
the polynomial and its derivative at the endpoints of the intervals, in order to

find isolating intervals for $A(x)$. In one case it is necessary to use a tangent construction to verify that an interval does not contain any roots.

The key to obtaining an algorithm using Fourier's theorem is to guarantee that the intervals that are tested do not contain any roots of the derivative. In this case, Fourier's theorem reduces to a combination of the intermediate value theorem and Rolle's theorem: if $\forall x \in [a, b]$, $A^{(1)}(x) \neq 0$, then the number of roots of $A(x)$ in $(a, b]$ is equal to $\mathrm{var}(\mathrm{sign}(A(a)), \mathrm{sign}(A^{(1)}(a))) - \mathrm{var}(\mathrm{sign}(A(b)), \mathrm{sign}(A^{(1)}(b)))$. In particular, if $\alpha_1' < \cdots, < \alpha_r'$ are the distinct roots of $A^{(1)}(x)$ then the intermediate value theorem can be applied to the intervals $(\alpha_i', \alpha_{i+1}']$ to find the isolating intervals for the simple roots $A(x)$. It is necessary to include the possibility of multiple roots since even if $A(x)$ is squarefree, $A^{(1)}(x)$ may not be. For the multiple roots it is necessary to determine if $A(\alpha_i') = 0$. The algorithm ROLLE recursively uses the derivative sequence and these ideas to find isolating intervals for $A(x)$.

$$L \leftarrow \mathrm{ROLLE}(A(x))$$

Isolate real roots using Rolle's theorem and the derivative sequence.
Input: $A(x)$ is an integral polynomial.
Output: $L = (I_1, m_1, ..., I_r, m_r)$ is a list of isolating intervals for the real roots of $A(x)$. m_j is the multiplicity of the root in I_j.

(1) [Initialize.] $B \leftarrow \mathrm{RootBound}(A(x))$; $L \leftarrow ()$.
(1) [Recursion.]
    if $\deg(A(x)) = 1$ then {
      $L := ((-B, B), 1)$; return }
    else { $L' \leftarrow \mathrm{ROLLE}(A'(x))$; $r' \leftarrow \mathrm{LENGTH}(L')/2$ }.
(2) [Compute $\mathrm{sign}(A(\alpha_i'))$ and apply intermediate value theorem.]
    Let $\alpha_1' < ... < \alpha_{r'}'$ be the distinct roots of $A'(x)$);
    $\alpha_0 \leftarrow -B$; $\alpha_{r'+1} \leftarrow B$;
    for $i = 1, ..., r' + 1$ do {
      if $sign(A(\alpha_i')) = 0$ then
        $L \leftarrow \mathrm{append}(L, I_i, m_i' + 1)$
      else if $\mathrm{sign}(A(\alpha_{i-1}')) \cdot \mathrm{sign}(A(\alpha_i)) < 0$ then
        $L \leftarrow \mathrm{append}(L, (\alpha_{i-1}, \alpha_i), 1)$ }

A practical improvement to the algorithm is obtained by using the primitive derivative sequence, which uses the primitive part of each derivative, rather than the derivative sequence, since this reduces the size of the coefficients.

The algorithm ROLLE requires the computation of the sign of a real algebraic number. In particular we must determine the sign of a polynomial at the real roots of its derivative where the roots of the derivative are given by isolating intervals. Let α' be a root of $A^{(1)}(x)$ and let $I' = (a, b]$ be an isolating interval for α'. If $C(x) = \mathrm{gsfd}(\gcd(A(x), A^{(1)}(x)))$, the greatest squarefree divisor of $A(x)$ and its derivative, then $A(\alpha') = 0$ if and only if $C(\alpha') = 0$. Since any root of $C(x)$ is a root of $A^{(1)}(x)$, the only possible root of $C(x)$ in the interval I' is α'. Moreover, since $C(x)$ is squarefree it only has simple roots and $C(\alpha') = 0$ if and only if $C(a)C(b) < 0$ or $C(b) = 0$.

If $A(\alpha') \neq 0$ then its sign can be computed by repeatedly bisecting I', always retaining the subinterval that contains α' until I' does not contain any roots of $A(x)$. If I' does not contain any roots of $A(x)$ then $\text{sign}(A(\alpha'))$ is equal to the sign of any number in I'. In particular $\text{sign}(A(\alpha')) = \text{sign}(A(b))$. The bisection process must eventually terminate since $A(\alpha') \neq 0$. Any algorithm which can determine the number of real roots in an interval can be used to test if I' contains any roots of $A(x)$. However, all that is required is an algorithm which can determine if there are any roots of $A(x)$ in I'.

The algorithm ROLLE produces isolating intervals $(\alpha_i', \alpha_{i+1}']$ which may not be rational intervals. Rational intervals can be produced as a side effect of the sign computation just described. Suppose $(\alpha_i', \alpha_{i+1}']$ is an isolating interval for a root, α, of $A(x)$, and $J_i = (a_i, b_i]$ is an isolating interval for α_i' which does not contain any roots of $A(x)$. Then $(b_i, a_{i+1}]$ is an isolating interval for α.

Since the algorithm ROLLE inductively uses the derivative, specialized algorithms can be designed to compute the sign of $A(x)$ at roots of its derivative. Rolle's theorem implies that between any two distinct roots of $A(x)$ there is a root of $A^{(1)}(x)$. This implies that there can be only zero, one, or two roots of $A(x)$ in an isolating interval for a simple root of $A^{(1)}(x)$. It further implies that between any two isolating intervals for $A(x)$ there is an isolating interval for $A^{(1)}(x)$ and hence the isolating intervals produced by the algorithm ROLLE are separated. The intervals $I_j = (a_j, b_j]$ and $I_k = (a_k, b_k]$ are separated if either $b_j < a_k$ or $b_k < a_j$. Since the isolating intervals for $A^{(1)}(x)$ are inductively separated from the isolating intervals for $A^{(2)}(x)$, $A^{(2)}(x)$ does not vanish on any isolating interval, I', for a simple root of $A^{(1)}(x)$, which implies, in this case, that $A^{(1)}(x)$ is monotone on I'. If $A^{(1)}(x)$ is monotone on the interval $I = (a, b]$ then the tangent to $A(x)$ at a and the tangent to $A(x)$ at b do not intersect $A(x)$ in the interior of I and they can be used to determine if $A(x)$ has any roots in I.

Furthermore, a bound on the number of roots of $A(x)$ in an isolating interval for $A^{(1)}(x)$ can be obtained using Fourier's theorem. In some cases this is enough information to determine the number of roots of $A(x)$ in the interval and the general purpose sign algorithm or tangent construction can be avoided. The case where the tangent construction is required corresponds to the case where Collins and Loos required the tangent construction.

The algorithm IPRIDS (Integral polynomial real root isolation using the derivative sequence) implements algorithm ROLLE, using Descartes' rule of signs and polynomial transformations (see the next section) to compute the sign of $A(x)$ at the roots of its derivative. The algorithm IPRIPDS (Integral polynomial real root isolation using the primitive derivative sequence) is the same as IPRIDS except the primitive derivative sequence is used. The algorithm IPRIDSF incorporates the use of the test using Fourier's theorem to IPRIPDS, and the algorithm IPRIDTC (Integral polynomial real root isolation using the derivative sequence and tangent construction) uses the tangent construction for sign computation.

3.3 The Coefficient Sign Variation Method

The coefficient sign variation method (Collins and Akritas 1976; Collins and Johnson 1989b), for polynomial real root isolation, is based on two special cases of Descartes' rule of signs and the use of polynomial transformations. Descartes rule (Theorem 4) can be applied to obtain an upper bound on the number of positive roots of a polynomial $A(x)$. An upper bound can be obtained on the number of roots of $A(x)$ in the interval (a, b) by first mapping the roots of $A(x)$ in the interval (a, b) to the positive roots of $A^*(x)$ and then applying Descartes' rule to $A^*(x)$. The interval (a, b) can be mapped to the interval $(0, \infty)$ using a linear fractional transformation, T, that maps the circle whose diameter is (a, b) to the right-half plane. The polynomial $A^*(x)$ is obtained by replacing x by $T^{-1}(x)$.

The polynomial transformations needed are homothetic transformations, $A(x) \rightarrow A(ax)$, polynomial translations, $A(x) \rightarrow A(x + b)$, and reciprocal transformations, $A(x) \rightarrow x^m A(1/x)$, where $m = \deg(A(x))$.

The approach we take isolates the positive and negative roots separately. The algorithm first decides whether $A(0) = 0$ by checking if $x|A(x)$. Then a positive root bound, b_p, is computed and the roots of $A(x)$ in the interval $(0, b_p)$ are isolated. The normalized polynomial $\tilde{A}_p(x) = A(b_p x)$, whose roots in the normalized interval $(0, 1)$ correspond to the positive roots of $A(x)$, is computed. Since the algorithms we use for computing root bounds return a root bound which is a power of two, a binary homothetic transformation can be used. The subalgorithm IPRINCS is then used to isolate the real roots of $\tilde{A}_p(x)$ in the interval $(0, 1)$. IPRINCS computes the number of variations of $\operatorname{var}(D(\tilde{A}_p(x)))$, where $D(A(x)) = (x + 1)^m A(1/(x + 1))$ $(m = \deg(A(x)))$ is the polynomial transformation that maps $A(x)$ to a polynomial whose positive roots correspond to the roots of $A(x)$ in the interval $(0, 1)$.

If $\operatorname{var}(D(\tilde{A}_p(x))) = 0$ then there are no roots in $(0, 1)$. If $\operatorname{var}(D(\tilde{A}_p(x))) = 1$ there is a single positive root of $\tilde{A}_p(x)$ and $(0, 1)$ is an isolating interval. If $\operatorname{var}(D(\tilde{A}_p(x))) > 1$, the interval $(0, 1)$ is bisected and IPRINCS recursively checks the left and right subintervals, using the bisection polynomials $\tilde{A}_1(x) = 2^m \tilde{A}(x/2)$ and $\tilde{A}_2(x) = \tilde{A}_1(x+1)$ for the left and right subintervals respectively. The roots of $\tilde{A}_1(x)$ in the interval $(0, 1)$ correspond to the roots of $\tilde{A}(x)$ in the interval $(0, 1/2)$, and the roots of $\tilde{A}_2(x)$ in the interval $(0, 1)$ correspond to the roots of $\tilde{A}(x)$ in the interval $(1/2, 0)$. After IPRINCS has isolated the roots of $\tilde{A}(x)$ in the interval $(0, 1)$, the positive roots of $A(x)$ are obtained by scaling the isolating intervals returned by IPRINCS.

After the positive roots have been isolated, the negative roots are isolated by isolating the positive roots of $A_n(x) = A(-x)$. A negative root bound, b_n, is computed and IPRINCS is used to isolate the roots of $\tilde{A}_n(x) = A_n(b_n x)$. The algorithms IPRICS and IPRINCS are listed below.

$$L \leftarrow \text{IPRICS}(A(x))$$

Algorithm IPRICS (Integral polynomial real root isolation, coefficient sign variation method).
Input: $A(x)$ is a squarefree integral polynomial.
Output: $L = (I_1, \ldots, I_r)$ is a list of isolating intervals for the real roots of $A(x)$.
$I_j = (a_j, b_j)$ is either a standard open or a one-point binary rational interval and $a_1 \le b_1 \le \cdots \le a_r \le b_r$.

(1) [Initialize and check if $A(0) = 0$.]
if $x | A(x)$ then $\{ L_0 \leftarrow ((0,0)); \quad A(x) = A(x)/x; \}$
else $L_0 \leftarrow ()$;
$L_p \leftarrow ()$; $L_n \leftarrow ()$.
(2) [Isolate positive roots.]
$b \leftarrow \text{IUPPRB}(A(x))$; $\tilde{A}_p \leftarrow A(b_p x)$;
$L_p \leftarrow \text{IPRINCS}(\tilde{A}_p(x), 0, 1)$;
Scale intervals in L_p by b_p.
(3) [Isolate negative roots.]
$A_n(x) \leftarrow A(-x)$; $b_n \leftarrow \text{IUPPRB}(A_n(x))$; $\tilde{A}_n(x) \leftarrow A_n(b_n x)$;
$L_n \leftarrow \text{IPRINCS}(\tilde{A}_n(x), 0, 1)$;
Scale intervals in L_n by $-b_n$.
(4) [Combine.] $L \leftarrow \text{concat}(L_n, L_0, L_p)$

$$L \leftarrow \text{IPRINCS}(A(x), I)$$

Algorithm IPRINCS (Integral polynomial real root isolation, normalized coefficient sign variation method).
Inputs: $A(x)$ is a squarefree integral polynomial. $I = (a, b)$ is a standard interval.
Output: $L = (I_1, \ldots, I_r)$ is a list of isolating intervals for $T(A(x))$ in the interval $(0,1)$, where T is the linear fractional transformation that maps (a, b) onto $(0,1)$. $I_j = (a_j, b_j)$ is either a standard open or one-point interval and $a_1 \le b_1 \le \cdots \le a_r \le b_r$.

(1) [Initialize and check if $A(a) = 0$.]
$a \leftarrow \text{LeftEndpoint}(I)$; $b \leftarrow \text{RightEndpoint}(I)$;
if $x | A(x)$ then $\{ L_0 \leftarrow ((a, a)); \quad A(x) \leftarrow A(x)/x \}$
else $L_0 \leftarrow ()$.
(2) [Base case.] $m \leftarrow \deg(A(x))$; $A^*(x) \leftarrow (x+1)^m A(1/(x+1))$;
if $\text{var}(A^*(x)) = 0$ then $\{ L \leftarrow L_0; \text{return} \}$; if $\text{var}(A^*(x)) = 1$ then $\{$
$L \leftarrow \text{concat}(L_0, ((a, b)))$; return $\}$.
(3) [Bisect.] $c \leftarrow (a+b)/2$;
$A_1(x) \leftarrow 2^m A(x/2)$; $A_2(x) \leftarrow A_1(x+1)$.
(4) [Left recursive call.]
$L_1 \leftarrow \text{IPRINCS}(A_1(x), (a, c))$.
(5) [Right recursive call.]
$L_2 \leftarrow \text{IPRINCS}(A_2(x), (c, b))$.

(6) [Combine.] $L \leftarrow \text{concat}(L_1, L_0, L_2)$

4 Computing Time Analysis

In this section we derive maximum computing time bounds for each of the real root isolation algorithms introduced in Section 3. Improved computing time bounds are obtained through the use of Davenport's Theorem 10 and its corollaries. Davenport (1985) first used this theorem in the analysis of an algorithm based on Sturm sequences.

The algorithm for which we were able to derive the best maximum computing bound is the algorithm based on polynomials transformations and Descartes' rule of signs, IPRICS. The next best bound is for the derivative sequence algorithm, and the worst bound is for the Sturm sequence algorithm. Even though these theorems are not indicative of typical computing times, they do suggest which algorithms are better in practice. We conclude by giving an example which shows that the computing time bound for the algorithm based on Descartes' rule of signs is the best possible.

The following notation will be used in the statements of the computing time bounds. The sum norm of the polynomial, $A(x)$, denoted by $|A(x)|_1$, is the sum of the absolute values of its coefficients, and the max norm of the polynomial, $A(x)$, denoted by $|A(x)|_\infty$, is the maximum of the absolute values of its coefficients. The β-length of an integer, denoted by L_β, is the number of digits in base β representation. A function f is dominated by a function g, denoted by $f \preceq g$ if $f = O(g)$. A function f is codominant with g, denoted by $f \sim g$ if $f \preceq g$ and $g \preceq f$. Since $L_{\beta_1} \sim L_{\beta_2}$ for any two integers β_1 and β_2 greater than one, we will ignore the radix when indicating length. For a more thorough discussion on these concepts see (Collins 1974a)

Theorem 11 (Computing Time of IPRIST**)** *Let* $m = \deg(A)$ *and* $|A|_1 = d$. *Then the computing time of* IPRIST(A) *is dominated by* $m^6 L(d)^3$.

Proof. The time to compute a Sturm sequence with IPPNPRS is dominated by $m^5 L(d)^2$ (see Heindel 1971 or Johnson 1991).

Associated with the algorithm IPRIST is a binary tree, whose nodes correspond to subintervals of the initial interval that is being searched for isolating intervals. The proof involves estimating the computing time associated with each node in the associated tree and obtaining a bound on the height and number of nodes in the tree.

At level h the endpoints of each interval are of the form $b/2^i$ with $i \leq h$, therefore the length of any endpoint is less than or equal to $L(b) + h$, which by Theorem 9 is dominated by $L(d) + h$. By Theorem 6, the height of the tree (the maximum value of h) is dominated by $mL(d)$.

An interval at level h is bisected if and only if it contains two or more real roots. Let k be the number of intervals that are bisected at level h. Since the length of any interval at level h is equal to $b/2^h$ an interval is bisected if and only if it contains two real roots whose separation is less than $b/2^h$. Choosing, γ_{2i-1} and γ_{2i}, for $1 \leq i \leq k$, to be two real roots in the i-th interval that needs to be bisected, we have

$$\prod_{i=1}^{k} |\gamma_{2i-1} - \gamma_{2i}| \leq b/2^h.$$

However, by Theorem 7 this product is greater than $c_0 d^{-c_1 m}$, for constants c_0 and c_1, which implies that $hk \preceq mL(md)$.

The cost of IPRIST at level h is dominated by the cost of the k evaluation of the Sturm sequence. Using the modification Horner's method discussed in Section 3.1, the cost of each evaluation at a point a is dominated by $m^3 L(d)L(a) + m^3 L(a)^2$ (see Johnson 1991), which since $L(a) \preceq L(d) + h$ is dominated by $m^3 L(d)^2 + m^3 L(d)h + m^3 L(d)^2 + m^3 h^2$. The k evaluations at level h are dominated $m^3 kL(d)^2 + hk(m^3 L(d)m^3 h)$, which is dominated by $m^4 L(d)^2 + m^4 L(d)^2 m^5 L(d)^2 \preceq m^5 L(d)^2$. Since there are at most $mL(d)$ levels the total computing time of IPRIST is dominated by $m^6 L(d)^3$. ∎

A better computing time bound can be obtained for the algorithm which uses the recurrence relation method for evaluating Sturm sequences.

Theorem 12 (Computing time of IPRIST using IPLREV.) *Let* $m = \deg(A)$ *and* $|A|_\infty = d$. *Then the computing time of* IPRIST(A) *is dominated by* $m^6 L(md)^3$.

Proof. The proof is the same as Theorem 11 except that the improved computing time bound to evaluate a Sturm sequence using IPLREV is used. ∎

The same technique used in the proof of Theorem 11 can be used to derive a computing time bound for IPRICS. However, unlike the Sturm sequence algorithm, complex roots have an effect on whether an interval is bisected or not, and the corollary of Davenport's theorem which deals with pairs of complex conjugate roots must be used.

Theorem 13 (Computing time of IPRICS) *Let* $m = \deg(A(x))$ *and* $|A(x)|_1 = d$. *Then the computing time of* IPRICS$(A(x))$ *is dominated by* $m^5 L(d)^2$.

Proof. As with IPRIST, there is a binary tree associated with IPRICS; however, in this case, in addition to an interval, each node is also associated with a polynomial – the transformed polynomial whose roots in the interval $(0, 1)$ correspond to the roots of the original polynomial in the interval associated with the node.

The cost associated with a node in the search tree is equal to the time needed to transform the associated polynomial so that Descartes' rule of signs may be applied, and if more than one variation is detected, the time to compute the two bisection polynomials. The polynomial transformations are dominated by the cost of performing a translation by one. The time to compute $P(x+1)$ is dominated by $p^3 + p^2 L(e)^2$, where $p = \deg(P(x))$ and $e = |P(x)|_\infty$ (see Johnson 1991).

The max norm of a polynomial associated with a node at level h is dominated by $2^{mh}d$. Therefore, the time to perform a polynomial translation by one of a polynomial associated with a node at level h is dominated by $m^3 h + m^2 L(d)$.

An interval associated with a node at level h has length equal to $b/2^h$, where b is the root bound used by the algorithm. By Theorem 7 an interval at level h is bisected only if it contains two or more real roots or there is a pair of complex conjugate roots in the circle of radius 2^{h+1} centered about the interval in question. Therefore an interval is bisected only if it contains two real roots whose separation is less than $b/2^h$ or if the surrounding circle contains a pair of complex conjugate roots whose separation is less than $b/2^{h-1}$.

Let k_1 denote the number of intervals at level h with two or more real roots, and let k_2 denote the number of nodes that are bisected due to a pair of complex conjugate roots in the surrounding circle. By Corollary 7 and 8, $hk_1 \preceq bymL(d)$ and $hk_2 \preceq bymL(d)$. Therefore, if k is the number of intervals that are bisected at level h, $hk \preceq mL(d)$.

The cost of the k translations at level h is dominated by $m^3 hk + km^2 L(d)$, which is dominated by $m^4 L(d)$. By Theorem 6 the height of the tree is dominated by $mL(d)$ and hence there are at most $mL(d)$ levels. Therefore, the total computing time of IPRICS is dominated by $m^5 L(d)^2$. ∎

The proof of the maximum computing time bound for the derivative sequence algorithms depends on the time required to compute the sign of a real algebraic number. In particular, we must bound the cost for computing the signs of all of the elements of the derivative sequence at the roots of their derivatives.

Theorem 14 (Computing Time of IPRIDS**)** *Let $A(x)$ be an integral polynomial with $\deg(A(x)) = m$ and $|A(x)|_1 = d$. Then the computing time of* IPRIDS$(A(x))$ *is dominated by $m^8 + m^6 \log(d)^2$.*

If instead the tangent construction is used to compute the signs of the polynomials at the roots of their derivatives, the maximum computing time increases.

Theorem 15 (Computing Time of IPRIDTC**)** *Let $A(x)$ be an integral polynomial with $\deg(A(x)) = m$ and $|A(x)|_1 = d$. Then the computing time of* IPRIDTC$(A(x))$ *is dominated by $m^9 + m^6 \log(d)^3$.*

Theorem 16 (Mignotte) *Let $A(x) = x^m - 2(ax-1)^2$, $m \geq 3$ and $a \geq 2$ be an integer, then $A(x)$ has two real roots in the interval $(1/a - h, 1/a + h)$ where $h = a^{-(m+2)/2}$.*

Proof. See (Mignotte 1982). ∎

Since var($A(x)$) = 3, Descartes' rule of signs implies that $A(x)$ has three positive roots, and since var($A(-x)$) = 1 if m is even and var($A(-x)$) = 0 if m is odd, $A(x)$ has one or zero negative roots depending on whether m is even or odd. Moreover, using the prime 2, Eisenstein's criteria shows that $A(x)$ is irreducible.

Theorem 17 (Maximum Computing Time) *Let* $A(x) = x^m - 2(ax - 1)^2$ *with* $m \geq 3$ *and* $a \geq 2$ *an integer. Then either the computing time of* IPRICS($A(x)$) *or the computing time of* IPRICS($A(-x + 1)$) *is codominant with* $m^5 \mathrm{L}(a)^2$.

Proof. By Theorem 16 the height of the tree associated with IPRICS of $A(x)$ is codominant with $m\mathrm{L}(a)$. The polynomial associated with node (l, n) (node n from the left on level l of a complete binary tree) is $2^{ml}A(x + n/2^l) = (x + n)^m - 2^{ml-1}a(x + n)^2 + 2^{ml+1}(x + n) - 2^{ml}$. This polynomial is dense and has max norm codominant with $\mathrm{L}(l)$. However, if n is small most of the coefficients will have small sizes. To guarantee that all of the coefficients have large sizes, we want to ensure that the search tree has nodes with large n. If all of the nodes corresponding to $A(x)$ are such that $n < 2^{l-1}$ the transformed polynomial $A(-x + 1)$ will have nodes with $n > 2^{l-1}$. Therefore, for either $A(x)$ or $A(-x + 1)$ the search tree will have nodes with polynomials that are dense and have coefficients codominant with $\mathrm{L}(l)$ which is codominant with $m\mathrm{L}(a)$. The computing time for the translation associated with such a node is $m^3 l + m^2 \mathrm{L}(a)^2$ which is codominant with $m^4 \mathrm{L}(a)$. Since there are at least $m\mathrm{L}(a)$ nodes, the total computing time is codominant with $m^5 \mathrm{L}(a)^2$. ∎

5 Empirical Computing Times

This section contains an empirical comparison of the Sturm sequence algorithm, the derivative sequence algorithm, and the coefficient sign variation method. All three algorithms were executed on a large number of examples from several classes of polynomials. These classes include Mignotte's example, polynomials with random coefficients, polynomials with varying number of random real roots, and orthogonal polynomials. Further examples such as random bivariate resultants and discriminants are discussed in (Johnson 1991). For almost all of these examples the coefficient sign variation method is far superior to the other two algorithms, even for orthogonal polynomials that have Sturm sequences with small coefficients. Nonetheless, there are examples for which the Sturm sequence algorithm is superior, namely for polynomials that have a pair of complex conjugate roots with very small imaginary part, but whose minimum real root separation is large. Finally, a modification of the derivative sequence method is superior to both the coefficient sign variation method and the method based on Sturm sequences for random sparse polynomials.

All of the timings in this paper are from (Johnson 1991) and were obtained on a SPARCstation 1+, with 64 Megabytes of memory and rated at 15.8 mips. The timings were obtained with the UNIX timing procedure "times" and are reported in milliseconds.

5.1   Random Polynomials and Average Computing Time

In this section we investigate the average performance of the three real root isolation algorithms described in this chapter. Each algorithm was executed for one thousand randomly generated input polynomials with degrees from 10 to 100 and coefficient sizes from 10 to 250 bits. The average execution times are calculated and compared. This comparison clearly shows that, for random polynomials, the coefficient sign variation method is far superior to the other two methods, with the Sturm sequence method significantly worse than the various derivative sequence methods.

In our experiments, one thousand squarefree polynomials with uniformly distributed random integral coefficients were generated, and were used as input for each of the three algorithms. For each degree, m, and coefficient size, k, 100 random integral polynomials, whose coefficients are uniformly distributed between $-2^k + 1$ and $2^k - 1$, were generated. The computing times for each algorithm are averaged over the 100 random polynomials for each degree and coefficient size specifications. All times are reported in milliseconds and are listed under the corresponding algorithm's name. For each of the three algorithms several variations are compared.

Besides recording the computing times, a variety of statistics were gathered that characterize the computing times of the various algorithms. These statistics are also averaged over the 100 polynomials for each degree and coefficient size. For the Sturm sequence algorithm, IPRIST, the length of the Sturm sequence, l, the average number of bits in the Sturm sequence, S, and the maximum number of bits in the Sturm sequence, S_∞, were computed. This information was also computed for the recurrence relation version of the Sturm sequence algorithm, IPRISTR. The average number of bits, D, were computed for the coefficient sign variation method, IPRICS. Information concerning the search trees, such as height, H, number of nodes, N, and number of leaf nodes, L, were recorded for both IPRIST and IPRICS.

The information characterizing the derivative sequence algorithms include the average number of bits in derivative sequence, the average number of real roots of the polynomials in the derivative sequence, and the average number of bisections per sign computation. The number of bits in the derivative sequence is compared to the number of bits in the primitive derivative sequence. For the improved algorithm IPRIDSF, which uses Fourier's theorem, the number of times $V_a - V_b$ is equal to zero, one, or two is recorded.

Table 2 reports on Sturm Sequence Calculation for 1000 randomly generated polynomials with degrees from 10 to 50 and coefficient sizes from 10 to 250 bits. For each degree and coefficient size equal to k, 100 random integral polynomials, whose coefficients are uniformly distributed between $-2^k + 1$ and $2^k - 1$, were generated. The average length of the Sturm sequence, the average number of bits in the primitive Sturm sequence (S) and quotient sequence (Q), the average maximum number of bits in the primitive Sturm sequence (S_m) and quotient sequence (Q_m), and computing times for IPPNPRS and IPPSQSEQ are recorded.

Table 2: Sturm Sequence Statistics for Random Polynomials

m	k	l	S	S_m	T_1	Q	Q_m	T_2
10	10	11	73.1	174.5	109.8	143.4	327.0	119.0
20	10	21	191.9	427.8	1460.2	380.8	833.1	1569.1
30	10	31	320.0	693.3	7064.4	636.7	1364.7	7547.3
40	10	41	455.3	975.4	22209.8	907.3	1927.7	22704.08
50	10	51	592.9	1257.3	54404.1	1182.3	2492.6	55256.5
10	50	11	342.4	815.6	1216.4	678.34	1529.17	1277.3
10	100	11	679.1	1615.4	4272.2	1347.18	3028.87	4429.19
10	150	11	1015.4	2415.7	9223.3	2015.30	4528.95	9713.84
10	200	11	1351.8	3215.3	16039.3	2683.65	6028.74	16923.05
10	250	11	1687.5	4013.5	24755.5	3350.35	7525.62	26018.95

T_1 is the time for IPPNPRS and T_2 is the time for IPPSQSEQ.

Table 3 reports the time for isolating the real roots after a Sturm sequence is computed. The times for IPIISS which uses IPLEV for evaluation of Sturm sequences and IPIISRS which uses IPLREV for evaluation of Sturm sequences are reported. In addition the average number of real roots, r, the average height of the search trees, H, the average number of nodes in the search trees, N, and the average number of leaf nodes, L, in the search tree are reported. This data was obtained from the same random polynomials that were used for the experiments reported in Table 3.

Table 4 reports computing times and search tree statistics for IPRICS. The table contains the same information concerning the search tree as was recorded for IPRIST. In addition, the average number of bits of the transformed polynomials was computed. B_1 is the average number of bits for IPRICS.

Table 5 compares the computing times of several variations of the derivative sequence algorithm. Algorithms IPRIDS, IPRIPDS, and IPRIDSF use Descartes' rule of signs to compute the sign of the polynomial at the roots of its derivative. IPRIDS uses the derivative sequence, and IPRIPDS and IPRIDSF use the primitive derivative sequence. D is the average number of bits in the derivative sequence and D' is the average number of bits in the primitive derivative sequence. The algorithm IPRIDSF uses Fourier's theorem to test if $A(x)$ has zero, one, or two roots in the isolating intervals for the roots of its de-

Table 3: Sturm Search Tree Statistics for Random Polynomials

m	k	r	H	N	L	IPIISS	IPIISRS
10	10	2.14	2.33	5.06	3.53	13.85	65.89
20	10	2.58	3.01	6.72	4.36	109.26	679.44
30	10	2.80	3.18	7.14	4.57	368.78	2569.21
40	10	2.92	3.32	7.76	4.88	933.14	14181.9
50	10	3.14	3.65	8.72	5.36	2041.13	14838.31
10	50	2.04	2.26	4.88	3.44	36.32	742.46
10	100	2.10	2.24	4.90	3.45	65.52	2650.99
10	150	1.92	2.11	4.56	3.28	87.66	5534.48
10	200	1.94	2.10	4.46	3.23	110.14	9537.05
10	250	1.96	1.94	4.10	3.05	127.47	14051.05

Table 4: IPRICS Search Tree Statistics for Random Polynomials

m	k	r	H	N	L	T_1	B_1
10	10	2.14	3.41	8.94	5.47	25.78	20.44
20	10	2.58	3.94	10.92	6.46	40.55	56.51
30	10	2.80	3.98	11.52	6.76	54.41	111.17
40	10	2.92	4.31	12.60	7.30	73.38	192.99
50	10	3.14	4.57	13.36	7.68	89.95	306.96
60	10	3.10	4.29	12.60	7.30	97.43	404.50
70	10	3.05	4.09	12.50	7.25	112.58	554.46
80	10	3.28	4.89	14.48	8.24	142.50	919.37
90	10	3.38	4.85	14.54	8.27	153.65	1201.48
100	10	3.74	5.10	15.20	8.60	175.39	1659.05
10	50	2.04	3.27	8.46	5.23	65.80	27.35
10	100	2.10	3.31	8.60	5.30	113.01	33.06
10	150	1.92	3.17	8.06	5.03	163.40	36.55
10	200	1.94	3.03	7.54	4.77	213.53	39.48
10	250	1.96	2.87	7.34	4.67	263.37	44.16

T_1 is the time for IPRICS.

rivative. The algorithm IPRIDTC also performs this test; however, if it needs to refine an isolating interval, this is done with the tangent construction. See Section 3.2 for more details concerning these variations.

Table 5: Comparison of Derivative sequence algorithms

n	k	D	T_1	D'	T_2	T_3	T_4
10	10	22.07	152.56	14.44	151.38	112.35	101.17
20	10	44.71	1118.18	21.47	921.82	644.03	589.33
30	10	71.12	4087.46	28.32	3345.96	2157.35	1945.32
40	10	100.23	10457.64	35.44	8428.08	4916.02	4396.89
50	10	131.08	23083.10	42.53	18445.36	9611.62	8479.45
60	10	163.64	43510.33	49.77	34231.45	16306.39	14237.53
70	10	197.18	78325.01	56.90	60643.33	26606.55	22774.57
10	50	62.10	258.67	54.50	266.21	227.41	235.33
10	100	112.13	337.77	104.47	366.05	324.19	362.47
10	150	162.18	400.22	154.57	448.17	412.82	486.83
10	200	212.08	506.37	204.53	570.20	524.71	653.28
10	250	262.06	583.84	254.43	671.64	623.31	792.25

T_1 is the time for IPRIDS, T_2 is the time for IPRIPDS, T_3 is the time for IPRIDSF, and T_4 is the time for IPRIDTC.

Tables 6 and 7 report statistics characterizing the performance of the algorithms IPRIDSF and IPRIDTC. The average number of real roots, r, the average number of real roots of the polynomials in the derivative sequences, r', and the average maximum number of real roots in the derivative sequence, r_0 are reported. The information indicates the number of sign computations and the maximum number of sign computations for any element of the derivative sequence. The number of bisections needed for the sign computations was also recorded. B is the average number of bisections per sign computation needed for each input polynomial. B_l is the average number of bisections per element of the derivative sequence. This is the number of bisections required to compute the signs of a polynomial at all of the roots of its derivative. B_∞ is the maximum of the number of bisections needed to compute the signs of a polynomial at any particular root of its derivative. V_0, V_1, and V_2 is the number of times Fourier's theorem indicated that there were mzero, one, or two roots of $A(x)$ in the isolating intervals for the roots of its derivative.

5.2 Polynomials with Varying Number of Real Roots

In the previous section we examined the behavior of the real root isolation algorithms for random polynomials. However, random polynomials may not be indicative of the polynomials one is interested in practice. Random poly-

Table 6: IPRIDSF Statistics for Random Polynomials

m	k	r	r'	r_0	B	B_l	B_∞	V_0	V_1	V_2
10	10	2.14	2.17	3.42	1.79	2.45	4.39	0.81	10.69	5.86
20	10	2.58	2.68	4.59	2.16	3.03	5.37	2.48	33.31	12.63
30	10	2.80	3.22	5.76	2.37	3.49	6.02	4.98	65.52	20.14
40	10	2.92	3.59	6.46	2.55	3.78	6.67	6.9	102.56	27.52
50	10	3.14	3.96	7.14	2.60	3.97	6.97	9.29	147.06	34.67
60	10	3.10	4.17	7.55	2.62	4.02	6.84	11.56	190.06	41.54
70	10	3.05	4.55	8.19	2.67	4.18	7.23	14.74	247.02	49.25
10	50	2.04	2.16	3.48	1.77	2.40	4.24	0.79	10.53	6.10
10	100	2.10	2.18	3.45	1.64	2.25	4.03	1.08	10.36	6.04
10	150	1.92	1.97	3.21	1.65	2.15	4.07	0.79	9.08	5.95
10	200	1.94	2.06	3.30	1.70	2.22	4.05	0.76	9.91	5.91
10	250	1.96	1.99	3.26	1.68	2.16	3.94	0.79	9.38	5.81

Table 7: IPRIDTC Statistics for Random Polynomials

n	k	r	r'	R_0	B	B_l	B_∞	V_0	V_1	V_2
10	10	2.14	2.17	3.42	1.76	2.45	4.48	0.80	10.69	5.85
20	10	2.58	2.68	4.59	2.16	3.10	5.54	2.48	33.32	12.62
30	10	2.80	3.22	5.76	2.35	3.54	6.15	4.98	65.53	20.13
40	10	2.92	3.59	6.46	2.54	3.84	6.85	6.9	102.53	27.51
50	10	3.14	3.96	7.14	2.60	4.02	7.09	9.30	147.04	34.66
60	10	3.10	4.17	7.55	2.62	4.07	6.99	11.58	190.03	41.53
70	10	3.05	4.55	8.19	2.67	4.22	7.33	14.76	246.95	49.24
10	50	2.04	2.16	3.48	1.74	2.44	4.33	0.79	10.53	6.10
10	100	2.10	2.18	3.45	1.63	2.28	4.20	1.08	10.36	6.04
10	150	1.92	1.97	3.21	1.62	2.17	4.17	0.79	9.08	5.95
10	200	1.94	2.06	3.30	1.68	2.26	4.16	0.76	9.91	5.91
10	250	1.96	1.99	3.26	1.66	2.21	4.09	0.79	9.38	5.81

nomials typically have a small number of real roots. In practice this may not be the case. In this section we compare the algorithms on polynomials with a specified number of randomly generated real roots. In this experiment 10 random polynomials of degree 20 with r real roots were generated. A random polynomial with r real roots and s pairs of complex conjugate roots is equal to

$$\left\{ \prod_{i=1}^{r} (a_i x + b_i) \right\} \left\{ \prod_{j=1}^{s} (c_j^2 x^2 - 2d_j c_j x + d_j^2 + e_j^2) \right\},$$

where a_i, b_i, c_j, d_j, and e_j are random integers uniformly distributed between $-2^k + 1$ and $2^k - 1$. In all of our experiments, $k = 5$. Since we compute the primitive part of the polynomial before isolating its real roots, we list the average number of bits, K, in the input polynomials. As the number of real roots increases, the average number of bits in the primitive parts decrease since there is a greater likelihood that a linear factor will have a non-trivial content. Each table in this section records the same information as was recorded in the section on random polynomials (Section 5.1).

Table 8: Sturm Sequence Statistics for Polynomials with Random Roots

n	k	r	l	S	S_m	T_1	Q	Q_m	T_2
20	88.4	0	21	402.65	838.30	4789.9	725.17	1641.10	4921.5
20	86.8	2	21	399.46	828.40	4690.1	720.38	1618.70	4845.0
20	84.6	4	21	396.13	811.40	4646.7	716.64	1597.70	4817.0
20	82.0	6	21	391.21	803.00	4604.9	709.93	1578.60	4733.6
20	79.4	8	21	384.58	788.50	4471.6	699.92	1548.10	4573.4
20	79.1	10	21	379.06	771.70	4373.3	689.10	1508.30	4490.0
20	76.3	12	21	355.79	707.00	3895.0	644.60	1386.20	4015.2
20	74.1	14	21	332.04	642.60	3483.5	599.23	1263.10	3575.0
20	74.0	16	21	302.71	558.70	2978.3	540.10	1092.30	3071.7
20	71.0	18	21	260.97	464.90	2321.9	459.36	900.60	2400.0
20	72.3	20	21	209.50	363.50	1616.7	353.04	692.30	1661.5

T_1 is the time for IPPNPRS and T_2 is the time for IPPSQSEQ.

5.3 Orthogonal Polynomials

A sequence of polynomials $f_n(x)$, with $\deg(f_n(x)) = n$, is called orthogonal on the interval $[a, b]$ with respect to the weight function $w(x) > 0$, if

$$\int_a^b f_m(x) f_n(x) w(x) dx = 0 \text{ for } m \neq n.$$

Table 9: Sturm Search Tree Statistics for Polynomials with Random Roots

n	k	r	H	N	L	IPIISS	IPIISRS
20	88.4	0	1.00	2.00	2.00	78.3	1303.3
20	86.8	2	2.40	4.80	3.40	134.8	1826.6
20	84.6	4	6.80	15.60	8.80	341.8	4131.6
20	82.0	6	7.10	22.00	12.00	463.4	5380.2
20	79.4	8	7.50	26.60	14.30	541.5	6180.0
20	79.1	10	7.80	32.00	17.00	636.6	7025.2
20	76.3	12	8.30	36.80	19.40	701.8	7028.4
20	74.1	14	8.10	42.00	22.00	763.5	6891.4
20	74.0	16	8.10	45.40	23.70	778.1	6096.7
20	71.0	18	8.30	49.60	25.80	783.3	4966.8
20	72.3	20	8.30	53.00	27.50	749.9	3446.7

Table 10: Descartes' Search Tree Statistics for Polynomials with Random Roots

n	k	r	H	N	L	T_1	B_1
20	88.4	0	4.80	10.80	6.40	86.6	110.22
20	86.8	2	5.20	13.20	7.60	103.3	107.27
20	84.6	4	6.50	18.80	10.40	146.5	101.85
20	82.0	6	6.60	21.20	11.60	165.0	95.57
20	79.4	8	6.80	25.00	13.50	190.1	91.03
20	79.1	10	7.10	26.80	14.40	198.4	88.24
20	76.3	12	7.30	31.00	16.50	225.2	85.01
20	74.1	14	7.30	34.40	18.20	244.8	82.03
20	74.0	16	7.50	35.20	18.60	248.3	79.94
20	71.0	18	7.40	37.80	19.90	263.4	77.45
20	72.3	20	7.40	42.00	22.00	288.1	77.13

T_1 is the time for IPRICS.

Table 11: Comparison of Derivative Sequence Algorithms for Polynomials with Random Roots

n	k	r	D	T_1	D'	T_2	T_3	T_4
20	88.4	0	75.06	999.9	50.84	876.8	694.5	616.7
20	86.8	2	74.31	1242.0	50.52	1115.1	814.8	718.4
20	84.6	4	73.50	1486.6	50.26	1361.8	891.7	794.9
20	82.0	6	72.94	1641.7	49.2	1475.1	898.4	818.5
20	79.4	8	71.76	1924.9	48.22	1720.1	991.8	933.4
20	79.1	10	71.47	2194.7	47.75	1963.7	1046.7	978.2
20	76.3	12	70.59	2530.0	46.88	2276.9	1129.9	1111.7
20	74.1	14	70.30	2798.1	46.78	2505.2	1203.2	1145.0
20	74.0	16	70.30	2953.4	46.71	2619.6	1235.1	1198.5
20	71.0	18	69.24	3228.4	45.52	2915.0	1321.6	1306.6
20	72.3	20	69.87	3571.8	45.94	3208.3	1401.7	1380.0

T_1 is the time for IPRIDS, T_2 is the time for IPRIPDS, T_3 is the time for IPRIDSF, and T_4 is the time for IPRIDTC.

Table 12: IPRIDSF Statistics for Random Polynomials

n	k	r	r'	R_0	B	B_l	B_∞	V_0	V_1	V_2
20	88.4	0	1.76	3.5	2.39	2.96	6.20	4.40	16.30	12.70
20	86.8	2	2.47	4.5	2.31	3.55	7.20	6.40	25.80	12.80
20	84.6	4	3.20	6.0	2.24	3.86	6.50	6.60	37.80	12.40
20	82.0	6	3.66	6.9	2.15	3.96	6.70	7.20	45.20	11.20
20	79.4	8	4.67	8.8	1.93	3.96	6.00	10.90	58.80	11.10
20	79.1	10	5.59	10.5	1.73	3.75	5.90	14.00	72.60	9.60
20	76.3	12	6.87	12.4	1.57	3.80	5.70	21.2	89.1	8.30
20	74.1	14	7.94	14.2	1.36	3.70	6.00	32.00	98.40	6.40
20	74.0	16	8.60	16.0	1.27	3.76	5.50	44.50	96.90	6.00
20	71.0	18	9.74	18.0	1.25	3.83	5.40	46.80	116.30	3.90
20	72.3	20	11.05	20.0	1.10	3.58	5.50	64.9	123.7	1.4

Orthogonal polynomials have many interesting properties. For an overview see (Abramowitz and Stegun 1970) and detailed exposition see (Szego 1939). The properties of interest to us are: (1) Orthogonal polynomials satisfy a recurrence relation of the form $f_{n+1}(x) = (a_n + b_n x)f_n(x) - c_n f_{n-1}(x)$, where a_n, b_n, and c_n are constants that can be computed. (2) The derivatives of a sequence of orthogonal polynomials are orthogonal. (3) All of the roots of an orthogonal polynomial are real, simple, and located in the interior of the interval of orthogonality. (4) The roots of $f_{n-1}(x)$ separate the roots of $f_n(x)$.

These properties can be used to show that the primitive Sturm sequence of a polynomial is related to the primitive derivative sequence. Moreover, since the derivatives satisfy a recurrence relation, the sizes of the coefficients in the primitive derivative sequence typically decrease instead of increasing. See (Johnson 1991) for a complete discussion. This explains the empirical observation in (Collins and Loos 1982b) that the sizes of the coefficients in the Sturm sequence of Legendre and Chebycheff polynomials are abnormally small.

An example of a sequence of orthogonal polynomials are the Chebycheff polynomials. The Chebycheff polynomials of the first kind are defined by $T_n(x) = \cos(n\theta)$, where $x = \cos(\theta)$, and the Chebycheff polynomials of the second kind are defined by $U_n(x) = \text{sign}(n\theta)/\sin(\theta)$. Using these definitions it is easy to verify that $T_n'(x) = nU_{n-1}(x)$. The Chebycheff polynomials satisfy the recurrence relation $T_{n+1}(x) = 2xT_n(x) - T_{n-1}(x)$, where $T_1(x) = x$ and $T_0(x) = 1$, and $U_{n+1}(x) = 2xU_n(x) - U_{n-1}(x)$, where $U_1(x) = 2x$ and $U_0(x) = 1$.

The empirical computing times in Table 13 shows that the Sturm sequence algorithm is faster for small n; however, when $n > 32$, the coefficient sign variation method is faster. Table 13 also lists the computing time for the derivative sequence based algorithm IPRIDSF. The computing times for the derivative sequence algorithms are considerably more expensive than the other two algorithms; however, the derivative sequence algorithms isolate all of the roots in the derivative sequence, which implies that as a side effect from isolating the roots of $T_n(x)$ the roots of $U_k(x)$ are also isolated for $k < n$.

Table 13: Timings for Chebycheff Polynomials

n	IPPNPRS	IPIISS	IPRICS	H	N	L	IPRIDSF
8	0	17	33	6	22	12	66
16	33	84	133	8	38	20	433
32	200	1117	867	11	74	38	4833
64	1383	14349	8017	13	138	70	62188

5.4 Exceptional Polynomials

In this section we examine the performance of the different root isolation algorithms on Mignotte's class of polynomials. We also discuss a related example for which the Sturm sequence algorithm is superior to the other two algorithms.

Tables 14 and 15 report the computing times and search tree information for the Sturm sequence algorithm IPRIST and the coefficient sign variation method IPRICS for the degree ten Mignotte polynomials $A(x) = x^{10} - (3^k x - 1)^2$, with k ranging from 5 to 9. H is equal to the height of the corresponding search trees, N is equal to the number of nodes in the search trees, and L is equal to the number of leaf nodes in the search trees. The computing times are in milliseconds and are listed in the column under the algorithms name. For this example, the tables show that the algorithms behave similarly in terms of the number of bisections they must perform. Initially the computing time of the Sturm sequence algorithm is faster than the coefficient sign variation method; however, as k increases, the coefficient sign variation eventually becomes faster. The reason for this is that the polynomial evaluations that the Sturm sequence algorithm uses become slower compared to the translation required by the coefficient sign variation method as the height of the tree increases.

Table 14: IPRIST Statistics for $x^m - (ax - 1)^2$

m	k	r	H	N	L	IPPNPRS	IPRIST
10	5	4	51	102	52	0	167
10	6	4	102	204	103	33	650
10	7	4	150	300	151	33	1567
10	8	4	201	402	202	33	1567
10	9	4	246	492	247	33	5183

Table 15: IPRICS Statistics for $x^m - (ax - 1)^2$

m	a	r	H	N	L	IPRICS
10	5	4	51	102	52	334
10	6	4	102	204	103	1000
10	7	4	150	300	151	1917
10	8	4	201	402	202	3366
10	9	4	246	492	247	5017

Let $A(x) = x^m + (ax - 1)^2$. These polynomials do not have any positive real roots since $A(x) > 0$ for $x > 0$. The polynomial $A(x)$ has a pair of complex conjugate roots close to $1/a$ since it is a perturbation of $(ax - 1)^2$ which has a multiple at $1/a$. The multiple root must transform into a pair of complex conjugate roots since $A(x)$ does not have any positive real roots. The derivative sequence algorithm and the coefficient sign variation method behave poorly on this example, while the Sturm sequence algorithm behaves well. The reason for the poor performance of the coefficient sign variation method is that it continues transforming the polynomial until the pair of complex conjugate roots have been sufficiently separated. Similarly, the derivative sequence algorithm must perform many bisections to determine the sign of $A(x)$ at the root of its derivative which is close to $1/a$, since $A(x)$ is nearly tangent to the x-axis at this root. The Sturm sequence algorithm, however, is not affected by complex roots and hence determines that there are no positive roots immediately. The Sturm sequence algorithm also benefits from that fact that $A(x)$ only has five polynomials in its Sturm sequence.

Tables 16 and 17 report the computing times and search tree information for the Sturm sequence algorithm IPRIST and the coefficient sign variation method IPRICS for the degree ten Mignotte polynomials $A(x) = x^{10}+(3^k x-1)^2$, with k ranging from 5 to 9. For this example, the Sturm sequence algorithm is always significantly faster than the coefficient sign variation method. The height and number of nodes for the coefficient sign variation method is similar to the previous example; however, the number of nodes is always equal to 2 for the Sturm sequence algorithm. The reason for this, as pointed out previously, is that the Sturm sequence is not affected by the pair of complex conjugate roots near $1/a$ as is the coefficient sign variation method.

Table 16: Sturm Algorithm Statistics for $x^m + (ax - 1)^2$

m	a	r	H	N	L	IPIISS	IPIISRS
10	3^5	0	1	2	2	0	0
10	3^6	0	1	2	2	0	17
10	3^7	0	1	2	2	0	33
10	3^8	0	1	2	2	0	33
10	3^9	0	1	2	2	0	50

5.5   Sparse Polynomials

A sparse polynomial is a polynomial with few nonzero coefficients. Let $A(x) = \sum_{i=1}^{t} a_i x^{e_i}$, where $e_1 < \cdots < e_t$ are non-negative integers. Ideally an algorithm whose inputs are sparse polynomials should have a computing time dependent on t rather than e_t. Mignotte's (Theorem 16) implies that the reciprocal of the root separation for sparse polynomials can be codominant with n. Therefore, any algorithm whose computing time is codominant with the reciprocal of the root separation can depend on n rather than t. Nonetheless, we still seek

Table 17: IPRICS Statistics for $x^m + (ax - 1)^2$

m	a	r	H	N	L	IPRICS
10	3^5	0	51	100	51	350
10	3^6	0	97	194	98	1084
10	3^7	0	146	292	147	2133
10	3^8	0	194	388	195	3550
10	3^9	0	240	480	241	5332

algorithms that take advantage of sparsity.

Algorithms based on Sturm sequences do not seem to be able to take advantage of sparsity since the Sturm sequence, computed with a PRS, of a sparse polynomial quickly suffers from coefficient fill in. Furthermore, the coefficient sign variation method also destroys sparsity since polynomial translations typically fill in missing coefficients. The only algorithm presented in this paper which can take advantage of sparsity is the derivative sequence method. The reason for this is (1) the derivative of a sparse polynomial is sparse, and (2) that the inductive algorithm ROLLE needn't use the entire derivative sequence – instead it may use a sequence of length t that we call the sparse derivative sequence. This sequence is defined by $A_1(x) = A(x)/x^{e_1}$, and for $2 \leq i \leq t$, $A_i(x) = A'_{i-1}(x)/x^{e_i - e_{i-1} - 1}$. In other words, each time ROLLE is called, the input polynomial, $A(x)$, is divided by the largest power of x dividing $A(x)$. To preserve sparsity when evaluating $A(x)$ at the roots of $A'(x)$, the tangent construction should be used.

Preliminary experiments indicate that a modified version of the derivative sequence method based on the sparse derivative sequence out performs both the Sturm sequence algorithm and the coefficient sign variation method for random sparse polynomials. Polynomials were generated in such a way that the coefficient of x^i, $0 \leq i \leq n$ is non-zero with a given probability. Non-zero coefficients were selected from a uniform distribution of integers with k bits.

Acknowledgments

The material in this paper is based on the author's Ph.D. thesis, done under the supervision of Professor G. E. Collins. The author wishes to thank Professor Collins for the influence he has had on this work. The preparation of this paper was partially supported by NSF grant CCR-9211016.

Sturm–Habicht Sequences, Determinants and Real Roots of Univariate Polynomials

L. González-Vega, T. Recio, H. Lombardi, and M.-F. Roy

1 Introduction

The real root counting problem is one of the main computational problems in Real Algebraic Geometry. It is the following: Let $\mathbb{D}$ be an ordered domain and $\mathbb{B}$ a real closed field containing $\mathbb{D}$. Find algorithms which for every $P \in \mathbb{D}[x]$ compute the number of roots of P in $\mathbb{B}$. More precisely we shall study the following problem.

Problem 1 *Let $\mathbb{D}$ be an ordered domain and $\mathbb{B}$ a real closed field containing $\mathbb{D}$. Find algorithms which for every P and Q, polynomials in $\mathbb{D}[x]$, compute*

$$\operatorname{card}(\{\alpha \in \mathbb{B}\,|\,P(\alpha) = 0,\ Q(\alpha) > 0\}) - \operatorname{card}(\{\alpha \in \mathbb{B}\,|\,P(\alpha) = 0,\ Q(\alpha) < 0\}).$$

When we make $Q = 1$, a solution to Problem 1 immediately gives a solution to the real root counting problem.

Several methods for solving these problems appear in the literature of Numerical Analysis, Computer Algebra and Real Algebraic Geometry. The first known algorithm solving the real root counting problem is due to C. Sturm (1835). He obtained the number of real roots in terms of the Sturm sequence, which is the Euclidean remainder sequence for P and its derivative modulo some sign changes. Sylvester (1853) generalized Sturm's method to Problem 1. Finally C. Hermite (1856) developed a general theory giving the number of real roots as the signature of a Hankel matrix (see Section 4).

The computation of the Sturm sequence has bad numerical properties. Using exact arithmetic for a polynomial with integer coefficients, the polynomials in the Sturm sequence have rational coefficients, whose size grows exponentially in the degree of the polynomials. Because of the denominators appearing in the Euclidean division process, in the case when the polynomial depends on parameters, the computation of the Sturm sequence has no good specialization properties, i.e., it has to be redone completely for special values of the parameters.

More recently R. Loos (1982b) studied the connection between subresultants and Euclidean remainder sequences. Contrary to the polynomials in the Euclidean remainder sequence, the subresultants are stable under specialization and the size of their coefficients are well controlled.

It is thus a natural idea to use subresultants for the real root counting problem, where one has to take care of signs. The method proposed by R. Loos (1982b), known as the negative polynomial remainder sequence method, uses the subresultant algorithm with some sign modifications. This method solves the problem of the growth of coefficients in the Sturm sequence. But again, in the case of a polynomial depending on parameters, the choice of the signs is not uniform and this method has no good specialization properties (Gonzàlez-Vega et al. 1990, 1994).

The method we present here is based on the Sturm–Habicht sequence, introduced by W. Habicht (1948). W. Habicht proved that the Sturm–Habicht sequence (to be introduced in Section 2) solves Problem 1 with P squarefree and $\gcd(P, Q) = 1$.

The main objectives of this paper are the following. First to present the Sturm–Habicht sequence and to show its algebraic properties which are easily derived from the Structure Theorem for subresultants (Section 2). Second to show that the results of W. Habicht can be extended for any two polynomials on any open interval (even if the endpoints are roots of polynomials in the sequence) (Section 3). Finally the connections between Sturm–Habicht sequences and Hermite's method are explained (Section 4). The last section is devoted to the presentation of some examples.

The Sturm–Habicht sequence has been studied in (Gonzàlez-Vega et al. 1989) and its properties were shown in (Gonzàlez-Vega et al. 1990, 1994) using its connection with the corresponding Sturm sequence. Here we shall derive the same properties using only an algebraic approach. Specialization properties of Sturm–Habicht sequence are studied more deeply in (Gonzàlez-Vega et al. 1990, 1994).

2   Algebraic Properties of Sturm–Habicht Sequences

Let $\mathbb{D}$ be an ordered domain. In this section we define the Sturm–Habicht sequence and we prove its structure theorem. First of all we introduce a slight modification of the usual definition of subresultant polynomials in order to get some control over the formal degrees of the polynomials defining the sequence (see Loos 1982b, for the classical definition).

Definition 1 *Let $P = a_p x^p + a_{p-1} x^{p-1} + \ldots + a_1 x + a_0$ and, $Q = b_q x^p + b_{q-1} x^{q-1} + \ldots + b_1 x + b_0$ be polynomials in $\mathbb{D}[x]$ and $p, q \in \mathbb{N}$ with $\deg(P) \leq p$*

and $\deg(Q) \leq q$. If $i \in \{0, \ldots, \min(p, q)\}$ we define the polynomial subresultant associated with P, p, Q and q of index i as follows:

$$\mathrm{Sres}_i(P, q, Q, q) = \sum_{j=0}^{i} M_j^i(P, Q)x^j,$$

where every $M_j^i(P, Q)$ is the determinant of the matrix built with the columns $1, 2, \ldots, p+q-2i-1$ and $p+q-i-j$ in the matrix:

$$m_i(P, p, Q, q) = \overbrace{\begin{pmatrix} a_p & \cdots & a_0 & & & \\ & \ddots & & \ddots & & \\ & & a_p & \cdots & a_0 \\ b_q & \cdots & b_0 & & \\ & \ddots & & \ddots & \\ & & b_q & \cdots & b_0 \end{pmatrix}}^{p+q-i} \left.\vphantom{\begin{pmatrix}a\\a\\a\end{pmatrix}}\right\} q-i \atop \left.\vphantom{\begin{pmatrix}a\\a\\a\end{pmatrix}}\right\} p-i$$

The determinant $M_i^i(P, Q)$ will be called i-th principal subresultant coefficient and will be noted $\mathrm{sres}_i(P, p, Q, q)$.

So every subresultant is a polynomial in $\mathbb{D}[x]$ whose coefficients are determinants of matrices constructed from the coefficients of the polynomials P and Q. The celebrated *subresultant theorem*, first proved by W. Habicht (1948) in the forties and reformulated by G. E. Collins (1967) and R. Loos (1982b) among others, provides the best known non-modular algorithm computing the subresultant sequence associated to two polynomials in $\mathbb{D}[x]$ and the greatest common divisor of such polynomials (it is the last subresultant in the sequence not identically 0). The proof of the version we present here corresponding to our subresultant definition can be found in (Gonzàlez-Vega et al. 1990, 1994).

In the statement of the *subresultant theorem* we use the following notation:

- if $P \in \mathbb{D}[x]$ then $\mathrm{lcof}(P)$ will be the leading coefficient of P and $\mathrm{coef}_k(P)$ the coefficient of x^k in P,
- if $P, Q \in \mathbb{D}[x]$ with $p = \deg(P) \geq q = \deg(Q)$ then:

$$\mathrm{Prem}(P, Q) = \mathrm{remainder}(\mathrm{lcof}(Q)^{p-q+1}P, Q).$$

Theorem 1 (Subresultant Theorem) *Let P, Q be polynomials in $\mathbb{D}[x]$ with $\deg(P) \leq p = q + 1$ and $\deg(Q) \leq q$, one of the last two inequalities being an equality. For every k in $\{0, \ldots, q\}$ let*

$$S_k = \mathrm{Sres}_k(P, p, Q, q) \text{ and } s_k = \mathrm{sres}_k(P, p, Q, q).$$

Then for every $j \in \{1, \ldots, q-1\}$ such that $s_{j+1} \neq 0$ and $\deg(S_j) = r \leq j$ we have the following equalities:

1. *if $r < j - 1$ then $S_{j-1} = \ldots = S_{r+1} = 0$,*
2. *if $r < j$ then $s_{j+1}^{j-r} S_r = \mathrm{lcof}(S_j)^{j-r} S_j$,*
3. *$(-s_{j+1})^{j-r+2} S_{r-1} = \mathrm{Prem}(S_{j+1}, S_j)$.*

We next define the Sturm–Habicht sequence associated with P and Q as the subresultant sequence of P and $P'Q$ modulo some specified sign changes.

Definition 2 *Let P and Q be polynomials in $\mathbb{D}[x]$ with $p = \deg(P)$ and $q = \deg(Q)$. If we write $v = p + q - 1$ and*

$$\delta_k = (-1)^{k(k+1)/2}$$

for every integer k, the Sturm–Habicht sequence associated to P and Q is defined as the list of polynomials $\{\mathrm{StHa}_j(P,Q)\}_{j=0,\ldots,v+1}$ where $\mathrm{StHa}_{v+1}(P,Q) = P$, $\mathrm{StHa}_v(P,Q) = P'Q$ and for every $j \in \{0, \ldots, v-1\}$:

$$\mathrm{StHa}_j(P,Q) = \delta_{v-j}\mathrm{Sres}_j(P, v+1, P'Q, v).$$

For every j in $\{0, \ldots, v+1\}$ the principal j-th Sturm–Habicht coefficient is defined as:
$$\mathrm{stha}_j(P,Q) = \mathrm{coef}_j(\mathrm{StHa}_j(P,Q)).$$

The definition of the Sturm–Habicht sequence through subresultants allows us to give a structure theorem for this new sequence. This theorem is used to give an algorithm for computing the Sturm–Habicht sequence and, most important, to prove that the Sturm–Habicht sequence can be used for real root counting as was the Sturm sequence. But the Sturm–Habicht sequence has nicer properties: nice specialization properties, good worst-case complexity bounds, etc. The next section will be devoted to showing such results using only the *Sturm–Habicht structure theorem.*

Theorem 2 (Sturm–Habicht Structure Theorem) *Let P, Q be polynomials in $\mathbb{D}[x]$ with $p = \deg(P)$ and $q = \deg(Q)$ and $v = p + q - 1$. For k in $\{0, \ldots, v\}$, let*
$$H_k = \mathrm{StHa}_k(P,Q) \text{ and } h_k = \mathrm{stha}_k(P,Q)$$

and let $h_{v+1} = 1$. Then for every $j \in \{1, \ldots, v\}$ such that $h_{j+1} \neq 0$ and $\deg(H_j) = r \leq j$ we have

1. *if $r < j - 1$ then $H_{j-1} = \ldots = H_{r-1} = 0$,*
2. *if $r < j$ then $h_{j+1}^{j-r} H_r = \delta_{j-r}\mathrm{lcof}(H_j)^{j-r} H_j$,*
3. *$h_{j+1}^{j-r+2} H_{r-1} = \delta_{j-r+2}\mathrm{Prem}(H_{j+1}, H_j)$.*

Proof. It is straightforward using the definition of $\mathrm{StHa}_k(P,Q)$, the subresultant theorem (Theorem 1) and taking care of the signs introduced by the δ_k's. ∎

3   Sturm–Habicht Sequences and Real Roots of Polynomials

Let $\mathbb{D}$ be an ordered domain and $\mathbb{B}$ its real closure. In order to establish the relation between the number of real zeros (zeros in $\mathbb{B}$) of a polynomial $P \in \mathbb{D}[x]$ and the polynomials in the Sturm–Habicht sequence of P and Q, with $Q \in \mathbb{D}[x]$, we introduce the following integers for every $\epsilon \in \{-1, 0, +1\}$ and $\alpha, \beta \in \mathbb{B}$ with $\alpha < \beta$:

- $c_\epsilon^{[\alpha,\beta]}(P;Q) = \mathrm{card}(\{\gamma \in (\alpha,\beta)/P(\gamma) = 0,\ \mathrm{sign}(Q(\gamma)) = \epsilon\})$,
- $c_\epsilon(P;Q) = \mathrm{card}(\{\alpha \in \mathbb{B}/P(\alpha) = 0,\ \mathrm{sign}(Q(\alpha)) = \epsilon\})$.

The principal aim of this section is to show how to use the polynomials in the Sturm–Habicht sequence of P and Q to compute $c_+(P;Q) - c_-(P;Q)$. In the case when $Q = 1$ we get the number of real zeros of P and when $Q = F^2$, with $F \in \mathbb{D}[x]$ we get the number of real zeros of P making F different from 0. The next definitions introduce several sign counting functions that we shall use to compute the integer $c_+(P;Q) - c_-(P;Q)$.

Definition 3 *Let $\{a_0, a_1, \ldots, a_n\}$ be a list of non zero elements in $\mathbb{B}$. $\mathbf{V}(\{a_0, a_1, \ldots, a_n\})$ is the number of sign variations in the list $\{a_0, a_1, \ldots, a_n\}$, that is the number of consecutive signs $\{+, -\}$ or $\{-, +\}$. $\mathbf{P}(\{a_0, a_1, \ldots, a_n\})$ is the number of sign permanences in the list $\{a_0, a_1, \ldots, a_n\}$, that is the number of consecutive signs $\{+, +\}$ or $\{-, -\}$.*

Example 1 $\mathbf{V}(\{1, -1, 2, 3, 4, -5, -2, 3\}) = 4$ *and* $\mathbf{P}(\{1, -1, 2, 3, 4, -5, -2, 3\}) = 3$.

Definition 4 *Let P and Q polynomials in $\mathbb{D}[x]$ and α an element of $\mathbb{B} \cup \{-\infty, +\infty\}$ with $P(\alpha) \neq 0$. We define the integer $\mathbf{W}_{\mathrm{StHa}}(P, Q; \alpha)$ in the following way.*

1. *We construct a list of polynomials $\{g_0, \ldots, g_s\}$ in $\mathbb{D}[x]$ obtained from $\{\mathrm{StHa}_j(P,Q)\}_{j=0,\ldots,v+1}$ by deleting the polynomials identically 0.*
2. *$\mathbf{W}_{\mathrm{StHa}}(P, Q; \alpha)$ is the number of sign variations in the list $\{g_0(\alpha), \ldots, g_s(\alpha)\}$ using the following rules for the groups of 0's.*

   - *Count 1 sign variation for the groups $[-, 0, +]$, $[+, 0, -]$, $[+, 0, 0, -]$ and $[-, 0, 0, +]$,*
   - *Count 2 sign variations for the groups $[+, 0, 0, +]$ and $[-, 0, 0, -]$.*

The Sturm–Habicht structure theorem implies that it is not possible to find more than two consecutive zeros in the sequence $\{g_0, \ldots, g_s\}$ and that the sign sequences $[+, 0, +]$, $[-, 0, -]$ cannot appear (see the proof of Theorem 3).

In the case when $\alpha \in \{-\infty, +\infty\}$, $\mathbf{W}_{\mathrm{StHa}}(P, Q; \alpha)$ coincides with the number of sign variations in the list of non-identically zero polynomials in $\{\mathrm{StHa}_j(P,Q)\}_{j=0,\ldots,v+1}$ evaluated at α.

$$\mathbf{W}_{\mathrm{StHa}}(P, Q; \alpha) = \mathbf{V}(\{g_0(\alpha), \ldots, g_s(\alpha)\})$$

Definition 5 *Let P and Q be polynomials in $\mathbb{D}[x]$ and α, β in $\mathbb{B} \cup \{-\infty, +\infty\}$ with $\alpha < \beta$. Define*

$$\mathbf{W}_{\mathrm{StHa}}(P, Q; \alpha, \beta) = \mathbf{W}_{\mathrm{StHa}}(P, Q; \alpha) - \mathbf{W}_{\mathrm{StHa}}(P, Q; \beta)$$

and

$$\mathbf{W}_{\mathrm{StHa}}(P, Q) = \mathbf{W}_{\mathrm{StHa}}(P, Q; -\infty) - \mathbf{W}_{\mathrm{StHa}}(P, Q; +\infty).$$

Theorem 3 *Let P and Q be polynomials in $\mathbb{D}[x]$ and α, β in $\mathbb{B} \cup \{-\infty, +\infty\}$ with $\alpha < \beta$ and $P(\alpha)P(\beta) \neq 0$. Then*

1. $\mathbf{W}_{\mathrm{StHa}}(P, Q; \alpha, \beta) = c_+^{[\alpha, \beta]}(P; Q) - c_-^{[\alpha, \beta]}(P; Q)$,
2. $\mathbf{W}_{\mathrm{StHa}}(P, Q) = c_+(P; Q) - c_-(P; Q)$.

Proof. Let v be equal to $p + q - 1$ with $\deg(P) = p$ and $\deg(Q) = q$. If $\alpha = \sigma_0 < \sigma_1 < \ldots < \sigma_s < \sigma_{s+1} = \beta$ are the real roots in $[\alpha, \beta]$ of the polynomials

$$H_j = \mathrm{StHa}_j(P, Q) \quad j \in \{0, \ldots, v + 1\}$$

we shall consider

- for every $i \in \{0, \ldots, s\}$ an element σ_i^+ between σ_i and σ_{i+1},
- for every $i \in \{1, \ldots, s+1\}$ an element σ_i^- between σ_{i-1} and σ_i.

We have the following equalities.

$$
\begin{aligned}
\mathbf{W}_{\mathrm{StHa}}(P, Q; \alpha, \beta) &= \mathbf{W}_{\mathrm{StHa}}(P, Q; \alpha) - \mathbf{W}_{\mathrm{StHa}}(P, Q; \beta) \\
&= \mathbf{W}_{\mathrm{StHa}}(P, Q; \sigma_0) - \mathbf{W}_{\mathrm{StHa}}(P, Q; \sigma_0^+) + \\
&\quad \sum_{i=1}^{s} \left[\mathbf{W}_{\mathrm{StHa}}(P, Q; \sigma_i^-) - \mathbf{W}_{\mathrm{StHa}}(P, Q; \sigma_i^+) \right] + \\
&\quad \mathbf{W}_{\mathrm{StHa}}(P, Q; \sigma_{s+1}^-) - \mathbf{W}_{\mathrm{StHa}}(P, Q; \sigma_{s+1})
\end{aligned}
$$

that reduce the proof to the study of the integers $\mathbf{W}_{\mathrm{StHa}}(P, Q; \sigma_i^-) - \mathbf{W}_{\mathrm{StHa}}(P, Q; \sigma_i^+)$, $\mathbf{W}_{\mathrm{StHa}}(P, Q; \sigma_0) - \mathbf{W}_{\mathrm{StHa}}(P, Q; \sigma_0^+)$ and $\mathbf{W}_{\mathrm{StHa}}(P, Q; \sigma_{s+1}^-) - \mathbf{W}_{\mathrm{StHa}}(P, Q; \sigma_{s+1})$ for every $i \in \{1, \ldots, s\}$.

We remark that the non-identically zero polynomials in the list $\{H_j : 0 \leq j \leq v + 1\}$ do not vanish on σ_i^- and σ_i^+ and that H_u, the last polynomial non-identically 0 in the sequence under consideration, is a greatest common divisor of P and $P'Q$. The proof of the theorem is obtained by proving the following lemmas. ∎

Lemma 1 *If $i \in \{1, \ldots, s\}$ and $P(\sigma_i) \neq 0$ then $\mathbf{W}_{\mathrm{StHa}}(P, Q; \sigma_i^-) - \mathbf{W}_{\mathrm{StHa}}(P, Q; \sigma_i^+) = 0$.*

Proof. We note first that the signs of H_{v+1}, H_u (because $P(\sigma_i) \neq 0$) and H_k (with $H_k(\sigma_i) \neq 0$) agree when evaluated in σ_i^- and σ_i^+:

$$\text{sign}(H_{v+1}(\sigma_i^-)) = \text{sign}(H_{v+1}(\sigma_i^+)),$$
$$\text{sign}(H_u(\sigma_i^-)) = \text{sign}(H_u(\sigma_i^+)),$$
$$\text{sign}(H_k(\sigma_i^-)) = \text{sign}(H_k(\sigma_i^+)).$$

So, we only need to know the behavior of the non-identically zero polynomials in the sequence when σ_i is a zero of some H_k.

The only three possibilities are the following.

(A$_1$) $k = v$ and $H_v(\sigma_i) = 0$.

In this case it is easy, applying Definition 2, to verify that

$$\begin{aligned} H_{v-1}(\sigma_i) &= -\text{lcof}(P'Q)^2 P(\sigma_i) \\ &= -\text{lcof}(P'Q)^2 H_{v+1}(\sigma_i), \end{aligned}$$

which implies (because $P(\sigma_i) \neq 0$) that

$$\begin{aligned} \mathbf{V}(H_{v+1}(\sigma_i^-), H_v(\sigma_i^-), H_{v-1}(\sigma_i^-)) &= \mathbf{V}(H_{v+1}(\sigma_i^+), H_v(\sigma_i^+), H_{v-1}(\sigma_i^+)) \\ &= 1. \end{aligned}$$

(A$_2$) $k \leq v - 1$ with $\deg(H_k) = k$, $\deg(H_{k+1}) = k + 1$ and $H_k(\sigma_i) = 0$.

In this case since $H_k(\sigma_i) = 0$ and $H_u(\sigma_i) \neq 0$, $H_{k+1}(\sigma_i) \neq 0$. So applying Identity (3) of Theorem 2 we get

$$\begin{aligned} &\text{lcof}(H_{k+1})^2 H_{k-1} = \delta_2 \text{Prem}(H_{k+1}, H_k) \\ \implies\ &\text{lcof}(H_{k+1})^2 H_{k-1}(\sigma_i) = -\text{lcof}(H_k)^2 H_{k+1}(\sigma_i) \\ \implies\ &\text{sign}(H_{k-1}(\sigma_i) H_{k+1}(\sigma_i)) = -1 \end{aligned}$$

which implies as in the last case

$$\begin{aligned} \mathbf{V}(H_{k+1}(\sigma_i^-), H_k(\sigma_i^-), H_{k-1}(\sigma_i^-)) &= \mathbf{V}(H_{k+1}(\sigma_i^+), H_k(\sigma_i^+), H_{k-1}(\sigma_i^+)) \\ &= 1. \end{aligned}$$

(A$_3$) $k \leq v - 1$ with $\deg(H_k) = r < k$, $\deg(H_{k+1}) = k + 1$ and $H_k(\sigma_i) = 0$.

As in the previous case we have $H_{k+1}(\sigma_i) \neq 0$. Applying Identity (3) of Theorem 2 we get

$$\text{lcof}(H_{k+1})^{k-r+2} H_{r-1}(\sigma_i) = \delta_{k-r+2} \text{lcof}(H_k)^{k-r+2} H_{k+1}(\sigma_i),$$

provided that $H_{k-1}(\sigma_i) \neq 0$, and using Identity (2) of Theorem 2 we obtain, with τ equal to σ_i^- or σ_i^+

$$\text{sign}(H_{r-1}(\tau) H_{k+1}(\tau)) = \frac{\delta_{k-r+2}}{\delta_{k-r}} \text{sign}(H_r(\tau) H_k(\tau)),$$

which implies directly that

$$\mathbf{V}(H_{k+1}(\tau), H_k(\tau), 0, \ldots, 0, H_r(\tau), H_{r-1}(\tau)) =$$
$$\begin{cases} 2 & \text{if } \operatorname{sign}(H_{r-1}(\sigma_i)H_{k+1}(\sigma_i)) = 1. \\ 1 & \text{if } \operatorname{sign}(H_{r-1}(\sigma_i)H_{k+1}(\sigma_i)) = -1. \end{cases}$$

So we conclude that, for σ_i satisfying $P(\sigma_i) \neq 0$, $\mathbf{W}_{\mathrm{StHa}}(P, Q; \sigma_i^-) - \mathbf{W}_{\mathrm{StHa}}(P, Q; \sigma_i^+) = 0$, which is what we wanted to show. ∎

Lemma 2 *If* $i \in \{1, \ldots, s\}$ *and* $P(\sigma_i) = 0$ *then* $\mathbf{W}_{\mathrm{StHa}}(P, Q; \sigma_i^-) - \mathbf{W}_{\mathrm{StHa}}(P, Q; \sigma_i^+) = \operatorname{sign}(Q(\sigma_i))$.

Proof. As a first step we define a new sequence of polynomials.

$$G_j = \frac{H_j}{H_u} \text{ for } j \in \{u, u+1, \ldots, v+1\},$$

and clearly we have $\mathbf{V}(G_{v+1}(\tau), \ldots, G_u(\tau)) = \mathbf{V}(H_{v+1}(\tau), \ldots, H_u(\tau))$ with $\tau = \sigma_i^+$ or $\tau = \sigma_i^-$. Moreover, observing carefully the relations appearing in Theorem 2 we obtain that the sequence G_j satisfies the same properties as the sequence H_j. So repeating the proofs we gave to prove (A$_2$) and (A$_3$) we conclude that if $G_v(\tau) \neq 0$ then $\mathbf{V}(G_v(\sigma_i^-), \ldots, G_u(\sigma_i^-)) - \mathbf{V}(G_v(\sigma_i^+), \ldots, G_u(\sigma_i^+)) = 0$, and if $G_v(\tau) = 0$ then $\mathbf{V}(G_v(\sigma_i^-), \ldots, G_u(\sigma_i^-)) - \mathbf{V}(G_v(\sigma_i^+), \ldots, G_u(\sigma_i^+)) = 0$.

Let

$$\begin{aligned} P(x) &= (x - \sigma_i)^e H(x) \text{ where } H(\sigma_i) \neq 0 \\ P'(x)Q(x) &= e(x - \sigma_i)^{e-1} H(x)Q(x) + (x - \sigma_i)^e H'(x)Q(x). \end{aligned}$$

We only need to study two cases.

(B$_1$) $\mathbf{V}(G_{v+1}(\sigma_i^-), G_v(\sigma_i^-)) - \mathbf{V}(G_{v+1}(\sigma_i^+), G_v(\sigma_i^+))$ when $G_v(\sigma_i) \neq 0$.
Since G_v is the quotient of $P'Q$ by the greatest common divisor of P and $P'Q$ and $G_v(\sigma_i) \neq 0$ then either $Q(\sigma_i) \neq 0$ or σ_i is a root of Q with multiplicity 1.
If $Q(\sigma_i) \neq 0$ then

$$\begin{aligned} G_{v+1}(x) &= (x - \sigma_i)\underline{H}(x) \text{ where } \underline{H}(\sigma_i) \neq 0, \\ G_v(x) &= e\underline{H}(x)Q(x) + (x - \sigma_i)U(x), \end{aligned}$$

which gives directly that

$$\mathbf{V}(G_{v+1}(\sigma_i^-), G_v(\sigma_i^-)) - \mathbf{V}(G_{v+1}(\sigma_i^+), G_v(\sigma_i^+)) =$$
$$\mathbf{V}(-\operatorname{sign}(\underline{H}(\sigma_i)), \operatorname{sign}(\underline{H}(\sigma_i)Q(\sigma_i))) - \mathbf{V}(\operatorname{sign}(\underline{H}(\sigma_i)), \operatorname{sign}(\underline{H}(\sigma_i)Q(\sigma_i)))$$
$$= \operatorname{sign}(Q(\sigma_i)).$$

If $Q(\sigma_i) = 0$, writing $Q(x) = (x - \sigma_i)q(x)$ with $q(\sigma_i) \neq 0$, then

$$
\begin{aligned}
G_{v+1}(x) &= \underline{H}(x) \text{ where } \underline{H}(\sigma_i) \neq 0 \\
G_v(x) &= e\underline{H}(x)Q(x) + (x - \sigma_i)V(x)
\end{aligned}
$$

which gives directly that

$$
\mathbf{V}(G_{v+1}(\sigma_i^-), G_v(\sigma_i^-)) - \mathbf{V}(G_{v+1}(\sigma_i^+), G_v(\sigma_i^+)) =
$$
$$
\mathbf{V}(\mathrm{sign}(\underline{H}(\sigma_i)), \mathrm{sign}(\underline{H}(\sigma_i)Q(\sigma_i))) - \mathbf{V}(\mathrm{sign}(\underline{H}(\sigma_i)), \mathrm{sign}(\underline{H}(\sigma_i)Q(\sigma_i)))
$$
$$
= 0 = \mathrm{sign}(Q(\sigma_i))a.
$$

(B_2) $\mathbf{V}(G_{v+1}(\sigma_i^-), G_v(\sigma_i^-), G_{v-1}(\sigma_i^-)) \; - \; \mathbf{V}(G_{v+1}(\sigma_i^+), G_v(\sigma_i^+), G_{v-1}(\sigma_i^+))$
when $G_v(\sigma_i) = 0$.

The condition $G_v(\sigma_i) = 0$ implies that the degree of Q must be greater than 1 which implies (looking at Definition 2) that

$$
H_{v-1} = -\mathrm{lcof}(P'Q)^2 P \quad \Longrightarrow \quad G_{v-1} = -\mathrm{lcof}(P'Q)^2 \underline{H}(x),
$$

with $\underline{H}(x)$ the same as in case (B_1). The equality

$$
G_{v+1}(\sigma_i)G_{v-1}(\sigma_i) = -\left[\mathrm{lcof}(P'Q)\underline{H}(\sigma_i)\right]^2 < 0
$$

provides directly that

$$
\mathbf{V}(G_{v+1}(\sigma_i^-), G_v(\sigma_i^-), G_{v-1}(\sigma_i^-)) - \mathbf{V}(G_{v+1}(\sigma_i^+), G_v(\sigma_i^+), G_{v-1}(\sigma_i^+))
$$
$$
= 0 = \mathrm{sign}(Q(\sigma_i)).
$$

From the equalities obtained in (B_1) and (B_2) we conclude that for every $i \in \{1, \ldots, s\}$ with $P(\sigma_i) = 0$ we have $\mathbf{V}(G_{v+1}(\sigma_i^-), G_v(\sigma_i^-), G_{v-1}(\sigma_i^-)) - \mathbf{V}(G_{v+1}(\sigma_i^+), G_v(\sigma_i^+), G_{v-1}(\sigma_i^+)) = \mathrm{sign}(Q(\sigma_i))$, which is what we wanted to show. ∎

Lemma 3 $\mathbf{W}_{\mathrm{StHa}}(P, Q; \sigma_0^-) - \mathbf{W}_{\mathrm{StHa}}(P, Q; \sigma_0^+) = \mathbf{W}_{\mathrm{StHa}}(P, Q; \sigma_{s+1}^-) - \mathbf{W}_{\mathrm{StHa}}(P, Q; \sigma_{s+1}^+) = 0$.

Proof. Clearly it is enough to study the case corresponding to σ_0. Given that $P(\sigma_0) \neq 0$, using the same strategy as in Lemma 1, it is enough to consider the two following cases.

(C_1) $k \leq v - 1$ with $\deg(H_k) = k$, $\deg(H_{k+1}) = k+1$ and $H_k(\sigma_i) = 0$. The only possibilities are:

	σ_0^-	σ_0^+	σ_0^-	σ_0^+	σ_0^-	σ_0^+	σ_0^-	σ_0^+
H_{k+1}	+	+	+	+	−	−	−	−
H_k	0	+	0	−	0	+	0	−
H_{k-1}	−	−	−	−	+	+	+	+
	$W = 1$		$W = 1$		$W = 1$		$W = 1$	

(C₂) $k \leq v - 1$ with $\deg(H_k) = r < k$, $\deg(H_{k+1}) = k + 1$ and $H_k(\sigma_i) = 0$. The only possibilities are:

	σ_0	σ_0^+	σ_0	σ_0^+	σ_0	σ_0^+	σ_0	σ_0^+
H_{k+1}	+	+	+	+	+	+	+	+
H_k	0	+	0	−	0	+	0	−
H_{k-1}	0	0	0	0	0	0	0	0
⋮	⋮		⋮		⋮		⋮	
H_{r+1}	0	0	0	0	0	0	0	0
H_r	0	−	0	+	0	+	0	−
H_{r-1}	+	+	+	+	−	−	−	−
	$W = 2$		$W = 2$		$W = 1$		$W = 1$	
	σ_0	σ_0^+	σ_0	σ_0^+	σ_0	σ_0^+	σ_0	σ_0^+
H_{k+1}	−	−	−	−	−	−	−	−
H_k	0	+	0	−	0	+	0	−
H_{k-1}	0	0	0	0	0	0	0	0
⋮	⋮		⋮		⋮		⋮	
H_{r+1}	0	0	0	0	0	0	0	0
H_r	0	+	0	−	0	+	0	−
H_{r-1}	+	+	+	+	−	−	−	−
	$W = 1$		$W = 1$		$W = 2$		$W = 2$	

The two tables in (C₁) and (C₂) allow us to conclude that

$$\mathbf{W}_{\text{StHa}}(P, Q; \sigma_0) - \mathbf{W}_{\text{StHa}}(P, Q; \sigma_0^+) = 0,$$

as we wanted to show. ∎

Another proof of this theorem can be found in (Gonzàlez-Vega et al. 1990, 1994) or (Lombardi 1989), based on the relations between the Sturm–Habicht sequence and the Sturm sequence. In (Habicht 1948) this theorem is proved when P is squarefree, P and Q are coprime and α and β are not zeros of a non identically zero polynomial in the Sturm–Habicht sequence associated with P and Q.

4   Sturm–Habicht Sequences and Hankel Forms

In this section we study how to compute the integer $c_+(P; Q) - c_-(P; Q)$ knowing only the principal Sturm–Habicht coefficients associated to P and Q. We shall also make the connection between the Sturm–Habicht sequence and the Hermite–Hankel form associated to P and Q.

Definition 6 *Let $a_0, a_1, \ldots, a_n$ be elements in $\mathbb{B}$, $a_0 \neq 0$, with the following distribution of zeros:*

$$\{a_0, a_1, \ldots, a_n\} =$$

$$\{a_0, \ldots, a_{i_1}, \overbrace{0, \ldots, 0}^{k_1}, a_{i_1+k_1+1}, \ldots, a_{i_2}, \overbrace{0, \ldots, 0}^{k_2}, a_{i_2+k_2+1}, \ldots, a_{i_3},$$

$$0, \ldots \ldots, 0, a_{i_{t-1}+k_{t-1}+1}, \ldots, a_{i_t}, \overbrace{0, \ldots, 0}^{k_t}\}$$

where all the a_i's that have been written are not 0. Let $i_0 + k_0 + 1 = 0$. Then

$$\mathbf{C}(\{a_0, a_1, \ldots, a_n\}) =$$

$$\sum_{s=1}^{t} (\mathbf{P}(\{a_{i_{s-1}+k_{s-1}+1}, \ldots, a_{i_s}\}) - \mathbf{V}(\{a_{i_{s-1}+k_{s-1}+1}, \ldots, a_{i_s}\})) +$$

$$\sum_{s=1}^{t-1} \varepsilon_{i_s},$$

where

$$\varepsilon_{i_s} = \begin{cases} 0 & \text{if } k_s \text{ is odd} \\ (-1)^{k_s/2} \operatorname{sign}(\frac{a_{i_s+k_s+1}}{a_{i_s}}) & \text{if } k_s \text{ is even.} \end{cases}$$

The next theorem gives a formula for $c_+(P; Q) - c_-(P; Q)$ based only on the sign determination of $p + 1$ determinants.

Theorem 4 *If P and Q are polynomials in $\mathbb{D}[x]$ with $p = \deg(P)$ then*

$$\mathbf{C}(\{\operatorname{stha}_p(P, Q), \ldots, \operatorname{stha}_0(P, Q)\}) = c_+(P; Q) - c_-(P; Q).$$

Proof. It is enough to prove the statement of the theorem when P is monic. Otherwise it suffices to apply the monic case to $P/\operatorname{lcof}(P)$. If all the polynomials in the Sturm–Habicht sequence associated to P and Q are regular (their index in the Sturm–Habicht sequence is equal to their degree) then it is clear that

$$
\begin{aligned}
&c_+(P; Q) - c_-(P; Q) \\
&= \mathbf{W}_{\mathrm{StHa}}(P, Q) = \mathbf{W}_{\mathrm{StHa}}(P, Q, -\infty) - \mathbf{W}_{\mathrm{StHa}}(P, Q, +\infty) \\
&= \mathbf{V}(\{(-1)^p, (-1)^v \operatorname{stha}_v(P, Q), \ldots, -\operatorname{stha}_1(P, Q), \operatorname{stha}_0(P, Q)\}) - \\
&\quad\; \mathbf{V}(\{1, \operatorname{stha}_v(P, Q), \ldots, \operatorname{stha}_1(P, Q), \operatorname{stha}_0(P, Q)\}) \\
&= \mathbf{P}(\{1, \operatorname{stha}_{p-1}(P, Q), \ldots, \operatorname{stha}_0(P, Q)\}) - \\
&\quad\; \mathbf{V}(\{1, \operatorname{stha}_{p-1}(P, Q), \ldots, \operatorname{stha}_0(P, Q)\}) \\
&= \mathbf{C}(\{\operatorname{stha}_p(P, Q), \ldots, \operatorname{stha}_0(P, Q)\}).
\end{aligned}
$$

This is a consequence of the fact that if all the polynomials in the sequence are regular then $v + 1 = p + q = p$ which implies that $q = 0$ and $v = p - 1$.

Problems arise when in the Sturm–Habicht sequence there appears some $\text{stha}_j(P,Q)$ which is equal to zero. There are two possibilities in this situation: (D_1) the index j such that $\text{stha}_j(P,Q) = 0$ is bigger than p or (D_2) smaller than p.

(D_1) We show that the structure of the sequence up to index p directly gives the result.

- If $q = \deg(Q) > 1$, then $\text{StHa}_{v+1}(P,Q) = P$, $\text{StHa}_v(P,Q) = P'Q$, $\text{StHa}_{v-1}(P,Q) = -\text{lcof}(P'Q)^2 P$, $\text{StHa}_{v-2}(P,Q) = \cdots = \text{StHa}_{p+1}(P,Q)$ $= 0$, and $\text{StHa}_p(P,Q) = \delta_{q-1}\text{lcof}(P'Q)^q P$, which allows us to conclude that

$$\mathbf{V}(\{\text{StHa}_j(P,Q)(-\infty)\}_{p \le j \le v+1}) - \mathbf{V}(\{\text{StHa}_j(P,Q)(+\infty)\}_{p \le j \le v+1}) = 0.$$

- If $q = \deg(Q) = 1$, then $\text{StHa}_{v+1}(P,Q) = \text{StHa}_{p+1}(P,Q) = P$, $\text{StHa}_v(P,Q) = \text{StHa}_p(P,Q) = P'Q$, which allows to conclude that

$$\mathbf{V}(\{\text{StHa}_j(P,Q)(-\infty)\}_{j=p+1,p}) - \mathbf{V}(\{\text{StHa}_j(P,Q)(+\infty)\}_{j=p+1,p}) = 0.$$

(D_2) In the second case,

$$\text{stha}_k(P,Q) \ne 0, \text{stha}_{k-1}(P,Q) = \cdots = \text{stha}_{k-h}(P,Q) = 0,$$
$$\text{stha}_{k-h-1}(P,Q) \ne 0,$$

and we need to prove the following equality.

$$\Delta \overset{\text{def}}{=} \mathbf{V}(\{\text{StHa}_k(P,Q)(-\infty), \text{StHa}_{k-1}(P,Q)(-\infty), 0, \ldots,$$
$$\text{StHa}_{k-h-1}(P,Q)(-\infty)\}) -$$
$$\mathbf{V}(\{\text{StHa}_k(P,Q)(+\infty), \text{StHa}_{k-1}(P,Q)(+\infty), 0, \ldots,$$
$$\text{StHa}_{k-h-1}(P,Q)(+\infty)\})$$
$$\overset{?}{=} \tau \overset{\text{def}}{=} \begin{cases} 0 & \text{if } h \text{ is odd} \\ (-1)^{h/2} \operatorname{sign}(\frac{\text{stha}_{k-h-1}(P,Q)}{\text{stha}_k(P,Q)}) & \text{if } h \text{ is even} \end{cases}$$

Applying (2) in Theorem 2 to this situation and denoting by ρ the leading coefficient of $\text{StHa}_{k-1}(P,Q)$ we get the following relation.

$$(\text{stha}_k(P,Q))^h \text{stha}_{k-h-1}(P,Q) = \delta_h \rho^{h+1}. \tag{1}$$

The equality denoted by $\overset{?}{=}$ is now obtained using Eq. 1 and considering the following three cases:

- if h is odd then $\tau = \Delta = 0$,
- if h is even and $h/2$ even then $\tau = \Delta = 1$,
- if h is even and $h/2$ odd then $\tau = \Delta = -1$.

Cases (D_1) and (D_2) provide the proof of the theorem. ∎

The "magic formula" appearing in Definition 6 requires some further explanation and will be better understood when we shall make the connection between Sturm–Habicht sequences and Hermite's method.

Let P and Q be in $\mathbb{D}[x]$ with $p = \deg(P)$ and $\alpha_1, \ldots, \alpha_p$ all the roots of P (counted with multiplicity) in an algebraic closure of $\mathbb{B}$. We define for every non negative integer k the generalized Newton sum as

$$s_k^{P,Q} = \sum_{i=1}^{p} \alpha_i^k Q(\alpha_i),$$

and the Hermite–Hankel form associated to P and Q as

$$B(P,Q) = \begin{pmatrix} s_0^{P,Q} & \cdots & s_{p-1}^{P,Q} \\ \vdots & & \vdots \\ s_{p-1}^{P,Q} & \cdots & s_{2p-2}^{P,Q} \end{pmatrix}.$$

For every j in $\{1, \ldots, p\}$ we denote by $b_j(P,Q)$ the j-th principal minor of the matrix $B(P,Q)$.

It is very easy to see that the signature of the matrix $B(P,Q)$ is the integer $c_+(P;Q) - c_-(P;Q)$ (see Gantmacher 1959 or Gonzàlez-Vega et al. 1990, 1994). On the other hand it is a classical (and non trivial) result in Hankel form theory (see Gantmacher 1959) that

$$\text{signature}(B(P,Q)) = \mathbf{C}(\{1, b_1(P,Q), \ldots, b_p(P,Q)\}). \tag{2}$$

So the mysterious quantity $\mathbf{C}$ appears naturally in the context of Hermite's method and solves Problem 1 too. This suggests connections between the Sturm–Habicht principal coefficients and the $b_j(P,Q)$. They will be given in the next theorem which proves immediately that

$$\mathbf{C}(\{1, b_1(P,Q), \ldots, b_p(P,Q)\}) = \mathbf{C}(\{\text{stha}_p(P,Q), \ldots, \text{stha}_0(P,Q)\}).$$

Using these connections between $\text{stha}_{p-j}(P,Q)$ and $b_j(P,Q)$, the fact that the signature of the matrix $B(P,Q)$ is $c_+(P;Q) - c_-(P;Q)$ and Eq. 2 we could give another proof of Theorem 4. We preferred to use the Sturm–Habicht structure theorem in order to illustrate that we can deduce everything we need from it.

Theorem 5 *Let P, Q be polynomials in $\mathbb{D}[x]$ with $p = \deg(P)$ and $\alpha_1, \ldots, \alpha_p$ all the roots of P (counted with multiplicity) in an algebraic closure, $\mathbb{B}$, of $\mathbb{D}$. For every j in $\{1, \ldots, p\}$ there exists $C_j \in \mathbb{B}$ such that $\text{sign}(C_i) = \text{sign}(C_k)$ for i and k in $\{1, \ldots, p\}$ and*

$$\text{stha}_{p-j}(P,Q) = C_j \cdot b_j(P,Q),$$

$$b_j(P,Q) = \sum_{\Delta \in \binom{[p]}{j}} \left[\prod_{t \in \Delta} Q(\alpha_t) \right] \left[\prod_{u < v \in \Delta} (\alpha_u - \alpha_v)^2 \right]$$

where $[p] = \{1, \ldots, p\}$ and

$$\binom{[p]}{j} = \{\{k_1, \ldots, k_j\} \subset [p]/k_1 < \cdots < k_j\}.$$

Proof. To prove the first equality we shall use the fact that every $s_k = s_k^{P,Q}$ can be obtained from the power series expansion at infinity of the rational function $P'Q/P$. In fact

$$\frac{P'Q}{P} = A(x) + \frac{R}{P} = A(x) + \frac{s_0}{x} + \frac{s_1}{x^2} + \cdots + \frac{s_i}{x^{i+1}} + \cdots \qquad (3)$$

where $A(x)$ is the quotient and $R(x)$ the remainder of the Euclidean division of $P'Q$ and P. If we write

$$P = \sum_{k=0}^{p} a_k x^k, \quad R = \sum_{k=0}^{p-1} c_k x^k, \quad Q = \sum_{k=0}^{q} b_k x^k, \quad \text{and} \quad P'Q = \sum_{k=0}^{v} d_k x^k,$$

then the formula in Eq. 3 gives the following relations between the c_k's, the a_k's and the s_k's:

$$c_k = \sum_{i=1}^{p} a_i s_{i-k-1} \quad \text{and} \quad 0 = \sum_{i=0}^{p} a_i s_{i-k-1} \text{ if } k < 0.$$

We now get the first equality. By definition

$$\text{stha}_{p-j}(P,Q) = \delta_{v-p+j} \begin{vmatrix} a_{v+1} & \cdots & \cdots & \cdots & a_* \\ & \ddots & & & \vdots \\ & & a_{v+1} & \cdots & a_{p-j} \\ d_v & \cdots & \cdots & \cdots & d_* \\ & \ddots & & & \vdots \\ & & d_v & \cdots & d_{p-j} \end{vmatrix}.$$

Performing Euclidean division this is equal to

$$\delta_{v-p+j} d_v^q a_p^q (-1)^{q(v-j)} \begin{vmatrix} a_p & \cdots & \cdots & \cdots & a_{p-2j+2} \\ & \ddots & & & \vdots \\ & & a_p & \cdots & a_{p-j} \\ c_{p-1} & \cdots & \cdots & \cdots & c_{p-2j+1} \\ & \ddots & & & \vdots \\ & & c_{p-1} & \cdots & c_{p-j} \end{vmatrix}.$$

Using the relations between the c_k's, the a_k's, and the s_k's the determinant in this equation factors into

$$
\begin{vmatrix}
1 & \dots & 0 & 0 & \dots & 0 \\
\vdots & & \vdots & \vdots & & \vdots \\
0 & \dots & 1 & 0 & \dots & 0 \\
s_0 & \dots & s_{j-2} & s_{j-1} & \dots & s_{2j-2} \\
\vdots & & \vdots & \vdots & & \vdots \\
0 & \dots & 0 & s_0 & \dots & s_{j-1}
\end{vmatrix}
\begin{vmatrix}
a_p & \dots & \dots & \dots & \dots & a_{p-2j+2} \\
 & \ddots & & & & \vdots \\
 & & a_p & \dots & \dots & a_{p-j} \\
 & & & \ddots & & \vdots \\
 & & & & \ddots & \vdots \\
 & & & & & a_p
\end{vmatrix}
$$

which implies that

$$
\mathrm{stha}_{p-j}(P,Q) = \left[(-1)^{pq} \delta_{q-1} p^q b_q^q a_p a_p^{2(q+j-1)} \right] b_j(P,Q) = C_j \cdot b_j(P,Q).
$$

To get the second equality we only need to decompose the principal minor with determinant $b_j(P,Q)$ as the product of two matrices of dimensions $j \times p$ and $p \times j$ and to apply the Cauchy–Binet formulas (see Gantmacher 1959) to compute its determinant.

$$
\begin{aligned}
b_j(P,Q) &= \left| \begin{pmatrix}
\alpha_1^0 Q(\alpha_1) & \alpha_2^0 Q(\alpha_2) & \dots & \alpha_p^0 Q(\alpha_p) \\
\alpha_1^1 Q(\alpha_1) & \alpha_2^1 Q(\alpha_2) & \dots & \alpha_p^1 Q(\alpha_p) \\
\vdots & \vdots & & \vdots \\
\alpha_1^{j-1} Q(\alpha_1) & \alpha_2^{j-1} Q(\alpha_2) & \dots & \alpha_p^{j-1} Q(\alpha_p)
\end{pmatrix} \right. \\
&\quad \left. \begin{pmatrix}
1 & \alpha_1 & \dots & \alpha_1^{j-1} \\
1 & \alpha_2 & \dots & \alpha_2^{j-1} \\
\vdots & \vdots & & \vdots \\
1 & \alpha_p & \dots & \alpha_p^{j-1}
\end{pmatrix} \right| \\
&= \sum_{\Delta \in \binom{[p]}{j}} \left[\prod_{t \in \Delta} Q(\alpha_t) \right] \left[\prod_{u < v \in \Delta} (\alpha_u - \alpha_v)^2 \right],
\end{aligned}
$$

which is what we wanted to show. ∎

5   Applications and Examples

An immediate application of Sturm–Habicht sequences are their good specialization properties: using the preceding results we can compute the number of real roots of a specific polynomial P of degree d in the following way.

The good specialization properties of Sturm–Habicht sequences provide the following algorithm for computing the number of real roots of a specific polynomial, P, of degree d.

(1) Compute the Sturm–Habicht sequence for a polynomial of same degree d with symbolic coefficients.

(2) Specialize the symbolic coefficients to the coefficients of P in the Sturm–Habicht sequence and evaluate the signs of the resulting principal Sturm–Habicht coefficients.

For practical purposes, this universal computation is inefficient and we would replace Step 1 by

(1') Compute the Sturm–Habicht sequence of P using the Sturm–Habicht theorem.

We hence need only d^2 algebraic operations and $d+1$ sign evaluations in $\mathbb{D}$ in order to compute the number of real roots. The specialization properties of the Sturm–Habicht sequence will be used in the following examples. In the first example, given a polynomial Q in x with coefficients depending on a parameter t, we can easily determine the conditions on t for the existence of a real root of Q.

Let

$$Q = x^8 - 2x^6 + tx^5 - x^2 - x + 1.$$

The Sturm–Habicht principal coefficients $(\mathrm{stha}_i(Q,1) = \mathbf{h}_i(t))$ are

$$
\begin{aligned}
h_8(t) &= 1 \\
h_7(t) &= 8 \\
h_6(t) &= 32 \\
h_5(t) &= 192 - 72t^2 \\
h_4(t) &= 1152 - 272t^2 - 135t^4 \\
h_3(t) &= 2160t^4 + 5720t^2 - 6720t - 34752 \\
h_2(t) &= 1215t^6 + 540t^5 + 51588t^4 + 113264t^3 - 584656t^2 - 32t + 1659104 \\
h_1(t) &= 24840t^7 + 19818t^6 + 352448t^5 + 1810496t^4 - 4725464t^3 - \\
&\quad 8804072t^2 + 36268784t + 18597192 \\
h_0(t) &= 110862t^7 - 84375t^8 - 799775t^6 - 3435148t^5 + 29105250t^4 - \\
&\quad 11266146t^3 - 187124876t^2 + 260682288t + 122141025
\end{aligned}
$$

So, considering the family of all the sign conditions $\{+,+,+,\epsilon_5,\dots,\epsilon_0\}$ ($\epsilon_i \in \{+,0,-\}$) such that $\mathbf{C}(\{+,+,+,\epsilon_5,\dots,\epsilon_0\}) > 0$ we get the necessary and sufficient conditions on t for Q to have a real root.

We finish with a more geometric example. Consider the surface in $\mathbb{R}^3$ defined by the equation

$$F = z^4 - 2xyz^3 + xz - x^2y + 1.$$

The computation of the Sturm–Habicht principal coefficients for F, considered as a polynomial in z, gives an easy way to determine the number of points of the surface over any point (x_0, y_0) in $\mathbb{R}^2$. In this case the Sturm–Habicht principal coefficients are $(\mathrm{stha}_i(Q, 1) = h_i(x, y))$

$$
\begin{aligned}
h_4(x, y) &= 1 \\
h_3(x, y) &= 4 \\
h_2(x, y) &= 12x^2 y^2 \\
h_1(x, y) &= x^2(96x^2 y^3 - 48y^2 - 36) \\
h_0(x, y) &= 864x^6 y^5 - 432x^8 y^6 + 184x^6 y^3 - 432x^4 y^4 - 24x^4 y^2 - 27x^4 - \\
&\quad 384x^2 y + 256
\end{aligned}
$$

and the information we look for is given by the expression

$$
\#\{\alpha \in \mathbb{R} : F(x_0, y_0, \alpha) = 0\} = \mathbf{C}(\{1, 4, 12x_0^2 y_0^2, h_1(x_0, y_0), h_0(x_0, y_0)\}).
$$

In the case $x_0 y_0 \neq 0$ the above expression reduces to

$$
\#\{\alpha \in \mathbb{R} : F(x_0, y_0, \alpha) = 0\} = 2 + \mathbf{C}(\{1, h_1(x_0, y_0), h_0(x_0, y_0)\}).
$$

Using Sturm–Habicht sequences, it is possible to perform Quantifier Elimination for formulas with only one quantifier and one polynomial sign in a combinatorial way (see González-Vega 1997).

Acknowledgments

L. González-Vega and T. Recio were partially supported by CICyT PB 92/0498/C02/01 (Geometría Real y Algoritmos). All of the authors were partially supported by Esprit/Bra 6846 (Posso).

Characterizations of the Macaulay Matrix and Their Algorithmic Impact

Georg Hagel

1 Introduction

Since the late eighties, the Macaulay matrix has recently received renewed attention in the field of quantifier elimination. Grigor'ev (1988) and Renegar (1992a, 1992b, 1992c) made it a central tool in their study of the complexity of quantifier elimination. They developed algorithms which have better complexities than Collins' quantifier elimination algorithm based on the CAD (Collins 1975), but until now Collins' CAD is much faster in practice.

Renegar (1992a, 1992b, 1992c) presents a method to construct a Macaulay matrix which is almost the same as the matrix introduced by Macaulay (1916) at the beginning of the century to get the solution set of a zero-dimensional ideal.

In this paper, we want to investigate the Macaulay matrix, the Macaulay determinant and the extraneous factor of (Macaulay 1916) to extensively characterize the entries of the Macaulay matrix and the extraneous factor. The fact that the homogenization method of Macaulay is in practice preferable to that proposed by Renegar is an easy consequence of our characterization of the Macaulay matrix.

In Section 3 we present the definitions of the Macaulay matrix by Renegar and Macaulay and general properties. In the subsequent sections we develop some formulae for the Macaulay matrix, the Macaulay determinant and the special case of the u-resultant computed by the Macaulay determinant. We also present a method for computing the u-resultant. After this we investigate the different methods of homogenization and their practical implications. Finally, we look at the extraneous factor of the Macaulay determinant. For more details and some of the proofs we refer the reader to (Hagel 1993).

2   Notation

Let F be a field and $f_1, f_2, \ldots, f_{n+1}$ be a set of polynomials in $F[x_1, x_2, \ldots, x_n]$ ($n \geq 1$) which is considered fixed. We denote by d_i the total degree of the polynomial f_i ($1 \leq i \leq n + 1$) and when we write deg or degree, we mean the total degree. Let x_{n+1} be a new variable, the variable of homogenization with which we bring the polynomials $f_1, f_2, \ldots, f_{n+1}$ into a homogeneous form by $F_i = x_{n+1}^{d_i} f_i(\frac{x_1}{x_{n+1}}, \frac{x_2}{x_{n+1}}, \ldots, \frac{x_n}{x_{n+1}}) \in F[x_1, x_2, \ldots, x_{n+1}](1 \leq i \leq n+1)$. This is what we call *degree-d_i-homogenization* instead of the *degree-d-homogenization*, which is defined later. Unless stated otherwise, we mean by homogenization the degree-d_i-homogenization.

A central concept is the one of a reduced polynomial: Let $f_1, f_2, \ldots, f_{n+1} \in F[x_1, x_2, \ldots, x_n]$ be polynomials of degree $d_1, d_2, \ldots, d_{n+1}$, respectively. A polynomial $P \in F[x_1, x_2, \ldots, x_{n+1}]$ is *reduced in x_i* for $1 \leq i \leq n + 1$, if no monomial of P is divisible by $x_i^{d_i}$. P is *reduced in* $\{x_{i_1}, x_{i_2}, \ldots, x_{i_k}\}$ with $1 \leq i_1, i_2, \ldots, i_k \leq n + 1$, if P is reduced in $x_{i_1}, x_{i_2}, \ldots, x_{i_k}$. We see that reducibility depends on the degree of the polynomials $f_1, f_2, \ldots, f_{n+1}$ and on their arrangement.

3   Definitions of the Macaulay Matrix

Here we present two different definitions of the Macaulay matrix. The first is given by Renegar (1992a, 1992b, 1992c) and the second by the original paper of Macaulay (1916).

Let V be the vector space of homogeneous polynomials in $F[x_1, x_2, \ldots, x_{n+1}]$ of degree $t := (\sum_{i=1}^{n+1} d_i) - n$ including the zero polynomial. The monomials of V form a basis B for the vector space V. For the elements of the basis B we define a linear transformation and extend it linearly:

Definition 1 *Let $\Psi : B \to V$ be the following transformation of the elements of the basis B: For $x_1^{\alpha_1} x_2^{\alpha_2} \cdots x_{n+1}^{\alpha_{n+1}} \in B$ let i denote the maximum index with $x_1^{\alpha_1} x_2^{\alpha_2} \cdots x_{n+1}^{\alpha_{n+1}}$ being reduced in $\{x_1, x_2, \ldots, x_{i-1}\}$. Then the image of $x_1^{\alpha_1} x_2^{\alpha_2} \cdots x_{n+1}^{\alpha_{n+1}}$ under Ψ is defined by*

$$\Psi(x_1^{\alpha_1} x_2^{\alpha_2} \cdots x_{n+1}^{\alpha_{n+1}}) := x_1^{\alpha_1} x_2^{\alpha_2} \cdots x_{i-1}^{\alpha_{i-1}} x_i^{\alpha_i - d_i} x_{i+1}^{\alpha_{i+1}} \cdots x_{n+1}^{\alpha_{n+1}} F_i.$$

By linear extension we get a linear transformation $\Psi : V \to V$.

The image of $x_1^{\alpha_1} x_2^{\alpha_2} \cdots x_{n+1}^{\alpha_{n+1}}$ is uniquely defined because i is the maximal index with $x_1^{\alpha_1} x_2^{\alpha_2} \cdots x_{n+1}^{\alpha_{n+1}}$ being reduced in $\{x_1, x_2, \ldots, x_{i-1}\}$. Therefore we have $\alpha_i \geq d_i$, and hence $\Psi(x_1^{\alpha_1} x_2^{\alpha_2} \cdots x_{n+1}^{\alpha_{n+1}}) \in V$.

Definition 2 *Let $M_R(F_1, F_2, \ldots, F_{n+1})$ denote the matrix representing Ψ with respect to the basis B. Then we call $M_R(F_1, F_2, \ldots, F_{n+1})$ the Macaulay matrix by Renegar and the Macaulay determinant by Renegar is defined by*

$$D_R(F_1, F_2, \ldots, F_{n+1}) := \det(M_R(F_1, F_2, \ldots, F_{n+1})).$$

The definition of Ψ, $M_R(F_1, F_2, \ldots, F_{n+1})$ and $D_R(F_1, F_2, \ldots, F_{n+1})$ is dependent on the input polynomials $f_1, f_2, \ldots, f_{n+1}$ and their arrangement. In addition, the entries of $M_R(F_1, F_2, \ldots, F_{n+1})$ are coefficients of $f_1, f_2, \ldots, f_{n+1}$ or 0. We extensively characterize $M_R(F_1, F_2, \ldots, F_{n+1})$ in Section 5.

The definition of the Macaulay matrix by Renegar based on the matrix of a linear transformation, is a slight generalization of the definition in (Renegar 1992a, 1992b, 1992c). At the beginning of the century Macaulay has defined this matrix in a slightly different way using different vector spaces:

- Let V, B and t be defined as above. Then V has dimension $\binom{t+n}{t}$.
- Let V_i be the vector space of all homogeneous polynomials of degree $t - d_i$ of $F[x_1, x_2, \ldots, x_{n+1}]$, reduced in $\{x_1, x_2, \ldots, x_{i-1}\}$, including the zero polynomial ($1 \leq i \leq n+1$). Then the monomials of V_i form a basis B_i of the vector space ($1 \leq i \leq n+1$). The dimension of V_i is calculated in Section 5.

We now also define the Macaulay matrix by Macaulay using the matrix of a linear transformation:

Definition 3 *Let* $\Phi : \bigoplus_{i=1}^{n+1} V_i \to V$ *denote the linear transformation defined by* $\Phi(g_1, g_2, \ldots, g_{n+1}) := \sum_{i=1}^{n+1} g_i F_i$, *with* $g_i \in V_i$. *Let* $M(F_1, F_2, \ldots, F_{n+1})$ *be the matrix representing* Φ *with respect to the canonical basis of the direct sum* $\bigoplus_{i=1}^{n+1} V_i$ *and the basis* B *of* V. *Then* $M(F_1, F_2, \ldots, F_{n+1})$ *is the Macaulay matrix of* $F_1, F_2, \ldots, F_{n+1}$ *and* $D(F_1, F_2, \ldots, F_{n+1}) := \det(M(F_1, F_2, \ldots, F_{n+1}))$ *the corresponding Macaulay determinant.*

To give a clear idea of what happens in the construction of the Macaulay matrix, we present another formula of the matrix: Let $b_1, b_2, \ldots, b_\mu$ be all elements of B and $B_i = \{b_{i1}, b_{i2}, \ldots, b_{i\mu_i}\}$ for $1 \leq i \leq n+1$. (We calculate the μ_i in Theorem 4.) Then the Macaulay matrix is $(\text{coeff}_{b_k}(b_{ij}F_i))_{1 \leq i \leq n+1, 1 \leq j \leq \mu_i, 1 \leq k \leq \mu}$. If we write on the head of each column of the matrix the corresponding $b_i \in B$ and on the left of each row the corresponding $b_{ij}F_i$ for $1 \leq i \leq n+1, 1 \leq j \leq \mu_i$, we get the so-called *bordered matrix* of (Macaulay 1916).

The proof of the following theorem can be found in (Hagel 1993).

Theorem 1 *Let* $f_1, f_2, \ldots, f_{n+1}$ *be polynomials in* $F[x_1, x_2, \ldots, x_n]$. *Then*

- $M_R(F_1, F_2, \ldots, F_{n+1}) = M(F_1, F_2, \ldots, F_{n+1})$.
- $D_R(F_1, F_2, \ldots, F_{n+1}) = D(F_1, F_2, \ldots, F_{n+1})$.

We find the definition of Macaulay, for example, in a paper of Hurwitz (1913) and in van der Waerden's (1950) book. In the papers of Habicht (1948) and González-Vega (1991) on elimination theory, the two authors define the Macaulay matrix without the homogeneous polynomials $F_1, F_2, \ldots, F_{n+1}$. They directly construct the Macaulay matrix by the inhomogeneous input polynomials. The construction of the Macaulay matrix can also be found in (Canny 1988; Lazard 1981). All these papers, with the exception of Canny's, are about elimination theory. Grigor'ev (1988) and Renegar (1992a, 1992b, 1992c) made the Macaulay matrix a central tool in quantifier elimination.

4   Extraneous Factor and First Properties of the Macaulay Determinant

For some well known properties of the Macaulay determinant and the correspondence between the Macaulay determinant and the resultant, we need the definition of a minor of $D(F_1, F_2, \ldots, F_{n+1})$.

Definition 4 *[Macaulay] We consider the Macaulay matrix in bordered form. The extraneous factor $A(F_1, F_2, \ldots, F_{n+1})$ is the minor obtained when deleting all columns of $M(F_1, F_2, \ldots, F_{n+1})$ for which the elements at the head of the column are reduced in any n of the variables $x_1, x_2, \ldots, x_{n+1}$, and all rows corresponding to products of F_i with monomials reduced in $\{x_{i+1}, x_{i+2}, \ldots, x_{n+1}\}$.*

With the extraneous factor we summarize some properties given by Macaulay (1903, 1916) and Hurwitz (1913):

Theorem 2 *Let $f_1, f_2, \ldots, f_{n+1} \in F[x_1, x_2, \ldots, x_n]$ be polynomials of degree $d_1, d_2, \ldots, d_{n+1}$, respectively. Let $F_1, F_2, \ldots, F_{n+1}$ be the corresponding homogeneous polynomials.*

- *The extraneous factor is independent of the coefficients of F_{n+1}.*
- *The extraneous factor depends only on the coefficients of $F_i(x_1, x_2, \ldots, x_n, 0)$ for $1 \leq i \leq n$.*
- *For the resultant $R(F_1, F_2, \ldots, F_{n+1})$ we have*

$$D(F_1, F_2, \ldots, F_{n+1}) = R(F_1, F_2, \ldots, F_{n+1}) A(F_1, F_2, \ldots, F_{n+1}) \ .$$

- *The resultant $R(F_1, F_2, \ldots, F_{n+1})$ is the gcd of all Macaulay determinants for all $(n+1)!$ arrangements of the $F_1, F_2, \ldots, F_{n+1}$.*
- *The resultant $R(F_1, F_2, \ldots, F_{n+1})$ is the gcd of all Macaulay determinants for all $(n+1)$ arrangements $F_1, F_2, \ldots, F_{i-1}, F_{i+1}, \ldots, F_{n+1}, F_i$ for $2 \leq i \leq n$ and $F_1, F_2, \ldots, F_{n+1}$ and $F_2, F_3, \ldots, F_{n+1}, F_1$.*
- *If the resultant is not identically zero, there exists an arrangement of the $F_1, F_2, \ldots, F_{n+1}$ for which the Macaulay determinant does not vanish.*

If we construct the Macaulay matrix for two input polynomials we get the well known Sylvester matrix. Of this matrix we know that its determinant is the resultant of the two polynomials. Algorithms to compute the determinant of the Sylvester matrix to get the resultant can be found in (Collins 1967; Loos 1982b).

5   Characterization of the Macaulay Matrix

We first calculate the dimensions of the vector spaces which occur in the definitions of the Macaulay matrix.

Theorem 3 *The Macaulay matrix is a $\mu \times \mu$ matrix with $\mu = \binom{\nu_1}{\nu_2}$ and $\nu_1 = \sum_{i=1}^{n+1} d_i$ and $\nu_2 = (\sum_{i=1}^{n+1} d_i) - n$.*

Proof.

$$\binom{\sum_{i=1}^{n+1} d_i}{(\sum_{i=1}^{n+1} d_i) - n}$$

is the number of monomials of degree $(\sum_{i=1}^{n+1} d_i) - n$ in $n + 1$ variables. ∎
We now calculate the dimensions μ_i of the vector spaces V_i ($1 \leq i \leq n + 1$), which are very important in the rest of the paper.

Theorem 4 *The dimension of the vector space* V_i ($2 \leq i \leq n+$)1 *and hence the number of rows in the Macaulay matrix with coefficients of the polynomial* F_i *is*

$$\mu_i = \sum_{\alpha_1=0}^{d_1-1} \sum_{\alpha_2=0}^{d_2-1} \cdots \sum_{\alpha_{i-1}=0}^{d_{i-1}-1} \binom{t - d_i - (\sum_{j=1}^{i-1} \alpha_j) + n + 1 - i}{t - d_i - \sum_{j=1}^{i-1} \alpha_j}$$

and for $i = 1$ we have $\binom{t-d_1+n}{t-d_1}$.

Proof. We prove the Theorem for $1 \leq i \leq n+1$: Let $m := x_1^{\alpha_1} x_2^{\alpha_2} \cdots x_{n+1}^{\alpha_{n+1}}$ be a monomial of degree $t-d_i$, reduced in $\{x_1, x_2, \ldots, x_{i-1}\}$. Then $x_i^{\alpha_i} x_{i+1}^{\alpha_{i+1}} \cdots x_{n+1}^{\alpha_{n+1}}$ has degree $t-d_i - \sum_{j=1}^{i-1} \alpha_j$. The number of monomials of degree $t-d_i - \sum_{j=1}^{i-1} \alpha_j$ in $x_i, x_{i+1}, \ldots, x_{n+1}$ is

$$\binom{t - d_i - (\sum_{j=1}^{i-1} \alpha_j) + n + 1 - i}{t - d_i - \sum_{j=1}^{i-1} \alpha_j}.$$

Hence we have this number of monomials m with exponent α_j in x_j for $1 \leq j \leq i-1$. Because monomials reduced in $\{x_1, x_2, \ldots, x_{i-1}\}$ have only exponents $0 \leq \alpha_j \leq d_j - 1$ in x_j for $1 \leq j \leq i - 1$, the proof is complete. ∎
For the dimension of the last vector space V_{n+1} in the definition of the Macaulay matrix we have a simpler formula than in Theorem 4.

Corollary 1 *The dimension of V_{n+1} and hence the number of rows in the Macaulay matrix with coefficients from the last input polynomial f_{n+1} is $\prod_{i=1}^{n} d_i$.*

A proof of this formula which differs from the proofs given in (Macaulay 1916) and (van der Waerden 1950) can be found in (Hagel 1993).

The number of zero entries and the number of non-zero entries of the Macaulay matrix can be bounded by:

Theorem 5 *With the μ_i of Theorem 4 we have:*

- *The maximal number of non-zero entries in the Macaulay matrix is*

$$\sum_{i=1}^{n+1} \mu_i \binom{d_i + n}{d_i}.$$

- *The number of zero entries of the Macaulay matrix is at least*

$$\left(\frac{\sum_{i=1}^{n+1} d_i}{(\sum_{i=1}^{n+1} d_i) - n} \right)^2 - \sum_{i=1}^{n+1} \mu_i \binom{d_i + n}{d_i}.$$

Proof. This follows directly from the fact that a homogeneous polynomial of degree d in $n + 1$ variables has at most $\binom{d+n}{d}$ monomials, and Theorem 4. ∎

If we have dense input polynomials, the number of non-zero entries in the Macaulay matrix is exactly the one given in the above theorem. If the input polynomials f_i ($1 \leq i \leq n + 1$) are sparse and we know the number τ_i of their monomials, the exact number of non-zero entries of the Macaulay matrix is $\sum_{i=1}^{n+1} \mu_i \tau_i$.

6 Characterization of the Macaulay Matrix, if It Is Used to Calculate the u-Resultant

Often we are interested in the calculation of the u-resultant by the Macaulay matrix (see, for example, Canny 1988; Lazard 1981; Macaulay 1916, Renegar 1992a, 1992b, 1992c, van der Waerden 1950). In this special case of the Macaulay matrix we have the following situation:

Let there be given n polynomials $f_1, f_2, \ldots, f_n \in F[x_1, x_2, \ldots, x_n]$ of degree $d_1, d_2, \ldots, d_n$, respectively. By the additional variable of homogenization x_{n+1} we get the homogeneous polynomials $F_1, F_2, \ldots, F_n \in F[x_1, x_2, \ldots, x_{n+1}]$ of degree $d_1, d_2, \ldots, d_n$, respectively. To these polynomials we add one linear polynomial

$$U := u_1 x_1 + u_2 x_2 + \cdots + u_{n+1} x_{n+1} \in F[x_1, x_2, \ldots, x_{n+1}][u_1, u_2, \ldots, u_{n+1}]$$

with indeterminate coefficients $u_1, u_2, \ldots, u_{n+1}$ and construct the Macaulay matrix. If the polynomial U is the last polynomial in the construction of the Macaulay matrix, we can divide the Macaulay matrix in two parts, namely M_1 and M_2: The upper part M_1, whose non-zero entries are only the coefficients of the input polynomials $f_1, f_2, \ldots, f_n$ (i.e., values in F), and the lower part M_2, whose non-zero entries are only the indeterminates $u_1, u_2, \ldots, u_{n+1}$. We characterize these parts by the following theorem which is a special case of the propositions of Section 5:

Theorem 6 *If the last polynomial in the construction of the Macaulay matrix is $F_{n+1} = U := u_1 x_1 + u_2 x_2 + \cdots + u_{n+1} x_{n+1}$, we have with the μ_i defined in Theorem 4:*

- *The upper part M_1 of the Macaulay matrix is an $m_1 \times \mu$ matrix with $\mu = \left(\frac{\sum_{i=1}^{n} d_i + 1}{(\sum_{i=1}^{n} d_i) + 1 - n} \right)$ and $m_1 = \mu - \prod_{i=1}^{n} d_i$.*
- *The lower part M_2 of the Macaulay matrix is an $m_2 \times \mu$ matrix with $m_2 = \prod_{i=1}^{n} d_i$, and μ as above.*

- *The maximal number of non-zero entries in M_1 is $\sum_{i=1}^{n} \mu_i \binom{d_i+n}{d_i}$.*
- *The number of non-zero entries in M_2 is $(n+1) \prod_{i=1}^{n} d_i$.*
- *The number of zero entries in M_2 is $(\mu - (n+1)) \prod_{i=1}^{n} d_i$.*

With this theorem we can see that a sparse representation of the matrix is beneficial if the Macaulay matrix is used to compute the u-resultant. We are able to present a special algorithm to calculate the u-resultant with respect to the structure of the Macaulay matrix: We first triangularize the upper part M_1 and then use an evaluation/interpolation algorithm to triangularize the rest of the matrix using the knowledge of a triangularized upper part.

This algorithm to compute the determinant of the Macaulay matrix to get the u-resultant is faster than the usual determinant algorithm. Because of the indeterminates $u_1, u_2, \ldots, u_{n+1}$ we take an evaluation/interpolation algorithm to calculate the determinant. But the upper part M_1 of the Macaulay matrix does not contain any indeterminates. So by the usual method we triangularize the upper part for every evaluation. But it suffices to triangularize this part once and then do the evaluation/interpolation. Hence we save at most $\left(\frac{(\prod_{i=1}^{n} d_i)+n}{\prod_{i=1}^{n} d_i} \right) - 1$ triangularizations of the upper part M_1 of the Macaulay matrix.

Further we can use our knowledge of a triangularized upper part of the Macaulay matrix to compute step (3) of the algorithm.

A separation of the numeric part and the symbolic part of the Macaulay matrix to compute the u-resultant can also be found in (Lazard 1981). Faster algorithms than the one presented here can be found for example in (Lazard 1983; Canny et al. 1989; Lakshman and Lazard 1991; Manocha and Canny 1992).

7   Two Sorts of Homogenization

In Section 2 we introduced the degree-d_i-homogenization. Here we introduce the degree-d-homogenization: Let $f_1, f_2, \ldots, f_{n+1} \in F[x_1, x_2, \ldots, x_n]$ be polynomials of degree $d_1, d_2, \ldots, d_{n+1}$, respectively. Let d be the maximum of $d_1, d_2, \ldots, d_{n+1}$. By *degree-$d$-homogenization* we denote the following homogenization:

$$F_i = x_{n+1}^d f_i \left(\frac{x_1}{x_{n+1}}, \frac{x_2}{x_{n+1}}, \ldots, \frac{x_n}{x_{n+1}} \right), \text{ for } 1 \leq i \leq n+1.$$

We see that all the F_i ($1 \leq i \leq n+1$) have the same degree d. This kind of homogenization to homogenize the input polynomials $f_1, f_2, \ldots, f_n$ was used by Renegar (1992a, 1992b, 1992c) to compute the u-resultant. The last polynomial is the polynomial U of the section above.

If the input polynomials have not the same degree, the Macaulay matrix constructed by the degree-d-homogenized polynomials F_i is much bigger than the one constructed by degree-d_i-homogenization. We will give an example:

Example 1 *Let $f_1, f_2, f_3 \in F[x_1, x_2, x_3]$ be of degree 4, 2, 2, respectively. We want to calculate the size of the Macaulay matrix used to compute the u-resultant by the two homogenizations.*

- *Degree-d_i-homogenization: The degrees of F_1, F_2, F_3 are 4, 2, 2, respectively, the degree of U is one. Then the Macaulay matrix is an 84×84 matrix and the u-resultant is a polynomial of degree 16.*
- *Degree-d-homogenization: The degrees of F_1, F_2, F_3 are 4, 4, 4, respectively, the degree of U is one. Then the Macaulay matrix is a 286×286 matrix and the u-resultant is a polynomial of degree 64.*

In this example we see that if the input polynomials have different degrees, the degree-d_i-homogenization is the better way to homogenize the polynomials to calculate the u-resultant, because of the size of the resulting matrix. If one is only interested in theoretical aspects, the degree-d-homogenization is favorable, because the formulae one gets with this method are simpler.

8   Characterization of the Matrix of the Extraneous Factor

The extraneous factor plays an important role in the calculation of the resultant and the u-resultant with the Macaulay matrix. In Theorem 2 we find the relationship between the extraneous factor and these resultants via the Macaulay determinant. Concerning the entries of the extraneous factor we have the following theorem:

Theorem 7 *The number of rows of the matrix of the extraneous factor, whose non-zero entries are the coefficients of the polynomials F_i, is*

$$\mu_i - (\prod_{j=1}^{i-1} d_j)(\prod_{j=i+1}^{n+1} d_j),$$

where the μ_i are the values defined in Theorem 4 and the d_i the degrees of the input polynomials f_i $(1 \leq i \leq n+1)$.

Proof. We prove this for $1 \leq i \leq n+1$. In the definition of the extraneous factor we remove from the μ_i rows of $M(F_1, F_2, \ldots, F_{n+1})$ whose non-zero entries consist of the coefficients of the F_i all the rows of the bordered matrix with corresponding monomials reduced in $\{x_1, x_2, \ldots, x_{i-1}\}$ and $\{x_{i+1}, x_{i+2}, \ldots, x_{n+1}\}$. Reduced in $\{x_1, x_2, \ldots, x_{i-1}\}$ because of the definition of the corresponding vector spaces V_i and reduced in $\{x_{i+1}, x_{i+2}, \ldots, x_{n+1}\}$ because of the definition of the extraneous factor. How many such monomials exist in the border of the matrix? These monomials have exponents in $\{0, 1, \ldots, d_j - 1\}$ in x_j for $1 \leq j \leq i-1$ and for $i+1 \leq j \leq n+1$. Because these monomials all have degree $t - d_i$, we have only one possibility for the exponent of x_i. So we get

$\mu_i - (\prod_{j=1}^{i-1} d_j)(\prod_{j=i+1}^{n+1} d_j)$ monomials, which are reduced in $\{x_1, x_2, \ldots, x_{i-1}\}$ and $\{x_{i+1}, x_{i+2}, \ldots, x_{n+1}\}$. ∎

That the matrix of the extraneous factor does not contain coefficients of the input polynomial f_{n+1} is a consequence of Corollary 1 and Theorem 7.

Now we look at the size of the matrix of the extraneous factor which follows from Theorem 2 and Theorem 7:

Corollary 2 *The matrix of the extraneous factor is an $l \times l$ matrix with*

$$l = \sum_{i=1}^{n} (\mu_i - (\prod_{j=1}^{i-1} d_j)(\prod_{j=i+1}^{n+1} d_j)),$$

where the μ_i are the values defined in Theorem 4.

The relation between the degree-d-homogenization and the extraneous factor is stated in the following theorem:

Theorem 8 *Let $f_1, f_2, \ldots, f_{n+1} \in F[x_1, x_2, \ldots, x_n]$ be polynomials, where the $f_1, f_2, \ldots, f_n$ are not all of the same degree. If we construct the Macaulay matrix with the degree-d-homogenized polynomials of $f_1, f_2, \ldots, f_n$, then the extraneous factor equals zero.*

Proof. We show that there is a row of zero entries in the matrix of the extraneous factor. If we construct the matrix of the extraneous factor, we remove all rows in the Macaulay matrix corresponding to products of F_i and monomials reduced in the set $\{x_{i+1}, x_{i+2}, \ldots, x_{n+1}\}$. That is, for every i, $1 \le i \le n$, we have one row corresponding to $x_{n+1}^{t-d} F_i$, where t is the degree of the monomials of the vector space V in the construction of the Macaulay matrix. Let f_j with $1 \le j \le n$ be a polynomial with degree $d_j < d$. Then every monomial of F_j has a factor $x_{n+1}^{\alpha_{n+1}}$ with $\alpha_{n+1} \ge 1$, because F_j is the degree-d-homogenization of f_j. Hence every monomial of $x_{n+1}^{t-d} F_i$ has a factor $x_{n+1}^{\beta_{n+1}}$ with $\beta_{n+1} \ge t - d + 1$. We show that there is no column in the matrix of the extraneous factor whose corresponding monomial on the head of the column of the bordered Macaulay matrix has a factor $x_{n+1}^{\beta_{n+1}}$ with $\beta_{n+1} \ge t - d + 1$.

Let m be such a monomial with a factor $x_{n+1}^{\beta_{n+1}}$ and $\beta_{n+1} \ge t - d + 1$. The degree of m is t. Hence $1/x_{n+1}^{\beta_{n+1}} m$ has degree $t - \beta_{n+1} \le t - (t - d + 1) = d - 1$. So m is reduced in $\{x_1, x_2, \ldots, x_n\}$, i.e., in n variables of $x_1, x_2, \ldots, x_{n+1}$. Hence this column shall be removed in the construction of the matrix of the extraneous factor. For this reason we have a row of zero entries in the matrix of the extraneous factor which corresponds to the row $x_{n+1}^{t-d} F_j$ of the bordered Macaulay matrix. ∎

An example can be found in (Hagel 1993). This theorem constitutes a more important argument in favor of using degree-d_i-homogenization instead of degree-d-homogenization in practical applications, as compared to the argument based on the size of the resulting matrix. If we use degree-d-homogenization with input polynomials not all of the same degree, and we want to calculate the resultant or the u-resultant with the help of the Macaulay matrix, Theorems 2 and 8 imply that the extraneous factor and the Macaulay determinant equal zero. Renegar, who used degree-d-homogenization in his work on the complexity of quantifier elimination, presents a procedure of Grigor'ev (1988) when the Macaulay determinant is identically zero. For this procedure we need some new indeterminates, what is undesirable in practice because of the explosion of computing time. Hence we should compute the Macaulay determinant using the degree-d_i-homogenization because in many cases we can avoid the difficulties with the new indeterminates.

9 Conclusion

In this paper we have presented a characterization of the Macaulay matrix, the Macaulay determinant and the extraneous factor and their implications for algorithms to compute them. Finally, we discussed two different homogenizations, the degree-d-homogenization and the degree-d_i-homogenization and their impact on the size of the Macaulay matrix and the time to compute the Macaulay determinant.

Acknowledgments

I wish to thank Prof. Loos for his supervision of my work.

Computation of Variant Resultants

Hoon Hong and J. Rafael Sendra

1 Introduction

In this paper, we give a new method for computing two variants of resultants, namely the trace resultants and the slope resultants. These two variants were introduced in (Hong 1993d) while devising quantifier elimination algorithms for a certain fragment of the elementary theory of the reals, where the input formulas are required to contain at least one quadratic polynomial equation. Hong (1993d) also gave a method for computing these two variant resultants. The method is based on expanding certain determinants whose entries come directly from the coefficients of the involved polynomials. In this paper, we provide a theoretical computing time analysis of the method. More importantly, we give a new and faster method, as well as modular algorithms for the computation of the variant resultants.

The main idea underlying the new method is that a linear recurrence sequence can be computed by one polynomial division. The polynomials in the division are obtained directly from the coefficients of the sequence, and correspond to the rational function associated with the recurrence (Sendra and Llovet 1992). In (Hong 1993d) certain recurrences for the variant resultants were derived. We apply the idea mentioned above on the recurrence formulas, obtaining a method for computing the variant resultants by one simple polynomial division (only quotient is needed) and several multiplications.

The structure of the paper is as follows. In Section 2, we give a precise definition of the variant resultants. In Section 3, we review the determinant based method. In Section 4, we describe the new method based on polynomial quotient computation. In Section 5, we analyze the problem modularly. In Section 6, we give theoretical computing time bounds for both the determinant based algorithms and the quotient based algorithms, and for the modular algorithms. In Section 7, we investigate the average computing times based on experimental results.

2   Problem Statement

In this section, we give a precise definition of the trace and the slope resultants and state the problem which will be tackled in the subsequent sections. Let I be an integral domain, and let $\bar{I}$ be the unique (up to isomorphism) algebraic closure of the quotient field of I. Consider two non-zero polynomials A and B over I of degree m and n such that

$$A = \sum_{i=0}^{m} a_i x^i = a_m \prod_{i=1}^{m} (x - \alpha_i),$$

$$B = \sum_{i=0}^{n} b_i x^i = b_n \prod_{i=1}^{n} (x - \beta_i)$$

where $\alpha_i, \beta_i \in \bar{I}$. We define a variant of the resultant, which we will call *trace resultant* and write as $\mathrm{tres}(A, B)$.

Definition 1 (Trace Resultant)

$$\mathrm{tres}(A, B) = a_m^n b_n \sum_{i=1}^{m} \prod_{j=1}^{n} (\alpha_i - \beta_j) = a_m^n \sum_{i=1}^{m} B(\alpha_i).$$

Intuitively, it is the average of the values of B on the roots of A, up to some constant factor.

Let u and v be two distinct indeterminates. Then it is easy to see that $u - v$ divides $B(u) - B(v)$ for any polynomial B. This makes possible the following definition:

Definition 2 (Slope) *Let B be a polynomial over I. Then the slope S_B of B is the polynomial over $I[u, v]$ such that*

$$S_B(u, v) = \frac{B(u) - B(v)}{u - v}.$$

Now we define the other variant of resultant which we will call *slope resultant* and write as $\mathrm{sres}(A, B)$.

Definition 3 (Slope Resultant)

$$\mathrm{sres}(A, B) = a_m^{n-1} \sum_{1 \le i < j \le m} S_B(\alpha_i, \alpha_j).$$

Intuitively, it is the average slope of B on the roots of A, up to some constant factor.

Now we are ready to state the problem: Devise algorithms which, given two polynomials, compute their trace and slope resultants.

3 Review of Determinant Based Method

Hong (1993d) showed that the variant resultants can be expressed as determinants of certain matrices whose entries come directly from the coefficients of the involved polynomials. This result immediately yields a method for computing the variant resultants, namely by expanding the determinants. In this section, we give a brief summary of the method. In Section 6, we will derive a theoretical computing time bound for the method. And in Section 7, we report empirical computing times. The following theorems were proved in (Hong 1993d).

Theorem 1 (Trace Resultant as Determinant)

$$\text{tres}(A, B) = \det(M)$$

where M is the $n + 1$ by $n + 1$ matrix defined by

$$M = \begin{bmatrix} a_m & a_{m-1} & \cdots & \cdots & \cdots & \\ & \cdots & \cdots & \cdots & \cdots & \cdots \\ & & a_m & a_{m-1} & a_{m-2} & 3 \cdot a_{m-3} \\ & & & a_m & a_{m-1} & 2 \cdot a_{m-2} \\ & & & & a_m & 1 \cdot a_{m-1} \\ b_n & b_{n-1} & \cdots & \cdots & b_1 & m \cdot b_0 \end{bmatrix}.$$

Precisely, the matrix is defined by

$$M_{i,j} = \begin{cases} a_{m-(j-i)} & \text{if } i < n+1 \text{ and } j < n+1, \\ b_{n-j+1} & \text{if } i = n+1 \text{ and } j < n+1, \\ (j-i)a_{m-(j-i)} & \text{if } i < n+1 \text{ and } j = n+1, \\ mb_0 & \text{if } i = n+1 \text{ and } j = n+1, \\ 0 & \text{otherwise.} \end{cases}$$

Theorem 2 (Slope Resultant as Determinant)

$$\text{sres}(A, B) = \det(M)$$

where M is the n by n matrix defined by

$$M = \begin{bmatrix} a_m & a_{m-1} & \cdots & \cdots & \cdots & \\ & \cdots & \cdots & \cdots & \cdots & \cdots \\ & & a_m & a_{m-1} & a_{m-2} & c_3^m \cdot a_{m-3} \\ & & & a_m & a_{m-1} & c_2^m \cdot a_{m-2} \\ & & & & a_m & c_1^m \cdot a_{m-1} \\ b_n & b_{n-1} & \cdots & \cdots & b_2 & c_m^m \cdot b_1 \end{bmatrix}$$

where $c_k^m = k(2m - k - 1)/2$. Precisely, the matrix is defined by

$$M_{i,j} = \begin{cases} a_{m-(j-i)} & \text{if } i < n \text{ and } j < n, \\ b_{n-j+1} & \text{if } i = n \text{ and } j < n, \\ c_{(j-i)}^m a_{m-(j-i)} & \text{if } i < n \text{ and } j = n, \\ c_m^m b_1 & \text{if } i = n \text{ and } j = n, \\ 0 & \text{otherwise.} \end{cases}$$

From these theorems, we trivially obtain the following algorithms.

TRES_D

Algorithm TRES_D (Trace Resultant by Determinant).

(1) Construct the matrix M in Theorem 1.
(2) Expand the determinant of M.

SRES_D

Algorithm SRES_D (Slope Resultant by Determinant).

(1) Construct the matrix M in Theorem 2.
(2) Expand the determinant of M.

4   Quotient Based Method

In this section, we give a new (better) method for computing the variant resultants. The main idea of the quotient based method is to use the relationship between linear recurrence sequences and rational functions.

Lemma 1 (Variants by P_k^m and Q_k^m)
$\mathrm{tres}(A, B) = a_m^n \sum_{k=0}^n b_k P_k^m$, where $P_k^m = \sum_{i=1}^m \alpha_i^k$.
$\mathrm{sres}(A, B) = a_m^{n-1} \sum_{k=0}^{n-1} b_{k+1} Q_k^m$, where $Q_k^m = \sum_{1 \le i < j \le m} \frac{\alpha_i^{k+1} - \alpha_j^{k+1}}{\alpha_i - \alpha_j}$.

Proof.

$$
\begin{aligned}
\mathrm{tres}(A, B) &= a_m^n \sum_{i=1}^m B(\alpha_i) \\
&= a_m^n \sum_{i=1}^m \sum_{k=0}^n b_k \alpha_i^k \\
&= a_m^n \sum_{k=0}^n b_k \sum_{i=1}^m \alpha_i^k \\
&= a_m^n \sum_{k=0}^n b_k P_k^m. \\
\mathrm{sres}(A, B) &= a_m^{n-1} \sum_{1 \le i < j \le m} \frac{B(\alpha_i) - B(\alpha_j)}{\alpha_i - \alpha_j} \\
&= a_m^{n-1} \sum_{1 \le i < j \le m} \frac{\sum_{k=0}^n b_k \alpha_i^k - \sum_{k=0}^n b_k \alpha_j^k}{\alpha_i - \alpha_j} \\
&= a_m^{n-1} \sum_{1 \le i < j \le m} \sum_{k=0}^n b_k \frac{(\alpha_i^k - \alpha_j^k)}{\alpha_i - \alpha_j} \\
&= a_m^{n-1} \sum_{k=0}^n b_k \sum_{1 \le i < j \le m} \frac{(\alpha_i^k - \alpha_j^k)}{\alpha_i - \alpha_j} \\
&= a_m^{n-1} \sum_{k=0}^n b_k Q_{k-1}^m.
\end{aligned}
$$

∎

Note that P_k^m and Q_k^m are symmetric polynomials in $\alpha_1, \ldots, \alpha_m$. Let s_k^m be the k-th elementary symmetric polynomial in $\alpha_1, \ldots, \alpha_m$, defined by $s_k^m = \sum_{1 \le i_1 < i_2 < \cdots < i_k \le m} \alpha_{i_1} \alpha_{i_2} \cdots \alpha_{i_k}$. Then by the fundamental theorem of symmetric polynomials, P_k^m and Q_k^m can be expressed as polynomials in $s_1^m, \ldots, s_m^m$. The following lemma gives recurrence formulas for computing such polynomials. The recurrence formula for P_k^m is due to Newton and well known, and that for Q_k^m is introduced and proved in (Hong 1993d).

Lemma 2 (Recurrence formulas for P_k^m and Q_k^m)

- $0 = P_k^m - s_1^m P_{k-1}^m + s_2^m P_{k-2}^m + \cdots + (-1)^m s_m^m P_{k-m}^m$, if $m < k$.

  $0 = P_k^m - s_1^m P_{k-1}^m + s_2^m P_{k-2}^m + \cdots + (-1)^{k-1} s_{k-1}^m P_1^m + (-1)^k s_k^m k$, if $m \geq k \geq 0$.
- $0 = Q_k^m - s_1^m Q_{k-1}^m + s_2^m Q_{k-2}^m + \cdots + (-1)^m s_m^m Q_{k-m}^m$, if $m < k$.

  $0 = Q_k^m - s_1^m Q_{k-1}^m + s_2^m Q_{k-2}^m + \cdots + (-1)^{k-1} s_{k-1}^m Q_1^m + (-1)^k s_k^m c_k^m$, if $m \geq k \geq 0$,

where $c_k^m = k(2m - k - 1)/2$.

The following lemma relates a linear recurrence formula with a rational function.

Lemma 3 (Recurrence Formulas and Rational Functions)

Let $f = (f_0, f_1, \ldots, f_m)$, $g = (g_0, g_1, \ldots, g_m)$ be two finite sequences, and let $h = (h_0, h_1, \ldots)$ be an infinite sequence, such that $f_m \neq 0$. Then we have

$$\begin{aligned} 0 &= f_m h_k + f_{m-1} h_{k-1} + \cdots + f_0 h_{k-m} && \text{if } m < k, \\ g_{m-k} &= f_m h_k + f_{m-1} h_{k-1} + \cdots + f_{m-k} h_0 && \text{if } m \geq k \geq 0 \end{aligned}$$

iff

$$h_0 x^0 + h_1 x^{-1} + \cdots = \frac{g_m x^m + g_{m-1} x^{m-1} + \cdots + g_0}{f_m x^m + f_{m-1} x^{m-1} + \cdots + f_0}.$$

Proof. Immediate by multiplying both sides of the rational function equation by its denominator and by equating the coefficients of equal powers of x. ∎

Theorem 3 (P_k^m and Q_k^m by Quotients)

$\sum_{k=0}^n P_k^m x^{n-k}$ is equal to the quotient $x^{n+1} A'/A$.
$\sum_{k=0}^{n-1} Q_k^m x^{n-1-k}$ is equal to the quotient $x^{n+1} A''/2A$.

Proof. First we rewrite the recurrence formulas for P_k^m and Q_k^m in Lemma 2 so that they conform to the pattern of the recurrence formula in Lemma 3. By multiplying the recurrence formulas in Lemma 2 by a_m, simplifying them using the relation $a_m(-1)^i s_i = a_{m-i}$, and manipulating the last terms suitably, we obtain

- $0 = a_m P_k^m + a_{m-1} P_{k-1}^m + \cdots + a_0 P_{k-m}^m$, if $m < k$.

  $(P_0^m - k)a_{m-k} = a_m P_k^m + a_{m-1} P_{k-1}^m + \cdots + a_{m-k} P_0^m$ if $m \geq k \geq 0$.
- $0 = a_m Q_k^m + a_{m-1} Q_{k-1}^m + \cdots + a_0 Q_{k-m}^m$ if $m < k$.

  $(Q_0^m - c_k^m)a_{m-k} = a_m Q_k^m + a_{m-1} Q_{k-1}^m + \cdots + a_{m-k} Q_0^m$ if $m \geq k \geq 0$.

By matching these against the recurrence formula in Lemma 3, we obtain for P_k^m

$$\begin{aligned} f_k &= a_k, \\ g_k &= (P_0^m - m - k)a_k = ka_k, \\ h_k &= P_k^m, \end{aligned}$$

and for Q_k^m

$$f_k = a_k,$$
$$g_k = (Q_0^m - c_{m-k}^m)a_k = \binom{k}{2}a_k,$$
$$h_k = Q_k^m.$$

By applying the relation between recurrence formulas and rational functions in Lemma 3, we obtain

$$P_0^m x^0 + P_1^m x^{-1} + \cdots = \frac{ma_m x^m + (m-1)a_{m-1}x^{m-1} + \cdots + a_1 x}{a_m x^m + a_{m-1}x^{m-1} + \cdots + a_0}$$
$$= xA'/A$$
$$Q_0^m x^0 + Q_1^m x^{-1} + \cdots = \frac{\binom{m}{2}a_m x^m + \binom{m-1}{2}a_{m-1}x^{m-1} + \cdots + \binom{2}{2}a_2 x^2}{a_m x^m + a_{m-1}x^{m-1} + \cdots + a_0}$$
$$= x^2 A''/2A.$$

By multiplying the both sides of the equations for P_k^m and Q_k^m respectively by x^n and x^{n-1}, we obtain

$$P_0^m x^n + P_1^m x^{n-1} + \cdots + P_n^m + \cdots = x^{n+1} A'/A$$
$$Q_0^m x^{n-1} + Q_1^m x^{n-2} + \cdots + Q_{n-1}^m + \cdots = x^{n+1} A''/2A$$

from which the theorem follows. ∎

Lemma 3 and Theorem 3 can be directly turned into algorithms, but such algorithms require working in the quotient field of the coefficient domain, making them inefficient. Fortunately we can easily bypass this problem by multiplying A' or A'' by appropriate powers of a_m before we carry out division (pseudo-division). Note that a_m divides the leading coefficients of A' and A''. Thus the appropriate powers are a_m^n and a_m^{n-1} respectively for A' and A''. Now we give algorithms based on these ideas. In the algorithm descriptions, $P_k = a_m^n P_k^m$ and $Q_k = a_m^{n-1} Q_k^m$.

TRES_Q

Algorithm TRES_Q (Trace Resultant by Quotient).

(1) Obtain $A'(x)$.
(2) Determine the quotient $\sum_{k=0}^{n} P_k x^{n-k}$ of $a_m^n x^{n+1} A'/A$.
(3) Compute $\sum_{k=0}^{n} b_k P_k$.

SRES_Q

Algorithm SRES_Q (Slope Resultant by Quotient).

(1) Obtain $A''(x)$.
(2) Determine the quotient $\sum_{k=0}^{n-1} Q_k x^{n-1-k}$ of $a_m^{n-1} x^{n+1} A''/2A$.
(3) Compute $\sum_{i=0}^{n-1} b_{k+1} Q_k$.

5 Modular Methods

In this section, we study the direct application of modular techniques to the computation of the variant resultants, following the ideas in (Collins 1971a) where a modular technique is applied to the computation of resultants.

Let $A, B \in \mathbb{Z}[x_1, \ldots, x_r]$, $d_i = \max\{\deg_{x_i}(A), \deg_{x_i}(B)\}$, and N the maximum max-norm of the coefficients of A and B. We denote by φ_m the modular homomorphism given by the positive integer m, and by $\psi_{j,a}$ the evaluation homomorphism that eliminates the variable x_j using the integer a.

Then, in computing by homomorphic images the variant resultants of A and B w.r.t. x_r, first the polynomials A and B are mapped to polynomials $A_i, B_i \in \mathbb{Z}_{m_i}[x_1, \ldots, x_r]$, by means of some lucky modular homomorphisms φ_{m_i} (m_i is a prime integer). Secondly, using recursively some lucky evaluation homomorphisms $\psi_{j,a_{i_j}}$, one obtains polynomials $\bar{A}_i, \bar{B}_i \in \mathbb{Z}_{m_i}[x_r]$. Once the problem has been reduced to univariate polynomials over finite fields, either algorithm TRES_D, SRES_D, TRES_Q, or SRES_Q is performed over $\mathbb{Z}_{m_i}[x_r]$. Finally, applying the inversion algorithms (interpolation and Chinese remainder), the variant resultant is determined. We will call the resulting algorithms: TRES_MD, SRES_MD, TRES_MQ, and SRES_MQ.

To analyze the lucky homomorphisms of the process, we observe that the order of the matrix M – in the determinant based methods – and the degree of the quotient – in the quotient based methods – depend on the degree of B w.r.t x_r. Thus, the lucky homomorphisms are those where the leading coefficient of B w.r.t. x_r is not mapped to zero. In order to determine the number of required modular and evaluation homomorphisms, one bounds the trace and slope resultant in norm and degree. For this propose, we consider the expression of the variant resultants given in theorems 1 and 2.

For the trace resultant, one has that the maximum max-norm of the entries of the matrix M (theorem 1) is bounded by $N d_r$. Hence, the max-norm of $\det(M)$ is bounded by $(d_r + 1)! \, (N \, (d_r + 1))^{d_r + 1}$. On the other hand, the degree of $\det(M)$ w.r.t. x_j is bounded by $(d_r + 1) \, d_j$. Therefore, for the modular homomorphisms, lucky primes $m_1, \ldots, m_s$ satisfying

$$\prod_{i=1}^{s} m_i > 2 \, (d_r + 1)! \, (N \, (d_r + 1))^{d_r + 1}$$

are taken. And, the number of lucky evaluation homomorphisms to eliminate the variable x_j is greater than $(d_r + 1) \, d_j + 1$.

Now we give algorithms for the modular computation of the trace resultant. First, we present the evaluation algorithm TRES_ED that determines the trace resultant of two polynomials $A, B \in \mathbb{Z}_m[x_1, \ldots, x_j][x_r]$ (m is a prime integer) by evaluating the variables $\{x_1, \ldots, x_j\}$, and using algorithm TRES_D over $\mathbb{Z}_m[x_r]$. In the design of TRES_ED the algorithm applies itself recursively. Secondly, we outline the modular algorithm TRES_MD that computes the trace resultant by determinants. In the inversion step of TRES_MD we use Garner's algorithm for Chinese remainder calculations.

TRES_ED

Algorithm TRES_ED (Trace Resultant by Evaluation Determinant).

(1) If $A, B \in \mathbb{Z}_m[x_r]$ apply TRES_D to A, B.
(2) Initialize $t = 1$, $V = 1$, TRES $= 0$.
(3) While $t \leq (d_r + 1)d_j + 1$ do
(3.1) Take a new $b \in \mathbb{Z}_m$ such that $\deg_{x_r}(\psi_{j,b}(A)) = d_r$.
(3.2) Apply TRES_ED to $\psi_{j,b}(A), \psi_{j,b}(B)$ obtaining $U(x_1, \ldots, x_{j-1})$.
(3.3) $W = \text{TRES} + (U - \text{TRES}(x_1, \ldots, x_{j-1}, b)) V(b)^{-1} V(x_j)$.
(3.4) Replace TRES by W, V by $V \cdot (x_j - b)$, and t by $t + 1$.
(4) Return TRES.

TRES_MD

Algorithm TRES_MD (Trace Resultant by Modular Determinant).

(1) Initialize $M = 1$, TRES $= 0$.
(2) While $M \leq 2(d_r + 1)! \left((d_r + 1) N \right)^{d_r+1}$ do
(2.1) Take a new prime m such that $\deg_{x_r}(\varphi_m(A)) = d_r$.
(2.2) Apply TRES_ED to $\varphi_m(A), \varphi_m(B)$ obtaining U.
(2.3) Apply Chinese remainders to $\{\text{TRES}, U\}$ with modulus $\{M, m\}$ obtaining W.
(2.4) Replace TRES by W, and M by $M \cdot m$.
(3) Return TRES.

For the slope resultant, one deduces that the required lucky primes $m_1, \ldots, m_s$ satisfy

$$\prod_{i=1}^{s} m_i > 2 d_r! (N d_r^2)^{d_r},$$

and the number of lucky evaluation homomorphisms needed to eliminate the variable x_j is greater than $d_r d_j + 1$. Thus analogous modular algorithms can be presented. Here, we describe the process using quotients, and we omit the corresponding algorithms SRES_ED and SRES_MD.

SRES_EQ

Algorithm SRES_EQ (Slope Resultant by Evaluation Quotient).

(1) If $A, B \in \mathbb{Z}_m[x_r]$ apply SRES_Q to A, B.
(2) Initialize $t = 1$, $V = 1$, SRES $= 0$.
(3) While $t \leq d_r d_j + 1$ do
(3.1) Take a new $b \in \mathbb{Z}_m$ such that $\deg_{x_r}(\psi_{j,b}(A)) = d_r$.
(3.2) Apply SRES_EQ to $\psi_{j,b}(A), \psi_{j,b}(B)$ obtaining $U(x_1, \ldots, x_{j-1})$.
(3.3) $W = \text{SRES} + (U - \text{SRES}(x_1, \ldots, x_{j-1}, b)) V(b)^{-1} V(x_j)$.

(3.4) Replace SRES by W, V by $V \cdot (x_j - b)$, and t by $t + 1$.
(4) Return SRES.

SRES_MQ

Algorithm SRES_MQ (Slope Resultant by Modular Quotient).

(1) Initialize $M = 1$, SRES $= 0$.
(2) While $M \leq 2d_r! \, (N \, d_r^2)^{d_r}$ do
(2.1) Take a new prime m such that $\deg_{x_r}(\varphi_m(A)) = d_r$.
(2.2) Apply SRES_EQ to $\varphi_m(A), \varphi_m(B)$ obtaining U.
(2.3) Apply Chinese remainders to $\{\text{SRES}, U\}$ with modulus $\{M, m\}$ obtaining W.
(2.4) Replace SRES by W, and M by $M \cdot m$.
(3) Return SRES.

6 Theoretical Computing Time Analysis

In this section, we derive asymptotic upper bounds of the computing times for the worst case of the algorithms given in the last three sections. In deriving bounds, we shall assume that all arithmetic operations on integers and polynomials are performed by classical algorithms. Furthermore, we also suppose that polynomials are represented recursively, that is, if $p(x_1, \ldots, x_r) = \sum_{i=0}^{k} p_i(x_1, \ldots, x_{r-1})x_r^i \in R[x_1, \ldots, x_r], p_k \neq 0$ and R a ring, then the representation of p is a list $(k, p_k^*, \ldots, p_0^*)$ where p_i^* is the representation of p_i. For computing determinants, we will use Bareiss' exact division algorithm (Bareiss 1968) .

Let $A, B \in \mathbb{Z}[x_1, \ldots, x_r]$, $d = \max\{\deg_{x_i}(A), \deg_{x_i}(B)\}_{1 \leq i \leq r}$ and ℓ the maximum bit length of the coefficients of A and B. We begin by recalling the following well-known asymptotic bounds for several basic polynomial operations.

Proposition 1 (Bounds for basic polynomial operations)

(1) *The computing time of adding A and B is bounded above by $O((d+1)^r \ell)$.*
(2) *The computing time of multiplying A and B is bounded above by $O((d+1)^{2r}\ell^2)$.*
(3) *The computing time of pseudo-dividing A by B with respect to x_r is $O((d+1)^{3r}\ell^2)$. Furthermore, the coefficient length of the pseudo-quotient is dominated by $O((d+1)(\ell + r\log d))$.*
(4) *The computing time of determinant by Bareiss' algorithm over $\mathbb{Z}[x_1, \ldots, x_r]$ is bounded above by $O(k^{2r+5}(d+1)^{2r}\ell^2)$, where k is the order of the matrix, ℓ the maximum bit length of the integer coefficients of the entries, and d the maximum of their degrees.*

Theorem 4 (Bound for Determinant Method) *The computing times for the worst cases for the algorithm* TRES_D *(Trace Resultant by Determinant) and for the algorithm* SRES_D *(Slope Resultant by Determinant) are bounded above by* $O((d+1)^{4r+5}(\ell + \log d)^2)$.

Proof. One basically has to analyze the computation of the determinant of a matrix. For both algorithms, the lengths of the integer coefficients of the entries in the matrices are dominated by $\ell + \log d$. Therefore, using Proposition 1 (4), one concludes that an upper bound for both algorithms is $O((d+1)^{4r+5}(\ell + \log d)^2)$. ∎

Theorem 5 (Bound for Quotient Method) *The computing times for the worst cases for the algorithm* TRES_Q *(Trace Resultant by Quotient) and for the algorithm* SRES_Q *(Slope Resultant by Quotient) are bounded above by* $O((d+1)^{3r}(\ell + \log d)^2)$.

Proof. We first observe that Step 1 in both algorithms requires $O((d+1)^r \ell \log d)$, and that the lengths of the integer coefficients of the polynomials $A'(X)$ and $A''(x)$ are dominated by $O(\ell + \log d)$. Hence one can derive same upper bounds for both algorithms. Let us then analyze Step 2 and Step 3. For Step 2, using the upper bound for pseudo-division given in Proposition 1 (3), one requires $O((d+1)^{3r}(\log d + \ell)^2)$. In Step 3, $(d+1)$ polynomial multiplications have to be performed. Thus, taking into account that the coefficient length of the pseudo-quotient computed in Step 2 is dominated by $O((d+1)(\ell + r \log d))$, one has that Step 3 is dominated in time by $O((d+1)^{2r+2}(\ell + r \log d)^2)$. Therefore an upper bound for the algorithms is $O((d+1)^{3r}(\ell + \log d)^2)$. ∎

Proposition 2 (Bounds for Evaluation Algorithms)

(1) *The computing time for the worst cases for the algorithm* TRES_ED *(Trace Resultant by Evaluation Determinant) and the algorithm* SRES_ED *(Slope Resultant by Evaluation Determinant) are bounded above by* $O(d^{2j+3})$.

(2) *The computing time for the worst cases for the algorithm* TRES_EQ *(Trace Resultant by Evaluation Quotient) and the algorithm* SRES_EQ *(Slope Resultant by Evaluation Quotient) are bounded above by* $O(d^{2j+2})$.

Proof. (1) We first observe that the number of iterations in Step 3 of both algorithms is dominated by $O(d^2)$, and that both procedures over $\mathbb{Z}_m[x_r]$ are bounded in time by $O(d^3)$. Hence, one can derive same upper bounds for both algorithms. Let T_j be the computing time for the worst case for algorithm TRES_ED when the input polynomials are given in $\mathbb{Z}_m[x_1, \ldots, x_j][x_r]$. Then, one basically has to analyze Step 3. In the k-th execution of the loop one has that: Step 3.1 requires $O(\alpha_k d^j)$, where α_k denotes the number of integers tested (since an integer is unlucky if the leading coefficient of A is mapped to zero, it follows that at most the number of integers tested in Step 3 is dominated by d^2); Step 3.2 is dominated by $d^{j+1} + T_{j-1}$; and Step 3.3 takes $O(kd^{2j-2})$. Hence $T_j = O(d^2 T_{j-1} + d^{2j+2})$, and taking into account that $T_0 = O(d^3)$ one concludes that $O(d^{2j+3})$ is an upper bound for the algorithms.

(2) The number of iterations of Step 3 in both algorithms is also $O(d^2)$, and both algorithms over $\mathbb{Z}_m[x_r]$ are bounded in time by $O(d^2)$. Hence, upper bounds for both processes can be derived simultaneously. Then, if T_j denotes now the computing time for the worst case for algorithm TRES_EQ when the input polynomials are given in $\mathbb{Z}_m[x_1, \ldots, x_j][x_r]$, one has that the recursive formula is $T_j = O(d^2 T_{j-1} + d^{2j+2})$. Therefore, since T_0 is dominated by d^2, one concludes that $O(d^{2j+2})$ is an upper bound for the algorithms. ∎

Theorem 6 (Bound for Modular Determinant Method) *The computing time for the worst cases for the algorithm* TRES_MD *(Trace Resultant by Modular Determinant) and the algorithm* SRES_MD *(Slope Resultant by Modular Determinant) are bounded above by* $O(d^{2r+2}(\ell^2 + \log d))$. ∎

Proof. Note that the number of iterations in Step 2 of both algorithms is dominated by $O(d(\ell + \log d))$. Then, taking into account Proposition 2 (1), it follows that one can derive same upper bounds for both algorithms. Hence, we analyze Step 2 in algorithm TRES_MD. In the k-th execution of the loop one has that: Step 2.1 requires $O(\alpha_k \ell d^{r-1})$, where α_k denotes the number of primes tested (since a prime is unlucky if it divides the content of the leading coefficient of A, it follows that at most the number of primes tested in Step 2 is dominated by $d(\ell + \log d)$); applying Proposition 2 (1), one deduces that Step 2.2 is dominated by $\ell d^r + d^{2r+1}$; and Step 2.3 takes $O(kd^{2r-2})$. Thus, Step 2 is dominated by $d^{2r+2}(\ell + \log d) + d^{2r}(\ell + \log d)^2$. And therefore, $O(d^{2r+2}(\ell^2 + \log d))$ is an upper bound for the algorithms. ∎

Theorem 7 (Bound for Modular Quotient Method) *The computing time for the worst cases for the algorithm* TRES_MQ *(Trace Resultant by Modular Quotient) and the algorithm* SRES_MQ *(Slope Resultant by Modular Quotient) are bounded above by* $O(d^{2r+1}(\ell^2 + \log d))$. ∎

Proof. One simply has to follow the previous proof in combination with Proposition 2 (2). ∎

7 Experiments

In this section, we investigate the average computing times of the algorithms (for small inputs). Since such study is difficult to carry out theoretically, we will base our study on experiments. Before we jump into implementation and experiment, we first need to find out what to measure. For this purpose, let us assume that the average computing times have similar forms as the worst case computing times obtained in the last section, namely:

$$T = c_1(d+1)^{c_2 r + c_3}(\ell + \log_2 d)^{c_4}$$

where c_1, c_2, c_3, and c_4 are some constants. While the constants c_2, c_3, and c_4 depend only on the algorithms used, the constant c_1 also depends on hardware, operating system, programming language, etc. Our goal is to determine the constants c_1, c_2, c_3 and c_4 for each algorithm.

In order to facilitate the measurements, let us take logarithm (base 2) of both sides, obtaining:

$$\log_2 T = \log_2 c_1 + (c_2 r + c_3)\log_2(d+1) + c_4 \log_2(\ell + \log_2 d).$$

Using the short-hands $T^* = \log_2 T, c_1^* = \log_2 c_1, d^* = \log_2(d+1), \ell^* = \log_2 \ell$, it can be rewritten as

$$T^* = c_1^* + (c_2 r + c_3)d^* + c_4 \log_2(2^{\ell^*} + \log_2(2^{d^*} - 1)).$$

From this, we obtain the following partial differentiations:

$$\frac{\partial T^*}{\partial r} = c_2 d^*$$

$$\frac{\partial T^*}{\partial \ell^*} = c_4 \frac{\ell}{\ell + \log_2 d}$$

$$\frac{\partial T^*}{\partial d^*} = c_2 r + c_3 + c_4 \frac{d+1}{d} \frac{1}{\log_e 2(\ell + \log_2 d)}.$$

By rearranging them, we obtain

$$c_2 = \frac{1}{d^*}\frac{\partial T^*}{\partial r}$$

$$c_4 = \frac{\ell + \log_2 d}{\ell}\frac{\partial T^*}{\partial \ell^*}$$

$$c_3 = \frac{\partial T^*}{\partial d^*} - c_2 r - c_4 \frac{d+1}{d}\frac{1}{\log_e 2(\ell + \log_2 d)}$$

$$c_1 = \frac{T}{(d+1)^{c_2 r + c_3}(\ell + \log_2 d)^{c_4}}.$$

Thus in order to determine the constants, we need to measure $\frac{\partial T^*}{\partial r}$, $\frac{\partial T^*}{\partial \ell^*}$, $\frac{\partial T^*}{\partial d^*}$, and T^* for several inputs characterized by d, ℓ, and r.

In order to measure them, we have implemented the algorithms TRES_D, SRES_D, TRES_Q, and SRES_Q in the C language on top of the computer algebra C library SACLIB (Buchberger et al. 1993). We have not yet implemented the modular algorithms. All the experiments were done on a DEC station

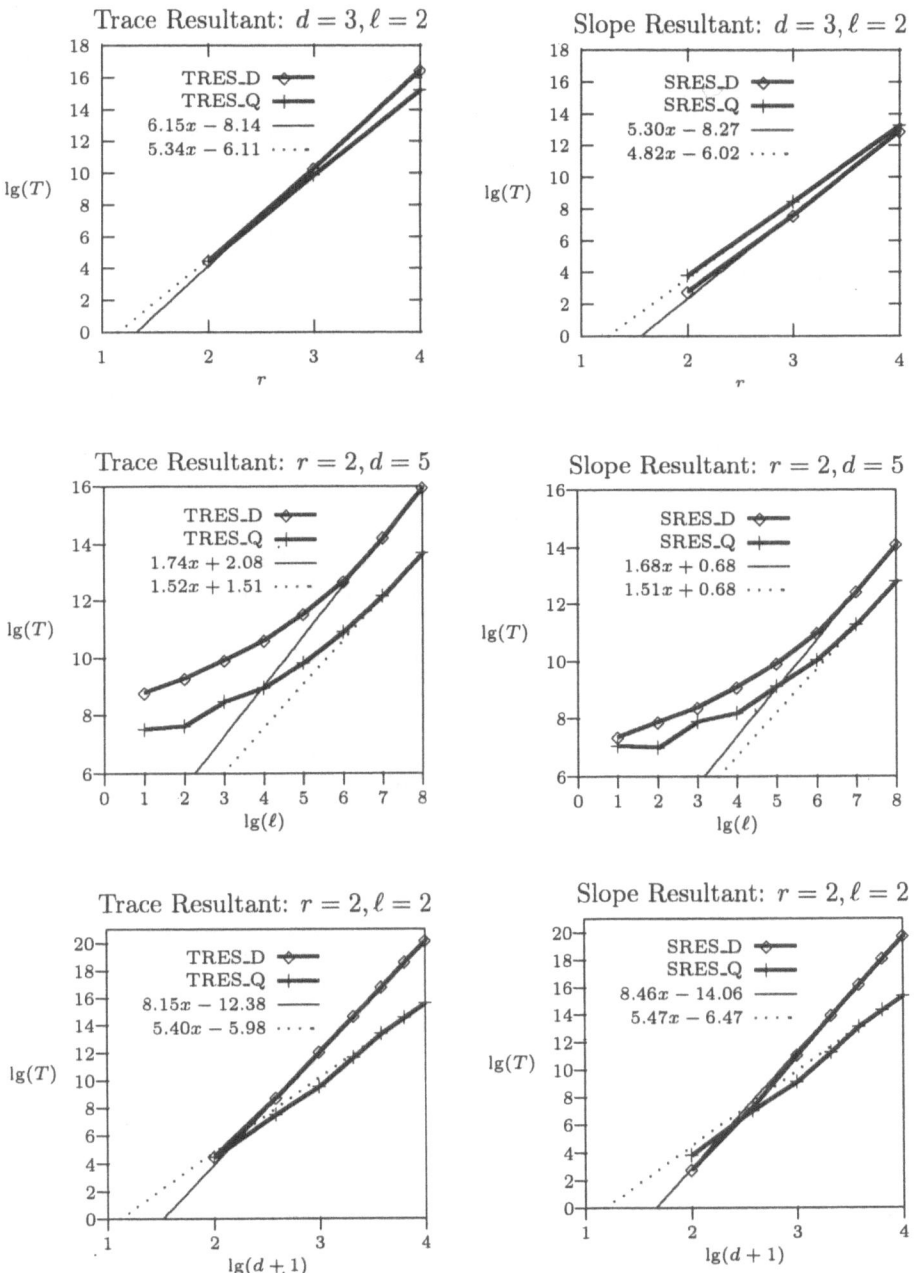

Figure 1: Experimental Results

5000/200 with 8 megabyte heap space for list processing. The input polynomials were randomly generated and were completely dense. All the computing times were measured in milliseconds.

In Figure 1, we give the experimental results. In order to help obtaining the partial derivatives, we plotted straight lines which have the same slopes as the experimental curves at the largest values of the horizontal axes. From these experimental measurements, we obtained the following values for c_1, c_2, c_3, and c_4.

	TRES_D	SRES_D	TRES_Q	SRES_Q
c_1	$2^{-15.13}$	$2^{-16.68}$	$2^{-8.39}$	$2^{-8.85}$
c_2	3.08	2.65	2.67	2.41
c_3	1.54	2.73	-0.33	0.26
c_4	1.76	1.69	1.53	1.52

It is interesting to note that in most cases, the values c_2, c_3 and c_4 are (significantly) smaller than the values predicted by theoretical analysis (for the worst cases). One might conclude from this that there is a big gap between average cases and worst cases, or that the theoretical estimation were too coarse (over-estimation).

Note that the values of c_1 for the determinant based methods are much smaller than that for the quotient based methods. This indicates that the determinant method is faster for small inputs. In fact, this phenomena is illustrated in the top two graphs in Figure 1. However for most inputs, the quotient based methods are faster due to the smaller values for c_2, c_3, and c_4. In fact, this sort of data can be used for guessing the trade-off point between the two kinds of algorithms. Such information can be implemented into a "driver" routine so that it can decide which algorithm should be called for a particular input.

Acknowledgments

Hoon Hong was partially supported by the ACCLAIM project of European Community Basic Research Action (ESPRIT 7195) and Austrian Science Foundation (P9374-PHY). J. Rafael Sendra partially supported by DGICYT A.I. Spain-Austria HU-007.

A New Algorithm to Find a Point in Every Cell Defined by a Family of Polynomials

Saugata Basu, Richard Pollack, and Marie-Françoise Roy

We consider s polynomials $P_1, \ldots, P_s$ in $k < s$ variables with coefficients in an ordered domain A contained in a real closed field R, each of degree at most d. We present a new algorithm which computes a point in each connected component of each non-empty sign condition over $P_1, \ldots, P_s$. The output is the set of points together with the sign condition at each point. The algorithm uses $s(s/k)^k d^{O(k)}$ arithmetic operations in A. The algorithm is nearly optimal in the sense that the size of the output can be as large as $s(O(sd/k))^k$.

Previous algorithms of Canny (1988, 1993a, 1993b) and Renegar (1992a, 1992b, 1992c) used $(sd)^{O(k)}$ operations. We use either these algorithms in the case $s = 1$ as a subroutine in our algorithm. As a bonus, our algorithm yields an independent proof of the bound on the number of connected components in all non-empty sign conditions (Pollack and Roy 1993) and also yields an independent proof of a theorem of Warren (1968) which is used in (Pollack and Roy 1993) as well as in many other problems of a combinatorial nature (see Alon (1995) for example).

1 Introduction

A sign condition for a set $\mathcal{P} = \{P_1, \ldots, P_s\}$, of s polynomials in k variables with coefficients in an ordered ring A, is specified by a sign vector $\sigma \in \{-1, 0, +1\}^s$. The sign condition σ is called *non-empty* (with respect to $\mathcal{P}$) if there is a point $x \in R^k$ (where R is a real closed field containing A and extending its order), such that

$$(\operatorname{sign} P_1(x), \ldots, \operatorname{sign} P_s(x)) = \sigma.$$

The *realization* of the sign condition σ in R is the set

$$\{x \in R^k | (\operatorname{sign} P_1(x), \ldots, \operatorname{sign} P_s(x)) = \sigma\}.$$

We will abuse language and have a sign condition denote the realization of the sign condition as well.

We say that a set $\mathcal{P} = \{P_1, \ldots, P_s\}$ of polynomials in k variables is in *general position* if no $k + 1$ have a common zero in R. (This is a weak notion of general position as we assume neither smoothness nor transversality.)

We will prove the following theorem

Theorem 1 *Let $\mathcal{P} = \{P_1, \ldots, P_s\}$ be a set of polynomials in $k < s$ variables each of degree at most d and each with coefficients in an ordered domain A contained in a real closed field R.*

There is an algorithm which outputs at least one point in each cell (semi-algebraically connected component) of every non-empty sign condition on $P_1, \ldots, P_s$ and provides the sign vector

$$(\mathrm{sign}(P_1(x)), \ldots, \mathrm{sign}(P_s(x)))$$

of $\mathcal{P}$ at each output point x.

The algorithm terminates after at most $\binom{O(s)}{k} sd^{O(k)} = (s/k)^k sd^{O(k)}$ arithmetic operations in A.

The idea behind the algorithm is that, if the polynomials are in general position, it is enough to find a point in every semi-algebraically connected component of the algebraic sets defined by a conjunction of less than $k + 1$ equations $P_i + e\epsilon$ (where $e \in \{0, 1, -1\}$ and ϵ is small enough) in order to find a test point in every semi-algebraically connected component of every non-empty sign condition over $\mathcal{P}$. Since the polynomials may not be in general position, we perturb the polynomials using three additional infinitesimals $\frac{1}{\Omega}, \delta', \delta$, such that the perturbed polynomials are in general position and their zero set is bounded. We can then use the method described above, and finally we let these infinitesimals go to zero in an appropriate way.

Previously, algorithms with complexity $(sd)^{O(k)}$, due to Canny (1988, 1993a, 1993b) and section 4 of Renegar (1992b), which test the emptyness of a sign condition, actually produced a point in every semi-algebraically connected component of that sign condition when it is not empty. In the case $s = 1$, it is easy to see, using Bezout's theorem, that these algorithms produce $O(d)^k$ points. We use any one of these algorithms as a subroutine in our algorithm. In our result, we consider an algebraic model in which we measure complexity in terms of the number of algebraic operations in the underlying ring A. We separate this complexity into a combinatorial part of $s\binom{O(s)}{k}$, involving s and k where precise bounds are obtained and an algebraic part, with complexity $d^{O(k)}$.

This idea of separating the complexity into a combinatorial part and an algebraic part already appears in (Canny 1993c) which has an algorithm which also constructs a point in every semi-algebraically connected component of every non-empty sign condition with complexity $s^{k+1}d^{O(k^2)}$ in the algebraic model (and $s^{k+1}d^{O(k)}$ randomized).

We are able to obtain as algebraic part of the complexity $d^{O(k)}$ by using the perturbation of Renegar (1992a, 1992b, 1992c) which uses only three additional infinitesimals.

We use the terminology and properties of infinitesimals, semi-algebraically connected components and paths in non-Archimedean extensions. A full discussion of these can be found in (Bochnak et al. 1987) but we offer a brief summary below.

The order relation on a real closed field R defines, as usual, the Euclidean topology on R^k. Semi-algebraic sets are finite unions of sets defined by a finite number of polynomial equalities and inequalities, and semi-algebric homeomorphisms are homeomorphism whose graph is semi-algebraic.

In particular we have the following elementary definitions and properties of semi-algebraic sets, over a real closed field R (see also Bochnak et al. 1987)

- A semi-algebraic set S is *semi-algebraically connected* if it is not the disjoint union of two non-empty closed semi-algebraic sets in S.
- A *semi-algebraically connected component* of a semi-algebraic set S is a maximal semi-algebraically connected subset of S.
- A semi-algebraic set has a finite number of semi-algebraically connected components.
- A *semi-algebraic path* between x and x' in R^k is a semi-algebraic subset γ, semi-algebraically homeomorphic to the unit interval of R through a semi-algebraic homeomorphism f_γ with $f_\gamma(0) = x$ and $f_\gamma(1) = x'$. A semi-algebraic path γ is *defined over A* if the graph of f_γ is described by polynomials with coefficients in A.
- A semi-algebraic set is semi-algebraically connected if and only if it is semi-algebraically path connected.
- In the case of real numbers, semi-algebraically connected components of semi-algebraic sets are ordinary connected components.
- Let S be a semi-algebraic set in R^k, and R' a real closed field containing R, we write $S_{R'}$ for the subset of R'^k defined by the same boolean combination of inequalities that define S.
- Let A be a subring of the real closed field R. A semi-algebraic set S is defined over A if it can be described by polynomials with coefficients in A.
- Let S be defined over $A \subset R$ and let R' be a real closed field extension of R. The semi-algebraically connected components of $S_{R'}$ are the extensions to R' of the semi-algebraically connected components of S.

We also need the following definitions and properties of ordered rings and Puiseux series. Again, fuller details can be found in (Bochnak et al. 1987). We suggest, in order to understand the following definitions as well as the proof of Proposition 1, that it is helpful to think of ϵ as positive and small enough.

- For a subring $A \subset R$ and a new variable ϵ, we define the order on $A[\epsilon]$ making ϵ infinitesimal and positive by saying that $P \in A[\epsilon]$ is positive if and only if the tail coefficient of P is positive. Similarly for $Q \in R(\epsilon)$ the order making

ϵ infinitesimal and positive is defined by saying that Q is positive iff the tail coefficients of the numerator and denominator of Q have the same sign. It follows that $P(\epsilon) > 0$ iff for $t \in R$ sufficiently small and positive, $P(t) > 0$.

- Let $R\langle \epsilon \rangle$ be the field of Puiseux series in ϵ with coefficients in R. Its elements are 0 and series of the form

$$\sum_{i \in \mathbf{Z}} a_i \epsilon^{i/q}.$$

with $q \in \mathbf{N}$, $a_{i_0} \neq 0$ for some $i_0 \in \mathbf{Z}$, $a_i \in R$, and $a_i = 0$ for $i < i_0$.

The field $R\langle \epsilon \rangle$ is real closed (Bochnak et al. 1987). An element x' of $R\langle \epsilon \rangle$ is *infinitesimal* (with respect to R) if and only if its absolute value is strictly smaller than any positive element in R. The element ϵ of $R\langle \epsilon \rangle$ is infinitesimal positive. The elements of $R\langle \epsilon \rangle$ bounded over R form a *valuation ring* denoted $V(\epsilon)$; the elements of $V(\epsilon)$ are 0 or Puiseux series

$$\sum_{i \in \mathbf{Z}} a_i \epsilon^{i/q}$$

with $a_i = 0$ for $i < 0$. We denote by eval the ring homomorphism from $V(\epsilon)$ to R which maps $\sum_{i \in \mathbf{Z}} a_i \epsilon^{i/q}$ (with $a_i = 0$ for $i < 0$) to a_0. If the Puiseux series is infinitesimal ($i_0 > 0$), it is mapped by eval to 0. We can think of eval as the evaluation of the Puiseux series at 0.

- A Puiseux series $f(\epsilon) \in R\langle \epsilon \rangle$ is algebraic over $A[\epsilon]$ if it is a root of a polynomial $P_\epsilon(X)$ with coefficients in $A[\epsilon]$. The root $f(\epsilon)$ is distinguished among the other roots of P, by its Thom encoding (see Coste and Roy 1988), using for example Algorithm RAN in (Roy and Szpirglas 1990a). For t sufficiently small and positive, we may replace ϵ by t in the polynomials which are the coefficients of $P_\epsilon(X)$ to obtain the polynomial $P_t(X)$ and (since inequalities in $A[\epsilon]$ will agree with the corresponding inequalities in $A[t]$ for t positive and small enough) $P_t(X)$ will have the same number of roots in R as $P_\epsilon(X)$ has in $R\langle \epsilon \rangle$, with the corresponding Thom encodings and the algebraic Puiseux series $f(\epsilon)$ thus defines a semi-algebraic function $f(t)$ for t small enough.

- Let $S(\epsilon)$ be a semi-algebraic set in $R\langle \epsilon \rangle^k$ defined over $A[\epsilon]$ and for $t \in R$ let $S(t)$ be the semi-algebraic set in R^k obtained by substituting t for ϵ. Let $P(S(\epsilon))$ be a property of the semi-algebraic set $S(\epsilon)$ in $R\langle \epsilon \rangle^k$ defined over $A[\epsilon]$ which is expressible by a first order formula $\Phi(\epsilon)(x_1, \ldots, x_k)$ with parameters in $A[\epsilon]$. Then for t positive and small enough $P(S(t))$ is satisfied.

This is an easy consequence of quantifier elimination and of the preceding lemma on the ways signs are computed in $A[\epsilon]$.

We will use this property to replace ϵ by t in a path defined over $A[\epsilon]$ to obtain a "neighboring" path defined over $A[t]$.

We can now be a little more precise about the output of our computation.

The coordinates of the points that we construct in every semi-algebraically connected component of a non-empty sign condition will not necessarily belong to the realization C of the sign condition in R but to its realization $C_{R\langle 1/\Omega,\delta\rangle}$ in some non-Archimedean real closed extension $R\langle 1/\Omega,\delta\rangle$ of the ordered domain $A[\Omega,\delta]$, where $1/\Omega,\delta$, are infinitesmals with $1/\Omega \gg \delta$. Nevertheless, whenever we output a point $x = (x_1,\ldots,x_k)$ what we actually output will be:

1. A univariate polynomial f with coefficients in $A[\Omega,\delta]$,
2. A root, say α of f which is distinguished among the other real roots of f by the sign vector $(\operatorname{sign} f(\alpha), \operatorname{sign} f'(\alpha), \ldots, \operatorname{sign} f^{\{\deg f\}})$. This sign vector is called the *Thom encoding* of α after Thom's lemma (see Coste and Roy 1988).
3. k rational functions $R_1,\ldots,R_k$ with coefficients in $A[\Omega,\delta]$ and $x_i = R_i(\alpha)$ for $i = 1,\ldots,k$.

All the computations will take place in these polynomial rings, and our complexity analysis counts the number of arithmetic operation in A (algebraic operations and sign determinations).

The fact that the sample points constructed may belong to a non-Archimedean real closed extension $R\langle 1/\Omega,\delta\rangle$ of R is quite unusual but is not a problem at all because of the relationship between semi-algebraically connected components in $R\langle 1/\Omega,\delta\rangle$ and in R for semi-algebraic sets defined over A (see the ninth bullet above).

We can answer any sign question concerning these sample points by algebraic computations and sign evaluations in A and the complexity is evaluated in A.

If we are only interested in the existential theory of reals, our algorithm outputs the set of non-empty sign conditions over R, even if the sample points belong to $R\langle 1/\Omega,\delta\rangle$.

The following proposition will easily assure the correctness of our algorithms (see also Grigor'ev and Vorobjov 1992, lemma 1 (b)).

Proposition 1 *Let C be a semi-algebraically connected component of a non-empty sign condition of the form $P_1 = \ldots = P_\ell = 0, P_{\ell+1} > 0,\ldots,P_s > 0$. Then we can find an algebraic set V in $R\langle\epsilon\rangle^k$ defined by equations $P_1 = \ldots = P_\ell = P_{i_1} - \epsilon = \ldots P_{i_m} - \epsilon = 0$, such that a semi-algebraically connected component of V is contained in $C_{R\langle\epsilon\rangle}$, where $R\langle\epsilon\rangle$ is the real closed field of Puiseux series over R in the variable ϵ.*

Proof. If C is closed, it is a semi-algebraically connected component of the algebraic set defined by $P_1 = \ldots = P_\ell = 0$. If not, we consider Γ, the set of all semi-algebraic paths γ in R going from some point $x(\gamma)$ in C to a $y(\gamma)$ in $\bar{C} \setminus C$ such that $\gamma \setminus \{y(\gamma)\}$ is entirely contained in C. For any $\gamma \in \Gamma$, there exists an $i > \ell$ such that P_i vanishes at $y(\gamma)$. Then on $\gamma_{R\langle\epsilon\rangle}$ there exists a point $z(\gamma,\epsilon)$ such that one of the $P_i - \epsilon$ vanishes at $z(\gamma,\epsilon)$ and that on the portion of the path between x and $z(\gamma,\epsilon)$ no such $P_i - \epsilon$ vanishes. We write I_γ for the set of indices between $\ell + 1$ and s such that $i \in I_\gamma$ if and only if $P_i(z(\gamma,\epsilon)) - \epsilon = 0$.

Now choose a path $\gamma \in \Gamma$ so that the set $I_\gamma = \{i_1, \ldots, i_m\}$ is maximal under set inclusion and let V be defined by $P_1 = \ldots = P_\ell = P_{i_1} - \epsilon = \ldots = P_{i_m} - \epsilon = 0$.

It is clear that at $z(\gamma, \epsilon)$, defined above, we have $P_{\ell+1} > 0, \ldots, P_s > 0$ and $P_j - \epsilon > 0$ for every $j \notin I_\gamma$. Let C' be the semi-algebraically connected component of V containing $z(\gamma, \epsilon)$. We shall prove that no polynomial $P_{\ell+1}, \ldots, P_s$ vanishes on this semi-algebraically connected component, and thus that C' is contained in $C_{R\langle\epsilon\rangle}$.

Let us suppose on the contrary that some new P_i ($i > l, i \notin I_\gamma$) vanishes on C', say at y_ϵ. We can suppose without loss of generality that y_ϵ is defined over $A[\epsilon]$. Take a semi-algebraic path γ_ϵ defined over $A[\epsilon]$ connecting $z(\gamma, \epsilon)$ to y_ϵ with $\gamma_\epsilon \subset C'$. Denote by $z(\gamma_\epsilon, \epsilon)$ the first point of γ_ϵ with $P_1 = \ldots = P_\ell = P_{i_1} - \epsilon = \ldots = P_{i_m} - \epsilon = P_j - \epsilon = 0$ for some new j not in I_γ.

For t in R small enough, the set γ_t (obtained by replacing ϵ by t in γ_ϵ) defines a semi-algebraic path from $z(\gamma, t)$ to $z(\gamma_\epsilon, t)$ contained in C. Replacing ϵ by t in the Puiseux series which give the coordinates of $z(\gamma_\epsilon, \epsilon)$ defines a path γ' containing $z(\gamma_\epsilon, \epsilon)$ from $z(\gamma_\epsilon, t)$ to $y = \mathrm{eval}(z(\gamma_\epsilon, \epsilon))$ (which is a point of $\bar{C} \setminus C$). Let us consider the new path γ^* consisting of the beginning of γ (up to the point z_t for which $P_{i_1} = \ldots, P_{i_m} = t$), followed by γ_t and then followed by γ'. Now the first point in γ^* such that there exists a new j with $P_j - \epsilon = 0$ is $z(\gamma_\epsilon, \epsilon)$ and thus $\gamma^* \in \Gamma$ with I_{γ^*} strictly larger than I_γ. This is impossible by the maximality of I_γ. ∎

2 Proof of the Theorem

2.1 The Algorithm

First we introduce some notation. Let $\Omega, \delta, \delta', \epsilon$ be new variables.

We order $A[\Omega]$, requiring that Ω be positive and bigger than any positive element of A. Similarly, we order $A[\Omega, \delta]$, (resp. $A[\Omega, \delta, \delta']$, $A[\Omega, \delta, \delta', \epsilon]$), such that δ is positive and smaller than any positive element of $A[\Omega]$, (resp. δ' positive and smaller than any positive element of $A[\Omega, \delta]$, ϵ positive and smaller than any positive element of $A[\Omega, \delta, \delta']$. In short, the ordering is $1/\Omega \gg \delta \gg \delta' \gg \epsilon$, where $a \gg b$ stands for, "b is positive and infinitesmal with respect to a".

Let $R\langle 1/\Omega \rangle$ be the field of Puiseux series in $1/\Omega$ with coefficients in R, $R\langle 1/\Omega, \delta \rangle$ the field of Puiseux series in δ with coefficients in $R\langle 1/\Omega \rangle$, $R\langle 1/\Omega, \delta, \delta' \rangle$ the field of real Puiseux series in δ' with coefficients in $R\langle 1/\Omega, \delta \rangle$, $R\langle 1/\Omega, \delta, \delta', \epsilon \rangle$ the field of real Puiseux series in ϵ with coefficients in $R\langle 1/\Omega, \delta, \delta' \rangle$.

Let $V(1/\Omega, \delta, \delta', \epsilon)$ be the valuation ring of elements of $R\langle 1/\Omega, \delta, \delta', \epsilon \rangle$ bounded by elements of $R\langle 1/\Omega, \delta \rangle$, and eval the corresponding map from $V(1/\Omega, \delta, \delta', \epsilon)$ to $R\langle 1/\Omega, \delta \rangle$ sending δ' and ϵ to 0.

We are now in a position to describe the algorithm for finding a point in every semi-algebraically connected component of every non-empty sign condition.

The algorithm is as follows:

(1) Replace the set $\mathcal{P}$ by the set $\mathcal{Q}$ of $4s+1$ polynomials $\{P_0 = x_1^2 + \ldots + x_k^2 - \Omega^2\} \cup_{i=1,\ldots,s} \{(1-\delta)P_i - \delta H_i, (1-\delta)P_i + \delta H_i, (1-\delta)P_i - \delta'\delta H_i, (1-\delta)P_i + \delta'\delta H_i\}$ where $H_i = (1 + \sum_{1 \leq j \leq k} i^j x_j^{d'})$ and d' is an even number bigger than the degree of the P_i.

(2) Enlarge the set of $4s + 1$ polynomials to the following set of $3(4s + 1)$ polynomials.

$$\{Q_{i,e}\}_{1 \leq i \leq s, e \in \{0,1,-1\}}$$

where ϵ is an infinitesimal positive element and $Q_{i,e} = Q_i + e\epsilon$.

(3) For every $(\ell \leq k)$-tuple of polynomials $Q_{i_1,e_1}, \ldots, Q_{i_\ell,e_\ell}$ consider $Q = Q_{i_1,e_1}^2 + \cdots + Q_{i_\ell,e_\ell}^2$, test if $Q = 0$ defines the empty set, and if it does not construct at least one point in every semi-algebraically connected component of the zero set of Q (using, say, the construction in section 4 of (Renegar 1992b)).

(The i-th coordinate of one of these points is a univariate rational function $R_i(\alpha)$ where α is defined over $A[\Omega, \delta, \delta', \epsilon]$. Recall items 1, 2, 3 of the introduction.) Apply the eval map, by letting $\epsilon, \delta' \to 0$, on the points satisfying $P_0 < 0$, and obtain points defined over $A[\Omega, \delta]$.

(4) Apply the eval map, by letting $\epsilon, \delta' \to 0$, on the points satisfying $P_0 < 0$, and obtain points defined over $A[\Omega, \delta]$.

(5) Characterize the coordinates of these points using Thom's lemma and determine at these points the sign of the other polynomials $P_i \in \mathcal{P}$ (using the algorithms RAN and RANSI in (Roy and Szpirglas 1990a)). Note that now all polynomials are univariate with coefficients in $A[\Omega, \delta,]$ (again, recall items 1, 2, 3 of the introduction).

In practical applications, it might be useful to check as a first step, whether the given polynomials are in general position. That is for each $(k + 1)$-tuple of polynomials $P_{i_1}, \ldots, P_{i_{k+1}}$, determine whether the polynomial $Q = P_{i_1}^2 + \ldots + P_{i_{k+1}}^2 = 0$ defines the empty set or not. If every such $(k + 1)$-tuple defines the empty set, then the polynomials are in general position. Thus we could do without the additional infinitesmals $1/\Omega$, δ and δ'. In this case, we omit Step 1 and Step 4 of the algorithm. As a result, we generate points in $R\langle\epsilon\rangle^k$, defined over $A[\epsilon]$, where $R\langle\epsilon\rangle$ is the field of Puiseux series in ϵ with coefficients in R. This would lead to a significant practical (though not asymptotic) improvement in the run-time of the algorithm.

2.2 Proof of the Correctness of the Algorithm

We first prove that the polynomials in $\mathcal{Q}$ are in general position. This is an easy consequence of the fact that the polynomials $(1-\delta)P_i + \delta H_i$ are in general position (Renegar 1992a, 1992b, 1992c lemma 3.11.1).

We recall this proof. If H is a polynomial we denote by H^h its homogenization with respect to the variable x_0.

Lemma 1 *The $H_i^h = (x_0^{d'} + \sum_{1 \le j \le k} i^j x_j^{d'})$ are in general position inside the complex projective space.*

Proof. Take $H = (x_0{}^{d'} + \sum_{1 \le j \le k} t^j x_j^{d'})$. If $k+1$ of the H_i^h (say $H_{j_1}^h, \ldots, H_{j_{k+1}}^h$) had a common zero $\bar{x}$, in the complex projective space, substituing this root in H would give a non zero univariate polynomial of degree at most k with $k+1$ distinct roots $(j_1, \ldots, j_{k+1})$, which is impossible. ∎

Since a common zero of a set of H_i would certainly produce a common zero of the corresponding set of H_i^h we have:

Corollary 1 *The polynomials H_i are in general position.*

Lemma 2 *Polynomials $(1 - \delta)P_i + \delta H_i$ are in general position.*

Proof. Let $Q_{i,t} = (1 - t)P_i + tH_i$. Let $Q_{i,t}^h$ denote the homogenization of $Q_{i,t}$ with respect to x_0. Let us consider the set T of t such that the $Q_{i,t}$ are in general position in the complex projective space, which means that no $k+1$ subset of the $Q_{i,t}^h$ has a common zero in the projective complex space. The set T contains 1, according to the preceding lemma, it contains an open interval containing 1, since being in general position in the complex projective space is a stable condition. It is also Zariski constructible, since it can be defined by a first order formula of the language of algebraically closed fields. So the transcendental element δ belongs to the extension of T, which proves the lemma. ∎

Let C be a semi-algebraically connected component of $P_1 = \ldots = P_\ell = 0, P_{\ell+1} > 0, \ldots, P_s > 0$. We shall prove that the extension of C to $R\langle 1/\Omega, \delta \rangle$, $C_{R\langle 1/\Omega, \delta \rangle}$, contains eval$(D_{R\langle 1/\Omega, \delta, \delta', \epsilon \rangle})$ for some bounded, semi-algebraically connected component, D of the set S defined in $R\langle 1/\Omega, \delta, \delta' \rangle^k$ by the sign conditions $P_0 < 0, -\delta'\delta H_1 < (1 - \delta)P_1 < \delta'\delta H_1, \ldots, -\delta'\delta H_\ell < (1 - \delta)P_\ell < \delta'\delta H_\ell, (1 - \delta)P_{\ell+1} > \delta H_{\ell+1}, \ldots, (1 - \delta)P_s > \delta H_s$.

Let x be a point of C, then x belongs to S. Now let D be the bounded semi-algebraically connected component of S containing x. We know that $D_{R\langle 1/\Omega, \delta, \delta', \epsilon \rangle}$ is again semi-algebraic and bounded. Since the image under eval of a bounded semi-algebraically connected set (defined over $A[\Omega, \delta, \delta', \epsilon]$) is again semi-algebraically connected (see Heintz et al. 1994b), it is clear that eval$(D_{R\langle 1/\Omega, \delta, \delta', \epsilon \rangle})$ is contained in $C_{R\langle 1/\Omega, \delta \rangle}$. ∎

The correctness of the algorithm is now an easy consequence of Proposition 1, since by Lemma 2, it suffices to consider sets of polynomials of size $< k + 1$. ∎

2.3 Complexity Analysis of the Algorithm

First note that, since we have introduced only four infinitesimals, and all the algebraic computations are done in the ordered ring $A[\Omega, \delta, \delta', \epsilon]$ using only linear algebra subroutines, the asymptotic complexity is not affected because of the introduction of these infinitesimals.

The total number of $\ell \leq k$-tuples to be examined is $\sum_{\ell \leq k} \binom{12s+3}{\ell} = (O(s/k))^k$.

For each of these ℓ-tuples, we test whether the zero set of the corresponding polynomial is empty or not, and if not compute a point in each of its semi-algebraically connected components.

For each subset of size ℓ for which the zero set of the corresponding polynomial is non-empty we produce at most $(O(d))^k$ points. So the total number of points constructed will be $(O(sd/k)^k$. The Thom characterization and sign evaluation of the other polynomials at these points is done with complexity $sd^{O(k)}$ (using the variant of the Ben-Or–Kozen–Reif algorithm that is algorithm RAN and RANSI in Roy and Szpirglas 1990a).

2.4 Generalizations

In a forthcoming paper (Basu et al. 1997) we show that

Theorem 2 *Let $\mathcal{V}$ be a k' dimensional variety which is the zero set of the polynomial $Q \in A[X_1, \ldots X_k]$ and Q has degree at most d. Let $\mathcal{P} = \{P_1, \ldots, P_s\}$ be a set of polynomials in $k < s$ variables each of degree at most d and each with coefficients in an ordered domain A contained in a real closed field R.*

There is an algorithm $\mathcal{A}$, which determines whether the family $\mathcal{P}$ is in general position with respect to $\mathcal{V}$ (which means that no $k' + 1$ polynomials in $\mathcal{P}$ have a common zero on $\mathcal{V}$) and if it is in general position, outputs at least one point in each cell (semi-algebraically connected component) in $\mathcal{V}$ of every sign condition of $\mathcal{P}$ which is non-empty in $\mathcal{V}$ and provides the signs of all the polynomials at each of these points.

In case the family is not in general position with respect to $\mathcal{V}$, there is an algorithm, $\mathcal{B}$, which gives the same output as algorithm $\mathcal{A}$.

Algorithm $\mathcal{A}$ terminates after at most $\binom{O(s)}{k'} sd^{O(k)} = s^{k'} sd^{O(k)}$ arithmetic operations in A while algorithm $\mathcal{B}$ terminates after at most $\binom{O(s)}{k'} sd^{O(k^2)} = s^{k'} sd^{O(k^2)}$ arithmetic operations in A.

Corollary 2 *The number of semi-algebraically connected components of all non-empty sign conditions defined by a family of s polynomials in $k < s$ variables, on a variety $\mathcal{V}$ of dimension k' with each polynomial of degree at most d is $s^{k'} O(d/k)^k$.*

In the case $k' = 0$ we get, being slightly more specific, an algorithm evaluating the signs of a family of s polynomials at a finite set of points defined by ℓ polynomials in time $sd^{O(k)}$

This is a multivariate version of the Ben-Or–Kozen–Reif algorithm improving the complexity of (Pedersen et al. 1993).

We plan to use the methods in this paper in order to describe a new quantifier elimination algorithm of single exponential complexity when the number of alternation of quantifiers is fixed. In this new algorithm the degrees of the polynomials in the quantifier free equivalent formula will be independent of the number of equations, which is new for single exponential methods, and the time complexity in terms of the number of input polynomials will be more precise than in (Heintz et al. 1990) and Renegar (1992a, 1992b, 1992c).

Acknowledgments

Saugata Basu and Richard Pollack were supported in part by NSF grant CCR-9122103. Marie-Françoise Roy was supported in part by the project ESPRIT-BRA 6846POSSO.

Local Theories and Cylindrical Decomposition

Daniel Richardson

1 Introduction

There are many interesting problems which can be expressed in the language of elementary algebra, or in one of its extensions, but which do not really depend on the coordinate system, and in which the variables can be restricted to an arbitrarily small neighborhood of some point. It seems that it ought to be possible to use cylindrical decomposition techniques to solve such problems, taking advantage of their special features. This article attempts to do this, but many unsolved problems remain.

Let $\mathbb{Q}$ be the rational numbers. Let $\mathbb{R}$ be the real numbers. The language of elementary algebra is the first order language appropriate to the structure $(\mathbb{R}, =, <, +, -, *, \mathbb{Q})$. Let $f_1(x), f_2(x), \ldots, f_k(x)$ be elementary functions defined (and analytic) around 0, such as for example, e^x or $\sin(x)$. An extension of the language of elementary algebra, in the context of this article, means the first order language appropriate to $(\mathbb{R}, =, \leq, +, -, *, f_1(x), \ldots, f_k(x), \mathbb{Q})$. The theory of the reals then means the set of all true sentences in this structure.

If S is a sentence in one of these languages, let $S(\epsilon)$ be S with all variables restricted to the interval $(-\epsilon, \epsilon)$. So $S(\epsilon)$ is obtained by changing each expression $(\forall v)A$ or $(\exists v)A$ to $(\forall v)((-\epsilon < v < \epsilon) \rightarrow A)$, or $(\exists v)(-\epsilon < v < \epsilon \wedge A)$ respectively. For fixed ϵ the set of true sentences of the form $S(\epsilon)$ is called the ϵ theory, $T(\epsilon)$.

If the limit, as $\epsilon \to 0$, of $T(\epsilon)$ exists, we can call this the local theory around the origin. To say that the limit exists means that for every sentence, S, the truth value of $S(\epsilon)$ is eventually constant for sufficiently small ϵ.

The work of Gabrielov (1968b), van den Dries (1988b) and Denef and van den Dries (1988) implies that these local theories will exist for extensions by elementary functions such as e^x and $\sin(x)$. We also know from their work that if Γ is any finite set of terms in the extension language, there exists a neighborhood of the origin and a cylindrical decomposition of that neighborhood which is sign invariant for Γ. In spite of this existence result, we only know how to compute the cylindrical decompositions in the algebraic case.

In the global situation, cylindrical decompositions do not necessarily exist. For example, if the extension is by $\sin(x)$, we can write a set of terms with so many critical points that their projection is dense.

The relationship between the global theories and their local counterparts is not quite clear. I expect that the local theories are all decidable, and that the local cylindrical decompositions are not very hard to compute. As a first step, we could try to find an easy way to compute cylindrical decomposition in the algebraic case in an infinitesimal neighborhood of the origin. In Section 2 this is attempted. The resulting algorithm looks easier than the standard one, but no relative complexity results are available. (An implementation of this algorithm has been written, in AXIOM, by Jetender Kang, at Bath, but not enough testing has yet been done to draw any conclusions.)

In Section 3 the methods of Section 2 are applied, also in the algebraic case, to neighborhoods of infinity. In Section 4 cylindrical decompositions local to the origin are found which are sign invariant for sets of polynomials in x, e^x, y, e^y. The problem of extending this to higher dimensions is briefly discussed.

2 Infinitesimal Sectors at the Origin

Assume the variables are ordered by importance: $x_1 \prec x_2 \prec \cdots \prec x_n$. Define S_η^n to be the subset of $\mathbb{R}^n$ with coordinates $(x_1, \ldots, x_n)$ satisfying $0 < x_1 < \eta \wedge -x_1 < x_2 < x_1 \wedge \cdots \wedge -x_1 < x_n < x_1$. S_η^n is called the n dimensional η sector at the origin. We will suppose $\eta > 0$ and study the limiting situation as $\eta \to 0$.

Definition 1 *There is only one cell in S_η^1, namely $(0, \eta)$. The dimension of $(0, \eta)$ is defined to be 1. Suppose we have defined cells and their dimension in S_η^n. Let C be a cell in S_η^n of dimension k, and let $X = (x_1, \ldots, x_n)$. Let $f : C \to \mathbb{R}$ be a continuous function, whose graph is contained in the sector, i.e.,*

$$(\forall X \in C)(-x_1 < f(X) < x_1).$$

Then $\{(X, y) : X \in C \wedge y = f(X)\} = C(f)$ is a cell of S_η^{n+1} of dimension k. If $C(f)$ and $C(g)$ are cells of this type, and

$$(\forall X \in C)(f(X) < g(X)),$$

then $\{(X, y) : X \in C \wedge f(X) < y < g(X)\} = C(f, g)$ is a $k+1$ dimensional cell of S_η^{n+1}.

Definition 2 *The base of a cell $C(f)$ or $C(f, g)$ is C.*

Cells of the form $C(f)$ are called *thin* cells, and cells of the form $C(f, g)$ are called *thick* cells.

Definition 3 *A stack of cells is a finite non-empty list of disjoint cells with the same base. The cells in the list are ordered in increasing size of the most important variable.*

For example, a typical stack with base C would be: $C(f_1, f_2), C(f_2), C(f_2, f_3),$ $\ldots, C(f_{k-1}, f_k)$. Note that according to the definitions above there is no such thing as an empty cell or an empty stack.

Definition 4 *A cylindrical decomposition of the cell in S_η^1 is the cell itself (there being only one of them anyway). A cylindrical decomposition of a cell C in S_η^{n+1} is a partition of C into finitely many disjoint stacks, the bases of which form a cylindrical decomposition of the base of C in S_η^n.*

If Γ is a set of functions, a sign condition on Γ is a conjunction of statements which specifies, for all f in Γ, either $f < 0$, or $f = 0$, or $f > 0$. If s is a sign condition on Γ, let $C_s^n(\eta) = \{X : X \in S_\eta^n \wedge s\}$.

Definition 5 *Suppose Γ is a subset of $\mathbb{Q}[x_1, \ldots, x_n]$, and Σ is a set of sign conditions on Γ. We will say that Σ specifies a cylindrical decomposition of S_η^n for Γ if there exists $\delta > 0$ such that for all $\eta > 0$ with $\eta < \delta$ the following conditions hold:*

1. *$\{C_s^n(\eta) : s \in \Sigma\}$ is a cylindrical decomposition of S_η^n and*
2. *each $C_s^n(\eta)$ is a single non-empty cell.*

Note that if Σ specifies a cylindrical decomposition for Γ, then each s in Σ must give a sign to each polynomial in Γ, and thus every polynomial in Γ must have constant sign on every $C_s^n(\eta)$.

Definition 6 *If Σ specifies a cylindrical decomposition for Γ of S_η^n, we will say that we know the stack structure of Σ if*

1. *we can partition $\{C_s^n(\eta) : s \in \Sigma\}$ into the stacks of the cylindrical decomposition,*
2. *we know the dimension of each cell $C_s^n(\eta)$, for s in Σ, and*
3. *within each stack we can list the cells in order of increasing size of the most important variable, x_n.*

Note that if x_n does not occur in Γ, but all the variables in Γ are below x_n then all the stacks in a cylindrical decomposition of S_η^n consist of a single cell. Also, conditions 2 and 3 together imply that we know for any cell $C_s^n(\eta)$ whether it is a thin cell or a thick cell; and if it is a thick cell, of the form $C(f, g)$, we can find the bottom and top bounding surfaces with the same base, $C(f)$ and $C(g)$. If $C(f, g)$ is not the top of its stack, then $C(g)$ is a cell described by a sign condition in Σ, and if $C(f, g)$ is the top of its stack then $C(g)$ is just $C(x_n - x_1)$. Similarly, we can find $C(f)$.

In the following we will suppose that $p(y)$ is a polynomial in $\mathbb{Q}[x_1, \ldots, x_n]$, and y is the most important of the variables which actually occur in $p(y)$, and $(x_1, \ldots, x_k)$ are the variables below y. Let

$$p(y) = a_d y^d + a_{d-1} y^{d-1} + \cdots + a_0$$

be a polynomial in $(x_1, \ldots, x_k)$ and y. Define the degree of $p(y)$ to be d, the leading coefficient to be a_d, and the reduct of $p(y)$ to be $a_{d-1} y^{d-1} + \cdots + a_0$. Also the derivative, $p'(y)$, is the partial derivative with respect to y. Also if $p(y)$ and $q(y)$ are two such polynomials, the remainder $\text{rem}(p, q, y)$ of $p(y)$ and $q(y)$ is the usual pseudo-remainder of p and q with respect to y. That is $\text{rem}(p, q, y)$ is obtained by considering p and q to be polynomials in y with coefficients in $\mathbb{Q}(x_1, \ldots, x_k)$, dividing q into p, and multiplying the remainder by a power of the leading coefficient of p, in order to clear denominators.

We will say that a polynomial is *small* if it is zero at the origin. If Γ is a set of polynomials, the *small part* of Γ is the subset which is zero at the origin.

The factored form of a polynomial p will mean $Cq_1 \cdots q_k$, where C is rational and as large as possible, and $q_1, \ldots, q_k$ are non-constant polynomials with integral coefficients with $p = Cq_1 \cdots q_k$, and k is as large as possible. If C is not 1, or k is not 1, the polynomial p will be said to be reducible. Otherwise, if the constant factor of p is 1, and its only non-constant factor is itself, p will be called irreducible.

Define the sector boundary terms to be $x_2 - x_1, x_2 + x_1, \ldots, x_n - x_1, x_n + x_1$.

Definition 7 *Suppose Δ is a set of polynomials in $x_1, \ldots, x_k, y$, and y is the most important of the variables which occur in Δ. We will say that Δ is closed, with respect to y, if*

(a) *Δ is closed under factorization in $\mathbb{Q}[x_1, \ldots, x_k, y]$,*
(b) *if $p(y)$ is in Δ, and $p(y)$ is irreducible and small, and has degree greater than zero in y, then the leading coefficient and derivative with respect to y of p is also in Δ; if the leading coefficient of $p(y)$ is also small, then the reduct of $p(y)$ is also in Δ*
(c) *if $p(y)$ and $q(y)$ are in Δ and are both small and irreducible, and have positive degree in y, then the remainder of $p(y)$ with respect to $q(y)$, $\text{rem}(p, q, y)$, is also in Δ.*

Definition 8 *If Γ is a set of polynomials in $x_1, \ldots, x_k, y$, and y is the largest of the variables, define $\text{cl}(\Gamma, y)$ to be the smallest set of polynomials containing Γ, and the sector boundary terms $x_2 - x_1, x_2 + x_1, \ldots, y - x_1, y + x_1$, and closed with respect to y. Let $\text{proj}(\Gamma, y)$ be that part of $\text{cl}(\Gamma, y)$ which does not contain y. Define Γ^* recursively by*

$$\Gamma^* = \Gamma$$

if only the least important of the variables occurs in Γ, and otherwise, if y is the most important variable in Γ,

$$\Gamma^* = \text{cl}(\Gamma, y) \cup \text{proj}(\Gamma, y)^*.$$

Example 1 *Let* $p = y^4 + (2x^2 + 4x)y^2 + x^4 - 4x^3$. *Let* $\Gamma = \{p\}$. *The sector boundary terms are* $y + x, y - x$. *The polynomial* p *is irreducible. Its leading coefficient is not zero at the origin, so we do not need to take the reduct of* p. $p_y = 4y(y^2 + x^2 + 2x)$. *Let* $q = y^2 + x^2 + 2x$. *The remainder of* p *with respect to* q *is a polynomial in* x, $-8x^3 - 4x^2$. *So all the terms in* $\mathrm{cl}(\Gamma, y)$ *involving* y *are* $p, p_y, y, q, y + x, y - x$. *All the terms in* x *are, of course, either reducible or not small, except for* x *itself.*

Remark 1 *It is claimed that* Γ^* *contains all the information needed to find a cylindrical decomposition for* Γ *for a sector near the origin. In the computation of* Γ^* *we take advantage of the fact that we are intending to work close to the origin; this means that we are able to throw away factors of polynomials which are not small. For example, if we had* $y^{3000} + y$, *we could cancel out the large factor* $y^{2999} + 1$, *since it has known sign near the origin.*

Define the leading variable of a polynomial to be the most important variable which occurs in it. Extend the ordering on variables to an ordering on polynomials as follows. $p \prec q$ if the leading variable of p is less important than the leading variable of q, or if they have the same leading variable but the degree of this leading variable is smaller in p than it is in q.

Sector Decomposition Algorithm

Input: finite $\Gamma^* \subset \mathbb{Q}[x_1, \ldots, x_n], n$.
Output: A set Σ of sign conditions on Γ^* which determine a cylindrical decomposition of S_η^n for Γ^*.

(1) We suppose that Γ^* given to us is closed under all the operations mentioned earlier. List the polynomials in Γ^*: $p_1, p_2, \ldots, p_n, \ldots$ in such a way that

  (a) if $p_i \prec p_j$ then $i < j$.
  (b) for each x_j, with $j > 1$ the first two polynomials, if there are any, in the list involving x_j are the sector boundary terms $(x_j - x_1), (x_j + x_1)$.
  (c) If p is in the list and can be factored, its non-constant factors all come before it in the list.

(2) Now proceed recursively up the list. Let $G_n = (p_1, \ldots, p_{n-1})$. Assume x_k is the most important variable in G_n. We make the following induction hypothesis:

  We have a set of sign conditions Σ_n which determine a cylindrical decomposition of S_η^k for G_n.

Now consider p_n. We suppose that the most important variable of p_n is y. So p_n can be written as

$$p_n(y) = a_d y^d + \cdots + a_0$$

and a_d is non-zero as a polynomial. What we need to do is extend the sign conditions in Σ_n to account for $p_n(y)$. Each sign condition in Σ_n will have either one extension or three extensions. There are several cases:

(0) y is the least important variable, x_1. In this case determine the sign of $p_n(y)$ on $(0, \eta)$, and extend each s in Σ_n accordingly. The sign may be found by repeatedly differentiating and evaluating at 0 until some non-zero quantity is found.

(i) y is not the same as x_k. In this case $p_n(y)$ and p_{n+1} must be the sector boundary terms $y - x_1$ and $y + x_1$. We extend each s in Σ_n to $(s, y - x_1 < 0)$

(ii) $p_n(y)$ is $y + x_1$. Extend each s in Σ_n to $(s, y + x_1 > 0)$.

(iii) $p_n(y)$ is not small. In this case extend each s in Σ_n according to the sign of $p_n(y)$ at the origin

(iv) $p_n(y)$ can be factored. Then the factors have already been given signs in each sign condition in Σ_n. Extend each sign condition by giving a sign to $p_n(y)$ which is consistent with the signs given to its factors.

(v) $p_n(y)$ is small and irreducible. This is the interesting case.
Divide Σ_n into stacks and do each stack independently. Divide each stack into thin cells and thick cells. This can be done since we know the stack structure and dimension.
(part a) All the thin cells are now done independently. Take a sign condition s in Σ_n which describes a thin cell

$$C_s^k(\eta) = \mathcal{C}(f).$$

We need to extend s to include the sign of $p_n(y)$ on this cell. Since this is a thin cell, the value of y is determined for each X in $\mathcal{C}$, and the way in which y is determined is defined by s. Therefore s must include an equality $q(y) = 0$, where $q(y)$ is a polynomial in y whose leading coefficient is not set to zero by s. The leading coefficient is in G_n, so its sign is fixed and non-zero in the cell. The remainder $\mathrm{rem}(p, q, y)$ is also in G_n. So we already have the sign of the remainder in s. Therefore in the cell $\mathcal{C}(f)$ the sign of $p_n(y)$ is determined by s. So in the extension of s we set the sign of $p_n(y)$ to be consistent with the information already in s.

(part b) All the thick cells are now done independently. Take a typical thick cell

$$C_s^k(\eta) = \mathcal{C}(f, g).$$

The leading coefficient of $p_n(y)$ is already in G_n. If this is set to zero by s, we set the sign of $p_n(y)$ in the extension of s to be the same as the sign of the reduct of $p_n(y)$, which is in G_n.

Suppose the leading coefficient of $p_n(y)$ is not set to zero in s. The derivative of $p_n(y)$ is in G_n, and its sign is therefore set in s. Thus $p_n(y)$ is monotone in y in the cell $\mathcal{C}(f, g)$.

Determine the signs of $p_n(y)$ on $\mathcal{C}(f)$ and $\mathcal{C}(g)$. If the thick cell we are considering is neither the top nor the bottom of its stack, we can find $\mathcal{C}(f)$ and $\mathcal{C}(g)$ as thin cells, since we assume we know the stack structure, and we have already found the signs of $p_n(y)$ on them. We can also find the sign of $p_n(y)$ at the top and bottom of the stack, by taking the remainder of $p_n(y)$ with the sector boundary terms $y - x_1$ and $y + x_1$. These remainders have signs set by s.

If the sign of $p_n(y)$ is the same on $C(f)$ and $C(g)$, or if one of the signs is zero and the other is non-zero, then s has just one extension and we set the sign of $p_n(y)$ in the extension to be consistent with the signs on its upper and lower boundaries, i.e., to be the same as one of the non-zero signs.

Suppose the signs of $p_n(y)$ differ on $C(f)$ and $C(g)$ one being plus and the other minus. Then s has three extensions

$$(s, p_n(y) < 0),$$
$$(s, p_n(y) = 0),$$
$$(s, p_n(y) > 0)$$

which will define three cells $C(f, h)$, $C(h)$, $C(h, g)$, and these are ordered in the stack in a way consistent with the signs on the boundaries, i.e., so that the new negative thick cell is next to the old negative thin cell, and the new positive thick cell is next to the old positive thin cell.

Theorem 1 *Given any finite set of polynomials $\Gamma \subset \mathbb{Q}[x_1, \ldots, x_k]$, and natural number k, the sector decomposition algorithm just described produces a set Σ of sign conditions on Γ^* which determines a cylindrical decomposition of S_η^k for Γ^*, and we can find the stack structure for this decomposition.*

Proof. Polynomials which are not small, i.e., not zero at the origin, will be said to be large.

We can list all the large polynomials in Γ^*, and find $\delta > 0$ so that if p is large and in Γ^*, then p does not change sign in S_δ^k. This is a consequence of continuity of polynomials.

The theorem can now be proved by induction, following the recursive definition of the algorithm. Using the notation given, our induction hypothesis is:

> We have a set Σ_n of sign conditions on G_n so that if $0 < \eta < \delta$, and x_k is at least as large as the largest variable in G_n then $\{C_s^k(\eta) : s \epsilon \Sigma_n\}$ is a cylindrical decomposition for G_n of S_η^k with known stack structure.

This hypothesis will be abbreviated *correct(n)*. We need to show

$$correct(n) \Rightarrow correct(n + 1).$$

The polynomial considered at the $n + 1$ stage is $p_n(y)$, which can be written as

$$a_d y^d + \cdots + a_0$$

with the leading coefficient a_d non-zero as a polynomial. There are five cases to consider, depending of which branch of the algorithm applies.

In case (i), $y \neq x_k$ and the dimension is raised by one. All the cells defined by Σ_n become bases of stacks. These stacks just have one cell in them, since we are confining attention to the interior of the sector. All the extensions from Σ_n to Σ_{n+1} are correct. In cases (i), (ii), (iii), and (iv) we evidently have $correct(n) \Rightarrow correct(n + 1)$.

Consider case (v). Here $y = x_k$. The action of the algorithm at this stage is to subdivide some of the cells in an existing cylindrical decomposition of S_η^k.

A sign condition $s+$ of Σ_{n+1} which is the same as s in Σ_n as far as G_n is concerned will be said to be an extension of s.

We observe:

1. Each sign condition s in Σ_n has either three extensions in Σ_{n+1},

$$(s, P_n(y) < 0), (s, p_n(y) = 0), (s, p_n(y) > 0)$$

or one extension, which is one of these.
2. $G_n^* = G_n$, and $G_{n+1}^* = G_{n+1}$.
3. The bases of the cells determined by Σ_{n+1} are the same as the bases of the cells determined by Σ_n.
4. The construction is independent on each stack, and within each stack the subdivisions only occur on some of the thick cells, and these subdivisions are made independently.
5. Since the sets $C_s^k(\eta)$, for s in Σ_{n+1} are determined by sign conditions on G_{n+1}, it must be that G_{n+1} is sign invariant on each of these sets.

We need to show that when single extensions are made to sign conditions s in Σ_n these extensions are correct, and that all the new sign conditions in Σ_{n+1} describe single non-empty cells.

First consider s in Σ_n which describes a thin cell $C_s^k(\eta) = \mathcal{C}(f)$. The condition s has only one extension $s+$ in Σ_{n+1}. The claim is that $s+$ is true on $\mathcal{C}(f)$, so that the cell described by the extension of s is the same as the cell described by s.

The leading coefficient, a_d of $p_n(y)$ is in G_n, so s must give a sign to a_d. If this sign is zero, a_d must be small, since otherwise the cell defined by s would be empty. Thus if the leading coefficient of $p_n(y)$ is set to zero by s, the reduct of $p_n(y)$ must be in G_n, and therefore the algorithm can refer to this to set the sign of $p_n(y)$. Next suppose that s does not set a_d to zero. We already know that $p_n(y)$ itself is small, since this is a precondition of case (v). We also know that $\mathcal{C}(f)$ must be described by $q(y) = 0$, where $q(y)$ is in G_n, together with its leading coefficient. Since $q(y) = 0$ does not define the empty set, $q(y)$ must be small. Thus the remainder of $p_n(y)$ with respect to $q(y)$ is in G_n. So the sign of $p_n(y)$ can be set correctly by the algorithm.

The above argument applies to all the thin cells. So we can suppose that all the extensions to the thin cells are done correctly.

It is only necessary to show that the thick cell extensions are also correct. This seems entirely clear from the remarks made in that section of the algorithm definition. ∎

The above construction can also be used to find cylindrical decompositions in small neighborhoods of the origin, rather then in sectors. This is done by adding an extra variable. For example, if we have a polynomial in $y_1, \ldots, y_n$, and we wish a cylindrical decomposition adapted to the polynomial in a sufficiently small neighborhood of the origin, we use variables $x_1, \ldots, x_{n+1}$, and, for $i \leq n$, define $x_{i+1} = y_i$.

Example 2 *Let* $\Gamma = \{x^3 y^4 - x^3 z^4 + y^3 z^4\}$. *The small irreducible part of* Γ^* *is then* $\{x, y - x, y + x, y, 2y + x, 4y + 3x, y^2 + xy + x^2, y^4 + xy^3 - x^4, z - x, z + x, z, (y^3 - x^3)z^4 + x^3 y^4\}$, *according to the Jetender Kang implementation at Bath. The cell construction algorithm produces 51 cells.*

Remark 2 *The above construction is unreasonably dominated by the idea of degree. It seems that to improve this algorithm this dominance should be questioned. It is obviously necessary to induct on some well ordered quantity. The idea of degree is familiar, but it is not very well related to geometric complexity of the zero set near the origin. We are looking for another idea, but have not yet found it. In the above, for example, we would like to replace the use of* $\mathrm{rem}(p, q, y)$ *with some polynomial in the ideal generated by* p *and* q *which minimizes geometric complexity at the origin.*

2.1   An Application

An important problem in practice is classification of singularities. The above technique does not solve this interesting problem, but it does seem to be a step in the right direction.

Suppose for example that we want to decide whether or not a polynomial which is zero at the origin has a maximum or a minimum at that point. That is, we want to decide

$$(\exists \epsilon)(\forall y_1, \ldots, y_k)(0 < |(y_1, \ldots, y_k)| < \epsilon \rightarrow (p \neq 0))$$

where p is in $\mathbb{Q}[y_1, \ldots, y_k]$ and is zero at the origin. Let $x_1 = \epsilon$ and $x_{i+1} = y_i$ for $i = 1, \ldots, k$. The problem is solved by decomposing the S_η^{k+1} sector for $\Gamma = \{p\}$.

It would be useful to solve the equivalent problem for functions defined by power series. The statement of the problem would be as follows. Define a computable real field to be a field of computable real numbers, in which the field operations are computable, and the zero equivalence problem is decidable. Suppose $F(x_1, \ldots, x_k)$ is an analytic function, i.e., has a convergent multivariate power series, around the origin, and is zero at the origin. Suppose also that we know the coefficients of the power series, in the sense that we have a computable embedding of them in a computable real field. (So, in particular, we can approximate each coefficient as closely as we wish by rationals.) Let $X = (x_1, \ldots, x_k)$. Let $d(X)$ be the distance of X from the origin. We suppose that for each n, we can find a polynomial $F_n(X)$, obtained by truncation of the power series, with coefficients in our computable real field, and a constant K_n so that, for sufficiently small X,

$$|F(X) - F_n(X)| < K_n d(X)^n.$$

The problem now is: Does there exist a neighborhood $N_\epsilon(0)$ of the origin in which $X = 0$ is the only solution of $F(X) = 0$? If there is such a neighborhood, find one.

The construction above can be used to give a semialgorithm to solve this problem. That is, it can be shown that if there is such an isolating neighborhood, this fact can be verified, by using the methods given above. The method is sketched below.

A punctured neighborhood of the origin is the neighborhood with the origin removed. Suppose $F(X) > 0$ in some punctured neighborhood of the origin. It can be shown that this implies that $F(X) > d(X)^k$ for some natural number k in some punctured neighborhood of the origin. This implies that $F_{k+1}(X) > d(X)^k$ in some punctured neighborhood of the origin.

Also, the cylindrical decomposition techniques work on polynomials with coefficients in a computable real field. So if it ever happens that $F_{k+1}(X) > d(X)^k$ for X sufficiently near the origin, then this can be verified. If this can be done for some k, this will reveal to us that $F(X) > 0$ for X sufficiently close to the origin, since $d(X)^k$ will be much bigger than $K_{k+1} d(X)^{k+1}$ for small X.

3   Neighborhoods of Infinity

The construction given above can also be made to work for large values of x, instead of for points in a sector near the origin. We only need to change the definition of cell.

The definition is slightly simpler than before. Define S_ζ^n to be the subset of $\mathbb{R}^n$ with coordinates $(x_1, \ldots, x_n)$ satisfying $\zeta < x_1$. $\check{S}_\zeta^n$ is called the n dimensional $x_1 > \zeta$ block. We will suppose ζ is large, and consider limiting behavior as ζ tends to infinity.

Definition 9 *A cell in S_ζ^1 is (ζ, ∞). The dimension of (ζ, ∞) is defined to be 1. Suppose we have defined cells and their dimension in $\mathbb{R}^n$. Let C be a cell in S_η^n of dimension k. Let $f : C \to \mathbb{R}$ be a continuous function. Then $\{(x, y) : x \in C \wedge y = f(x)\} = C(f)$ is a cell of S_ζ^{n+1} of dimension k. If $C(f)$ and $C(g)$ are cells of this type, and*

$$(\forall x \in C)(f(x) < g(x))$$

then $\{(x, y) : x \in C \wedge f(x) < y < g(x)\} = C(f, g)$ is a $k + 1$ dimensional cell of S_ζ^{n+1}.

The set cl(Γ) is defined as before except that we do not put in any of the sector boundary terms, which were used before, and we need to apply the operations to all the irreducible polynomials, not just to small ones.

The decomposition theorem can now be proved as before. However, the result is less interesting, since if the first variable does not occur in our set Γ we get nothing different from the usual cylindrical decomposition.

4 Exponential Polynomials in Two Variables

In this section we consider polynomials in $\mathbb{Q}[x, e^x, y, e^y]$. We will call x, e^x, y, e^y *basic terms* and order them: $x \prec e^x \prec y \prec e^y$. We will normally write our exponential polynomials as polynomials in the highest occurring basic term with coefficients which are similarly written as polynomials in the lower basic terms.

Differentiation does not necessarily reduce a term to zero after finitely many steps, so we define a false-derivative operation. Let t be one of the variables. Let

$$p(t) = a_n(t)e^{nt} + \cdots + a_1(t)e^t + a_0(t).$$

Define $D_t(p) = p_t/e^{kt}$, where the natural number k is chosen as large as possible so that e^{kt} divides p_t. Usually k will be zero and $D_t(p)$ will be the ordinary derivative. Each time this differentiation is done, the degree of the trailing coefficient $a_0(t)$ will go down, until eventually it is reduced to zero and at that point division by e^t will become possible.

The operation D_t will be called false differentiation with respect to t. After a finite number of D_t operations, a polynomial in t and e^t, say $p(t, e^t)$, will be reduced to a non-zero constant. This number of operations is analogous to degree, and will be called the *index* of $p(t, e^t)$. So the D_t operator reduces the index.

We also observe that $D_t(p)$ has the same sign as p_t, so it has many of the useful properties of the derivative. In particular, roots of $p(t, e^t)$ are separated by roots of $D_t(p)$. For discussion of false derivatives, see (Richardson 1992).

We now need to define cl(Γ), where Γ is a set of exponential polynomials. If neither y nor e^y occur in Γ, i.e., if the only variable is x, we let cl(Γ) = Γ. If the most important basic term in Γ is y, the closure is defined just as before. Suppose the most important basic term is e^y. Polynomials in Γ can be written as polynomials in e^y with coefficients in lower terms. So leading coefficient, degree, and reduct can be defined just as before. For terms with degree zero in e^y, differentiation with respect to y has the usual meaning. For polynomials with degree greater than zero in e^y replace differentiation with the false derivative operation defined above.

We now come to the real difficulty, which is remainders. We can take remainders between two polynomials which both have degree zero in e^y, or between two polynomials which both have positive degree in e^y. But we don't know how to take the remainder of $p(y, e^y)$ after division by $q(y)$. So the remainder operation has to be undefined in this case.

Let Γ be a finite set of polynomials in $\mathbb{Q}[x, e^x, y, e^y]$. We now define Γ^* as before, in terms of cl(Γ). (Note that except when differentiating, x, e^x, y, e^y are treated as if they were four unrelated variables; so, for example, if Γ is a set of polynomials involving x, e^x, y, e^y, proj(Γ, e^y) will be a set of polynomials which may involve x, e^x, and y.)

Exponential Sector Decomposition Algorithm

Input: A finite set $\Gamma^* \subset \mathbb{Q}[x, e^x, y, e^y]$.
Output: A set Σ of sign conditions which determine a cylindrical decomposition of S^2_η for Γ^*.

(1) We list Γ^* in four blocks, first polynomials in x, then polynomials in x and e^x, then polynomials in x, e^x and y, and finally polynomials in x, e^x, y, e^y. As before we make sure the sector boundary terms are first in each block, and factors come before products; subject to that, each block is ordered by index.

(2) We suppose, for an induction hypothesis, that we have constructed our set of sign conditions for G_n, the first n polynomials in the list. Everything now works as before, with one exception. This happens in part (a) of step (v).

The problem occurs in the following case. Our sign condition s in Σ_n describes a thin cell $(0, \eta)(f)$, and f is determined by some inequalities and an equality $q(y) = 0$ in s. However, $p_n(y) = p(y, e^y)$ involves e^y, but there are no equalities involving e^y in s. What we need here is a way to determine the sign of $p(y, e^y)$ on $(0, \eta)(f)$. We can suppose that p is zero at the origin.

There are various ways of solving this problem. The method below, due essentially to Khovanskii, is described because there seems to be some hope of generalizing it.

We now form another set of polynomials Γ_w. This is obtained from Γ^* in two steps as follows.

1. Form Γ_1 by replacing every instance of e^y in Γ^* by $2w + 1$. This particular linear function is used so that if $-x < y < x$ and $2w + 1 = e^y$ then $-x < w < x$, for x small and positive. We want to have this condition in order to apply the methods of Section 2. Order the variables $x \prec y \prec w$.

2. Form Γ_2 as the set of polynomials

$$R = p_y q_x - p_x q_y + (2w + 1)q_x p_w/2$$

where $q(y)$ is small and irreducible and in Γ_1 with most important variable y, and $p(y, 2w + 1)$ is small and irreducible and in Γ_1 with most important variable w. The polynomial R is obtained by solving

$$q_y y' + q_x = 0,$$
$$p_w w' + p_y y' + p_x = 0$$

for y' and w', substituting the result into

$$(2w+1)' - (2w+1)y',$$

and clearing denominators.

Let $\Gamma_w = \Gamma_1 \cup \Gamma_2$. Using the methods of Section 2, we find a cylindrical decomposition of S_η^3, with coordinate variables (x, y, w), sign invariant for Γ_w, ordering the basic terms $x \prec e^x \prec y \prec w$.

(Note that, using false derivatives, we can find the sign of any polynomial in x and e^x in $(0, \eta)$ for small η.)

Now return to our construction of sign conditions for Γ^*.

We have a thin cell $(0, \eta)(f)$ determined by sign condition s, and s has in it an equality $q(y) = 0$, where the leading coefficient of $q(y)$ is not set to zero in s. We are trying to determine the sign of $p(y, e^y)$ on this cell. (A sign is determined since the cell is an analytic curve whose limit is the origin.)

At this point we refer to our other cylindrical decomposition, for Γ_w. In this other cylindrical decomposition we also have $(0, \eta)(f)$, and above $(0, \eta)(f)$, in the w dimension, we have a stack of cells on which, simultaneously,

$$p(y, 2w+1) = 0, \quad q_y y' + q_x = 0 \text{ and } p_w w' + p_y y' + p_x = 0.$$

Say $w = h_1(x)$ or $\cdots$ or $w = h_k(x)$, with $h_1(x) < \cdots < h_k(x)$. On each of these cells the sign of $(2w+1)' - (2w+1)y'$ is known. Note that w' is the derivative of w as a function of x implicitly defined by $q(y) = 0, p(y, 2w+1) = 0$. On an initial segment of the stack, we will have $(2w+1)' < (2w+1)y'$. Suppose we have $(2w+1)' < (2w+1)y'$ on $w = h_i(x)$, but not on $h_{i+1}(x)$.

It might happen that $(2w+1)' \equiv (2w+1)y'$ when $w = h_{i+1}(x)$. In that case $p(y, e^y) = 0$.

The other possibility is that $(2w+1)' > (2w+1)y'$ when $w = h_{i+1}(x)$. Then we know that $w = (e^y - 1)/2$ lies in $(0, \eta)(f)(h_i, h_{i+1})$. We can read off the sign of $p(y, 2w+1)$ in this region. This now tells us the sign of $p(y, e^y)$ on $(0, \eta)(f)$. (The only purpose of the auxiliary decomposition was to discover these signs.)

Now consider a cell of the form $(0, \eta)(f, g)$, and irreducible polynomial $p_n(y) = p(y, e^y)$. Since G_n contains the false derivative of $p(y, e^y)$, it cannot happen that $p(y, e^y)$ has two roots on $(0, \eta)(f, g)$. In fact $p(y, e^y)$ must be monotone in y in this cell. Since we are doing the thin cells first, we already know the signs of $p(y, e^y))$ on the boundaries $(0, \eta)(f)$ and $(0, \eta)(g)$. If the signs on the boundaries are the same, we do not need to refine the cell. If the signs are different there is an implicitly defined function $h(x)$ so that $f(x) < h(x) < g(x)$, and $p(y, e^y)$ is zero on $(0, \eta)(h)$. So we refine our cylindrical decomposition in this case by replacing the cell $(0, \eta)(f, g)$ by three cells $(0, \eta)(f, h), (0, \eta)(h), (0, \eta)(h, g)$. Since $(0, \eta)(f, g)$ was described by a sign condition on G_n, the new cells are described by a sign condition on G_{n+1}. The only new aspect of this is that the function $h(x)$ may not be algebraic.

4.1   Exponential Polynomials in Higher Dimensions

I believe the above construction can be made to work for exponential polynomials in any number of variables, but I don't know how to do it in general. The problems which occur can be seen in the three dimensional case. Suppose we just went ahead with the construction as described above for polynomials in x, e^x, y, e^y, z, e^z. It is fairly clear that no problems occur for polynomials which do not involve e^z. Suppose we have a set Γ of terms, closed under some operations such as false derivatives, remainders, leading coefficients and reducta as applied to small and irreducible terms, and we already have a cylindrical decomposition sign invariant for that part of Γ which does not involve e^z. Let $\mathcal{C}(f)$ be a cell, and let $p(z, e^z) = a(z)e^z + b(z)$ be one of our terms. We want to find the sign of this term on $\mathcal{C}(f)$. If $\mathcal{C}(f)$ has dimension one, we can find the sign using a more elaborate version of the construction given above. Suppose, however, that the dimension of $\mathcal{C}(f)$ is two. So underneath $\mathcal{C}(f)$ is a sector in the xy plane. $\mathcal{C}(f)$ is the zero set of some term $q(z)$. At this point our lack of a remainder operation between $p(z, e^z)$ and $q(z)$ becomes quite serious. It may happen that $p(z, e^z)$ changes sign on $\mathcal{C}(f)$. This seems to force subdivision of the sector in the xy plane. This may happen also on other sheets of the zero set of $q(z)$ above the sector. So our lack of a remainder may mean that the problem we solved in the xy plane was not sufficient to allow successful lifting to three dimensions.

There are serious difficulties here, but I am sure they can, eventually, be overcome. It may be that such problems will call forth a re-evaluation of the cylindrical decomposition technique, or even a change in the definition. In the case discussed above, for example, it seems tempting just to subdivide the cells independently in three dimensional space, and forget about the projection to the xy plane. If we are mainly interested in the topology of the zero set, why do we need the projection?

Acknowledgments

This paper was written with the support of CEC, ESPRIT BRA contract 6846 "POSSO."

A Combinatorial Algorithm Solving Some Quantifier Elimination Problems

Laureano González-Vega

1 Introduction

The main problem in Computational Real Algebraic Geometry is the development of efficient Quantifier Elimination algorithms. It is well known (Davenport and Heintz 1988) that the general problem of quantifier elimination cannot be solved in polynomial time. Therefore the only way to attack this problem is to consider specific cases where efficient algorithms can be applied. By efficient we do not mean "polynomial time". Instead we are looking for algorithms, methods and criteria that can be used to solve specific quantifier elimination problems involving low degree polynomials in a reasonable amount of time. This strategy has already been investigated by Hong (1993d) for formulas with quadratic polynomial constraints and by Heintz, Roy, and Solernó (1993) for specific inputs for which the general Quantifier Elimination algorithm presented in (Heintz et al. 1990) is efficient.

The main tool used to achieve this goal is the good specialization properties of Sturm–Habicht sequences introduced in (Gonzàlez-Vega et al. 1989, 1990, 1994, 1997) and (Lombardi 1989). The properties of Sturm–Habicht sequences are presented in Section 2 and used in Section 3 to perform quantifier elimination for some specific formulas in an efficient way. Section 3 provides algorithms performing quantifier elimination for the following formulas:

$$\mathbb{H}_n : \ \forall x \, (x^n + a_{n-1} x^{n-1} + \ldots + a_1 x + a_0 > 0),$$
$$\mathbb{E}_n : \ \exists x \, (x^n + a_{n-1} x^{n-1} + \ldots + a_1 x + a_0 = 0),$$
$$\mathbb{L}_n : \ \exists x \, (x^n + a_{n-1} x^{n-1} + \ldots + a_1 x + a_0 < 0).$$

The algorithms presented have two distinct parts. The first one, purely algebraic, computes a finite family of polynomials in $\mathbb{Z}[a_0, \ldots, a_{n-1}]$ through the computation of one or several Sturm–Habicht sequences. The second one, purely combinatorial, considers all the possible sign conditions over the constructed polynomials, and selects, using the rule shown in Section 2, only those making the considered formula true. There exists a third step in the algorithm, currently performed by hand, which simplifies the resulting quantifier free formula by deleting trivially empty conditions and by using the well known simplification rules:

$$< \cup > \longrightarrow \neq, \ < \cup = \longrightarrow \leq, \ > \cup = \longrightarrow \geq .$$

The result obtained after performing the algorithm in Section 3 is a quantifier free formula for $\mathbb{H}_n$, $\mathbb{E}_n$ or $\mathbb{L}_n$ represented as a finite union of basic semialgebraic sets: sets defined as the intersection of a finite family of polynomial sign conditions (see Bochnak et al. 1987 for example). Some of these sets, possibly all, can be empty and in this sense the algorithm proposed in Section 3 has a philosophy similar to Hormander's algorithm (see chapter I in Bochnak et al. 1987).

The last section discusses extensions of the methods presented in Section 3 to more complicated formulas, and presents future improvements that will lead to more efficient algorithms for these types of quantifier elimination problems. Finally we remark that the strategies presented in this paper cannot be extended in an efficient way to solve the general quantifier elimination problem: the lower bounds for the general problem cannot be avoided.

2 Sturm–Habicht Sequence

The definition and main properties of Sturm–Habicht sequences are presented in this section. Proofs of the theorems in this section can be found in (Gonzàlez-Vega et al. 1989, 1990, 1994, 1997) and (Lombardi 1989). We begin with a minor modification to the usual definition of the subresultant polynomial which allows some control over the formal degrees of the polynomials defining the sequence (see (Loos 1982b) or (Collins 1967) for the classical definition).

Let $\mathbb{D}$ be an ordered integral domain.

Definition 1 *Let P, Q be polynomials in $\mathbb{D}[x]$ and $p, q \in \mathbb{N}$ with $\deg(P) \leq p$ and $\deg(Q) \leq q$:*

$$P = \sum_{k=0}^{p} a_k x^k \quad Q = \sum_{k=0}^{q} b_k x^k.$$

If $i \in \{0, \ldots, \min(p, q)\}$ we define the polynomial subresultant associated to P, p, Q and q of index i as follows:

$$\mathrm{Sres}_i(P, q, Q, q) = \sum_{j=0}^{i} M_j^i(P, Q) x^j$$

where $M_j^i(P, Q)$ is the determinant of the matrix containing columns $1, 2, \ldots,$ $p + q - 2i - 1$ and $p + q - i - j$ of the matrix:

$$m_i(P, p, Q, q) = \overbrace{\begin{pmatrix} a_p & \cdots & a_0 & & & \\ & \ddots & & \ddots & & \\ & & a_p & \cdots & a_0 \\ b_q & \cdots & b_0 & & \\ & \ddots & & \ddots & \\ & & b_q & \cdots & b_0 \end{pmatrix}}^{p+q-i} \left.\begin{matrix}\\ \\ \\ \\ \\ \\\end{matrix}\right\} \begin{matrix} q - i \\ \\ p - i \end{matrix}$$

The determinant $M_i^i(P, Q)$ will be called i-th principal subresultant coefficient and will be denoted by $\mathrm{sres}_i(P, p, Q, q)$.

The next definition defines the Sturm–Habicht sequence associated with P and Q as the subresultant sequence of P and $P'Q$ modulo some sign changes.

Definition 2 *Let P, Q be polynomials in $\mathbb{D}[x]$ with $p = \deg(P)$ and $q = \deg(Q)$. If we write $v = p + q - 1$ and*

$$\delta_k = (-1)^{k(k+1)/2}$$

for every integer k, the Sturm–Habicht sequence associated with P and Q is defined as the list of polynomials $\{\mathrm{StHa}_j(P, Q)\}_{j=0,\ldots,v+1}$ where $\mathrm{StHa}_{v+1}(P, Q) = P$, $\mathrm{StHa}_v(P, Q) = P'Q$ and for every $j \in \{0, \ldots, v - 1\}$:

$$\mathrm{StHa}_j(P, Q) = \delta_{v-j}\mathrm{Sres}_j(P, v + 1, P'Q, v).$$

For every j in $\{0, \ldots, v + 1\}$ the principal j-th Sturm–Habicht coefficient is defined as:

$$\mathrm{stha}_j(P, Q) = \mathrm{coef}_j(\mathrm{StHa}_j(P, Q)).$$

Let $\mathbb{D}$ be an ordered integral domain and $\mathbb{B}$ its real closure. To establish the relation between the real zeros (zeros in $\mathbb{B}$) of a polynomial $P \in \mathbb{D}[x]$ and the polynomials in the Sturm–Habicht sequence of P and Q, with $Q \in \mathbb{D}[x]$, we introduce the following integer number for every $\epsilon \in \{-1, 0, +1\}$:

$$c_\epsilon(P; Q) = \mathrm{card}(\{\alpha \in \mathbb{B} / P(\alpha) = 0, \; \mathrm{sign}(Q(\alpha)) = \epsilon\}).$$

The polynomials in the Sturm–Habicht sequence of P and Q provide an efficient method of computing the integer number $c_+(P; Q) - c_-(P; Q)$. In the case when $Q = 1$ we are computing the number of real zeros of P and when $Q = F^2$, with $F \in \mathbb{D}[x]$ we are computing the number of real zeros, α, of P such that $F(\alpha) \neq 0$. The following definitions introduce several sign counting functions that we shall use to compute the integer number $c_+(P; Q) - c_-(P; Q)$.

Definition 3 *Let* $\{a_0, a_1, \ldots, a_n\}$ *be a list of nonzero elements in* $\mathbb{B}$. *Define* $\mathbf{V}(\{a_0, a_1, \ldots, a_n\})$ *to be the number of sign variations in* $\{a_0, a_1, \ldots, a_n\}$, *and* $\mathbf{P}(\{a_0, a_1, \ldots, a_n\})$ *to be the number of sign permanences in* $\{a_0, a_1, \ldots, a_n\}$.

Definition 4 *Let* $a_0, a_1, \ldots, a_n$ *be elements in* $\mathbb{B}$ *with* $a_0 \neq 0$ *and suppose that we have the following distribution of zeros:*

$$\{a_0, a_1, \ldots, a_n\} = \{a_0, \ldots, a_{i_1}, \overbrace{0, \ldots, 0}^{k_1}, a_{i_1+k_1+1}, \ldots, a_{i_2}, \overbrace{0, \ldots, 0}^{k_2},$$

$$a_{i_2+k_2+1}, \ldots, a_{i_3}, 0, \ldots, 0, a_{i_{t-1}+k_{t-1}+1}, \ldots, a_{i_t}, \overbrace{0, \ldots, 0}^{k_t}\}$$

where all the a_i*'s that have been written are not zero. We define* $i_0 + k_0 + 1 = 0$ *and*

$$\mathbf{C}(\{a_0, a_1, \ldots, a_n\}) =$$

$$\sum_{s=1}^{t} \left(\mathbf{P}(\{a_{i_{s-1}+k_{s-1}+1}, \ldots, a_{i_s}\}) - \mathbf{V}(\{a_{i_{s-1}+k_{s-1}+1}, \ldots, a_{i_s}\})\right) + \sum_{s=1}^{t-1} \varepsilon_{i_s}$$

where

$$\varepsilon_{i_s} = \begin{cases} 0 & \text{if } k_s \text{ is odd,} \\ (-1)^{k_s/2} \, \mathrm{sign}(\frac{a_{i_s+k_s+1}}{a_{i_s}}) & \text{if } k_s \text{ is even.} \end{cases}$$

Theorem 1 *If* P *and* Q *are polynomials in* $\mathbb{D}[x]$ *with* $p = \deg(P)$ *then*

$$\mathbf{C}(\{\mathrm{stha}_p(P, Q), \ldots, \mathrm{stha}_0(P, Q)\}) = c_+(P; Q) - c_-(P; Q).$$

In particular we have that for the case $Q = 1$ the number of real roots of P is determined exactly by the signs of the $p - 1$ determinants $\mathrm{stha}_i(P, 1)$ (the first two ones are $\mathrm{lcof}(P)$ and $p \cdot \mathrm{lcof}(P)$). Moreover, as these determinants are the formal leading coefficients of the subresultant sequence for P and P' (modulo some sign changes) we get also as a by-product, very useful in the Section 3, the greatest common divisor of P and P':

$$\mathrm{StHa}_i(P, 1) = \gcd(P, P') \Longleftrightarrow \begin{cases} \mathrm{stha}_0(P, 1) = \cdots = \mathrm{stha}_{i-1}(P, 1) = 0, \\ \mathrm{stha}_i(P, 1) \neq 0. \end{cases}$$

The definition of Sturm–Habicht sequence through determinants allows us to perform computations dealing with real roots in a generic way: if P and Q are two polynomials with parametric coefficients whose degrees do not change after specialization we can compute the Sturm–Habicht sequence for P and Q without specializing the parameters and the result is always good after specialization (modulo the condition over the degrees). This behavior is studied carefully in (Gonzàlez-Vega et al. 1989) where the Sturm–Habicht sequence definition is modified in order to get good specialization properties even in cases where the degrees change after specialization. This is not true using Sturm sequence, use of denominators, or negative polynomial remainder sequences, with fixed degrees for P and Q the sequence does not always have the same number of elements (see Loos 1982b). This observation is the key to the methods introduced in Section 3.

To quote finally that the most efficient way of computing the $\mathrm{stha}_j(P,Q)$'s is through Subresultant algorithm which requires $\mathbf{O}(\deg(P)(\deg(P)+\deg(Q)))$ arithmetic operations in $\mathbb{D}$. In the cases, $\mathbb{D} = \mathbb{Z}$ or $\mathbb{D}$ a polynomial domain, integer size and degrees of intermediate results are well bounded (Gonzàlez-Vega et al. 1989, 1990, 1994, 1997).

3   The Algorithms

This section is devoted to showing how to perform quantifier elimination for some formulas using properties of Sturm–Habicht sequences. The algorithms proposed depend heavily on the type of formulas to be considered. Let n be a positive integer number. The generic monic polynomial with degree n will be denoted by

$$P_n(\underline{a}, x) = x^n + a_{n-1}x^{n-1} + \ldots + a_1 x + a_0.$$

The proof of the next proposition is an easy consequence of Theorem 1.

Proposition 1 *Let n be a positive even integer and*

$$\mathbb{H}_n : \ \forall x\,(P_n(\underline{a}, x) > 0),$$
$$\mathbb{E}_n : \ \exists x\,(P_n(\underline{a}, x) = 0).$$

Then

$$\mathbb{H}_n \ \Longleftrightarrow \ \mathbf{C}(\mathrm{stha}_n(P_n, 1), \ldots, \mathrm{stha}_0(P_n, 1)) = 0$$
$$\mathbb{E}_n \ \Longleftrightarrow \ \mathbf{C}(\mathrm{stha}_n(P_n, 1), \ldots, \mathrm{stha}_0(P_n, 1)) > 0.$$

Conditions equivalent to $\mathbb{H}_n$ and $\mathbb{E}_n$ are obtained as a union of basic semi-algebraic sets by considering all the 3^{n-2} possible sign conditions over the polynomials $\mathrm{stha}_i(P_n, 1)$ and keeping those which imply $C = 0$ (for $\mathbb{H}_n$) and $C > 0$ (for $\mathbb{E}_n$).

Example 1 *For the first nontrivial case, $n = 4$, the formula $\mathbb{H}_4$ is equivalent to the union of the following 9 semialgebraic basic sets.*

$$[S_2 > 0, S_1 < 0, S_0 > 0] \cup [S_2 < 0, S_1 > 0, S_0 > 0] \ \cup$$
$$[S_2 < 0, S_1 < 0, S_0 > 0] \cup [S_2 < 0, S_1 = 0, S_0 > 0] \ \cup$$
$$[S_2 < 0, S_1 = 0, S_0 < 0] \cup [S_2 = 0, S_1 > 0, S_0 < 0] \ \cup$$
$$[S_2 = 0, S_1 < 0, S_0 > 0] \cup [S_2 = 0, S_1 = 0, S_0 > 0] \ \cup$$
$$[S_2 < 0, S_1 = 0, S_0 = 0]$$

where

$$S_2 \;=\; 3a_3^2 - 8a_2$$

$$S_1 \;=\; 2a_2^2 a_3^2 - 8a_2^3 + 32a_2 a_0 + a_1 a_2 a_3 - 12a_3^2 a_0 - 6a_1 a_3^3 - 36a_1^2$$

$$S_0 \;=\; -27a_1^4 - 4a_3^3 a_1^3 + 18a_2 a_3 a_1^3 - 6a_3^2 a_0 a_1^2 + 144a_2 a_0 a_1^2 + a_2^2 a_3^2 a_1^2 -$$
$$\quad 4a_2^3 a_1^2 - 192a_3 a_0^2 a_1 + 18a_0 a_2 a_3^3 a_1 - 80a_0 a_2^2 a_3 a_1 + 256a_0^3 -$$
$$\quad 27a_3^4 a_0^2 + 144a_2 a_3^2 a_0^2 - 128a_2^2 a_0^2 - 4a_2^3 a_3^2 a_0 + 16a_2^4 a_0.$$

The previous description for $\mathbb{H}_4$ *can be reduced to*

$$\begin{aligned}
\mathbb{H}_4 \;=\; & [S_2 < 0, S_1 \neq 0, S_0 > 0] \cup [S_2 = 0, S_1 \leq 0, S_0 > 0] \cup \\
& [S_2 > 0, S_1 < 0, S_0 > 0] \cup [S_2 = 0, S_1 > 0, S_0 < 0] \cup \\
& [S_2 < 0, S_1 = 0] \\
\;=\; & [S_2 < 0, S_1 \neq 0, S_0 > 0] \cup [S_2 = 0, S_1 \leq 0, S_0 > 0] \cup \\
& [S_2 > 0, S_1 < 0, S_0 > 0] \cup [S_2 < 0, S_1 = 0].
\end{aligned}$$

The last simplification is due to the following fact:

$$S_2 = 0 \quad \Longrightarrow \quad a_2 = \frac{3a_3^2}{8} \quad \Longrightarrow \quad S_1 = -(16a_1 + a_3^3)^2 \leq 0.$$

For $\mathbb{E}_4$ *we get the union of the following 16 semialgebraic basic sets:*

$$\begin{aligned}
& [S_2 > 0, S_1 > 0, S_0 > 0] \cup [S_2 > 0, S_1 > 0, S_0 < 0] \cup \\
& [S_2 > 0, S_1 < 0, S_0 < 0] \cup [S_2 < 0, S_1 < 0, S_0 < 0] \cup \\
& [S_2 > 0, S_1 = 0, S_0 > 0] \cup [S_2 > 0, S_1 = 0, S_0 < 0] \cup \\
& [S_2 = 0, S_1 > 0, S_0 > 0] \cup [S_2 = 0, S_1 < 0, S_0 < 0] \cup \\
& [S_2 = 0, S_1 = 0, S_0 < 0] \cup [S_2 > 0, S_1 > 0, S_0 = 0] \cup \\
& [S_2 > 0, S_1 < 0, S_0 = 0] \cup [S_2 < 0, S_1 < 0, S_0 = 0] \cup \\
& [S_2 = 0, S_1 > 0, S_0 = 0] \cup [S_2 = 0, S_1 < 0, S_0 = 0] \cup \\
& [S_2 > 0, S_1 = 0, S_0 = 0] \cup [S_2 = 0, S_1 = 0, S_0 = 0].
\end{aligned}$$

The same strategy used to simplify $\mathbb{H}_4$ *can be used to obtain*

$$\begin{aligned}
\mathbb{E}_4 \;=\; & [S_2 = 0, S_0 \leq 0] \cup [S_2 > 0, S_1 < 0, S_0 \leq 0] \cup \\
& [S_2 < 0, S_1 < 0, S_0 \leq 0] \cup [S_2 > 0, S_1 \geq 0].
\end{aligned}$$

We remark finally that some of the sets appearing in the description of $\mathbb{H}_4$ *and* $\mathbb{E}_4$ *may be empty.*

The next formula to be considered is:

$$\mathbb{L}_n : \; \exists x \, (x^n + a_{n-1} x^{n-1} + \ldots + a_1 x + a_0 < 0).$$

Quantifier elimination for this formula requires a more detailed study than that needed for the formulas $\mathbb{H}_n$ and $\mathbb{E}_n$. First we introduce the following definition.

Definition 5 *Let n be a positive integer number. The set of $\underline{a}$ in $\mathbb{R}^{n+1}$ such that the polynomial $P_n(\underline{a}, x)$ has a real root with odd multiplicity will be denoted by $\mathcal{I}_n$.*

Proposition 2 *Let n be a positive even integer number and $\mathbb{L}_n$ the formula $\exists x \, (P_n(\underline{a}, x) < 0)$. Then*

$$\mathbb{L}_n \iff \underline{a} \in \mathcal{I}_n.$$

Proof. If $\underline{a}_0 \in \mathcal{I}_n$ then $\mathbb{L}_n$ is true. By considering the values of x close to a real root of $P_n(\underline{a}_0, x)$ with odd multiplicity, we see that if $a \in \mathcal{I}_n$ then $\mathbb{L}_n$ is true. If $\underline{a}_0 \notin \mathcal{I}_n$, then for all x, $P_n(\underline{a}_0, x) > 0$. ∎

The construction of a description for $\mathcal{I}_n$ is more complicated than in the cases studied before. We shall use the following notation for the multiplicity of a root. If P is a polynomial in $\mathbb{D}[x]$ then we denote the multiplicity of α in P by the symbol $\text{mult}(P, \alpha)$. With this notation

$$P(\alpha) \neq 0 \iff \text{mult}(P, \alpha) = 0 \text{ and } P(\alpha) = 0 \iff \text{mult}(P, \alpha) > 0.$$

First we perform a decomposition of the set $\mathcal{I}_n$ in the following terms:

$$\mathcal{I}_n = \bigcup_{j=1}^{n/2} \{\underline{a} \in \mathbb{R}^{n+1} : \exists \alpha \in \mathbb{R} \, (\text{mult}(\alpha, P_n(\underline{a}, x)) = 2j - 1)\}$$

$$= \bigcup_{j=1}^{n/2} \mathcal{I}_n^j = \bigcup_{j=1}^{n/2-1} \mathcal{I}_n^j,$$

since $\mathcal{I}_n^{n/2} = \mathcal{I}_n^1$. If j is an element of $\{1, \ldots, n/2 - 1\}$ then $\mathcal{I}_n^j$ is obtained in the following terms:

$$\begin{aligned}
\mathcal{I}_n^j &= \{\underline{a} \in \mathbb{R}^{n+1} : \exists \alpha \in \mathbb{R} \, (\text{mult}(\alpha, P_n(\underline{a}, x)) = 2j - 1)\} \\
&= \{\underline{a} \in \mathbb{R}^{n+1} : \exists \alpha \in \mathbb{R} \, (P_n(\underline{a}, \alpha) = P_n^{(1)}(\underline{a}, \alpha) = \cdots = P_n^{(2j-2)}(\underline{a}, \alpha) = 0 \\
&\quad \text{and } P_n^{(2j-1)}(\underline{a}, \alpha) \neq 0)\} \\
&= \{\underline{a} \in \mathbb{R}^{n+1} : \exists \alpha \in \mathbb{R} \, (\sum_{k=0}^{2j-2} (P_n^{(k)}(\underline{a}, \alpha))^2 = 0 \text{ and } P_n^{(2j-1)}(\underline{a}, \alpha) \neq 0)\}.
\end{aligned}$$

To simplify the description of every $\mathcal{I}_n^j$ we introduce the notation

$$R_n^j(\underline{a}, x) = \sum_{k=0}^{2j-2} (P_n^{(k)}(\underline{a}, x))^2,$$

which allows us to write

$$\begin{aligned}
\mathcal{I}_n^j &= \{\underline{a} \in \mathbb{R}^{n+1} : \exists \alpha \in \mathbb{R} \, (R_n^j(\underline{a}, \alpha) = 0 \text{ and } P_n^{(2j-1)}(\underline{a}, \alpha) \neq 0)\} \\
&= \{\underline{a} \in \mathbb{R}^{n+1} : \mathbf{V}_{\text{StHa}}(R_n^j(\underline{a}, x), (P_n^{(2j-1)}(\underline{a}, x))^2) > 0\}.
\end{aligned}$$

We have obtained the proof of the following theorem.

Theorem 2 *Let n be a positive even integer. Then the set $\mathcal{I}_n$ can be described as the following union.*

$$\mathcal{I}_n = \bigcup_{j=1}^{n/2-1} \{\underline{a} \in \mathbb{R}^{n+1} : \mathbf{V}_{\text{StHa}}(R_n^j, (P_n^{(2j-1)})^2) > 0\}.$$

This theorem allows to give a description for $\mathbb{L}_n$ similar to the one obtained in Proposition 1 for the formulas $\mathbb{H}_n$ and $\mathbb{E}_n$.

Corollary 1 *Let n be a positive even integer. Then $\mathbb{L}_n$ is equivalent to*

$$\underline{a} \in \bigcup_{j=1}^{n/2-1} \{\underline{a} \in \mathbb{R}^{n+1} : \mathbf{C}(\text{stha}_{2n}(R_n^j, (P_n^{(2j-1)})^2), \ldots, \text{stha}_0(R_n^j, (P_n^{(2j-1)})^2)) > 0\}.$$

Proof. Use Proposition 2, Theorem 2 and Theorem 1. ∎

There exists a description of the set $\mathcal{I}_n^1$ simpler than the general one previously presented. In fact $\mathcal{I}_1$ is the set of elements in $\underline{a} \in \mathbb{R}^{n+1}$ such that $P_n(\underline{a}, x)$ has a simple real root and the general description for the semialgebraic sets $\mathcal{I}_n^j$ produces the following result.

$$\begin{aligned}
\mathcal{I}_n^1 &= \{\underline{a} \in \mathbb{R}^{n+1} : \exists \alpha \in \mathbb{R} \, (R_n^1(\underline{a}, \alpha) = 0 \text{ and } P_n^{(1)}(\underline{a}, \alpha) \neq 0)\} \\
&= \{\underline{a} \in \mathbb{R}^{n+1} : \exists \alpha \in \mathbb{R} \, (P_n(\underline{a}, \alpha)) = 0, \, P_n^{(1)}(\underline{a}, \alpha) \neq 0)\} \\
&= \{\underline{a} \in \mathbb{R}^{n+1} : \mathbf{V}_{\text{StHa}}(P_n(\underline{a}, x), (P_n^{(1)}(\underline{a}, x))^2) > 0\}.
\end{aligned} \tag{1}$$

But even this last description is redundant because once we have assigned signs to the polynomials $\text{stha}_j(P_n, 1)$ we know the number of real roots of $P_n(\underline{a}, x)$ when $\underline{a}$ satisfies the given sign conditions and, very importantly, the greatest common divisor of $P_n(\underline{a}, x)$ and $P_n^{(1)}(\underline{a}, x)$. This remark implies the following description for the semialgebraic set $\mathcal{I}_n^1$.

Proposition 3 *Let n be an even positive integer number. Then*

$$\mathcal{I}_n^1 = \bigcup_{i=0}^{n-2} \left\{ \underline{a} \in \mathbb{R}^n : \begin{array}{l} \text{stha}_0(P_n, 1) = \cdots = \text{stha}_{i-1}(P_n, 1) = 0, \\ \text{stha}_i(P_n, 1) \neq 0, \text{ and} \\ \mathbf{V}_{\text{StHa}}(P_n, 1) - \mathbf{V}_{\text{StHa}}(\text{StHa}_i(P_n, 1), 1) > 0 \end{array} \right\}.$$

It is worth noting that this description for the set $\mathcal{I}_n^1$ is very useful for small n's but produces a combinatorial explosion in the number of cases for large n's, since we must consider, for every i and for every sign condition over the $\text{stha}_j(P_n, 1)$ (with $j \geq i$), all the possible sign conditions over the polynomials $\text{stha}_k(\text{StHa}_i(P_n, 1), 1))$. So in general what is proposed for the semialgebraic set $\mathcal{I}_n^1$ is the formula appearing in Eq. 1.

Example 2 *We study the first nontrivial case, $n = 4$, using the two approaches discussed here. In the first one, using Proposition 2 and the properties of $\mathcal{I}_4^1$ we get the following equivalence.*

$$\mathbb{L}_4 \iff \underline{a} \in \mathcal{I}_4 = \mathcal{I}_4^1.$$

The description of the set $\mathcal{I}_4^1$ is obtained by means of the formula in Proposition 3 which provides the following description for $\mathcal{I}_4^1$.

$$[S_1 < 0, S_0 < 0] \cup [S_2 > 0, S_1 = 0, S_0 \neq 0] \cup [S_2 > 0, S_1 > 0] \cup$$
$$[S_2 = 0, S_1 = 0, S_0 < 0] \cup [S_2 > 0, S_1 = 0, S_0 = 0, H_0 \leq 0].$$

The polynomials S_i, $i = 0, 1, 2$, appearing in the description of $\mathcal{I}_4^1$ are the same as those in Example 1 and

$$H_0 = (8a_2 - 3a_3^2)(a_3^2 a_2^2 - 128a_2 a_0 - 4a_1 a_3 a_2 + 48a_3^2 a_0 + 36a_1^2 - 3a_1 a_3^3).$$

In the second one the description of $\mathcal{I}_4^1$ is obtained using the formula:

$$\mathcal{I}_4^1 = \{(a_3, a_2, a_1, a_0) \in \mathbb{R}^4 : \mathbf{V}_{\text{StHa}}(P_4(\underline{a}, x), (P_4^{(1)}(\underline{a}, x))^2) > 0\}.$$

Computation of the Sturm–Habicht sequence for $P_4(\underline{a}, x)$ and $(P_4^{(1)}(\underline{a}, x))^2)$ provides the following results.

$$
\begin{aligned}
T_3 =\ & -a_3^6 - 32a_2 a_0 + 12a_3^2 a_0 - 18a_3^2 a_2^2 + 8a_2 a_3^4 - 28a_1^2 + 8a_2^3 - 10a_1 a_3^3 + \\
& 36a_1 a_3 a_2
\end{aligned}
$$

$$
\begin{aligned}
T_2 =\ & 1664a_2^3 a_3 a_0 a_1 - 549a_2^2 a_3^4 a_1^2 - 279a_3^2 a_1^4 + 576a_0^2 a_1 a_3^3 + 504a_0^2 a_3^4 a_2 - \\
& 1536a_0^2 a_3^2 a_2^2 - 2688a_2^2 a_1^2 a_0 - 2304a_3 a_1^3 a_0 + 986a_3^3 a_1^3 a_2 - 90a_1 a_3^7 a_0 + \\
& 1512a_1^4 a_2 - 918a_1^2 a_3^4 a_0 - 3072a_0^2 a_3 a_2 a_1 - 2048a_2 a_0^3 + 2048a_2^2 a_0^2 + \\
& 768a_3^2 a_0^3 - 2496a_2^2 a_3^3 a_0 a_1 + 4512a_2^3 a_1^2 a_0 a_2 - 2352a_3 a_1^2 a_2^2 + 64a_2^7 - \\
& 276a_2^3 a_3^4 a_0 - 480a_2^5 a_3 a_1 + 596a_2^3 a_3^2 a_1^2 - 78a_2^3 a_3^5 a_1 + 784a_2^4 a_3^3 a_0 + \\
& 882a_2 a_3^5 a_0 a_1 - 640a_2^5 a_0 + 336a_2^4 a_3^3 a_1 + 6a_2^2 a_3^7 a_1 - 48a_2^6 a_3^2 + \\
& 30a_2^2 a_3^6 a_0 + 126a_2 a_3^6 a_1^2 - a_2^4 a_3^6 - 9a_1^2 a_3^8 - 63a_0^2 a_3^6 + 12a_2^5 a_3^4 - \\
& 102a_1^3 a_3^5 + 640a_2^4 a_1^2 + 4608a_0^2 a_1^2
\end{aligned}
$$

$$
\begin{aligned}
T_1 =\ & -5736a_1^4 a_3^6 a_2 a_0 - a_1^2 a_3^6 a_2^6 - 2208a_1 a_3^3 a_2^7 a_0 - 61440a_1 a_3^3 a_0^4 a_2 - \\
& 48a_1^2 a_2^8 a_3^2 + 2628a_1 a_3^7 a_0^4 a_2^3 + 39324a_1^4 a_2^2 a_0 a_3^4 - 77760a_1^6 a_2 a_0 - \\
& 13608a_1^6 a_3^2 a_2^2 + 8424a_1^6 a_3^2 a_0 - 6516a_1^6 a_2 a_3^4 + 10a_1^3 a_2^4 a_3^7 + \cdots
\end{aligned}
$$

$$
\begin{aligned}
T_0 =\ & 594864a_1^9 a_0 a_3 a_2^2 - 150174a_1^9 a_0 a_2 a_3^3 + 3888a_1^9 a_0 a_3^5 + 64a_1^9 a_3^9 + \\
& 8748a_1^{10} a_2^3 + 1296a_1^{10} a_3^6 + \cdots,
\end{aligned}
$$

and the description of $\mathcal{I}_4^1$ is given by

$$[T_3 > 0, T_2 > 0, T_1 > 0, T_0 \geq 0] \cup [T_3 < 0, T_2 \neq 0, T_1 > 0, T_0 \geq 0] \cup$$
$$[T_3 = 0, T_2 \geq 0, T_1 > 0, T_0 \geq 0] \cup [T_3 < 0, T_2 = 0, T_1 \geq 0, T_0 \geq 0] \cup$$
$$[T_3 < 0, T_2 = 0, T_1 < 0, T_0 \leq 0] \cup [T_3 = 0, T_2 < 0, T_1 < 0, T_0 \leq 0] \cup$$
$$[T_3 < 0, T_2 < 0, T_1 \leq 0].$$

Using the fact that T_0 is the resultant of P_4 and $(P_4')^3$ it is possible to replace T_0 with a polynomial involving fewer terms. The multiplicative property of resultants implies that

$$T_0 = -4^6 \text{Resultant}(P_4, (P_4')^3) = -4^6 \text{Resultant}(P_4, P_4')^3 = -4^6 (S_0)^3,$$

where S_0 is the polynomial introduced in Example 1. This equality allows us to replace T_0 by $-S_0$ in the previous expression for the semialgebraic set $\mathcal{I}_4^1$.

4   Conclusions

Several methods for solving some specific quantifier elimination problems have been presented. Attention has been focused on atomic formulas involving only one polynomial sign condition and one quantifier over a single variable. The methods presented obtain, as output, a finite union of possibly empty semialgebraic basic sets, using the following algebraic and combinatorial steps:

1. The algebraic part of the method is performed by means of the computation of one or several Sturm–Habicht sequences, depending on the particular formula considered.
2. The first step of the combinatorial part proceeds by assigning all the possible sign conditions to the polynomials obtained in the algebraic part and selecting those for which the considered formula is true.
3. The second step of the combinatorial part reduces the quantifier free formula obtained by deleting some trivially empty sets and by decreasing the number of unions using the following well known rules

$$< \cup > \longrightarrow \neq, \quad < \cup = \longrightarrow \leq, \quad > \cup = \longrightarrow \geq . \tag{2}$$

The first two steps have been implemented in Maple. Currently the second step of the combinatorial stage is performed by hand. This is done for two reasons. The first reason is related to the way some trivially empty sets are deleted. Some deletions depend on the structure of the computed polynomials. See Example 1 where the vanishing of one polynomial implies a linear dependence between one variable and the others allowing us to discard one of the obtained sets. The second reason is that application of the rules in Eq. 2 does not commute with the ordering between the polynomials involved. By visual inspection we decide which is the good ordering, i.e., the one providing a smaller number of unions. Currently this step of the method is performed using the ordering appearing in the Sturm–Habicht sequence considered.

The proposed methods work efficiently (in a few minutes) for the general case (considering the general polynomial $P_n(\underline{a}, x)$) for degrees up to $n = 10$, and in the non-generic case (considering polynomial coefficients not all different from 0) for degrees up to $n = 14$. As expected the most time consuming part of the algorithm corresponds (for these low degrees) to the algebraic part of the method (the computation of the Sturm–Habicht sequences needed).

The extension of these methods to a general quantifier elimination method is straightforward but does not provide an efficient algorithm. The method is inherently exponential (the number of cases to be considered is at least 3^n). Dealing with more complicated formulas implies that we consider too many cases to manage in a similar way to the one presented in Section 3. For example if we are interested in the elimination of the existential quantifier in the formula

$$\exists x \qquad P_n(\underline{a}, x) = 0,\ Q_1(\underline{a}, x) > 0, \ldots, Q_m(\underline{a}, x) > 0,$$

the natural extension of the results in Section 3 is the equality "á la Ben-Or, Kozen and Reif" (see Ben-Or et al. 1986 or Korkina and Kushnirenko 1985):

$$\#\{\alpha \in \mathbf{R} : P_n(\underline{a}_0, x) = 0, Q_1(\underline{a}_0, x) > 0, \ldots, Q_m(\underline{a}_0, x) > 0\}$$
$$= \frac{1}{2^m} \sum_{j=1}^{2^m} \mathbf{V}_{\mathrm{StHa}}(P_n(\underline{a}_0, x), \prod_{i=1}^{m} Q_i^{\epsilon_j^i}(a_0, x))$$

with ϵ_j denoting all the elements in the set $\{1, 2\}^m$ ($\epsilon_j^i = 1$ or $\epsilon_j^i = 2$). This implies the computation of 2^m Sturm–Habicht sequences and the consideration of $2^m 3^n$ different cases.

Nonetheless, we expect practical improvements to our combinatorial methods, allowing us to solve more general problems. We will attempt to obtain practical improvements by investigating the following:

1. the use of multiplicative Sturm–Habicht (or subresultant) properties to reduce the size of the polynomials involved (similar to the reduction made in Example 2),
2. the implication of the vanishing of a Sturm–Habicht principal coefficient (it was seen in Example 1 that the vanishing of a Sturm–Habicht principal coefficient can imply the factorization of the next Sturm–Habicht principal coefficients),
3. the development of algorithms determining which sign conditions give some specific character to the functional $\mathbf{C}$ without considering all possibilities,
4. the use of rewrite rules, based on three valued logic minimization as in (Hong 1992e), the simplification of unions of basic semialgebraic sets over the same set of polynomials, as performed by hand in the examples of Section 3.

Acknowledgments

Laureano González-Vega was partially supported by CICyT PB 92/0498/C02/01 (Geometría Real y Algoritmos) and Esprit/Bra 6846 (Posso).

A New Approach to Quantifier Elimination for Real Algebra

V. Weispfenning

1 Introduction

Quantifier elimination for the elementary formal theory of real numbers is a fascinating area of research at the intersection of various field of mathematics and computer science, such as mathematical logic, commutative algebra and algebraic geometry, computer algebra, computational geometry and complexity theory. Originally the method of quantifier elimination was invented (among others by Th. Skolem) in mathematical logic as a technical tool for solving the decision problem for a formalized mathematical theory. For the elementary formal theory of real numbers (or more accurately of real closed fields) such a quantifier elimination procedure was established in the 1930s by A. Tarski, using an extension of Sturm's theorem of the 1830s for counting the number of real zeros of a univariate polynomial in a given interval. Since then an abundance of new decision and quantifier elimination methods for this theory with variations and optimizations has been published with the aim both of establishing the theoretical complexity of the problem and of finding methods that are of practical importance (see Arnon 1988a and the discussion and references in Renegar 1992a, 1992b, 1992c for a comparison of these methods). For subproblems such as elimination of quantifiers with respect to variables, that are linearly or quadratically restricted, specialized methods have been developed with good success (see Weispfenning 1988; Loos and Weispfenning 1993; Hong 1992d; Weispfenning 1997).

The theoretical worst-case complexity of the quantifier elimination problem for the reals is by now well-established (see Renegar 1992a, 1992b, 1992c); surprisingly the corresponding asymptotic lower bound is already valid for the elimination of linear quantifiers (see Weispfenning 1988). Of course, this asymptotic complexity is established only up to undetermined multiplicative constants that may vary extremely for the competing quantifier elimination methods. The Collins procedure and its optimizations (see Collins and Hong 1991) remain up to now the only completely implemented quantifier elimination procedure for the full elementary theory of reals. It remains doubtful (see Hong 1991b), whether for practical use in problems of small to moderate size future implementations of the asymptotically better procedures will improve on this procedure.

It has meanwhile become apparent that a wealth of problems, e.g., in geometry, algebra, analysis and robotics, can be formulated as quantifier elimination problems (see Collins and Hong 1991; Lazard 1988; Hong 1992e). Hence the search for practicable quantifier elimination methods for the full elementary theory or fragments thereof is of great importance.

The purpose of this note is to outline a new quantifier elimination procedure for the elementary theory of reals that differs from most of the known ones among others by the fact that it tends to eliminate whole *blocks* of quantifiers instead of one quantifier at a time. In contrast to the method in (Renegar 1992a, 1992b, 1992c), this is achieved not by a reduction of a quantifier-block to a single quantifier, but by dealing with the whole block at once, provided the equations on the variables of this block define a zero-dimensional ideal over the parameters. Only if this condition fails, the block has to be split into two or more smaller subblocks. By analogy with the corresponding problem for algebraically closed fields (see Weispfenning 1992), this feature may yield a good performance in some practical examples (compare Section 6). In contrast to the Collins–Hong method of quantifier elimination via partial cylindrical algebraic cell decomposition, our method eliminates quantifiers "from inside to outside". So in principle (except for additional simplification routines) the method depends only on the number of coefficients of the polynomials in the input formulas with respect to the free variables and not on the actual number of free variables.

At present, it is too early to estimate the performance of the method in general; implementation has, however, begun at the University of Passau.

The method is based on an exciting new method for counting the number of real joint zeroes of multivariate polynomials with side conditions that was found recently independently by Becker and Wörmann (1991) and by Pedersen et al. (1993), based on ideas of Hermite and Sylvester. The method uses quadratic forms; it applies, however, only to zeros of zero-dimensional ideals (i.e., polynomial systems that have only finitely many complex zeros). It is this restriction that has to be overcome in order to apply the method for quantifier elimination; moreover, the method has to be extended uniformly in arbitrary real parameters. The clue to the solution of these two problems is the use of *comprehensive Gröbner bases* that have proved to be of great value already for complex quantifier elimination. The combination of these two tools reduces the quantifier elimination problem to a combinatorial problem of the following type: Find a quantifier-free formula that expresses the fact that the number of sign changes in a sequence $a_1, \ldots, a_D$ of variables has some prescribed value. While this problem is easy to solve in principle, it seems hard to find "short" formulas for this purpose.

In the following, I will first recall the basic facts concerning quantifier elimination, the real zero counting using quadratic forms and comprehensive Gröbner basis, and then outline the main steps that combine these techniques into a quantifier elimination procedure. Finally some examples computed with interactive use of Maple will illustrate the method.

2   The Quantifier Elimination Problem for the Elementary Theory of the Reals

An *atomic formula* is an expression of the form $f(X_1, \ldots, X_n) \, \rho \, g(X_1, \ldots, X_n)$, where $f, g \in \mathbb{Q}[X_1, \ldots, X_n]$ and ρ is one of the relations $=, \leq, <$. *Formulas* are obtained from atomic formulas by means of the propositional operators $\wedge, \vee, \neg$ and quantification $\exists x_i, \forall x_i$ over variables x_i, together with appropriate use of parentheses. The *quantifier elimination problem* asks for an algorithm that on input of a formula φ outputs a quantifier-free formula φ' (i.e., a propositional combination of atomic formulas) such that φ and φ' are equivalent in the ordered field $\mathbb{R}$ of real numbers (i.e., yield the same truth value for any assignment of real numbers to unquantified variables).

By a well-known and easy algorithm, any formula can be rewritten as an equivalent *prenex formula*, i.e., a formula beginning with a string of quantifiers followed by a quantifier-free formula. Any string $\exists x_1 \ldots \exists x_k$ or $\forall x_1 \ldots \forall x_k$ of similar quantifiers is called a *quantifier block*. In order to solve the quantifier elimination problem, it suffices by recursion on the number of quantifiers in a prenex formula to handle input formulas of the form $\exists x(\varphi)$, where φ is quantifier-free (notice that $\forall x(\varphi)$ is equivalent to $\neg \exists x(\neg \varphi)$). In a similar but more efficient way, the quantifier elimination problem is solved by recursion on the number of quantifier blocks provided one can handle input formulas of the form $\exists x_1 \ldots \exists x_k(\varphi)$, where φ is quantifier-free. Using the fact that φ can be put into disjunctive normal form and that disjunctions commute with existential quantifiers, it suffices therefore to handle input formulas of the form

$$(*) \qquad \exists x_1 \ldots \exists x_k \left(\bigwedge_{i=1}^{m} f_i(x_1, \ldots, x_n) = 0 \wedge \bigwedge_{i=1}^{m'} h_i(x_1, \ldots, x_n) > 0 \right)$$

with $1 \leq k \leq n$, $f_i, h_i \in \mathbb{Q}[x_1, \ldots, x_n]$.

3   Counting Real Zeros Using Quadratic Forms

Let F be a finite subset of $R = \mathbb{R}[x_1, \ldots, x_n]$ such that the ideal $I = Id(F)$ generated by F is zero-dimensional (i.e., has only finitely many complex zeros). Then the residue class ring $A = R/I$ is finite-dimensional as $\mathbb{R}$-vector-space and an explicit basis of A consisting of residue classes of terms in R can be computed from a Gröbner basis G of I (see Becker and Weispfenning 1993). Let $h \in R$ and let $(t_1, \ldots, t_d)$ be an ordered $\mathbb{R}$-basis of A. Then the maps $a \mapsto t_i t_j h a$ are endomorphisms φ_{ijh} of A as $\mathbb{R}$-vector-space ($1 \leq i \leq j \leq d$). The matrix

μ_{ijh} of φ_{ijh} with respect to the given basis can be computed using polynomial reduction with respect to G. Let $m_{ijh} = \text{trace}(\mu_{ijh})$ and let $M_h = (m_{ijh})$ be the resulting real symmetric $(d \times d)$ matrix.

Put $\rho_h = \text{rank}(M_h)$ and $\sigma_h = \text{signature}(M_h)$ which is equal to the number of positive eigenvalues of M_h minus the number of negative eigenvalues of M_h.

Notice that all eigenvalues of M_h are real; so the number of positive eigenvalues of M_h, and hence the signature of M_h can be computed by applying Descartes' rule of signs to the characteristic polynomial χ_h of M_h. Moreover, in the univariate case $n = 1$, $x_1 = x$, M_h is a Hankel matrix (i.e., $m_{ij} = m_{kl}$ for $i+j = k+l$), provided the residues of $1, x, x^2, \ldots, x^{d-1}$ are taken as a basis of A.

Then the following result was proved in (Becker and Wörmann 1991) and in (Pedersen et al. 1993):

Theorem 1
$$\#\{c \in \mathbb{R}^n \mid F(c) = 0\} = \sigma_1$$

$$\#\{c \in \mathbb{R}^n \mid F(c) = 0 \wedge h(c) > 0\} - \#\{c \in \mathbb{R}^n \mid F(c) = 0 \wedge h(c) < 0\} = \sigma_h$$

$$\#\{c \in \mathbb{R}^n \mid F(c) = 0 \wedge h(c) \neq 0\} = \sigma_{h^2}$$

$$\#\{c \in \mathbb{C}^n \mid F(c) = 0 \wedge h(c) \neq 0\} = \rho_h.$$

Consequently, the numbers

$$Z_h^+ = \#\{c \in \mathbb{R}^n \mid F(c) = 0 \wedge h(c) > 0\},$$

$$Z_h^- = \#\{c \in \mathbb{R}^n \mid F(c) = 0 \wedge h(c) < 0\}$$

$$Z_h^0 = \#\{c \in \mathbb{R}^n \mid F(c) = 0 \wedge h(c) = 0\}$$

are the unique solutions of the following system of linear equations:

$$\begin{pmatrix} 1 & 1 & 1 \\ 1 & -1 & 0 \\ 1 & 1 & 0 \end{pmatrix} \begin{pmatrix} Z_h^+ \\ Z_h^- \\ Z_h^0 \end{pmatrix} = \begin{pmatrix} \sigma_1 \\ \sigma_h \\ \sigma_{h^2} \end{pmatrix}.$$

Using a technique introduced in (Ben-Or et al. 1986) this counting of real zeros of I with one side-condition given by h is extended to finitely many side conditions roughly as follows:

Let $\mathbf{h} = (h_1, \ldots h_m) \in R^m$ and let $\epsilon \in \{1, -1, 0\}^m$ be a generalized sign vector. Then we put $Z_{\mathbf{h}}^\epsilon = \#\{c \in \mathbb{R}^n \mid F(c) = 0 \wedge \bigwedge_{i=1}^m \text{sign}(h_i(c)) = \epsilon(i)\}$.

Using the m-th tensor power (Kronecker power) J of the 3×3-matrix

$$I = \begin{pmatrix} 1 & 1 & 1 \\ 1 & -1 & 0 \\ 1 & 1 & 0 \end{pmatrix}$$

considered above, one obtains a system of linear equations

$$J \cdot Z = \sigma,$$

where Z is a column vector of length 3^m with entries of the form $Z_{\mathbf{h}}^{\varepsilon}$ ordered in decreasing lexicographical order with respect to the order $1 > -1 > 0$ on the components, and σ is a column vector of the same length with entries of the form $\sigma_{h_{\delta}}$, where δ has entries $\delta_i \in \{0, 1, 2\}$ for $1 \le i \le 3^m$, $h_{\delta} = \prod_{i=1}^{m} h_i^{\delta_i}$, and the entries $\sigma_{h_{\delta}}$ of σ are ordered increasingly with respect to the inverse lexicographical order on the subscript δ.

Since $\det(I) = 2$, a recursion on m shows that $\det(J)$ is a power of 2, in particular $\det(J) > 0$.

The first entry of the column vector Z is $Z_{\mathbf{h}}^{\mathbf{1}}$, where $\mathbf{1} = (1, \dots, 1)$. Let D_j $(1 \le j \le 3^m)$ be the cofactors in the Laplace expansion of $\det(J)$ with respect to the first column and let $\sigma = (\sigma_1, \dots, \sigma_{(3^m)})^t$ be the right-hand side of the system $J \cdot Z = \sigma$ above.

Then by Cramer's rule, we have

$$Z_{\mathbf{h}}^{\mathbf{1}} = \det(J)^{-1} \cdot \sum_{j=1}^{3^m} (-1)^{j+1} \sigma_j D_j.$$

As a consequence, we note that

Corollary 1

$$Z_{\mathbf{h}}^{\mathbf{1}} > 0 \iff \sum_{j=1}^{3^m} (-1)^{j+1} \sigma_j D_j > 0.$$

Moreover, by the theorem above and the fact that the complex variety of F has at most $d = \dim_{\mathbb{C}}(\mathbb{C}[X_1, \dots, X_n]/Id(F))$ many points, one concludes that for each j,

$$\sum_{\varepsilon \in \{1, -1, 0\}^m} Z_{\mathbf{h}}^{\varepsilon} \le d$$

and

$$|\sigma_j| \le \#\{c \in \mathbb{R}^n \mid F(c) = 0\} \le d.$$

This fact yields the following structural property of the linear system

$$J \cdot x = s:$$

If the column vector s on the right-hand side is of the form σ as considered so far, then at most d entries of the unique solution x are $\ne 0$. Consequently, these entries are uniquely determined by a suitable subsystem consisting of d equations only; moreover, s has at most 3^d pairwise different entries.

There is another observation concerning the linear system $J \cdot x = \sigma$ that will be of crucial importance for our quantifier elimination algorithm:

Consider the 2×3 submatrix I^* of I consisting of the second and third row of I. The tensor-power $J^* = I^{*m}$ is a $(2^m \times 3^m)$-submatrix of $J = I^m$ that coincides up to certain zero columns with the $(2^m \times 2^m)$-matrix considered in (Ben-Or et al. 1986). The corresponding linear subsystem

$$J^* \cdot Z = \sigma^* \quad \text{of} \quad J \cdot Z = \sigma$$

has as entries of σ^* all $\sigma_{(h_\delta)}$ with $\delta_i \in \{1,2\}$ only.

By induction on m one verifies easily that the sum of all rows of J^* yields the row-vector $(2^m, 0, \ldots, 0)$. As a consequence we obtain for any non-negative solution Z of $J^* Z = \sigma^*$:

$$2^m \cdot Z_{\mathbf{h}}^1 = \sum_{\delta \in \Delta} \sigma_{(h_\delta)},$$

where the sum ranges over the set Δ of all δ with entries $\delta_i \in \{1,2\}$.

So we obtain

$$Z_{\mathbf{h}}^1 = 0 \quad \Longleftrightarrow \quad \sum_{\delta \in \Delta} \sigma_{(h_\delta)} = 0 \quad \Longleftrightarrow \quad \sum_{\delta \in \Delta} \sigma_{(h_\delta)} < 2^m.$$

Recall that $\sigma_{(h_\delta)}$ is the signature of the matrix M_{h_δ}. Define for a univariate real polynomial $f(X)$ the type $\tau(f)$ as the number of positive zeros of f minus the number of negative zeros of f – both counted with multiplicities. Then $\sigma_{(h_\delta)}$ is by definition the type $\tau(\chi_\delta)$ of the characteristic polynomial χ_δ of M_{h_δ}. Using the obvious equation

$$\tau(f \cdot g) = \tau(f) + \tau(g)$$

for univariate real polynomials f, g, we can now rewrite the equivalence above as follows

$$Z_{\mathbf{h}}^1 = 0 \quad \Longleftrightarrow \quad \tau(\chi) = 0 \quad \Longleftrightarrow \quad \tau(\chi) < 2^m,$$

where $\chi = \prod_{\delta \in \Delta} \chi_\delta$, or equivalently

$$Z_{\mathbf{h}}^1 > 0 \quad \Longleftrightarrow \quad \tau(\chi) \neq 0 \quad \Longleftrightarrow \quad \tau(\chi) \geq 2^m.$$

Alternatively, we have the two implications

$$Z_{\mathbf{h}}^1 = 0 \Rightarrow \tau(\chi) = 0 \ , \ \tau(\chi) < 2^m \Rightarrow Z_{\mathbf{h}}^1 = 0.$$

4   Comprehensive Gröbner Bases

For the basic facts on Gröbner bases we refer to (Becker and Weispfenning 1993). Let $R = \mathbb{Q}[U_1, \ldots, U_m, X_1, \ldots X_n]$ and fix a term-order $<$ on the set T of terms in $X_1, \ldots, X_n$. Let $I = I(\mathbf{U}, \mathbf{X})$ be an ideal in R and let $G = G(\mathbf{U}, \mathbf{X})$ be a finite subset of I. Then G is a *comprehensive Gröbner basis* of I if for every m-tuple $(a_1, \ldots, a_m)$ of elements in some extension field K of $\mathbb{Q}$, $G(\mathbf{a}, \mathbf{X})$ is a Gröbner basis of $I(\mathbf{a}, \mathbf{X})$ in $K[\mathbf{X}]$ with respect to the term-order $<$. For every finite $F \subseteq R$ one can compute a comprehensive Gröbner basis G of $I = Id(F)$. From G one can compute mutually exclusive quantifier-free formulas $\varphi_1, \ldots, \varphi_s$ in $U_1, \ldots, U_m$, whose disjunction is true, corresponding numbers $d_1, \ldots, d_s \in \{-1, \ldots, n\}$ and corresponding subsets $\mathcal{Y}_1, \ldots, \mathcal{Y}_s$, of $\{X_1, \ldots, X_n\}$ such that in every extension field K of $\mathbb{Q}$ and all $(a_1, \ldots, a_m)$ in K, if $\varphi_i(\mathbf{a})$ holds true in K, then the ideal $I(\mathbf{a}, \mathbf{X})$ has dimension d_i and $\mathcal{Y}_i$ is a maximal set of independent variables modulo $I(\mathbf{a}, \mathbf{X})$ (see Weispfenning 1992).

Moreover, we remark that G immediately yields a quantifier-free formula $\varphi'(\mathbf{U})$ containing no order relations, such that $\varphi'(\mathbf{U})$ is equivalent in the field $\mathbb{C}$ of complex numbers to the formula $\exists x_1 \ldots \exists x_n (F(\mathbf{U}, x_1, \ldots, x_n) = 0)$.

Corresponding algorithms have been implemented at the University of Passau in the computer algebra systems ALDES/SAC-2, MAS and AXIOM.

5   Steps of the Quantifier Elimination Method

By Section 2, it suffices to consider input formulas φ of the form

$$(*)\qquad \exists x_1 \ldots \exists x_k (\bigwedge_{i=1}^{m} f_i(x_1, \ldots, x_n) = 0 \wedge \bigwedge_{i=1}^{m'} h_i(x_1, \ldots, x_n) > 0)$$

with $1 \le k \le n$, $f_i, h_i \in \mathbb{Q}[x_1, \ldots, x_n]$.

To begin with, we describe two *preprocessing steps* that depend only on m' and some integer d $(0 \le d \le n)$, whose intended meaning is the dimension of the residue class ring to be considered. These steps can be quite complex for large m' and d, but have to be executed only once for every pair (m', d). The results can be stored in a file to be called upon, whenever needed.

Step P1: For a monic univariate polynomial of degree d, $\chi(X) = X^d + \sum_{i=0}^{d-1} a_i X$ with indeterminate coefficients $a_0, \ldots, a_{d-1}$ (that is implicitly intended to have only real zeros) one writes down quantifier-free formulas $N_{d,j}(\mathbf{a})$ $(0 \le j \le d)$ that are valid in $\mathbb{R}$ if and only if the number of sign-variations in the sequence $(1, a_{d-1}, \ldots, a_0)$ (and hence by Descartes' rule the number of positive zeros of χ) equals j. Using these formulas, one constructs quantifier-free formulas $S_d(\mathbf{a})$ that are valid in $\mathbb{R}$ if and only if the type $\tau(\chi)$ of χ, i.e., the number of positive zeros of χ minus the number of negative zeros of χ (both counted with multiplicities) equals 0.

In the following we list such formulas S_d for $1 \le d \le 6$:

$$
\begin{aligned}
S_1(a_0) \quad &: \quad a_0 = 0; \\
S_2(a_0, a_1) \quad &: \quad a_1 = 0 \ \vee \ a_0 < 0; \\
S_3(a_0, a_1, a_2) \quad &: \quad a_0 = 0 \ \wedge \ (a_2 = 0 \ \vee \ a_1 < 0); \\
S_4(a_0, \ldots, a_3) \quad &: \quad [a_0 = a_1 = 0 \ \wedge \ (a_3 = 0 \ \vee \ a_2 < 0)] \ \vee \\
&\qquad [a_0 > 0 \ \wedge \ (a_2 < 0 \ \vee \ a_1 \cdot a_3 < 0))];
\end{aligned}
$$

$$S_5(a_0, \ldots, a_4) \quad : \quad a_0 = 0 \,\wedge\, S_4(a_1, \ldots, a_4);$$
$$S_6(a_0, \ldots, a_5) \quad : \quad [a_0 = a_1 = 0 \,\wedge\, S_4(a_2, \ldots, a_5)] \,\vee$$
$$[a_0 < 0 \,\wedge\, (((a_2 > 0 \vee a_3 a_1 < 0) \wedge$$
$$(a_4 < 0 \vee a_5 a_3 < 0)) \vee (a_5 a_1 > 0 \wedge a_4 a_2 < 0))].$$

In order to construct these formulas, one uses among others the fact that $\tau(\chi) = 0$ implies that the number of sign changes in the coefficients of $\chi(X)$ equals the number of sign changes in the coefficients of $\chi(-X)$. For the given short form of S_6 containing only 16 atomic subformulas I am indebted to A. Dolzmann.

Step P2: To every vector $\delta \in \{1, 2\}^{m'} = \Delta$ we assign a monic univariate polynomial of degree d, $\chi_\delta(X) = X^d + \sum_{i=0}^{d-1} a_{\delta, i} X^i$ with indeterminate coefficients $a_{\delta, 0}, \ldots, a_{\delta, d-1}$. We let $\chi(X) = \prod_{\delta \in \Delta} \chi_\delta(X) = X^D + \sum_{i=0}^{D-1} a_i X^i$, $(D = d \cdot 2^{m'})$ and let $S_D(\mathbf{a})$ be the formula of step P1 applied to the coefficients of χ.

Next we describe the *main steps* of the quantifier elimination algorithm. It accepts as input formulas φ of the form $(*)$ above and outputs a quantifier-free formula φ' equivalent to φ in $\mathbb{R}$.

Step M1: If $m = 0$, i.e., no equations are present, we may by a finite case distinction and by recursion on k adjoin equations of the type $\frac{\partial g}{\partial x_k} = 0$, where g is a product of at most two polynomials occurring in the inequalities.

This uses the elementary argument that with respect to the distinguished variable x_k,

$$(**) \qquad \exists x_k (\bigwedge_{i=1}^{m'} h_i(x_1, \ldots, x_n) > 0 \wedge \bigwedge_{i=1}^{m'} I_i \neq 0),$$

where I_i is the initial of g_i with respect to x_k, is equivalent to

$$(***) \qquad \left(\bigwedge_{i=1}^{m'} I_i \neq 0 \wedge \bigwedge_{i=1}^{m'} h_i(x_1, \ldots, x_{k-1}, \infty, x_{k+1}, \ldots, x_n) > 0 \right) \vee$$

$$\left(\bigwedge_{i=1}^{m'} I_i \neq 0 \wedge \bigwedge_{i=1}^{m'} h_i(x_1, \ldots, x_{k-1}, -\infty, x_{k+1}, \ldots, x_n) > 0 \right) \vee$$

$$\bigvee_{j \in C} \exists x_k \left(\bigwedge_{i=1}^{m'} I_i \neq 0 \wedge \bigwedge_{i=1}^{m'} h_i(x_1, \ldots, x_n) > 0 \,\wedge \right.$$

$$\left. \frac{\partial(h_j)}{\partial x_k}(x_1, \ldots, x_n) = 0 \right) \vee$$

$$\bigvee_{j, j' = 1 \ldots m', j < j'} \exists x_k \left(\bigwedge_{i=1}^{m'} I_i \neq 0 \wedge \bigwedge_{i=1}^{m'} h_i(x_1, \ldots, x_n) > 0 \,\wedge \right.$$

$$(h_j - h_{j'})(x_1, \ldots, x_n) = 0\Big),$$

where $C = \{i | 1 \leq i \leq m', \ \deg_{x_k}(h_i) \geq 2\}$.

This is the case, because if $(**)$ holds for given real values of $x_1, \ldots, x_{k-1}$, $x_{k+1}, \ldots, x_n$ on a maximal bounded interval J, then at the endpoints of J at least one of the h_i's vanishes. So either some h_j vanishes at both endpoints of J, and hence h_j has a local maximum in J; or there are two $h_j, h_{j'}$ with different indices, such that each vanishes at a different endpoint of J. Then by the intermediate value theorem, $h_j - h_{j'}$ must have a zero in J.

Notice that we do not adjoin the equation

$$\frac{\partial \prod_{i=1}^{m'} h_i}{\partial x_k}(x_1, \ldots, x_n) = 0$$

which is the usual procedure that is correct too, in order to keep the degree of the equation as low as possible, even at the expense of creating a large disjunction.

The first two disjunctive parts of $(***)$ can be easily expressed by a quantifier-free formula on the I_i's only. Each of the remaining parts contains the required non-trivial equation.

Step M2: If $m > 0$, we regard $x_{k+1}, \ldots, x_n$ as parameters, and compute a comprehensive Gröbner basis G of $I = Id(f_1, \ldots, f_m)$ with respect to the main variables $x_1, \ldots, x_k$ and the quantifier-free formulas $\varphi_i(x_{k+1}, \ldots, x_n)$ together with d_i and $\mathcal{Y}_i \subseteq \{x_1, \ldots, x_k\}$ as described in Section 4.

Step M3: Next we form a disjunction of quantifier-free formulas ψ_i over the indices i: For those indices i, where $d_i = -1$, we put $\psi_i = false$. This is legitimate, since $\varphi_i(x_{k+1}, \ldots, x_n)$ asserts that I has not even a complex zero in $x_1, \ldots, x_k$. The indices i, for which $d_i = 0$, are handled using the preprocessing steps and the methods of Section 3:

With the help of G, we find (for each such index i separately) a basis of the residue ring consisting of d terms in $x_1, \ldots, x_k$.

If $m' = 0$, we compute the matrix M_1 and its characteristic polynomial $\chi_1 = X^d + \sum_{i=0}^{d-1} a_{1,i} X^i$. The corresponding quantifier-free formula ψ_i is $\psi_i = \varphi_i \wedge \neg S_d(\mathbf{a})$, where S_d is defined as in step P1.

If $m' > 0$ we compute for all functions $\delta : \{1, \ldots, m'\} \longrightarrow \{1, 2\}$ the matrix M_{h_δ} with $h_\delta = \prod_{i=1}^{m'} h_i^{\delta_i}$ and its characteristic polynomial $\chi_\delta = X^d + \sum_{i=0}^{d-1} a_{\delta,i} X^i$.

Next we call the two preprocessing steps for the given values of m' and d and the given coefficient-tuples $\mathbf{a}_\delta$ of χ_δ to obtain the quantifier-free formula $S_D(\mathbf{a})$.

The corresponding quantifier-free formulas ψ_i are now taken as $\psi_i = \varphi_i \wedge \neg S_D(\mathbf{a})$.

Step M4: The indices i, for which $d_i > 0$, require a recursive call of the quantifier elimination algorithm:

If $d_i = n$, one applies the algorithm to the new input formula

$$\varphi_\sim = \exists x_1 \ldots \exists x_k (\bigwedge_{i=1}^{m'} h_i(x_1, \ldots, x_n) > 0)$$

to get its quantifier-free equivalent $(\varphi_\sim)'$ and defines ψ_i by $\psi_i = \varphi_i \wedge (\varphi_\sim)'$.

If $0 < d_i < n$, one renumbers the variables so that $\mathcal{Y}_i = \{x_{k'+1}, \ldots, x_k\}$ one applies the algorithm to the new input formula

$$\varphi_\sim = \exists x_1 \ldots \exists x'_k (\bigwedge_{i=1}^{m} f_i(x_1, \ldots, x_n) = 0 \wedge \bigwedge_{i=1}^{m'} h_i(x_1, \ldots, x_n) > 0)$$

to get its quantifier-free equivalent $(\varphi_\sim)'$.

Next one applies the quantifier-elimination algorithm to the new input formula $\exists x_{k'+1} \ldots \exists x_k ((\varphi_\sim)')$ to get its quantifier-free equivalent $(\varphi_\sim)''$. Finally one defines ψ_i by $\psi_i = \varphi_i \wedge (\varphi_\sim)''$. The output formula φ' of the algorithm is now the disjunction $\bigvee_i \psi_i$.

Termination of the recursion is guaranteed by the following argument: In case $0 < d_i < n$, the number of main variables decreases in the recursive call. In case $d_i = n$, the number of main variables remains the same and step M1 will produce an equation of the form $h'_j = 0$ or $h_i - h_j = 0$. So if the same case $d_i = n$ is entered again in the recursive call, then at least one of the h_i will be a constant with respect to some main variable or one of h_i, h_j can be dropped. This fact guarantees that step M1 cannot be applied infinitely often with respect to the same main variable. So finally there will be a decrease in the number of main variables too.

Correctness of the algorithm follows from the correctness of its ingredients discussed in Sections 3 and 4.

The algorithm described above is obviously in a first, preliminary state that has to be optimized significantly during implementation. In particular, the preprocessing step P1 has to optimized in order to produce formulas S_D as small as possible. Moreover, the computation of comprehensive Gröbner bases should be combined with polynomial factorization in order to reduce the dimension d of the residue class algebras to be considered and thus the size of D in the formulas S_D.

The main bottleneck for performance of the main algorithm is the possible recursion occurring for some input formulas, in case the dimension of the ideal to be considered is greater than zero in the quantified variables. There are, however, many interesting input formulas for which this situation will not occur, among them well-known benchmark examples such as the quartic problem (see Collins and Hong 1991; Lazard 1988; Hong 1992e) or the determination of the image of parameterization of a real algebraic variety such as the Whitney umbrella or the Enneper surface (see Cox et al. 1992).

6   Examples

The following examples were computed interactively with the computer algebra system Maple, following the general procedure described in Section 5. They will give a first impression of the power and the bottlenecks of the method.

6.1   A Quadratic Polynomial

In this toy example, we want to determine the number of real zeros of x^2+p+q in dependence of p and q following the general method.

An easy computation yields

$$M = M_1 = \begin{pmatrix} 2 & -p \\ -p & p^2 - 2q \end{pmatrix},$$

and hence

$$\chi = \chi_1 = \det(xE - M) = x^2 + Ax + D$$

with $A = (2q - p^2 - 2)$, $D = p^2 - 4q$.

The number of sign-changes of the coefficients of χ is 1 if $D < 0$ or $(D = 0$ and $A < 0)$, 2 if $(A < 0$ and $D > 0)$, and 0 otherwise.

Moreover, $D = 0 \Rightarrow A < 0$ and $D > 0 \Rightarrow A < 0$. So we get by Descartes' rule of signs, as expected

$$\sigma = \sigma_1 = \begin{cases} 0 & \text{if } D < 0, \\ 1 & \text{if } D = 0, \\ 2 & \text{if } D > 0. \end{cases}$$

In particular, we get the equivalence

$$\exists x(x^2 + px + q = 0) \iff D \geq 0.$$

6.2   The Quartic Problem

This well-known test problem (see Lazard 1988) asks for necessary and sufficient conditions on the coefficients of a univariate real polynomial f of degree 4 to be positive semidefinite. By a linear transformation of the variable, we may assume that f is of the form $f = x^4 + px^2 + qx + r$. So we require quantifier elimination for the formula

$$\forall x(f(x) \geq 0),$$

or equivalently for its negation

$$\varphi = \exists x(f(x) < 0).$$

Since $f(\pm\infty) > 0$, φ, the adjunction of an equation by step (M1), yields the equivalent formula

$$\exists x(f'(x) = 0 \wedge f(x) < 0)$$

with $f' = 4x^3 + 2px + q$.

Another way of obtaining an equivalent formula with a non-trivial equation for this special type of problem was noticed by L. Gonzales-Vega: It suffices to say that f has at least one simple zero. This yields the formula

$$\exists x (f(x) = 0 \wedge f'(x) \neq 0).$$

Using the first alternative, we proceed as follows: Taking the residues of $1, x, x^2$ as basis of the residue algebra $A = \mathbb{R}[x]/Id(f')$, we compute normal forms of the residues of $f', xf', x^2 f', x^3 f', x^4 f', x^5 f', x^6 f', f'^2, xf'^2, \ldots, x^6 f'^2$ in A by division with remainder and obtain the following trace matrices for $h = f$ (notice the Hankel structure):

$$M_h = \begin{pmatrix} 3r - \frac{1}{2}p^2 & -\frac{9}{8}pq & -\frac{9}{16}q^2 - pr + \frac{1}{4}p^3 \\ -\frac{9}{8}pq & -\frac{9}{16}q^2 - pr + \frac{1}{4}p^3 & -\frac{3}{4}qr + \frac{11}{16}qp^2 \\ -\frac{9}{16}q^2 - pr + \frac{1}{4}p^3 & -\frac{3}{4}qr + -\frac{11}{16}qp^2 & -\frac{1}{8}p^4 + \frac{9}{16}pq^2 + \frac{1}{2}p^2 r \end{pmatrix},$$

$$M_{h^2} = \begin{pmatrix} m_{11} & m_{12} & m_{13} \\ m_{21} & m_{22} & m_{23} \\ m_{31} & m_{32} & m_{33} \end{pmatrix},$$

with

$$m_{11} = 3r^2 - \frac{9}{8}pq^2 - p^2 r + \frac{1}{8}p^4 \; ; \; m_{12} = m_{21} = -\frac{27}{64}q^3 + \frac{17}{32}qp^3 - \frac{9}{4}qpr,$$

$$m_{13} = m_{22} = m_{31} = \frac{51}{64}q^2 p^2 - \frac{9}{8}q^2 r - pr^2 + \frac{1}{2}p^3 r - \frac{1}{16}p^5,$$

$$m_{23} = m_{32} = -\frac{19}{64}qp^4 - \frac{3}{4}qr^2 + \frac{63}{128}pq^3 + \frac{11}{8}qp^2 r,$$

$$m_{33} = \frac{27}{256}q^4 - \frac{17}{32}q^2 p^3 + \frac{1}{2}p^2 r^2 + \frac{9}{8}q^2 pr + \frac{1}{32}p^6 - \frac{1}{4}p^4 r.$$

This yields the following characteristic polynomials:

$$\chi_h = x^3 + a_h x^2 + b_h x + c_h,$$

with

$$c_h = \frac{1}{4096}\left(27 q^2 + 8 p^3\right)$$
$$\left(16 p^4 r - 4 p^3 q^2 - 128 p^2 r^2 + 144 pq^2 r + 256 r^3 - 27 q^4\right),$$

$$b_h = -\frac{81 q^4 p}{256} - \frac{9 q^2 r^2}{16} - \frac{p^4 r}{8} + \frac{3 q^2 p^2 r}{16} + \frac{p^5 r}{4} + \frac{p^2 r^2}{2} - \frac{27 q^2 r}{16} + \frac{5 rp^3}{4} -$$
$$\frac{p^5}{8} - \frac{67 q^2 p^4}{256} - \frac{p^7}{32} - 3 pr^2 - \frac{81 q^4}{256} - \frac{p^3 r^2}{2} + \frac{9 pq^2 r}{16} - \frac{63 q^2 p^2}{64},$$

$$a_h = -\frac{9 pq^2}{16} + pr + \frac{p^2}{2} - \frac{p^3}{4} - 3r + \frac{p^4}{8} + \frac{9 q^2}{16} - \frac{p^2 r}{2},$$

$$\chi_{h^2} = x^3 + a_{h^2}x^2 + b_{h^2}x + c_{h^2},$$

with

$$
c_{h^2} \;=\; \frac{1}{1048576}\left(27\,q^2 + 8\,p^3\right)
$$
$$
\left(16\,p^4 r - 4\,p^3 q^2 - 128\,p^2 r^2 + 144\,pq^2 r + 256\,r^3 - 27\,q^4\right)^2,
$$

$$
\begin{aligned}
b_{h^2} \;=\; & -\frac{27\,q^2 p^2 r^2}{64} + \frac{27\,q^4 p^2 r}{64} - \frac{9\,q^2 p^3 r^2}{16} + \frac{9\,q^2 pr^3}{8} + \frac{57\,rq^2 p^4}{64} + \frac{q^2 p^5 r}{64} + \\
& \frac{99\,q^4 p^3 r}{512} - \frac{81\,rq^4 p}{128} - \frac{13\,p^5 r^2}{16} + \frac{5\,p^3 r^3}{2} + \frac{rp^7}{8} - \frac{27\,q^2 r^3}{8} - \\
& \frac{459\,p^3 q^4}{1024} - \frac{115\,p^6 q^2}{1024} - \frac{p^9}{128} - \frac{729\,q^6}{4096} - \frac{9\,q^2 p^4 r^2}{16} + \frac{63\,q^2 p^6 r}{256} - \\
& \frac{81\,q^4 r^2 p}{128} + \frac{3\,q^2 r^3 p^2}{8} - \frac{81\,q^6 p^2}{512} - \frac{141\,q^4 p^5}{1024} - \frac{123\,q^2 p^8}{4096} - \frac{243\,q^6 r}{2048} - \\
& \frac{p^3 r^4}{2} - \frac{3\,p^7 r^2}{16} + \frac{p^5 r^3}{2} + \frac{rp^9}{32} - \frac{9\,q^2 r^4}{16} - \frac{99\,q^4 p^4}{4096} - \frac{q^2 p^7}{512} - \frac{243\,q^4 r^2}{256} - \\
& \frac{243\,q^6 p}{2048} + \frac{p^2 r^4}{2} - \frac{p^4 r^3}{4} + \frac{p^6 r^2}{32} - \frac{p^{11}}{512} - 3\,r^4 p,
\end{aligned}
$$

$$
\begin{aligned}
a_{h^2} \;\dot{=}\; & p^2 r + \frac{9\,pq^2}{8} - \frac{p^4}{8} - \frac{p^6}{32} + \frac{p^5}{16} - \frac{9\,pq^2 r}{8} + \frac{17\,p^3 q^2}{32} - \\
& \frac{51\,q^2 p^2}{64} - \frac{p^2 r^2}{2} - \frac{27\,q^4}{256} - 3\,r^2 - \frac{rp^3}{2} + \frac{p^4 r}{4} + \frac{9\,q^2 r}{8} + pr^2.
\end{aligned}
$$

Let $\chi = \chi_h \cdot \chi_{h^2} = X^6 + a_5 X^5 + a_4 X^4 + a_3 X^3 + a_2 X^2 + a_1 X + a_0$. Then the output formula $\varphi'(p, q, v)$ equals $\neg S_6(a_0, \ldots, a_5)$. Written down explicitly, this formula contains 16 atomic subformulas, each containing a polynomial in (p, q, r). Notice that a simplification of the polynomials in this formula is possible using the factorization of the coefficients of χ_h and χ_{h^2}.

Following the second alternative, we proceed as follows: Taking the residues of $1, x, x^2, x^3$ as basis of the residue algebra $A = \mathbb{R}[x]/Id(f)$, we compute normal forms of the residues of $f'^2, x f'^2, x^2 f'^2, \ldots, x^9 f'^2$ in A by division with remainder and obtain the following trace matrix for $h^2 = f'^2$ (notice again the Hankel structure):

$$
M_{h^2} = \begin{pmatrix}
m_{11} & m_{12} & m_{13} & m_{14} \\
m_{21} & m_{22} & m_{23} & m_{24} \\
m_{31} & m_{32} & m_{33} & m_{34} \\
m_{41} & m_{42} & m_{43} & m_{44}
\end{pmatrix},
$$

with

$$m_{11} = 28\,q^2 - 8\,p^3 + 32\,pr,$$
$$m_{12} = m_{21} = 80\,qr - 36\,qp^2,$$

$$m_{13} = m_{22} = m_{31} = 64\,r^2 - 48\,p^2 r - 54\,pq^2 + 8\,p^4,$$

$$m_{14} = m_{23} = m_{32} = m_{41} = -144\,qpr - 27\,q^3 + 44\,qp^3,$$

$$m_{24} = m_{33} = m_{42} = 56\,rp^3 - 108\,q^2 r - 96\,pr^2 + 90\,q^2 p^2 - 8\,p^5,$$

$$m_{34} = m_{43} = -144\,qr^2 + 228\,qp^2 r + 81\,pq^3 - 52\,qp^4,$$

$$m_{44} = 144\,p^2 r^2 + 306\,pq^2 r - 64\,r^3 - 64\,p^4 r + 27\,q^4 - 134\,p^3 q^2 + 8\,p^6.$$

The characteristic polynomial of M_{h^2} is

$$\chi_{h^2} = x^4 + a_{h^2} x^3 + b_{h^2} x^2 + c_{h^2} x + d_{h^2},$$

with

$$
\begin{aligned}
d_{h^2} \;=\;& \left(16\,p^4 r - 4\,p^3 q^2 - 128\,p^2 r^2 + 144\,pq^2 r + 256\,r^3 - 27\,q^4\right)^3, \\
c_{h^2} \;=\;& -(16\,p^4 r - 4\,p^3 q^2 - 128\,p^2 r^2 + 144\,pq^2 r + 256\,r^3 - 27\,q^4) \\
& (3744\,q^2 p^2 r^2 - 96\,q^2 p^2 r - 720\,rq^2 p^4 - 2560\,r^3 p^2 - 2000\,q^2 rp^3 + \\
& 108\,q^4 r - 1728\,pq^2 r - 3402\,rq^4 p + 54\,q^4 p + 16\,q^2 p^4 - 32\,p^5 r + \\
& 256\,p^3 r^2 + 1152\,q^2 r^2 + 304\,p^3 q^2 + 1024\,p^2 r^2 - 320\,p^4 r + 352\,p^5 r^2 - \\
& 1280\,p^3 r^3 + 756\,q^4 + 32\,p^6 - 32\,rp^7 - 3456\,q^2 r^3 + 216\,p^3 q^4 + \\
& 16\,p^6 q^2 + 729\,q^6 + 1600\,p^4 r^2 - 1024\,r^3 + 1024\,r^4 + 336\,q^2 p^5 + \\
& 882\,q^4 p^2 + 1536\,r^4 p + 32\,p^8 - 384\,p^6 r - 512\,pr^3 + 2112\,q^2 pr^2), \\
b_{h^2} \;=\;& -11520\,q^2 p^2 r^2 + 1620\,q^4 p^2 r + 2688\,q^2 p^2 r - 1664\,q^2 p^3 r^2 + \\
& 2304\,q^2 pr^3 + 8608\,rq^2 p^4 + 3072\,r^3 p^2 + 528\,q^2 p^5 r + 6588\,q^4 p^3 r + \\
& 128\,q^2 rp^3 - 3024\,q^4 r - 288\,rq^4 p + 6144\,pr^5 - 1512\,q^4 p - 640\,q^2 p^4 + \\
& 640\,p^5 r - 2048\,p^3 r^2 - 4608\,q^2 r^2 - 7168\,p^5 r^2 + 13312\,p^3 r^3 - 64\,p^7 + \\
& 1600\,rp^7 - 8704\,q^2 r^3 - 4076\,p^3 q^4 - 1424\,p^6 q^2 - 128\,p^9 - 702\,q^6 - \\
& 4096\,r^5 - 768\,p^4 r^2 - 4096\,r^4 - 80\,q^2 p^5 - 396\,q^4 p^2 - 17088\,q^2 p^4 r^2 + \\
& 7136\,q^2 p^6 r - 12312\,q^4 r^2 p + 14976\,q^2 r^3 p^2 - 4131\,q^6 p^2 - 3852\,q^4 p^5 - \\
& 912\,q^2 p^8 - 2916\,q^6 r - 17408\,p^3 r^4 - 5504\,p^7 r^2 + 14720\,p^5 r^3 + 960\,rp^9 - \\
& 13824\,q^2 r^4 - 648\,q^4 p^4 - 64\,q^2 p^7 - 9936\,q^4 r^2 - 1458\,q^6 p + 3072\,p^2 r^4 - \\
& 768\,p^4 r^3 + 64\,p^6 r^2 - 64\,p^{11} - 8192\,r^4 p + 64\,p^6 r + 2048\,pr^3 + 768\,q^2 pr^2, \\
a_{h^2} \;=\;& 48\,p^2 r - 28\,q^2 + 8\,p^5 - 27\,q^4 + 64\,r^3 + 54\,pq^2 + 8\,p^3 + 96\,pr^2 - 56\,rp^3 - \\
& 64\,r^2 + 134\,p^3 q^2 + 64\,p^4 r - 90\,q^2 p^2 - 8\,p^6 - 144\,p^2 r^2 - 8\,p^4 + 108\,q^2 r - \\
& 32\,pr - 306\,pq^2 r.
\end{aligned}
$$

The corresponding output formula $\varphi'(p, q, v)$ equals $\neg S_4(d_{h^2}, c_{h^2}, b_{h^2}, a_{h^2})$. Written down explicitly, this formula contains only 7 atomic subformulas, each containing a polynomial in (p, q, r). Again a simplification of the polynomials in this formula is possible by dropping the exponent 3 in the factorization of d_{h^2} and exploiting the factorization of c_{h^2}.

As a comparison we note that corresponding output in Collins' original algorithm contains 401 atomic subformulas, the latest optimization thereof by Hong (1992e) has 5 atomic subformulas, while the ingenious (but not automatically obtained) solution in (Lazard 1988) has only 3 atomic subformulas.

6.3 The Whitney Umbrella

Consider the map $\omega : \mathbb{R}^2 \longrightarrow \mathbb{R}^3$ given by

$$x = uv, \quad y = v, \quad z = u^2.$$

The Whitney umbrella U is the smallest real variety containing the image of this map (see Cox et al. 1992, p. 133). By computing a Gröbner basis G of $I = Id(x - uv, y - v, z - u^2)$ with respect to a term-order in which x, y, z are lexicographically small in relation to u, v, one finds the elimination ideal $J = I \cap \mathbb{R}[x, y, z]$ to be generated by $f = y^2 z - x^2$. Hence U is implicitly given by $f = 0$.

We are trying to find the image of ω inside U, i.e., a quantifier-free formula $\varphi'(x, y, z)$ equivalent in $\mathbb{R}$ to

$$\varphi(x, y, z) = \exists u \exists v (x = uv \ \wedge \ y = v \ \wedge \ z = u^2).$$

To do so, we compute a comprehensive Gröbner basis H of I in $\mathbb{Q}[x, y, z][u, v]$ with respect to the total-degree-inverse-lex order on u, v.

$$H = \{y^2 z - x^2, yu - x, v - y, u^2 - z\}.$$

From H we automatically get the following quantifier-free equivalent to φ in the field $\mathbb{C}$ of complex numbers:

$$\varphi''(x, y, z) = y^2 z - x^2 - 0 \ \wedge \ (y = 0 \Longrightarrow x = 0),$$

which is equivalent to

$$y^2 z - x^2 - 0.$$

Moreover, inspection of H shows that a corresponding formula for the reals can be taken as

$$\varphi'(x, y, z) = y^2 z - x^2 \ \wedge \ (y = 0 \implies z \geq 0).$$

Next, we describe, how such a real elimination formula is produced automatically by our method:

For the residue class ring $A = \mathbb{Q}[x, y, z][u, v]/I$ for real values x, y, z we have to distinguish the cases $y = 0$ and $y \neq 0$. In the first case $A = A_1$ is 2-dimensional with ordered basis $(1, u)$, in the second case $A = A_2$ is one-dimensional with ordered basis (1). So the second case $y \neq 0$ produces the characteristic polynomial $\chi_2 = X - 1$, and so the signature is 1, and the quantifier-free equivalent to φ is *true*.

In the first case, we compute the matrix

$$M = \begin{pmatrix} 2 & 0 \\ 0 & 2z \end{pmatrix}$$

with characteristic polynomial

$$\chi_1 = (X - 2)(X - 2z) = X^2 - (2z + 2) + 4z.$$

So the signature σ of M is determined as follows

$$\sigma = \begin{cases} 2 & \text{for } z > 0, \\ 1 & \text{for } z = 0, \\ 0 & \text{for } z < 0. \end{cases}$$

In particular in this case, the quantifier-free equivalent to φ is $z \geq 0$.
As a consequence, the final quantifier-free formula equivalent to φ is

$$\varphi' \;=\; y^2 z - x^2 = 0 \;\wedge\; [(y \neq 0 \;\wedge\; true) \;\vee\; (y = 0 \;\wedge\; z \geq 0)]$$

which is equivalent to the formula obtained by inspection above.

6.4   The X-Axis Ellipse Problem

The problem is a special case of the general ellipse problem (see Lazard 1988).
It asks for a quantifier-free formula $\varphi'(a, b, c)$ that is necessary and sufficient
for the ellipse $\frac{(x-c)^2}{a^2} + \frac{y^2}{b^2} = 1$ with main axis a, b and midpoint $(c, 0)$ to lie
inside the closed unit circle $x^2 + y^2 \leq 1$ (compare Hong 1992e).
By the geometric nature of the problem, it suffices to consider the case
$a, b, c > 0$ and $a < b$. The problem amounts then to an instance of real
quantifier elimination applied to the following input formula $\varphi(a, b, c)$:

$$\forall x \forall y (b^2 (x - c)^2 + a^2 y^2 = a^2 b^2 \;\Rightarrow\; x^2 + y^2 \leq 1).$$

Passing to the negation of φ, we obtain:

$$\exists x \exists y (b^2 (x - c)^2 + a^2 y^2 - a^2 b^2 = 0 \;\wedge\; a^2 x^2 + a^2 y^2 - a^2 > 0).$$

In this formula, the variable y can be easily eliminated "by hand" (within
the field of reals) by eliminating the monomial $a^2 y^2$:

$$\exists x (a^2 b^2 - b^2 (x - c)^2 \geq 0 \;\wedge\; a^2 x^2 + a^2 b^2 - b^2 (x - c)^2 - a^2 > 0).$$

This formula is of the form

$$\exists x (h_1 \geq 0 \;\wedge\; h_2 > 0)$$

with
$$h_1 = -b^2 x^2 + 2b^2 cx + b^2(a^2 - c^2),$$
$$h_2 = (a^2 - b^2)x^2 + 2b^2 cx + b^2(a^2 - c^2) - a^2.$$

An application of step M1 yields:

$$\exists x(h_1' = 0 \ \wedge \ h_1 \geq 0 \ \wedge \ h_2 > 0) \ \vee$$
$$\exists x(h_2' = 0 \ \wedge \ h_1 \geq 0 \ \wedge \ h_2 > 0) \ \vee$$
$$\exists x(h_1 - h_2 = 0 \ \wedge \ h_1 \geq 0 \ \wedge \ h_2 > 0).$$

Since h_1' and h_2' are linear with non-vanishing highest coefficients, quantifier elimination for the first two formulas is easy. The third formula reduces to

$$\exists x(x^2 - 1 = 0 \ \wedge \ h_1 \geq 0 \ \wedge \ h_2 > 0).$$

By factoring $x^2 - 1 = (x + 1)(x - 1)$ one obtains in this way (under the given hypothesis on a, b, c and with mild simplification) the following simple output formula:

$$b^2 + c^2 > 1 \ \vee \ ((a^2 - b^2)^2 \geq a^2 c^2 \ \wedge \ b^4 + b^2 c^2 + a^2 - a^2 b^2 - b^2 > 0) \ \vee \ a^2 > (c-1)^2.$$

For comparison: Collins' original algorithm failed for this problem, (Hong 1992e) yields an output formula containing 5 atomic subformulas.

References

Abhyankar, S. S. (1990): Algebraic geometry for scientists and engineers. American Mathematical Society, Providence, Rhode Island (Mathematical Surveys and Monographs, vol. 35).

Abramowitz, M. and Stegun, I. A. (1970): Handbook of mathematical functions. Dover Publishing, Inc., New York.

Achieser, N. I. (1956): Theory of approximation. Frederick Ungar, New York.

Aiba, A., Sakai, K., Sato, Y., Hawley, D. J., and Hasegawa, R. (1988): Constraint logic programming language CAL. In Proc. of International Conf. on Fifth Generation Computer Systems (FGCS-88), ICOT, Tokyo. pp. 263–276.

Alon, N. (1995): Tools from higher algebra. In Handbook of Combinatorics. Elsevier, Amsterdam New York.

Arnborg, S. and Feng, H. (1986): Algebraic decomposition of regular curves. In: Char, B. W. (Ed.), Proc. of 1986 Symp. on Symbolic and Algebraic Computation. pp. 53–55.

Arnborg, S. and Feng, H. (1988): Algebraic decomposition of regular curves. J. Symb. Comput. 5(1,2): 131–140.

Arnol'd, V. I. and Oleinik, O. (1979): Topology of real algebraic manifolds. Vest. Mosk. Univ. Mat. 34: 7–17.

Arnon, D. S. (1979): A cellular decomposition algorithm for semi-algebraic sets. In: Ng, E. W. (Ed.), Proc. International Symp. on Symbolic and Algebraic Manipulation (EUROSAM '79). Springer, Berlin Heidelberg New York Tokyo, pp. 301–315 (Lect. Notes Comput. Sci., vol. 72).

Arnon, D. S. (1981): Algorithms for the geometry of semi-algebraic sets. Ph. D. thesis, Comput. Sci. Dept., University of Wisconsin-Madison. Tech. Rep. 436.

Arnon, D. S. (1983): Topologically reliable display of algebraic curves. In: Tanner, P. (Ed.), Proc. ACM SIGGRAPH '83. Association for Computing Machinery, New York, pp. 219–227.

Arnon, D. S. (1988a): A bibliography of quantifier elimination for real closed fields. J. Symb. Comput. 5(1,2): 267–274.

Arnon, D. S. (1988b): A cluster-based cylindrical algebraic decomposition algorithm. J. Symb. Comput. 5(1,2): 189–212.

Arnon, D. S. (1988c): Geometric reasoning with logic and algebra. Artificial

Intelligence 37(1-3): 37–60.

Arnon, D. S. and Buchberger, B. (1988): Algorithms in real algebraic geometry. Academic Press, London.

Arnon, D. S. and McCallum, S. (1982): Cylindrical algebraic decomposition by quantifier elimination. In Proc. of the European Computer Algebra Conference (EUROCAM '82). Springer, Berlin Heidelberg New York Tokyo, pp. 215–222 (Lect. Notes Comput. Sci., vol. 144).

Arnon, D. S. and McCallum, S. (1984): A polynomial-time algorithm for the topological type of a real algebraic curve – extended abstract. Rocky Mountain J. Math. 14: 849–852.

Arnon, D. S. and McCallum, S. (1988): A polynomial-time algorithm for the topological type of a real algebraic curve. J. Symb. Comput. 5(1,2): 213–236.

Arnon, D. S. and Mignotte, M. (1988): On mechanical quantifier elimination for elementary algebra and geometry. J. Symb. Comput. 5(1,2): 237–260.

Arnon, D. S. and Smith, S. (1983): Towards mechanical solution of the Kahan ellipse problem I. In: van Hulzen, J. A. (Ed.), Proc. EUROCAL '83, European Computer Algebra Conf. Springer, Berlin Heidelberg New York Tokyo, pp. 36–44 (Lect. Notes Comput. Sci., vol. 162).

Arnon, D. S., Collins, G. E., and McCallum, S. (1982): Cylindrical algebraic decomposition I: The basic algorithm. Tech. Rep. CSD-427, Computer Science Dept., Purdue University.

Arnon, D. S., Collins, G. E., and McCallum, S. (1984a): Cylindrical algebraic decomposition I: The basic algorithm. SIAM J. Comput. 13(4): 865–877. This volume, pp. 136–151.

Arnon, D. S., Collins, G. E., and McCallum, S. (1984b): Cylindrical algebraic decomposition II: An adjacency algorithm for the plane. SIAM J. Comput. 13(4): 878–889. This volume, pp. 152–165.

Arnon, D. S., Collins, G. E., and McCallum, S. (1988): Cylindrical algebraic decomposition, III: An adjacency algorithm for three-dimensional space. J. Symb. Comput. 5(1,2): 163–187.

Artin, E. and Schreier, O. (1926): Algebraische Konstruktion reeller Körper. Abh. Math. Sem. Hamburg 5: 83–115.

Ax, J. and Kochen, S. (1965): Diophantine problems over local fields, I,II. Am. J. Math. 87: 605–648.

Ax, J. and Kochen, S. (1966): Diophantine problems over local fields III. Ann. Math. 83: 305–309.

Bajaj, C. (1987): Compliant motion planning with geometric models. In Proc. Third ACM Symp. Computational Geometry. pp. 171–180.

Bareiss, E. H. (1968): Sylvester's indentity and multistep integer-preserving Gaussian elimination. Math. Comput. 22: 565–578.

Bartholomeus, M. and Man, H. D. (1985): Presto-II: Yet another logic minimizer for programmed logic arrays. In Proc. Int. Symp. Circ. Syst.

Barvinok, A. I. (1993): Feasibility testing for systems of real quadratic equations. Discrete Comput. Geom. 10: 1–13.

Barwise, J. (Ed.) (1977): Handbook of mathematical logic. North-Holland, Amsterdam (Studies in Logic and the Foundations of Mathematics).

Basu, S., Pollack, R., and Roy, M.-F. (1994): On the combinatorial and algebraic complexity of quantifier elimination. In Proc. 34th IEEE Symp. on Foundations of Computer Science. pp. 632–641.

Basu, S., Pollack, R., and Roy, M.-F. (1995): Computing a set of points meeting every cell defined by a family of polynomials on a variety. In: Goldberg, K. R., Halperin, D., Latombe, J.-C., and Wilson, R. H. (Eds.), Algorithmic Foundations of Robotics. A. K. Peters, Boston.

Basu, S., Pollack, R., and Roy, M.-F. (1997): On computing a set of points meeting every semi-algebraically connected component of a family of polynomials on a variety. J. Complex. 13(1): 28–37.

Becker, E. (1986): On the real spectrum of a ring and its application to semi-algebraic geometry. Bull. Am. Math. Soc. N. S. 15: 19–60.

Becker, E. and Wörmann, T. (1991): On the trace formula for quadratic forms and some applications. In Proc. RAGSQUAD, Berkeley, 1991.

Becker, T. and Weispfenning, V. (1993): Gröbner bases – a computational approach to commutative algebra. in cooperation with H. Kredel. Springer, Berlin Heidelberg New York Tokyo (Graduate Texts in Mathematics).

Ben-Or, M., Kozen, D., and Reif, J. H. (1986): The complexity of elementary algebra and geometry. J. Comput. Syst. Sci. 32(2): 251–264.

Ben-Or, M., Feig, E., Kozen, D., and Tiwari, P. (1988): A fast parallel algorithm for determining all roots of a polynomial with real roots. SIAM J. Comput. 17(6): 1081–1092.

Benedetti, R. and Risler, J.-J. (1990): Real algebraic and semi-algebraic sets. Hermann, Paris (Actualités Mathématiques).

Berman, L. (1977): Precise bounds for Presburger arithmetic and the reals with addition (preliminary report). In Proc. 18th IEEE Symp. on Foundations of Computer Science. pp. 95–99.

Berman, L. (1980): The complexity of logical theories. Theor. Comput. Sci. 11: 71–77.

Bernays, P. and Hilbert, D. (1934): Grundlagen der Mathematik. J. Springer, Berlin (vol. 1).

Bernays, P. and Hilbert, D. (1939): Grundlagen der Mathematik. J. Springer, Berlin (vol. 2).

Birkhoff, G. (1948): Lattice theory. American Mathematical Society, New York (American Mathematical Society Colloquium publications, vol. 25).

Blum, L., Shub, M., and Smale, S. (1989): On a theory of computation and complexity over the real numbers: NP-completeness, recursive functions and universal machines. Bull. Am. Math. Soc. 21(0): 1–46.

Bochnak, J., Coste, M., and Roy, M.-F. (1987): Géometrie algébrique réelle. Springer, Berlin Heidelberg New York Tokyo.

Bochner, S. and Martin, W. T. (1948): Several complex variables. Princeton University Press, Princeton.

Böge, W. (1980): Decision procedures and quantifier elimination for elementary real algebra and parametric polynomial nonlinear optimization. Manuscript in preparation.

Böge, W. (1986): Quantifier elimination for real closed fields. In Proc. AAECC-3. Springer, Berlin Heidelberg New York Tokyo (Lect. Notes Comput. Sci., vol. 229).

Brayton, R., McMullen, C., Hachtel, G. D., and Sangiovanni-Vincentelli, A. (1984): Logic minimization algorithms for VLSI synthesis. Kluwer, Dordrecht.

Brown, S. S. (1978a): Bonds on transfer principles for algebraically closed and complete discrete valued fields. American Mathematical Society, Providence (Mem. AMS, vol. 204).

Brown, W. S. (1971): On Euclid's algorithm and the computation of polynomial greatest common divisors. J. ACM 18(4): 478–504.

Brown, W. S. (1978b): The subresultant PRS algorithm. ACM Trans. Math. Softw. 4: 237–249.

Brown, W. S. and Traub, J. F. (1971): On Euclid's algorithm and the theory of subresultants. J. ACM 18(4): 505–514.

Brownawell, D. (1987): Bounds for the degrees of the nullstellensatz. Ann. Math. 126(2): 577–591.

Brumfiel, G. (1979): Partially ordered rings and semi-algebraic geometry. Cambridge University Press, Cambdridge (London Math. Soc. Lect. Notes, vol. 37).

Buchberger, B. (1965): An algorithm for finding a basis for the residue class ring of a zero-dimensional polynomial ideal. Ph. D. thesis, Universität Innsbruck, Institut für Mathematik. In German.

Buchberger, B. and Hong, H. (1991): Speeding-up quantifier elimination by Groebner bases. Tech. Rep. 91-06.0, Research Institute for Symbolic Computation, Johannes Kepler University, Linz, Austria.

Buchberger, B., Collins, G. E., and Loos, R. (Eds.) (1982): Computer algebra: Symbolic and algebraic computation (second ed.). Springer, Wien New York.

Buchberger, B., Collins, G. E., and Kutzler, B. (1988): Algebraic methods for geometrical reasoning. Annu. Rev. Comput. Sci. 3: 85–119.

Buchberger, B., Collins, G., Encarnación, M., Hong, H., Johnson, J., Krandick, W., Loos, R., and Neubacher, A. (1993): SACLIB 1.1 user's guide. Tech. Rep. 93-19, Research Institute for Symbolic Computation, Johannes Kepler University, Linz, Austria.

Bündgen, R. (1991): Term completion versus algebraic completion. Ph. D. thesis, Wilhelm-Schickard-Institut, Universität Tübingen.

Burnside, W. S. and Panton, A. W. (1899): The theory of equations. Dublin University Press, Dublin (vol. 1).

Canny, J. F. (1987): A new algebraic method of robot motion planning and real geometry. In Proc. 28th IEEE Symp. on Foundations of Computer Science. pp. 39–48.

Canny, J. F. (1988): Some algebraic and geometric computations in PSPACE.

In Proc. 20th Annual ACM Symp. on the Theory of Computing. Association for Computing Machinery, New York, pp. 460–467.

Canny, J. F. (1989): The complexity of robot motion planning. MIT Press, Cambridge, Mass.

Canny, J. F. (1990): Generalized characteristic polynomials. J. Symb. Comput. 9: 241–250.

Canny, J. F. (1991a): Computing roadmaps of general semi-algebraic sets. In: Mattson, H. F., Mora, T., and Rao, T. R. N. (Eds.), Proc. Ninth International Symp. on Applied Algebra, Algebraic Algor. and Error Correcting Codes. Springer, Berlin Heidelberg New York Tokyo, pp. 94–107 (Lect. Notes Comput. Sci., vol. 539).

Canny, J. F. (1991b): An improved sign determination algorithm. In: Mattson, H. F., Mora, T., and Rao, T. R. N. (Eds.), Proc. Ninth International Symp. on Applied Algebra, Algebraic Algor. and Error Correcting Codes. Springer, Berlin Heidelberg New York Tokyo, pp. 108–117 (Lect. Notes Comput. Sci., vol. 539).

Canny, J. F. (1993a): Computing roadmaps of general semi-algebraic sets. Comput. J. 36: 504–514.

Canny, J. F. (1993b): Improved algorithms for sign and existential quantifier elimination. Comput. J. 36: 409–418.

Canny, J. F. (1993c): Some practical tools for algebraic geometry. Technical Report in Spring school on Robot Motion Planning, PROMOTION ESPRIT.

Canny, J. F., Kaltofen, E., and Lakshman Y.N. (1989): Solving systems of non-linear polynomial equations faster. In Proc. ACM-SIGSAM 1989 International Symp. on Symbolic and Algebraic Computation. Association for Computing Machinery, New York, pp. 121–128.

Canny, J. F., Grigor'ev, D. Y., and Vorobjov, N. N. (1992): Finding connected components in subexponential time. Appl. Algebra Eng. Commun. Comput. 2: 217–238.

Carnap, R. (1937): The logical syntax of language. Harcourt Brace, New York.

Caviness, B. F. and Collins, G. E. (1972): Symbolic mathematical computation in a Ph.D. computer sciences program. SIGCSE Bull. 4(1): 19–23.

Caviness, B. F. and Collins, G. E. (1976): Algorithms for Gaussian integer arithmetic. In: Jenks, R. D. (Ed.), Proc. of 1976 ACM Symp. on Symbolic and Algebraic Computation. Association for Computing Machinery, pp. 36–45.

Cegielski, P. (1980): La theorie elementaire de la multiplication. C. R. Acad. Sci. Paris, ser. A 290: 935–938.

Chazelle, B. (1985): Fast searching in a real algebraic manifold with applications to geometric complexity. In Proc. CAAP '85. Springer, Berlin Heidelberg New York Tokyo, pp. 145–156 (Lect. Notes Comput. Sci., vol. 185).

Chistov, A. L. (1986a): Algorithm of polynomial complexity for factoring polynomials and finding the components of varieties in subexponential time. J. Soviet Math. 34: 1838–1882.

Chistov, A. L. (1986b): Polynomial complexity of the Newton–Puiseux algorithm. In Mathematical Foundations of Computer Science. Springer, Berlin Heidelberg New York Tokyo, pp. 247–255 (Lect. Notes Comput. Sci., vol. 233).

Chistov, A. L. (1989): Polynomial complexity algorithms for computational problems in the theory of algebraic curves. Zap. Nauchn. Semin. Leningrad. Otdel. Mat. Inst. Steklov (LOMI) 176: 127–150. In Russian.

Chistov, A. L. and Grigor'ev, D. Y. (1984): Complexity of quantifier elimination in the theory of algebraically closed fields. In: Chytil, M. P. and Koubek, V. (Eds.), Proc. 11th Symp. Math. Foundations of Comput. Sci. Springer, Berlin Heidelberg New York Tokyo, pp. 17–31 (Lect. Notes Comput. Sci., vol. 176).

Chou, S.-C. (1984): Proving elementary geometry theorems using Wu's method. In Automated Theorem Proving after 25 Years. American Mathematical Society, Providence, RI, pp. 243–286 (AMS Contemporary Mathematics, vol. 29).

Chou, S.-C. (1985): Proving and discovering geometry theorems using Wu's method. Tech. Rep. 49, Inst. for Computing Sci., Austin, Texas.

Chou, S.-C. (1986): A collection of geometry theorems proved mechanically. Tech. Rep. 50, Inst. for Computing Sci., Austin, Texas.

Chou, S.-C. (1988): Mechanical geometry theorem proving. K. Reidel, Dordrecht.

Chou, S.-C., Gao, X.-S., and McPhee, N. (1989): A combination of Ritt-Wu's method and Collins' method. Tech. Rep. TR-89-28, Comput. Sci. Dept., The University of Texas at Austin.

Chou, S.-C., Gao, X.-S., and Arnon, D. S. (1992): On the mechanical proof of geometry theorems involving inequalities. Adv. Comput. Res. 6: 139–181.

Chou, T. J. and Collins, G. E. (1982): Algorithms for the solution of systems of linear diophantine equations. SIAM J. Comput. 11(4): 687–708.

Church, A. (1936): An unsolvable problem of elementary number theory. Am. J. Math. 58: 345–363.

Cohen, J. (1990): Constraint logic programming languages. Commun. ACM 33(7): 52–68.

Cohen, P. J. (1969): Decision procedures for real and p-adic fields. Commun. Pure Appl. Math. 22(2): 131–151.

Collins, G. E. (1954): Distributivity and an axiom of choice. J. Symb. Logic 19(4): 275–277.

Collins, G. E. (1955): The modeling of new foundations in Zermelo type set theories. Ph. D. thesis, Cornell University.

Collins, G. E. (1956): The Tarski decision procedure. In Proc. 11th Annual ACM Meeting. Association for Computing Machinery, New York.

Collins, G. E. (1960a): A method for the overlapping and erasure of lists. Commun. ACM 3(12): 655–657.

Collins, G. E. (1960b): Tarski's decision method for elementary algebra. In Proc. Summer Institute for Symbolic Logic. Institute for Defense Analyses,

Princeton, NJ, pp. 64–70.

Collins, G. E. (1966a): PM, a system for polynomial manipulation. Commun. ACM 9(8): 578–589.

Collins, G. E. (1966b): Polynomial remainder sequences and determinants. Am. Math. Monthly 73(7): 708–712.

Collins, G. E. (1967): Subresultants and reduced polynomial remainder sequences. J. ACM 14(1): 128–142.

Collins, G. E. (1969a): Computing multiplicative inverses in GF(p). Math. Comput. 23(105): 197–200.

Collins, G. E. (1969b): Computing time analyses for some arithmetic and algebraic algorithms. In: Tobey, R. G. (Ed.), Proc. 1968 Summer Institute on Symbolic Math. Computation. IBM Boston Programming Center, Cambridge, Mass., pp. 195–231.

Collins, G. E. (1971a): The calculation of multivariate polynomial resultants. J. ACM 18(4): 515–532.

Collins, G. E. (1971b): The SAC-1 system: An introduction and survey. In: Petrick, S. R. (Ed.), Proc. ACM Symp. on Symbolic and Algebraic Manipulation. Association for Computing Machinery, New York, pp. 144–152.

Collins, G. E. (1973a): Computer algebra of polynomials and rational functions. Am. Math. Monthly 80(7): 725–755.

Collins, G. E. (1973b): Efficient quantifier elimination for elementary algebra. Abstract presented at Symp. on Complexity of Sequential and Parallel Algorithms, Carnegie-Mellon University.

Collins, G. E. (1974a): The computing time of the Euclidean algorithm. SIAM J. Comput. 3(1): 1–10.

Collins, G. E. (1974b): Quantifier elimination for real closed fields by cylindrical algebraic decomposition – preliminary report. SIGSAM Bull. 8(3): 80–90.

Collins, G. E. (1975): Quantifier elimination for the elementary theory of real closed fields by cylindrical algebraic decomposition. In: Brakhage, H. (Ed.), Automata Theory and Formal Languages. Springer, Berlin Heidelberg New York Tokyo, pp. 134–183 (Lect. Notes Comput. Sci., vol. 33). This volume pp. 85–121.

Collins, G. E. (1976): Quantifier elimination for real closed fields by cylindrical algebraic decomposition – a synopsis. SIGSAM Bull. 10(3): 10–12.

Collins, G. E. (1977): Infallible calculation of polynomial zeros to specified precision. In: Rice, J. R. (Ed.), Mathematical Software III. Academic Press, Inc., New York, San Francisco and London.

Collins, G. E. (1979): Factoring univariate integral polynomials in polynomial average time. In: Ng, E. W. (Ed.), Proc. of 1979 Symp. on Symbolic and Algebraic Computation. Springer, Berlin Heidelberg New York Tokyo, pp. 317–329 (Lect. Notes Comput. Sci., vol. 72).

Collins, G. E. (1982a): Factorization in cylindrical algebraic decomposition. In: Calmet, J. (Ed.), Computer Algebra EUROCAM '82, European Computer Algebra Conf. Springer, Berlin Heidelberg New York Tokyo, pp. 212–214 (Lect. Notes Comput. Sci., vol. 144).

Collins, G. E. (1982b): Quantifier elimination for real closed fields: A guide to the literature. In: Buchberger, B., Collins, G. E., and Loos, R. (Eds.), Computer Algebra: Symbolic and Algebraic Computation (Second ed.). Springer, Wien New York, pp. 79–82.

Collins, G. E. (1997): Quantifier elimination by cylindrical algebraic decomposition – twenty years of progress. In this volume. Springer, pp. 8–23.

Collins, G. E. and Akritas, A. G. (1976): Polynomial real root isolation using Descartes' rule of signs. In: Jenks, R. D. (Ed.), Proc. of 1976 ACM Symp. on Symbolic and Algebraic Computation. Association for Computing Machinery, New York, pp. 272–275.

Collins, G. E. and Halpern, J. D. (1970): On the interpretability of arithmetic in set theory. Notre Dame J. Formal Logic 11(4): 477–483.

Collins, G. E. and Hong, H. (1990): Computing coarsest squarefree bases and multiplicities. Tech. Rep. OSU-CISRC-5/90 TR13, Comput. Sci. Dept., The Ohio State University.

Collins, G. E. and Hong, H. (1991): Partial cylindrical algebraic decomposition for quantifier elimination. J. Symb. Comput. 12(3): 299–328. This volume pp. 174–200.

Collins, G. E. and Horowitz, E. (1974): The minimum root separation of a polynomial. Math. Comput. 28(126): 589–597.

Collins, G. E. and Johnson, J. R. (1989a): The probability of relative primality of Gaussian integers. In: Gianni, P. (Ed.), Proc. of 1988 International Symp. on Symbolic and Algebraic Computation. Springer, Berlin Heidelberg New York Tokyo, pp. 252–258 (Lect. Notes Comput. Sci., vol. 385).

Collins, G. E. and Johnson, J. R. (1989b): Quantifier elimination and the sign variation method for real root isolation. In Proc. ACM-SIGSAM 1989 International Symp. on Symbolic and Algebraic Computation. Association for Computing Machinery, New York, pp. 264–271.

Collins, G. E. and Krandick, W. (1992): An efficient algorithm for infallible polynomial complex root isolation. In: Wang, P. S. (Ed.), Proc. International Symp. on Symbolic and Algebraic Computation. Association for Computing Machinery, New York, pp. 189–194.

Collins, G. E. and Krandick, W. (1993): A hybrid method for high precision calculation of polynomial real roots. In: Bronstein, M. (Ed.), Proc. 1993 International Symp. on Symbolic and Algebraic Computation. Association for Computing Machinery, New York, pp. 47–52.

Collins, G. E. and Loos, R. (1976): Polynomial real root isolation by differentiation. In: Jenks, R. D. (Ed.), Proc. of 1976 ACM Symp. on Symbolic and Algebraic Computation. Association for Computing Machinery, New York, pp. 15–25.

Collins, G. E. and Loos, R. (1982a): The Jacobi symbol algorithm. SIGSAM Bull. 16(1): 12–16.

Collins, G. E. and Loos, R. (1982b): Real zeros of polynomials. In: Buchberger, B., Collins, G. E., and Loos, R. (Eds.), Computer Algebra: Symbolic and Algebraic Computation (Second ed.). Springer, Wien New York, pp. 83–94.

Collins, G. E. and Loos, R. G. K. (1990): Specifications and index of SAC-2 algorithms. Tech. Rep. WSI-90-4, Wilhelm-Schickard-Institut für Informatik, Universität Tübingen.

Collins, G. E. and Musser, D. R. (1977): Analysis of the Pope-Stein division algorithm. Inf. Process. Lett. 6(5): 151–155.

Collins, G. E., Mignotte, M., and Winkler, F. (1982): Arithmetic in basic algebraic domains. In: Buchberger, B., Collins, G. E., and Loos, R. (Eds.), Computer Algebra: Symbolic and Algebraic Computation (Second ed.). Springer, Wien New York, pp. 189–220.

Collins, G. E., Johnson, J. R., and Küchlin, W. (1992): Parallel real root isolation using the coefficient sign variation method. In: Zippel, R. E. (Ed.), Proc. Computer Algebra and Parallelism. Springer, Berlin Heidelberg New York Tokyo, pp. 71–88.

Cooper, D. C. (1972): Theorem-proving in arithmetic without multiplication. In Machine Intelligence. University of Edinburgh Press, Edinburgh, pp. 91–100 (vol. 7).

Coste, M. (1989): Effective semi-algebraic geometry. In: Boissonnat, J.-D. and Laumond, J.-P. (Eds.), Geometry and Robotics. Springer, Berlin Heidelberg New York Tokyo (Lect. Notes Comput. Sci., vol. 391).

Coste, M. and Roy, M. F. (1988): Thom's lemma, the coding of real algebraic numbers and the computation of the topology of semi-algebraic sets. J. Symb. Comput. 5(1,2): 121–129.

Cox, D., Little, J., and O'Shea, D. (1992): Ideals, varieties, and algorithms. Springer, Berlin Heidelberg New York Tokyo (Undergraduate Texts in Mathematics).

Cucker, F. (1993): On the complexity of quantifier elimination: The structural approach. Comput. J. 36(5): 400–408.

Cucker, F. and Roy, M.-F. (1990): A theorem on random polynomials and some consequences in average complexity. J. Symb. Comput. 10(5): 405–409.

Cucker, F., Lanneau, H., Mishra, B., Pedersen, P., and Roy, M.-F. (1992): Real algebraic numbers are in NC. Appl. Algebra Eng. Commun. Comput. 3: 79–98.

Davenport, J. (1986): A piano movers problem. SIGSAM Bull. 20(76): 15–17.

Davenport, J. H. (1985): Computer algebra for cylindrical algebraic decomposition. Tech. Rep. 85–11, The Royal Inst. of Tech., Dept. of Numerical Analysis and Computing Sci., S-100 44, Stockholm, Sweden.

Davenport, J. H. and Heintz, J. (1988): Real quantifier elimination is doubly exponential. J. Symb. Comput. 5(1,2): 29–35.

Delzell, C. (1982): A finiteness theorem for open semi-algebraic sets, with applications to Hilbert's 17th problem. In Ordered Fields and Real Algebraic Geometry. American Mathematical Society, Providence, RI, pp. 79–97 (AMS Contemporary Mathematics, vol. 8).

Delzell, C. (1984): A continuous, constructive solution to Hilbert's 17th problem. Invent. Math. 76: 365–384.

Denef, J. (1984): The rationality of the Poincaré series associated to the p-adic points on a variety. Invent. Math. 77: 1–23.

Denef, J. (1986): p-adic semialgebraic sets and cell decomposition. J. Reine Angew. Math. 369: 154–166.

Denef, J. and van den Dries, L. (1988): p-adic and real subanalytic sets. Ann. Math. 128: 79–138.

Dershowitz, N. (1979): A note on simplification orderings. Inf. Process. Lett. 9(5): 212–215.

Dershowitz, N. (1987): Termination of rewriting. J. Symb. Comput. 3: 69–116.

Descartes, R. (1969): Géométrie 1637. In A Source Book in Mathematics. Harvard University Press, Cambridge, Mass., pp. 90–93.

Dickmann, M. A. (1985): Applications of model theory to real algebraic geometry: a survey. In Methods in Mathematical Logic Proc. 1983. Springer, Berlin Heidelberg New York Tokyo, pp. 76–150 (Lect. Notes Math., vol. 1130).

Dubhashi, D. (1993): Quantifier elimination for p-adic fields. Comput. J. 36(5): 419–426.

Dubois, D. W. (1969): A nullstellensatz for ordered fields. Ark. Math. 8: 111–114.

Dubois, D. W. (Ed.) (1982): Ordered fields and real algebraic geometry. American Mathematical Society, Providence, RI (AMS Contemporary Mathematics, vol. 8).

Dubois, D. W. (Ed.) (1984): Ordered Fields and Real Algebraic Geometry, Proc. of the AMS Boulder Conference. Rocky Mountain J. Math. 14(4).

Dubois, D. W. and Efroymson, G. (1970): Algebraic theory of real varieties. In Studies and Essays Presented to Y.H. Chen for His 60th Birthday. Math. Res. Center Nat. Taiwan University, Taipei, pp. 107–135.

Encarnación, M. J. (1994): On a modular algorithm for computing gcds of polynomials over algebraic number fields. In Proc. 1994 International Symp. on Symbolic and Algebraic Computation. pp. 58–65.

Ershov, Y. (1967): Fields with a solvable theory. Sov. Math. Dokl. 8: 575–576.

Ershov, Y. (1980): Multiply valued fields. Sov. Math. Dokl. 22: 63–66.

Ershov, Y., Lavrov, I. A., Taimonov, A. D., and Taitslin, M. A. (1965): Elementary theories. Russian Math. Surveys 20(4): 35–105.

Ersov, J. L. (1965): On elementary theories of local fields. Algebra Logika 4: 5–30.

Ferrante, J. and Rackoff, C. (1973): A decision procedure for the first order theory of real addition with order. Tech. Rep. Projcet MAC Technical Memorandum 33, M.I.T., Cambridge, Mass.

Ferrante, J. and Rackoff, C. (1975): A decision procedure for the first order theory of real addition with order. SIAM J. Comput. 4(1): 69–76.

Ferrante, J. and Rackoff, C. W. (1979): The computational complexity of logical theories. Springer, Berlin Heidelberg New York Tokyo (Lect. Notes Math., vol. 718).

Fischer, M. and Rabin, M. (1974a): Super-exponential complexity of Presburger arithmetic. In Complexity of Computation. pp. 27–41 (AMS-SIAM

Proc., vol. 7). This volume pp. 122–135.

Fischer, M. and Rabin, M. (1974b): Super-exponential complexity of Presburger arithmetic. Tech. Rep. MAC Tech. Memo. 43, M.I.T.

Fitchas, N., Galligo, A., and Morgenstern, J. (1990a): Algorithmes rapides en séquentiel et parallèle pour l'élimination des quantificateurs en géométrie élémentaire. In Séminaire sur les Structures Algébriques Ordonnées, Vol. 1. pp. 103–145 (Publ. Math. Univ. Paris VII, vol. 32).

Fitchas, N., Galligo, A., and Morgenstern, J. (1990b): Precise sequential and parallel complexity bounds for quantifier elimination over algebraically closed fields. J. Pure Appl. Algebra 67: 1–14.

Forster, W. (1992): Some computational methods for systems of nonlinear equations and systems of polynomial equations. J. Global Optimization 2: 317–356.

Fürer, M. (1982): The complexity of Presburger arithmetic with bounded quantifier alternation depth. Theor. Comput. Sci. 18: 105–111.

Gabrielov, A. (1968a): Projections of semi-algebraic sets. Funkts. Analiz. Prilozh. 2(4): 18–30.

Gabrielov, A. (1968b): Projections of semianalytic sets. Funct. Anal. Appl. 2: 282–291.

Gabrielov, A. (1993): Existential formulas for analytic functions. Tech. Rep. 93–60, Math. Sci. Inst., Cornell University, Ithaca, NY.

Gabrielov, A. and Vorobjov, N. (1994): Complexity of stratification of semi-Pfaffian sets. Tech. Rep. 94-4, Math. Sci. Inst., Cornell University, Ithaca, NY.

Galligo, A., Heintz, J., and Morgenstern, J. (1987): Parallelism and fast quantifier elimination over algebraically (and real) closed fields. In Proc. of the FCT '87 Conf., Kazan USSR. Springer, Berlin Heidelberg New York Tokyo (Lect. Notes Comput. Sci.).

Gantmacher, F. R. (1959): The theory of matrices. Chelsea Publishing Co., New York (vol. I and II).

Gao, X.-S. and Chou, S.-C. (1994): A zero structure theorem for differential parametric systems. J. Symb. Comput. 16(6): 585–595.

Garey, M. R. and Johnson, D. S. (1979): Computers and intractability. W. H. Freeman, San Francisco.

Gelenter, H., Hanson, J. R., and Loveland, D. W. (1960): Empirical explorations of the geometry theorem proving machine. In Proc. Western Joint Computer Conf. pp. 143–147.

Giusti, M. and Heintz, J. (1991): La détermination des points isolés et de la dimension d'une variété algébrique peut se faire en temps polynomial. Preprint, Ecole Polytechnique, France et Univesidad de Buenos Aires, Argentina.

Gödel, K. (1931): Über formal unentscheidbare Sätze der Principia Mathematica und verwandter Systeme I. Monatsh. Math. Phys. 38: 173–198.

Goldhaber, J. K. and Ehrlich, G. (1970): Algebra. Macmillan Co., New York.

González-Vega, L. (1991): Determinantal formulae for the solution set of zero-dimensional ideals. J. Pure Appl. Algebra 76: 57–80.

González-Vega, L. (1997): A combinatorial algorithm solving some quantifier elimination problems. This volume, pp. 365–375.

Gonzàlez-Vega, L., Lombardi, H., Recio, T., and Roy, M.-F. (1989): Sturm-Habicht sequences. In Proc. ACM-SIGSAM 1989 International Symp. on Symbolic and Algebraic Computation. Association for Computing Machinery, New York, pp. 136–146.

Gonzàlez-Vega, L., Lombardi, H., Recio, T., and Roy, M.-F. (1990): Sous-résultants et spécialisation de la suite de Sturm I. Inform. Théor. Appl. 24(6): 561–588.

Gonzàlez-Vega, L., Lombardi, H., Recio, T., and Roy, M.-F. (1994): Sous-résultants et spécialisation de la suite de Sturm II. Inform. Théor. Appl. 27(1): 1–24.

Gonzàlez-Vega, L., Lombardi, H., Recio, T., and Roy, M.-F. (1997): Determinants and real roots of univariate polynomials. This volume, pp. 300–316.

Gournay, L. and Risler, J.-J. (1993): Construction of roadmaps in semi-algebraic sets. Appl. Algebra Eng. Commun. Comput. 4: 239–252.

Gózalez-López, M. J. and Recio, T. (1993): Path tracking in motion planning. Comput. J. 36(5): 515–524.

Grigor'ev, D. Y. (1986): Factorization of polynomials over a finite field and the solution of systems of algebraic equations. J. Soviet Math. 34: 1762–1803.

Grigor'ev, D. Y. (1987): Computational complexity in polynomial algebra. Proc. Int. Congr. Mathematicians 2: 1452–1460.

Grigor'ev, D. Y. (1988): The complexity of deciding Tarski algebra. J. Symb. Comput. 5(1,2): 65–108.

Grigor'ev, D. Y. and Vorobjov, N. N. (1988): Solving systems of polynomial inequalities in subexponential time. J. Symb. Comput. 5(1,2): 37–64.

Grigor'ev, D. Y. and Vorobjov, N. N. (1992): Counting connected components of a semi-algebraic set in subexponential time. Comput. Complex. 2: 133–186.

Grigor'ev, D. Y., Heintz, J., Roy, M.-F., Solernó, P., and Vorobjov, N. (1990): Comptage des composantes connexes d'un ensemble semi-algebrique en temps simplement exponentiel. C. R. Acad. Sci. Paris, Sér. I 312: 879–882.

Grigor'ev, D. Y., Karpinski, M., and Vorobjov, N. (1994): Lower bounds on testing membership to a polyhedron by algebraic decision trees. In Proc. of 26th ACM Symp. on Theory of Computing.

Gudkov, D. A. (1974): The topology of real projective algebraic varieties. Russian Math. Surveys 29: 1–79.

Gunning, R. C. and Rossi, H. (1965): Analytic functions of several complex variables. Prentice-Hall, Englewood Cliffs.

Habicht, V. W. (1948): Eine Verallgemeinerung des Sturmschen Wurzelzählverfahrens. Comm. Math. Helvetici 21: 99–116.

Hagel, G. (1993): Characterizations of the Macaulay matrix. Tech. Rep. 93-5, Wilhelm-Schickard-Institut, Universität Tübingen.

Hartshorne, R. (1977): Algebraic geometry. Springer, Berlin Heidelberg New

York Tokyo.

Heindel, L. E. (1970): Algorithms for exact polynomial root calculation. Ph. D. thesis, University of Wisconsin.

Heindel, L. E. (1971): Integer arithmetic algorithms for polynomial real zero determination. J. ACM 18(4): 533–548.

Heintz, J. (1983): Definability and fast quantifier elimination over algebraically closed field. Theor. Comput. Sci. 24: 239–277.

Heintz, J. and Wüthrich, H. (1975): An efficient quantifier elimination algorithm for algebraically closed fields of any characteristic. SIGSAM Bull. 9(4): 11.

Heintz, J., Roy, M.-F., and Solernó, P. (1989a): On the complexity of semialgebraic sets. In: Ritter, G. X. (Ed.), Proc. IFIP. North-Holland, Amsterdam, pp. 293–298.

Heintz, J., Roy, M.-F., and Solernó, P. (1989b): Sur la complexité du principe de Tarski-Seidenberg. C. R. Acad. Sci. Paris 309: 825–830.

Heintz, J., Roy, M.-F., and Solernó, P. (1990): Sur la complexité du principe de Tarski-Seidenberg. Bull. Soc. Math. France 118: 101–126.

Heintz, J., Krick, T., Roy, M.-F., and Solernó, P. (1991): Geometric problems solvable in single exponential time. In: Sakata, S. (Ed.), Proc. Eighth International Conf. on Applied Algebra, Algebraic Algor. and Error Correcting Codes: AAECC-8 Tokyo 1990. Springer, Berlin Heidelberg New York Tokyo, pp. 11–23 (Lect. Notes Comput. Sci., vol. 508).

Heintz, J., Krick, T., Slissenko, A., and Solernó, P. (1991): Construction of a shortest path around semi-algebraic obstacles in the plane (in Russian). Teorya Slojnosti Buchislenii, Academia Nauk Leningrad 5: 200–232.

Heintz, J., Recio, T., and Roy, M.-F. (1991): Algorithms in real algebraic geometry and applications to computational geometry. In: Goodman, J., Pollack, R., and Steiger, W. (Eds.), Discrete and Computational Geometry: Papers from the DIMACS Special Year. American Mathematical Society/Association for Computing Machinery, Providence/New York, pp. 137–164 (DIMACS Series in Discrete Mathematics and Theoretical Computer Science, vol. 6).

Heintz, J., Roy, M.-F., and Solernó, P. (1991a): Description des composante connexes d'un ensemble semi-algébrique en temps simplement exponentiel. C. R. Acad. Sci. Paris 313: 161–170.

Heintz, J., Roy, M.-F., and Solernó, P. (1991b): Single exponential path finding in semi-algebraic sets. part I: The case of a regular bounded hypersurface. In: Sakata, S. (Ed.), Proc. Eighth International Conf. on Applied Algebra, Algebraic Algor. and Error Correcting Codes: AAECC-8 Tokyo 1990. Springer, Berlin Heidelberg New York Tokyo (Lect. Notes Comput. Sci., vol. 508).

Heintz, J., Krick, T., Slissenko, A., and Solernó, P. (1993): Une borne inférieure pour la construction de chemins polygonaux dans $\mathbf{R}^n$. Tech. Rep. PUBLIM Preprint Series, Département de Mathématiques, Université de Limoges.

Heintz, J., Roy, M.-F., and Solernó, P. (1993): On the theoretical and practical complexity of the existential theory of the reals. Comput. J. 36(5): 427–431.

Heintz, J., Roy, M.-F., and Solernó, P. (1994a): Description of the connected components of a semialgebraic set in single exponential time. Discrete Comput. Geom. 11: 121–140.

Heintz, J., Roy, M.-F., and Solernó, P. (1994b): Single exponential path finding in semi-algebraic sets. part II: The general case. In: Bajaj, C. L. (Ed.), Algebraic Geometry and Its Applications. Springer, Berlin Heidelberg New York Tokyo, pp. 467–481.

Herbrand, J. (1930): Recherches sur la thérie de la démonstration. Warsaw (Prace Towarzystwa Naukowego Warszawskiego, WydziałIII).

Hermite, C. (1853): Remarques sur le théorème de Sturm. C. R. Acad. Sci. Paris 36: 52–54.

Hermite, C. (1856): Sur le nombre des racines d'une équation algébrique comprise entre des limites données. J. Reine Angew. Math. 52: 39–51.

Hilbert, D. (1930): Grundlagen der Mathematik. Leipzig and Berlin.

Hilton, H. (1932): Plane algebraic curves. Clarendon Press, Oxford.

Hironaka, H. (1975): Triangulations of algebraic sets. AMS Symp. Pure Math. 29: 165–185.

Hohn, F. E. (1966): Applied boolean algebra. Macmillan, New York.

Holthusen, C. (1974): Vereinfachungen für Tarskis Entscheidungsverfahren der Elementaren Reellen Algebra. Ph. D. thesis, Universität Heidelberg.

Hong, H. (1989): An improvement of the projection operator in cylindrical algebraic decomposition. Tech. Rep. OSU-CISRC-12/89 TR55, Dept. of Computer Science, The Ohio State University.

Hong, H. (1990a): An improvement of the projection operator in cylindrical algebraic decomposition. In: Watanabe, S. and Nagata, M. (Eds.), Proc. International Symp. on Symbolic and Algebraic Computation. Association for Computing Machinery, New York, pp. 261–264. This volume, pp. 166–173.

Hong, H. (1990b): Improvements in CAD-based quantifier elimination. Ph. D. thesis, The Ohio State University, Columbus, Ohio.

Hong, H. (1991a): Collision problems by an improved CAD-based quantifier elimination algorithm. Tech. Rep. 91-05, Research Institute for Symbolic Computation, Johannes Kepler University, Linz, Austria.

Hong, H. (1991b): Comparison of several decision algorithms for the existential theory of the reals. Tech. Rep. 91-41.0, Research Institute for Symbolic Computation, Johannes Kepler University, Linz, Austria.

Hong, H. (1991c): Parallelization of quantifier elimination on a workstation network. Tech. Rep. 91-55.0, Research Institute for Symbolic Computation, Johannes Kepler University, Linz, Austria.

Hong, H. (1992a): Half resultant. Tech. Rep. 92-51, Research Institute for Symbolic Computation, Johannes Kepler University, Linz, Austria.

Hong, H. (1992b): Inter-reduction of two polynomials is unique. Tech. Rep. 92-01.0, Research Institute for Symbolic Computation, Johannes Kepler University, Linz, Austria.

Hong, H. (1992c): Non-linear real constraints in constraint logic programming. In: Kirchner, H. and Levi, G. (Eds.), Algebraic and Logic Programming: Third International Conf. Springer, Berlin Heidelberg New York Tokyo, pp. 201–212 (Lect. Notes Comput. Sci., vol. 632).

Hong, H. (1992d): Quantifier elimination for formuals constrained by quadratic equations. Tech. Rep. 92-53, Research Institute for Symbolic Computation, Johannes Kepler University, Linz, Austria.

Hong, H. (1992e): Simple solution formula construction in cylindrical algebraic decomposition based quantifier elimination. In: Wang, P. S. (Ed.), Proc. International Symp. on Symbolic and Algebraic Computation. Association for Computing Machinery, New York, pp. 177–188. This volume pp. 201–219.

Hong, H. (1992f): Slope resultant. Tech. Rep. 92-52, Research Institute for Symbolic Computation, Johannes Kepler University, Linz, Austria.

Hong, H. (Ed.) (1993a): Computational quantifier elimination. Oxford University Press. Special issue of Comput. J. 36(5).

Hong, H. (1993b): Heuristic search strategies for cylindrical algebraic decomposition. In: Calmet, J. and Campbell, J. A. (Eds.), Proc. of Artificial Intelligence and Symbolic Math. Computing. Springer, Berlin Heidelberg New York Tokyo, pp. 152–165 (Lect. Notes Comput. Sci., vol. 737).

Hong, H. (1993c): Parallelization of quantifier elimination on a workstation network. In: Cohen, G., Mora, T., and Moreno, O. (Eds.), Applied Algebra, Algebraic Algor. and Error Correcting Codes - 10th International Symp. Springer, Berlin Heidelberg New York Tokyo, pp. 170–179 (Lect. Notes Comput. Sci., vol. 673).

Hong, H. (1993d): Quantifier elimination for formulas constrained by quadratic equations. In: Bronstein, M. (Ed.), Proc. 1993 International Symp. on Symbolic and Algebraic Computation. Association for Computing Machinery, New York, pp. 264–274.

Hong, H. (1993e): Quantifier elimination for formulas constrained by quadratic equations via slope resultants. Comput. J. 36(5): 439–449.

Hong, H. (1993f): RISC-CLP(Real): Constraint logic programming over real numbers. In: Benhamou, F. and Colmerauer, A. (Eds.), Constraint Logic Programming: Selected Research. MIT Press, Cambridge, Mass.

Hong, H. (1994): Topology analysis of plane algebraic curves by a hybrid method: Numeric, interval, and algebraic. In: Nesterov, V. and Musaev, E. (Eds.), Interval 94. St. Petersburg, pp. 109–114.

Hong, H. (1996): An efficient method for analyzing the topology of plane real algebraic curves. Math. Computers Simulation 42: 571–582.

Hong, H. (1997): Heuristic search and pruning in polynomial constraint satisfaction. Annals of Math. and AI 19(3,4): 319–334.

Hong, H. and Küchlin, W. (1990): Termination proof of term rewrite system by cylindrical algebraic decomposition. Unpublished notes.

Hong, S. J., Cain, R. G., and Ostapko, D. L. (1974): MINI: A heuristic approach for logic minimization. IBM J. Res. Dev. 18(5): 443–458.

Hörmander, L. (1983): The analysis of partial differential operators: Differential operators with constant coefficients. Springer, Berlin Heidelberg New York Tokyo (vol. 2).

Huet, G. and Oppen, D. (1980): Equations and rewrite rules. In Formal Language Theory. Perspectives and Open Problems. Academic Press, New York, pp. 349–405.

Hurwitz, A. (1913): Über die Trägheitsformen eines algebraischen Moduls. Ann. Mat., Ser. III 20: 113–151.

Jacobson, N. (1964): Lectures in abstract algebra i, ii, iii. D. Van Nostrand, Princeton, NJ.

Jacobson, N. (1974): Basic algebra I, II. W. H. Freeman, San Francisco.

Jebelean, T. (1993): An algorithm for exact division. J. Symb. Comput. 15(2): 169–180.

Johnson, J. R. (1991): Algorithms for polynomial real root isolation. Ph. D. thesis, Dept. of Comput. and Information Sci., The Ohio State University. Tech. Rep. OSU-CISRC-8/91-TR21.

Johnson, J. R. (1992): Real algebraic number computation using interval arithmetic. In: Wang, P. S. (Ed.), Proc. International Symp. on Symbolic and Algebraic Computation. Association for Computing Machinery, New York, pp. 195–205.

Kahan, W. (1975): Problem no. 9: An ellipse problem. SIGSAM Bull. 9(35): 11.

Kahn, P. J. (1979): Counting types of rigid frameworks. Invent. Math. 55: 297–308.

Kalkbrener, M. and Stifter, S. (1987): Some examples for using quantifier elimination for the collision problem. Tech. Rep. 87-22, Research Institute for Symbolic Computation, Johannes Kepler University, Linz, Austria.

Kaltofen, E. (1982): Polynomial factorization. In: Buchberger, B., Collins, G. E., and Loos, R. G. K. (Eds.), Computer Algebra: Symbolic and Algebraic Computation. Springer, Wien New York, pp. 95–114.

Kaplan, W. (1952): Advanced calculus. Addison-Wesley, Reading, Mass.

Kaplan, W. (1966): Introduction to analytic functions. Addison-Wesley, Reading, Mass.

Kaplansky, I. (1977): Hilbert's problems - preliminary edition (manuscript). Math. Dept., University of Chicago, Chicago, IL.

Kapur, D. (1986): Geometry theorem proving using Hilbert's nullstellensatz. In: Char, B. W. (Ed.), Proc. of 1986 Symp. on Symbolic and Algebraic Computation. Association for Computing Machinery, New York, pp. 202–208.

Kapur, D. (1988): A refutational approach to geometry theorem proving. Artificial Intelligence J. 37(1–3): 61–93.

Khovanskii, A. G. (1991): Fewnomials. American Mathematical Society, Providence (Translations of Mathematical Monographs, vol. 88).

Kleene, S. C. (1935–1936): General recursive functions of natural numbers. Math. Ann. 112: 727–742.

Knuth, D. E. (1969): The art of computer programming. Addison-Wesley, Reading, Mass. (vol. 2, Seminumerical Algorithms).

Knuth, D. E. (1981): The art of computer programming. Addison-Wesley, Reading, Mass. (vol. 2, Seminumerical Algorithms (Second ed.)).

Knuth, D. E., Larrabee, T., and Roberts, P. M. (1989): Mathematical writing. Mathematical Association of America, Washington, D.C. (MAA Notes, vol. 14).

Kochen, S. (1975): The model theory of local fields. In ISILC Logic Conference: proceedings of the Summer Institute. Springer, Berlin Heidelberg New York Tokyo (Lect. Notes Comput. Sci., vol. 499).

Koopman, B. and Brown, A. (1932): On the covering of analytic loci by complexes. Trans. Am. Math. Soc. 34: 231–251.

Korkina, E. I. and Kushnirenko, A. G. (1985): Another proof of the Tarski-Seidenberg theorem. Sib. Math. Zh. 26(5): 94–98.

Kozen, D. and Yap, C. K. (1985): Algebraic cell decomposition in NC. In 26th Annual IEEE Symp. on Foundations of Comput. Sci. (FOCS). pp. 515–521.

Krandick, W. and Johnson, J. R. (1994): A multiprecision floating point and interval package for implementing hybrid symbolic- numeric algorithms. In Proc. International Conf. on Interval and Computer-Algebraic Methods in Sci. and Eng. pp. 122–125.

Kreisel, G. (1975): Some uses of proof theory for finding computer programs. Colloq. Int. CNRS 249: 123–134.

Kreisel, G. (1977): From foundations to science; justifying and unwinding proofs. In Set theory. Foundations of Mathematics. pp. 63–72 (Recueil des travaux de l'Institut Mathématique, Nouvelle serié, vol. 2).

Kreisel, G. and Krivine, J. L. (1967): Elements of mathematical logic (model theory). North-Holland, Amsterdam.

Küchlin, W. (1982): An implementation and investigation of the Knuth-Bendix completion algorithm. Master's thesis, Informatik I, Universität Karlsruhe, Karlsruhe, Germany. (Reprinted as Rep. 17/82.).

Kuratowski, C. and Tarski, A. (1931): Les opérations logiques et les ensembles projectifs. Fund. Math. 17: 240–248.

Kutzler, B. (1988): Algebraic approaches to automated geometry theorem proving. Ph. D. thesis, Johannes Kepler University, Linz, Austria.

Kutzler, B. and Stifter, S. (1986a): Automated geometry theorem proving using Buchberger's algorithm. In: Char, B. W. (Ed.), Proc. of 1986 Symp. on Symbolic and Algebraic Computation. Association for Computing Machinery, New York, pp. 209–214.

Kutzler, B. and Stifter, S. (1986b): On the application of Buchberger's algorithm to automated geometry theorem proving. J. Symb. Comput. 2(4): 389–397.

Lakshman, Y. N. and Lazard, D. (1991): On the complexity of zero-dimensional algebraic systems. In Proc. MEGA'90. Birkhäuser, Basle, pp. 217–225.

Lam, Y. T. (1984): Introduction to real algebra. Rocky Mountain J. Math. 14: 767–814.

Langemyr, L. (1990): The cylindrical algebraic decomposition algorithm and multiple algebraic extensions. In Proc. 9th IMA Conf. on the Math. of Surfaces.

Langemyr, L. and McCallum, S. (1989): The computation of polynomial greatest common divisors over an algebraic number field. J. Symb. Comput. 8(5): 429–428.

Langford, C. H. (1926,1927a): Some theorems on deducibility. Ann. Math. 28: 16–40.

Langford, C. H. (1926,1927b): Theorems on deducibility (second paper). Ann. Math. 28: 459–471.

Lankford, D. (1979): On proving term rewriting systems are Noetherian. Tech. Rep. MTP-3, Dept. of Math., Louisiana Tech. University, Ruston, LA.

Lauer, M. (1977): A solution to Kahan's problem (SIGSAM problem no. 9). SIGSAM Bull. 11: 16–28.

Lazard, D. (1981): Resolution des systemes d'equations algebriques. Theor. Comput. Sci. 15: 77–110.

Lazard, D. (1983): Gröbner bases, Gaussian elimination and resolution of systems of algebraic equations. In Proc. EUROCAL '83, European Computer Algebra Conf. Springer, Berlin Heidelberg New York Tokyo, pp. 146–156 (Lect. Notes Comput. Sci., vol. 162).

Lazard, D. (1988): Quantifier elimination: Optimal solution for two classical examples. J. Symb. Comput. 5: 261–266.

Lazard, D. (1990): An improved projection for cylindrical algebraic decomposition. Unpublished manuscript.

Lazard, D. (1994): An improved projection for cylindrical algebraic decomposition. In: Bajaj, C. L. (Ed.), Algebraic Geometry and Its Applications. Springer, Berlin Heidelberg New York Tokyo.

Liska, R. and Steinberg, S. (1993): Applying quantifier elimination to stability analysis of difference schemes. Comput. J. 36(5): 497–503.

Lloyd, N. G. (1978): Degree theory. Cambridge University Press, Cambridge.

Lombardi, H. (1989): Algebre elémentaire en temps polynomial. Ph. D. thesis, University of Nice.

Loos, R. G. K. (1975): A constructive approach to algebraic numbers. Unpublished manuscript.

Loos, R. G. K. (1976): The algorithm description language ALDES (Report). SIGSAM Bull. 10(1): 15–39.

Loos, R. G. K. (1982a): Computing in algebraic extensions. In: Buchberger, B., Collins, G. E., and Loos, R. (Eds.), Computer Algebra: Symbolic and Algebraic Computation (Second ed.). Springer, Wien New York, pp. 173–188.

Loos, R. G. K. (1982b): Generalized polynomial remainder sequences. In: Buchberger, B., Collins, G. E., and Loos, R. G. K. (Eds.), Computer Algebra: Symbolic and Algebraic Computation. Springer, Wien New York, pp. 115–137.

Loos, R. G. K. and Weispfenning, V. (1993): Applying linear quantifier elimination. Comput. J. 36: 450–462.

Löwenheim, L. (1915): Über Möglichkeiten im Relativkalkül. Math. Ann. 76: 447–470.

Macaulay, F. S. (1903): Some formulae in elimination. Proc. London Math. Soc. Ser. 1 35: 3–27.

Macaulay, F. S. (1916): The algebraic theory of modular systems. Cambridge University Press, Cambridge.

MacIntyre, A. (1976): On definable subsets of p-adic fields. J. Symb. Logic 41: 605–610.

MacIntyre, A. (1977): Model completeness. In: Barwise, J. (Ed.), Handbook of Mathematical Logic. North-Holland, Amsterdam, pp. 139–180 (Studies in Logic and the Foundations of Mathematics).

MacIntyre, A. (1986): Twenty years of p-adic model theory. In: Paris, J. B., Wilkie, A. J., and Wilmers, G. M. (Eds.), Logic Colloquium '84. North-Holland, Amsterdam, pp. 121–153.

MacIntyre, A., McKenna, K., and van den Dries, L. (1983): Elimination of quantifiers in algebraic structures. Adv. Math. 47: 74–87.

Mahler, K. (1964): An inequality for the discriminant of a polynomial. Michigan Math. J. 11(3): 257–262.

Manocha, D. (1993): Efficient algorithms for multipolynomial resultant. Comput. J. 36(5): 485–496.

Manocha, D. and Canny, J. F. (1992): Multipolynomial resultants and linear algebra. In: Wang, P. S. (Ed.), Proc. International Symp. on Symbolic and Algebraic Computation. Association for Computing Machinery, New York, pp. 158–167.

Marden, M. (1949): The geometry of the zeroes of a polynomial in a complex variable. American Mathematical Society, Providence, R.I. (Mathematical Surveys, vol. 3).

Massey, W. S. (1978): Homology and cohomology theory. Dekker, New York, NY.

McCallum, S. (1979): Constructive triangulation of real curves and surfaces. Master's thesis, Math. Dept., University of Sydney, Australia.

McCallum, S. (1984): An improved projection operation for cylindrical algebraic decomposition. Ph. D. thesis, University of Wisconsin-Madison.

McCallum, S. (1987): Solving polynomial strict inequalities using cylindrical algebraic decomposition. Tech. Rep. 87-25.0, RISC-LINZ, Johannes Kepler University, Linz, Austria.

McCallum, S. (1988): An improved projection operation for cylindrical algebraic decomposition of three-dimensional space. J. Symb. Comput. 5(1,2): 141–161.

McCallum, S. (1993): Solving polynomial strict inequalities using cylindrical algebraic decomposition. Comput. J. 36(5): 432–438.

McCluskey, E. J. (1956): Minimization of boolean functions. Bell System Tech. J. 35: 1417–1444.

McKinsey, J. C. C. (1943): The decision problem for some classes of sentences without quantifiers. J. Symb. Logic 8: 61–76.

McPhee, N. (1990): Mechanically proving geometry theorems with a combination of Wu's method and Collins' method. Unpublished Ph.D. proposal.

Meserve, B. (1955): Decision methods for elementary algebras. Am. Math. Monthly 62: 1–8.

Meyer, A. R. (1973): Weak monadic second order theory of successor is not elementary-recursive. Tech. Rep. Project MAC Technical Report Memorandum 38, M.I.T.

Mignotte, M. (1974): An inequality about factors of polynomials. Math. Comput. 28(128): 1153,1157.

Mignotte, M. (1982): Some useful bounds. In: Buchberger, B., Collins, G. E., and Loos, R. (Eds.), Computer Algebra: Symbolic and Algebraic Computation (Second ed.). Springer, Wien New York, pp. 259–263.

Mignotte, M. (1986): Computer versus paper and pencil. In Proc. CALSYF. pp. 63–69 (vol. 4).

Milnor, J. (1964): On Betti numbers of real varieties. Proc. Am. Math. Soc. 15(2): 275–280.

Minc, H. and Marcus, M. (1964): A survey of matrix theory and matrix inequalities. Allyn and Bacon Inc., Boston.

Mishra, B. (1993): Algorithmic algebra. Springer, Berlin Heidelberg New York Tokyo.

Mishra, B. and Pedersen, P. (1990): Computation with sign-representations of real algebraic numbers. In Proc. International Symp. on Symbolic and Algebraic Computation. Association for Computing Machinery, New York, pp. 120–126.

Monk, L. (1974): Elementary recursive decision procedure for th($r, +, *$). Manuscript, Math. Dept. University of California at Berkeley.

Monk, L. (1975): Elementary recursive decision theories. Ph. D. thesis, University of California at Berkeley.

Montaña, J. L. and Pardo, L. M. (1993): Lower bounds for arithmetic networks. Appl. Algebra Eng. Commun. Comput. 4: 1–24.

Müller, F. (1978): Ein exakter Algorithmus zur nichtlinearen Optimierung für beliebige Polynome mit mehreren Veränderlichen. Verlag Anton Hain, Meisenheim am Glan.

Mulmuley, K. (1994): Lower bounds for parallel linear programming and other problems. In Proc. 26th Annual ACM Symp. on Theory Comput. pp. 603–614.

Musser, D. R. (1971): Algorithms for polynomial factorization. Ph. D. thesis, University of Wisconsin Computer Sciences Dept. Tech. Rep. 134.

Musser, D. R. (1975): Multivariate polynomial factorization. J. ACM 22(2): 291–308.

Neff, A. C. (1994): Specified precision polynomial root isolation is in NC. J. Comput. Syst. Sci. 48: 429–463.

Oppen, D. C. (1973): Elementary bounds for Presburger arithmetic. In Proc. 5th ACM Symp. on Theory of Computing. pp. 34–37.

Pan, V. (1989): Fast and efficient parallel evaluation of the zeros of a polynomial having only real zeros. Comput. Math. Appl. 17(11): 1475–1480.

Pedersen, P. (1989): Computational semi-algebraic geometry. Tech. Rep. 212, Courant Inst., New York University.

Pedersen, P. (1991a): Counting real zeroes. Ph. D. thesis, Courant Inst., New York University.

Pedersen, P. (1991b): Multivariate Sturm theory. In: Mattson, H. F., Mora, T., and Rao, T. R. N. (Eds.), Proc. Ninth International Symp. on Applied Algebra, Algebraic Algor. and Error Correcting Codes. Springer, Berlin Heidelberg New York Tokyo, pp. 318–332 (Lect. Notes Comput. Sci., vol. 539).

Pedersen, P., Roy, M.-F., and Szpirglas, A. (1993): Counting real zeroes in the multivariate case. In: Eyssette, F. and Galligo, A. (Eds.), Proc. MEGA'92: Computational Algebraic Geometry. Birkhäuser, Basle, pp. 203–224 (Progress in Math., vol. 109).

Pieri, M. (1908): La geometria elementare instituita sulle nozioni di 'punto' e 'sfera'. Mem. Mat. Fis. Soc. Ital. Sci. 15: 345–450.

Pollack, R. and Roy, M.-F. (1993): On the number of cells defined by a set of polynomials. C. R. Acad. Sci. Paris 316: 573–577.

Post, E. L. (1921): Introduction to a general theory of elementary propositions. Am. J. Math. 43: 163–185.

Presburger, M. (1930): Über die Vollständigkeit eines gewissen Systems der Arithmetik ganzer Zahlen, in welchem die Addition als einzige Operation hervortritt. In Comptes-rendus du I Congrès des Mathèmathiciens des Pays Slaves. Warsaw, pp. 92–101,395.

Prestel, A. (1984): Lectures on formally real fields. Springer, Berlin Heidelberg New York Tokyo (Lect. Notes Math., vol. 1093).

Prill, D. (1986): On approximation and incidence in cylindrical algebraic decompositions. SIAM J. Comput. 15: 972–993.

Quine, W. V. O. (1952): The problem of simplifying truth functions. Am. Math. Monthly 59: 521–531.

Quine, W. V. O. (1955): A way to simplify truth functions. Am. Math. Monthly 62: 627–631.

Quine, W. V. O. (1959): On cores and prime implicants of truth functions. Am. Math. Monthly 66: 755–760.

Rabin, M. (1977): Decidable theories. In: Barwise, J. (Ed.), Handbook of Mathematical Logic. North-Holland, Amsterdam, pp. 595–630 (Studies in Logic and the Foundations of Mathematics).

Rackoff, C. (1974): Complexity of some logical theories. Ph. D. thesis, M.I.T., Dept. of Electrical Engineering.

Reddy, C. R. and Loveland, D. W. (1978): Presburger arithmetic with bounded quantifier alternation. In Proc. ACM Symp. Theory of Comput. Association for Computing Machinery, New York, pp. 320–325.

Rees, E. G. (1983): Notes on geometry. Springer, Berlin Heidelberg New York Tokyo.

Reif, J. (1979): Complexity of the mover's problem and generalizations. In IEEE Symp. on Foundations of Computer Science. pp. 421–427.

Renegar, J. (1987): On the worst-case arithmetic complexity of approximating zeros of polynomials. J. Complex. 3: 90–113.

Renegar, J. (1988): A faster PSPACE algorithm for the existential theory of the reals. In IEEE Symp. on Foundations of Computer Science. pp. 291–295.

Renegar, J. (1991): Recent progress on the complexity of the decision problem for the reals. In: Goodman, J., Pollack, R., and Steiger, W. (Eds.), Discrete and Computational Geometry: Papers from the DIMACS Special Year. American Mathematical Society/Association for Computing Machinery, Providence/New York, pp. 287–308 (DIMACS Series in Discrete Math. and Theoretical Comput. Sci., vol. 6). This volume, pp. 220–241.

Renegar, J. (1992a): On the computational complexity and geometry of the first order theory of the reals I. J. Symb. Comput. 13: 255–300.

Renegar, J. (1992b): On the computational complexity and geometry of the first order theory of the reals II. J. Symb. Comput. 13: 301–328.

Renegar, J. (1992c): On the computational complexity and geometry of the first order theory of the reals III. J. Symb. Comput. 13: 329–352.

Renegar, J. (1992d): On the computational complexity of approximating solutions for real algebraic formulae. SIAM J. Comput. 21: 1008–1025.

Richardson, D. (1991): Towards computing non-algebraic cylindrical decompositions. In: Watt, S. M. (Ed.), Proc. 1991 International Symp. on Symbolic and Algebraic Computation. Association for Computing Machinery, New York, pp. 247–255.

Richardson, D. (1992): Computing the topology of a bounded non-algebraic curve in the plane. J. Symb. Comput. 14: 619–643.

Risler, J. J. (1988): Some aspects of complexity in real algebraic geometry. J. Symb. Comput. 5(1,2): 109–119.

Robinson, A. (1956): Complete theories. North-Holland, Amsterdam.

Robinson, A. (1958): Relative model-completeness and the elimination of quantifiers. Dialectica 12: 349–407.

Robinson, A. (1974a): A decision method for elementary algebra and geometry – revisited. AMS Symp. Pure Math. 25: 139–152.

Robinson, A. (1974b): Introduction to model theory and the metamathematics of algebra. North-Holland, Amsterdam.

Rosenkranz, G. (1979): Eine dem kombinierten Vorzeichenverhalten zweier Polynome äquivalente quantorenfreie Formel. Ph. D. thesis, Universität Heidelberg.

Rosser, B. (1936): Extensions of some theorems of Gödel and Church. J. Symb. Logic 1: 87–91.

Roy, M.-F. (1990): Computation of the topology of a real algebraic curve. Asterisque 192: 17–33.

Roy, M.-F. (1992): Géométrie algébrique réelle et robotique: la complexité du démenagement des pianos. Gaz. Math. (51): 75–96.

Roy, M.-F. and Szpirglas, A. (1990a): Complexity of computation on real algebraic numbers. J. Symb. Comput. 10(1): 39–51.

Roy, M.-F. and Szpirglas, A. (1990b): Complexity of the computation of cylindrical decomposition and topology of real algebraic curves using Thom's lemma. In Real Analytic and Algebraic Geometry: Proc. of the Conf. held in Trento, Italy, October 3–7, 1988. Springer, Berlin Heidelberg New York Tokyo, pp. 223–236 (Lect. Notes Math., vol. 1420).

Roy, M.-F. and Vorobjov, N. N. (1994): Finding irreducible components of some real transcendental varieties. Comput. Complex. 4: 107–132.

Rubald, C. M. (1974): Algorithms for polynomials over a real algebraic number field. Ph. D. thesis, University of Wisconsin Computer Sciences Dept. Tech. Rep. 206.

Rudell, R. (1986): Multiple-Valued Logic Minimization for PLA Synthesis. Tech. Rep. UCB/ERL M86/65, University of California, Berkeley, USA.

Rump, S. (1979): Polynomial minimum root separation. Math. Comput. 33(145): 327–336.

Runge, C. (1898–1904): Separation und Approximation der Wurzeln. In Encyklopädie der mathematischen Wissenschaffen mit Einschluss ihrer Anwendungen. B. G. Teubner, Leipzig, pp. 404–448 (vol. 1).

Sasao, T. (1982): Comparison of minimization algorithms for multiple-valued expressons. In Proc. 12th Int. Symp. on Multivalued Logic.

Saunders, B. D., Lee, H. R., and Abdali, S. K. (1989): A parallel implementation of the cylindrical algebraic decomposition algorithm. In Proc. ACM-SIGSAM 1989 International Symp. on Symbolic and Algebraic Computation. Association for Computing Machinery, New York, pp. 298–307.

Scarpellini, B. (1984): Complexity of subcases of Presburger arithmetic. Trans. Am. Math. Soc. 284: 203–218.

Schönhage, A. and Strassen, V. (1971): Schnelle Multiplikation großer Zahlen. Computing 7: 281–292.

Schuette, K. and van der Waerden, B. L. (1953): Das Problem der dreizehn Kugeln. Math. Ann. 125: 325–334.

Schwartz, J. and Sharir, M. (1981): On the 'piano movers' problem. I. The case of a two-dimensional rigid polygonal body moving amidst polygonal barriers. Tech. Rep. 39, Department of Computer Science, Courant Institute of Mathematical Sciences.

Schwartz, J. and Sharir, M. (1983a): Mathematical problems and training in robotics. Notices Am. Math. Soc. 30(5): 475–477.

Schwartz, J. and Sharir, M. (1983b): On the 'piano movers' problem II: General techniques for computing topological properties of real algebraic manifolds. Adv. Appl. Math. 4: 298–351.

Schwartz, J. and Sharir, M. (1988): A survey of motion planning and related geometric algorithms. Artificial Intelligence J. 37(1-3): 157–169.

Scrowcroft, P. and van den Dries, L. (1988): On the structure of semialgebraic sets of p-adic fields. J. Symb. Logic 53(4): 1138–1164.

Seidenberg, A. (1954): A new decision method for elementary algebra. Ann. Math. 60(2): 365–374.

Sendra, J. R. and Llovet, J. (1992): An extended polynomial GCD algorithm using Hankel matrices. J. Symb. Comput. 13(1): 25–39.

Sharir, M. and Schorr, A. (1984): On shortest paths in polyhedral spaces. In Proc. 16th ACM Symp. on Theory of Computing (STOC). pp. 144–153.

Shoenfield, J. (1976): Quantifier elimination in fields. In Proc. Third Latin-Amer. Symp. Math. Logic. Campinas. North-Holland, Amsterdam, pp. 243–252 (Stud. Logic Found. Math., vol. 89).

Shub, M. and Smale, S. (1993): Complexity of Bezout's theorem II: Volumes and probabilities. In: Eyssette, F. and Galligo, A. (Eds.), Proc. MEGA'92: Computational Algebraic Geometry. Birkhäuser, Basle, pp. 267–285 (Progress in Math., vol. 109).

Simanyi, P. (1983): POP reference manual. Tech. Rep., University of California, Berkeley, CA.

Skolem, T. (1919): Untersuchungen über die Axiome des Klassenkalküls und über Produktations- und Summationsprobleme, welche gewisse Klassen von Aussagen betreffen. Oslo (Skrifter utgit av Videnskapsselskapet i Kristiania, I. klasse).

Skolem, T. (1970): Logisch-kombinatorische Untersuchungen über die Erfüllbarkeit. In: Fenstad, J. E. (Ed.), Th. Skolem, Selected Works in Logic. Universitetsforlaget, Oslo.

Smoryński, C. (1981): Skolem's solution to a problem of Frobenius. Math. Intell. 3(3): 123–132.

Sontag, E. (1985): Real addition and the polynomial hierarchy. Inf. Process. Lett. 20: 115–120.

Specker, E. and Strassen, V. (Eds.) (1976): Komplexität von Entscheidungsproblemen: Ein Seminar. Springer, Berlin Heidelberg New York Tokyo (Lect. Notes Comput. Sci., vol. 43).

Stockmeyer, L. J. (1974): The complexity of decision problems in automata theory and logic. Tech. Rep. Project MAC Technical Rep. 133, M.I.T., Cambridge, Mass.

Struth, M. (1980): Vorzeichenbestimmung für Polynome auf der Zellenzerlegung nach Collins. Master's thesis, Universität Heidelberg.

Sturm, C. (1835): Mémoire sur la résolution des équations numériques. Inst. France Sci. Math. Phys. 6.

Sylvester, J. J. (1853): On a theory of syzygetic relations of two rational integral functions, comprising an application to the theory of Sturm's function. In The Collected Mathematical Papers of James Joseph Sylvester, vol. 1. Chelsea Publ., New York, pp. 429–586 (1993).

Szego, G. (1939): Orthogonal polynomials. American Mathematical Society, New York (American Mathematical Society Colloquim, vol. 23).

Tarski, A. (1931): Sur les ensebles dèfinissables de nombres rèels I. Fund. Math. 17: 210–239.

Tarski, A. (1933): Einige Betrachtungen über die Begriffe der ω-Widerspruchsfreiheit und der ω-Vollstädigkeit. Monatsh. Math. Phys. 40: 97–112.

Tarski, A. (1936): Der Wahrheitsbegriff in den formalisierten Sprachen. Stud.

Philos. 1: 261–405.

Tarski, A. (1939): New investigations on the completeness of deductive theories. J. Symb. Logic 4: 176.

Tarski, A. (1949): Arithmetical classes and types of algebraically closed fields and real-closed fields. Bull. Am. Math. Soc. 55: 64.

Tarski, A. (1951): A decision method for elementary algebra and geometry. University of California Press, Berkeley. This volume pp. 24–84.

Tarski, A. (1959): What is elementary geometry? In: L. Henkin, P. Suppes, A. T. (Ed.), Proc. of an International Symp. on the Axiomatic Method, with Special Reference to Geometry and Physics. North-Holland, Amsterdam, pp. 16–29 (Studies in Logic and the Foundations of Mathematics).

Toffalori, C. (1975): Eliminazione dei quantificatori per certe teorie di coppie di campi. Boll. Un. Mat. Ital. A 15: 159–166.

Trager, B. M. (1976): Algebraic factoring and rational function integration. In Proc. of 1976 ACM Symp. on Symbolic and Algebraic Computation. pp. 219–226.

Uspensky, J. V. (1948): Theory of equations. McGraw-Hill Book Company, New York.

van den Dries, L. (1982): Remarks on Tarski's problem concerning $(\mathbf{R}, +, \cdot, \exp)$. In Logic Coll. '82. pp. 97–121 (Stud. Logic Found. Math., vol. 112).

van den Dries, L. (1985): The field of reals with a predicate for the powers of two. Manu. Math. 54: 187–195.

van den Dries, L. (1986): A generalization of the Tarski-Seidenberg theorem and some nondefinability results. Bull. Am. Math. Soc. 15: 189–193.

van den Dries, L. (1988a): Alfred Tarski's elimination theory for real closed fields. J. Symb. Logic 53: 7–19.

van den Dries, L. (1988b): On the elementary theory of restricted elementary functions. J. Symb. Logic 53: 796–808.

van der Waerden, B. L. (1929): Topologische Begründung des Kalküls der abzählenden Geometrie. Math. Ann.: 337–362.

van der Waerden, B. L. (1937): Moderne algebra. Springer, Berlin (vol. I).

van der Waerden, B. L. (1949): Modern algebra. Frederick Ungar, New York (vol. I).

van der Waerden, B. L. (1950): Modern algebra. Frederick Ungar, New York (vol. II).

van der Waerden, B. L. (1970): Algebra. Frederick Ungar, New York (vol. I and II).

Vorobjov, N. (1984): Bound of real roots of a system of algebraic equations. In Notes of Sci. Seminars. Leningrad Branch of the Steklof Math. Inst., pp. 7–19 (vol. 137). In Russian [English transl., *J. Soviet Math.* 34 (1986), 1750–1762].

Vorobjov, N. (1991): Deciding consistency of exponential polynomials in subexponential time. In: Mora, T. and Traverso, C. (Eds.), Effective Methods in Algebraic Geometry. Birkhäuser, Basle, pp. 490–500.

Vorobjov, N. (1992a): The complexity of deciding consistency of systems of polynomial in exponent inequalities. J. Symb. Comput. 13: 139–173.

Vorobjov, N. (1992b): Effective stratification of regular real algebraic varieties. In: Coste, M., Mahe, L., and Roy, M.-F. (Eds.), Real Algebraic Geometry. Springer, Berlin Heidelberg New York Tokyo, pp. 403–415 (Lect. Notes Math., vol. 1524).

Walker, R. J. (1978): Algebraic curves. Springer, Berlin Heidelberg New York Tokyo.

Wang, H. (1981): Popular lectures on mathematical logic. Science Press/Van Nostrand Reinhold, Beijing/New York.

Warren, H. E. (1968): Lower bounds for approximation of nonlinear manifolds. Trans. Am. Math. Soc. 133: 167–178.

Weber, H. (1895): Lehrbuch der Algebra. Braunschweig (vol. 1).

Weispfenning, V. (1971): Elementary theories of valued fields. Ph. D. thesis, University of Heidelberg.

Weispfenning, V. (1975): Model-completeness and elimination of quantifiers for subdirect products of structures. J. Algebra 36: 252–277.

Weispfenning, V. (1976): On the elementary theory of Hensel fields. Ann. Math. Logic 10: 59–93.

Weispfenning, V. (1983): Quantifier elimination and decision procedures for valued fields. In: Muller, G. H. and Richter, M. M. (Eds.), Models and Sets: Logic Colloquium '83 Aachen. Springer, Berlin Heidelberg New York Tokyo, pp. 419–472 (Lect. Notes Math., vol. 1103).

Weispfenning, V. (1984): Aspects of quantifier elimination in algebra. In: Heldermann, V. (Ed.), Universal Algebra and its Links with Logic, Algebra, Combinatorics, and Computer Science, Proc. 25, Arbeitstagung über Allgemeine Algebra. Berlin, pp. 85–105.

Weispfenning, V. (1988): The complexity of linear problems in fields. J. Symb. Comput. 5: 3–27.

Weispfenning, V. (1992): Comprehensive Gröbner bases. J. Symb. Comput. 14: 1–29.

Weispfenning, V. (1994): Quantifier elimination for real algebra – the cubic case. In Proc. 1994 International Symp. on Symbolic and Algebraic Computation. pp. 258–263.

Weispfenning, V. (1997): Quantifier elimination for real algebra – the quadratic case and beyond. Appl. Algebra Eng. Commun. Comput. 8: 85–101.

Wernecke, W. (1983): Über die elementare Theorie separabel abgeschlossener Körper. Ph. D. thesis, University of Heidelberg.

Wheeler, W. (1979): Amalgamation and quantifier elimination for theories of fields. Proc. Am. Math. Soc. 77: 243–250.

Whitney, H. (1957): Elementary structure of real algebraic varieties. Ann. Math. 66(3): 545–556.

Whitney, H. (1972): Complex analytic varieties. Addison-Wesley, Reading, Mass.

Wilkie, A. J. (1994): Some model completeness results for expansions of the ordered field of real numbers by Pfaffian functions. Preprint, Mathematics Institute, Oxford.

Wilson, G. (1978): Hilbert's sixteenth problem. Topology 17: 53–73.

Wu, W.-t. (1986): Basic principles of mechanical theorem proving in elementary geometries. J. Automated Reason. 2: 221–252.

Wüthrich, H. (1976): Ein Entscheidungsverfahren für die Theorie der reell-abgeschlossenen Körper. In Komplexitaet von Entscheidungsproblemen: ein Seminar. Springer, Berlin Heidelberg New York Tokyo, pp. 138–162 (Lect. Notes Comput. Sci., vol. 43).

Wüthrich, H. (1977): Ein schnelles Quantoreneliminationsverfahren für die Theorie der algebraisch abgeschlossenen Körper. Ph. D. thesis, Philosophischen Fakultät II, Universität Zürich.

Zariski, O. (1965): Studies in equisingularity II. Am. J. Math. 87(4): 972–1006.

Zariski, O. (1975): On equimultiple subvarieties of algebroid hypersurfaces. Proc. Nat. Acad. Sci. 72(4): 1425–1426.

Zharkov, A. Y. and Blinkov, Y. A. (1993): Involutive bases of zero-dimensional ideals. Unpublished manuscript.

Index

Texts and Monographs in Symbolic Computation

Alfonso Miola, Marco Temperini (eds.)

Advances in the Design
of Symbolic Computation Systems

1997. 39 figures. X, 259 pages.
Soft cover DM 98,–, öS 682,–
ISBN 3-211-82844-3

New methodological aspects related to design and implementation of symbolic computation systems are considered in this volume aiming at integrating such aspects into a homogeneous software environment for scientific computation. The proposed methodology is based on a combination of different techniques: algebraic specification through modular approach and completion algorithms, approximated and exact algebraic computing methods, object-oriented programming paradigm, automated theorem proving through methods à la Hilbert and methods of natural deduction. In particular the proposed treatment of mathematical objects, via techniques for method abstraction, structures classification, and exact representation, the programming methodology which supports the design and implementation issues, and reasoning capabilities supported by the whole framework are described.

 SpringerWienNewYork

Sachsenplatz 4-6, P.O.Box 89, A-1201 Wien, Fax +43-1-330 24 26, e-mail: order@springer.at, Internet: http://www.springer.at
New York, NY 10010, 175 Fifth Avenue • D-14197 Berlin, Heidelberger Platz 3 • Tokyo 113, 3-13, Hongo 3-chome, Bunkyo-ku

Texts and Monographs in Symbolic Computation

Franz Winkler

Polynomial Algorithms in Computer Algebra

1996. 13 figures. VIII, 270 pages.
Soft cover DM 89,–, öS 625,–. ISBN 3-211-82759-5

The book gives a thorough introduction to the mathematical underpinnings of computer algebra. The subjects treated range from arithmetic of integers and polynomials to fast factorization methods, Gröbner bases, and algorithms in algebraic geometry. The algebraic background for all the algorithms presented in the book is fully described, and most of the algorithms are investigated with respect to their computational complexity. Each chapter closes with a brief survey of the related literature.

Jochen Pfalzgraf, Dongming Wang (eds.)

Automated Practical Reasoning

Algebraic Approaches

With a Foreword by Jim Cunningham

1995. 23 figures. XI, 223 pages.
Soft cover DM 108,–, öS 755,–. ISBN 3-211-82600-9

This book presents a collection of articles on the general framework of mechanizing deduction in the logics of practical reasoning. Topics treated are novel approaches in the field of constructive algebraic methods (theory and algorithms) to handle geometric reasoning problems, especially in robotics and automated geometry theorem proving; constructive algebraic geometry of curves and surfaces showing some new interesting aspects; implementational issues concerning the use of computer algebra systems to deal with such algebraic methods.
Besides work on nonmonotonic logic and a proposed approach for a unified treatment of critical pair completion procedures, a new semantic modeling approach based on the concept of fibered structures is discussed; an application to cooperating robots is demonstrated.

 SpringerWienNewYork

Sachsenplatz 4-6, P.O.Box 89, A-1201 Wien, Fax +43-1-330 24 26, e-mail: order@springer.at, Internet: http://www.springer.at
New York, NY 10010, 175 Fifth Avenue • D-14197 Berlin, Heidelberger Platz 3 • Tokyo 113, 3-13, Hongo 3-chome, Bunkyo-ku

Texts and Monographs in Symbolic Computation

Wen-tsün Wu

Mechanical Theorem Proving in Geometries

Basic Principles

Translated from the Chinese by Xiaofan Jin and Dongming Wang

1994. 120 figures. XIV, 288 pages.
Soft cover DM 98,–, öS 686,–. ISBN 3-211-82506-1

This book is a translation of Professor Wu's seminal Chinese book of 1984 on Automated Geometric Theorem Proving. The translation was done by his former student Dongming Wang jointly with Xiaofan Jin so that authenticity is guaranteed. Meanwhile, automated geometric theorem proving based on Wu's method of characteristic sets has become one of the fundamental, practically successful, methods in this area that has drastically enhanced the scope of what is computationally tractable in automated theorem proving. This book is a source book for students and researchers who want to study both the intuitive first ideas behind the method and the formal details together with many examples.

Bernd Sturmfels

Algorithms in Invariant Theory

1993. 5 figures. VII, 197 pages.
Soft cover DM 65,–, öS 455,–. ISBN 3-211-82445-6

J. Kung and G.-C. Rota, in their 1984 paper, write: "Like the Arabian phoenix rising out of its ashes, the theory of invariants, pronounced dead at the turn of the century, is once again at the forefront of mathematics."
The book of Sturmfels is both an easy-to-read textbook for invariant theory and a challenging research monograph that introduces a new approach to the algorithmic side of invariant theory. The Groebner bases method is the main tool by which the central problems in invariant theory become amenable to algorithmic solutions. Students will find the book an easy introduction to this "classical and new" area of mathematics. Researchers in mathematics, symbolic computation, and computer science will get access to a wealth of research ideas, hints for applications, outlines and details of algorithms, worked out examples, and research problems.

 SpringerWienNewYork

Sachsenplatz 4-6, P.O.Box 89, A-1201 Wien, Fax +43-1-330 24 26, e-mail: order@springer.at, Internet: http://www.springer.at
New York, NY 10010, 175 Fifth Avenue • D-14197 Berlin, Heidelberger Platz 3 •Tokyo 113, 3-13, Hongo 3-chome, Bunkyo-ku

Springer-Verlag
and the Environment

WE AT SPRINGER-VERLAG FIRMLY BELIEVE THAT AN international science publisher has a special obligation to the environment, and our corporate policies consistently reflect this conviction.

WE ALSO EXPECT OUR BUSINESS PARTNERS – PRINTERS, paper mills, packaging manufacturers, etc. – to commit themselves to using environmentally friendly materials and production processes.

THE PAPER IN THIS BOOK IS MADE FROM NO-CHLORINE pulp and is acid free, in conformance with international standards for paper permanency.